BIG IDEAS MATH®

Integrated Mathematics III

Resources by Chapter

- Family Communication Letters
- Start Thinking/Warm Up/Cumulative Review Warm Up
- Extra Practice (A and B)
- Enrichment and Extension
- Puzzle Time
- Cumulative Review

BIG IDEAS LEARNING®

Erie, Pennsylvania

Photo Credits

Cover Image Mix3r/Shutterstock.com

69 ©iStockphoto.com/mlevy

ISBN 13: 978-1-68033-093-9
ISBN 10: 1-68033-093-4

23456789-VLP-19 18 17 16 15

Contents

About the Resources by Chapter

Family Communication Letters (English and Spanish)

The Family Communication Letters provide a way to quickly communicate to family members how they can help their student with the material of the chapter. They make the mathematics less intimidating and provide suggestions for helping students see mathematical concepts in common activities.

Start Thinking/Warm Up/Cumulative Review Warm Up

Each Start Thinking/Warm Up/Cumulative Review Warm Up includes two options for getting the class started. The Start Thinking questions provide students with an opportunity to discuss thought-provoking questions and analyze real-world situations. The Warm Up questions review prerequisite skills needed for the lesson. The Cumulative Review Warm Up questions review material from earlier lessons or courses.

Extra Practice

The Extra Practice exercises provide additional practice on the key concepts taught in the lesson. There are two levels of practice provided for each lesson: A (basic) and B (average).

Enrichment and Extension

Each Enrichment and Extension extends the lesson and provides a challenging application of the key concepts.

Puzzle Time

Each Puzzle Time provides additional practice in a fun format in which students use their mathematical knowledge to solve a riddle. This format allows students to self-check their work.

Cumulative Review

The Cumulative Review includes exercises covering concepts and skills from the current chapter and previous chapters.

Chapter 1

Name ______________________________ Date __________

Chapter 1 Geometric Modeling

Dear Family,

In this chapter, your student will learn about geometric modeling. Geometric figures are everywhere we look and very much a part of everything we do. The basic figures your student has learned about are now being combined and used to make more complex figures, including 3-dimensional figures.

Engineers who have knowledge of geometric relationships help designers develop solutions to problems. Basic sketches are made, but may eventually need to be converted to computer models that will be more accurate for the designers to use. Computer-aided design, or CAD, programs have simplified this process for designers.

With your student, use the Internet to gain a better understanding of what a CAD program is by explaining the following topics.

- Write a paragraph summarizing the benefits of CAD programs.
- What is the most common CAD program used by engineers or designers?
- What type of post-secondary education is needed to become proficient in using CAD programs?
- Make a list of employers in your area that hire people with experience using CAD programs.

If you and your student have the opportunity, visit a local college or trade school that teaches students how to use CAD programs. Have your student research the use of CAD programs and their applications in real-world settings.

- What are some ways a company can benefit from having employees who are well-versed in CAD programs?
- What is the market for future employment in this field?
- What type of classes should students take in high school if they are interested in pursuing this career path?

May your student come to appreciate his or her knowledge of geometric figures!

Nombre ______________________________ Fecha________

Capítulo 1 Representación geométrica

Estimada familia:

En este capítulo, su hijo aprenderá sobre la representación geométrica. Las figuras geométricas están en cualquier parte que miremos y en una gran parte de todo lo que hacemos. Las figuras básicas que su hijo ha aprendido ahora se combinan y se usan para hacer figuras más complejas, entre ellas, figuras tridimensionales.

Los ingenieros que tienen conocimiento sobre las relaciones geométricas ayudan a los diseñadores a desarrollar soluciones a los problemas. Se hacen bosquejos básicos, pero eventualmente quizás necesiten convertirse en modelos por computadora que serán más precisos para el uso de los diseñadores. Los programas de diseño asistido por computadora, o CAD (por sus siglas en inglés), han simplificado este proceso para los diseñadores.

Con su hijo, usen el Internet para comprender mejor qué es un programa CAD mediante la explicación de los siguientes temas.

- Escriban un párrafo que resuma los beneficios de los programas CAD.
- ¿Cuál es el programa CAD más común que usan los ingenieros o diseñadores?
- ¿Qué clase de educación postsecundaria se necesita para ser competente en el uso de los programas CAD?
- Hagan una lista de empleadores en su área que contraten personas con experiencia en el uso de programas CAD.

Si usted o su hijo tienen la oportunidad, visiten una universidad local o escuela de formación profesional donde se enseñe a los estudiantes cómo usar los programas CAD. Pida a su hijo que investigue el uso de los programas CAD y sus aplicaciones en situaciones de la vida real.

- ¿Cuáles son algunas maneras en que una compañía puede beneficiarse si tiene empleados que estén capacitados para usar programas CAD?
- ¿Cuál es el mercado para el empleo a futuro en este campo?
- ¿Qué tipo de clases deberían tomar los estudiantes en la escuela preparatoria si les interesa seguir una carrera en esta área?

¡Ojalá su hijo llegue a apreciar su conocimiento de las figuras geométricas!

1.1 Start Thinking

The population density of a county is a measure of how many people live within a given area.

$$\text{Population density} = \frac{\text{number of people}}{\text{area of land}}$$

The approximate population densities of four counties in New York State in 2010 are shown below. Find the population density of each county in people per square mile. Then order the population densities from greatest to least.

Monroe: $\dfrac{744{,}000 \text{ people}}{657 \text{ square miles}}$

Rockland: $\dfrac{312{,}000 \text{ people}}{174 \text{ square miles}}$

Suffolk: $\dfrac{1{,}493{,}000 \text{ people}}{912 \text{ square miles}}$

Westchester: $\dfrac{949{,}000 \text{ people}}{431 \text{ square miles}}$

1.1 Warm Up

Find the indicated measures.

1. area of a circle with radius 9 centimeters
2. area of a circle with diameter 32 inches
3. radius of a circle with an area of 49π miles

1.1 Cumulative Review Warm Up

Solve the inequality. Graph the solution, if possible.

1. $|x| \geq 2$
2. $|n| < 1.5$
3. $|3t + 2| > 8$
4. $|4s - 6| + 10 < 4$

Name____ Date__________

1.1 Practice A

In Exercises 1–4, find the indicated measure.

1. The state of South Dakota has a population of about 814,000 people. Find the population density in people per square mile.

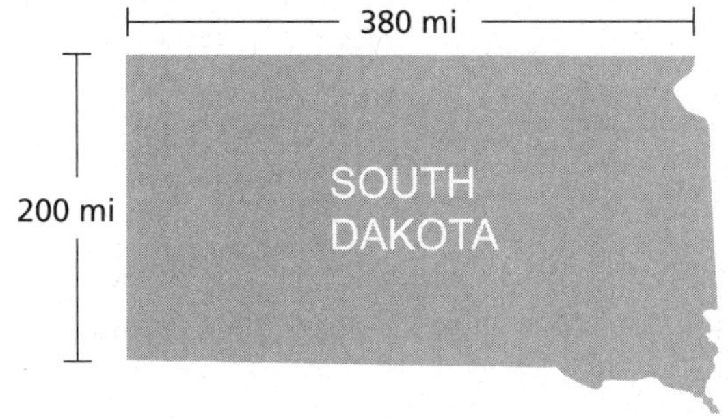

2. About 1.2 million people live in a circular region with a 6-mile radius. Find the population density in people per square mile.

3. A circular region with a 15-mile diameter has a population density of about 5000 people per square mile. Find the number of people who live in the region.

4. About 52,000 people live in a circular region with a population density of about 115 people per square mile. Find the radius of the region.

5. A rectangular swimming pool of length ℓ and width w has a perimeter of 64 feet.

   a. Write an expression that represents the area of the swimming pool in terms of w.

   b. Use your expression from part (a) to determine the dimensions of the pool that maximize the area. What do you notice?

In Exercises 6 and 7, describe how the change affects the surface area of the right prism or right cylinder.

6. tripling all the linear dimensions

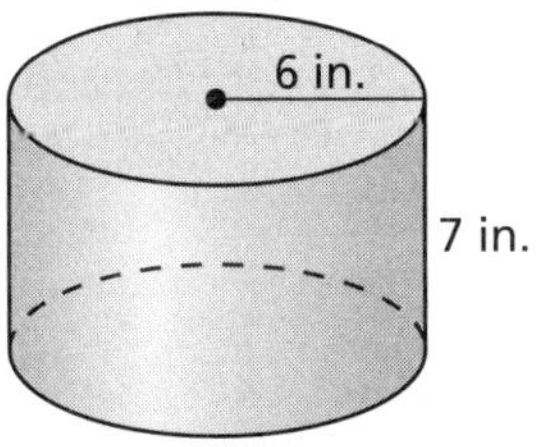

7. multiplying all the linear dimensions by $\frac{1}{4}$

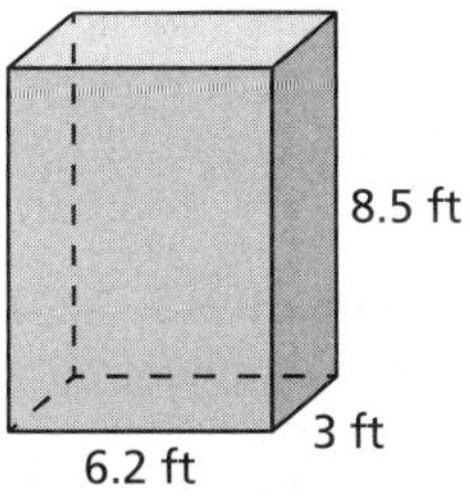

Name ______________________________ Date __________

1.1 Practice B

In Exercises 1–4, find the indicated measure.

1. A state park has an area of 112 acres. The table shows the estimated park populations for several animals. Find the population density in animals per acre for each animal.

Animal	Otter	Raccoon	Fox	Bobcat
Population	35	186	9	3

2. A city park is triangular with a base length of 4 blocks and a height of 7 blocks. During an evening concert, its population density is about 54 people per square block. Find the number of people in the park that evening.

3. About 150,000 people live in a circular region with a population density of about 1578 people per square mile. Find the radius of the region.

4. About 1.75 million people live in a circular region with a population density of about 5050 people per square mile. Find the radius of the region.

In Exercises 5 and 6, describe how the change affects the surface area of the right prism or right cylinder.

5. doubling the diameter

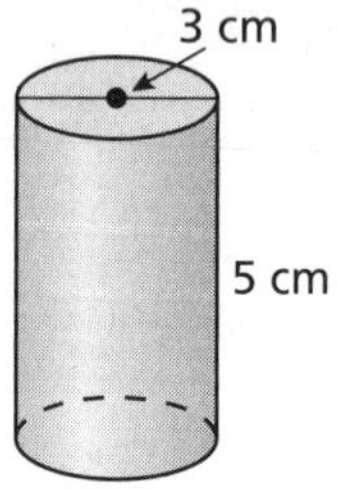

6. multiplying the base edge by 2 and the height by $\frac{1}{3}$

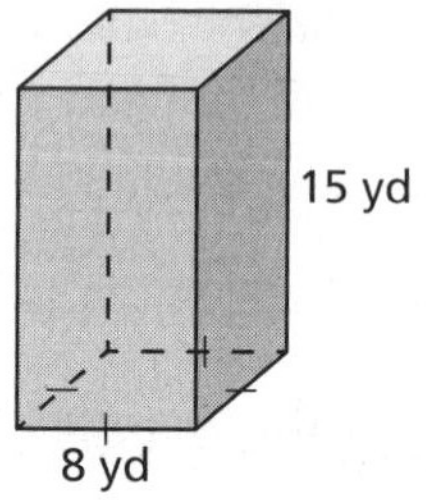

7. A baseball with a 2.9-inch diameter has a layer of leather on its surface.

   a. Does a softball with a diameter that is $\frac{4}{3}$ times the diameter of the baseball need $\frac{4}{3}$ times the amount of leather? Explain.

   b. What is the radius of a softball that uses four-thirds of the amount of leather used to cover the 2.9-inch baseball?

Name____ Date__________

1.1 Enrichment and Extension

Optimization Problems

1. A rancher has 600 feet of fencing to enclose two adjacent rectangular corrals, as shown in the diagram.

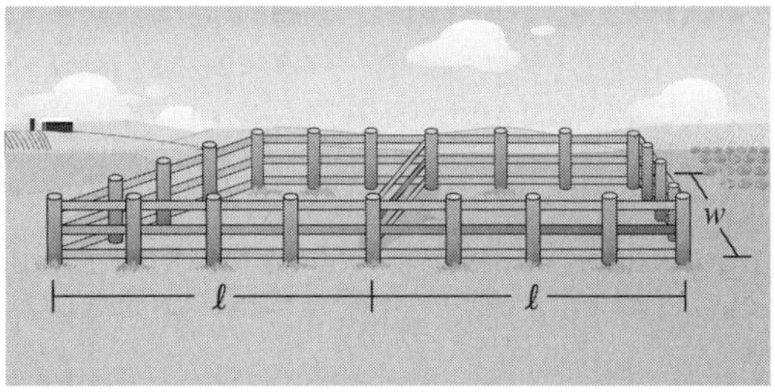

   a. Write an expression that represents the combined area of the two corrals in terms of ℓ.

   b. Use the *table* feature of a graphing utility and your expression from part (a) to determine the dimensions of the two corrals that maximize the area.

2. A manufacturer wants to design an open box having a square base and a surface area of 108 square inches, as shown in the diagram.

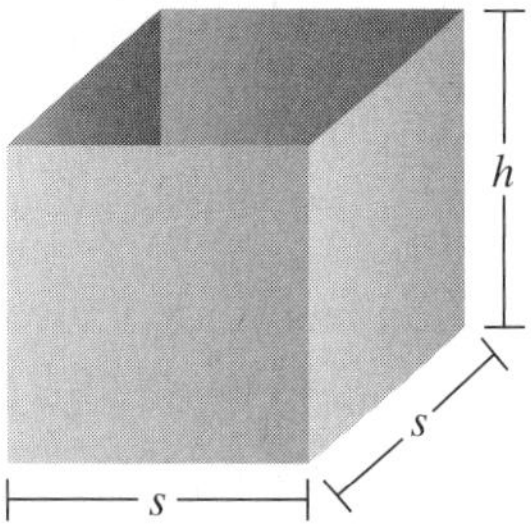

   a. Write an expression that represents the volume of the box in terms of s.

   b. Use the *table* feature of a graphing utility and your expression from part (a) to determine the dimensions of the box that maximizes the volume.

3. A rectangular page is to contain 24 square inches of print. The margins at the top and bottom of the page are $1\frac{1}{2}$ inches, and the margins on the left and right are 1 inch, as shown in the diagram.

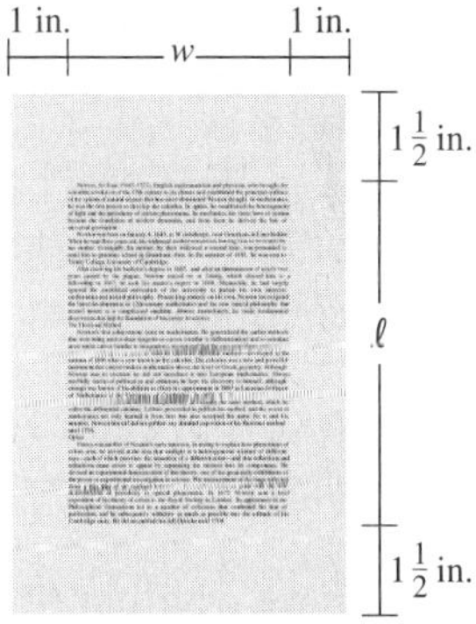

   a. Write an expression that represents the area of the page in terms of ℓ.

   b. Use the *table* feature of a graphing utility and your expression from part (a) to determine the dimensions of the printed area that minimizes the area of the page.

4. A product engineer wants the volume of a cylindrical can to be 20π cubic inches. The material for the top and bottom of the can costs \$10 per square inch, and the material for the side costs \$8 per square inch.

   a. Write an expression that represents the cost of making the can in terms of the radius r of the base of the can.

   b. Use the *table* feature of a graphing utility and your expression from part (a) to determine the dimensions of the can that result in the most economical can.

Name ______________________________ Date __________

1.1 Puzzle Time

What Kind Of Coat Can You Put On When Wet? A ...

Write the letter of each answer in the box containing the exercise number.

1. The state of Pennsylvania is approximately rectangular with a length of 280 miles and a width of 160 miles. The population is about 12.7 million people. Find the population density in people per square mile.

2. A rectangular corn field is about 50 feet long and 36 inches wide. The field yields about 108 plants. Find the population density of the field in plants per square foot.

3. A circular region has a population of about 235,000 people and a population density of about 5650 people per square mile. Find the radius of the region to the nearest tenth of a mile.

4. A circular region has a population of about 850,000 people and a population density of about 1254 people per square mile. Find the radius of the region to the nearest tenth of a mile.

5. About 1.64 million people live in a circular region with a 25-mile radius. Find the population density in people per square mile.

6. About 42,000 people live in a circular region with a 13-kilometer radius. Find the population density in people per square kilometer.

7. A professional basketball court of length ℓ and width w has a perimeter of 288 feet. Write an expression that represents the area of the basketball court in terms of w.

8. Use your expression from Exercise 7 to determine the width (in feet) of the field that maximizes the area.

Answers

N. about 1

P. $144w - w^2$

F. 3.6

A. 14.7

T. about 28.3

C. false

T. 835

A. 72

O. true

O. true

I. about 79

Determine whether the statement is *true* or *false*.

9. Doubling all the linear dimensions of a rectangular prism quadruples the surface area.

10. Tripling the height of a right cylinder will result in a right cylinder with 9 times the surface area.

11. A ball with a diameter that is $\frac{1}{3}$ times the diameter of another ball needs $\frac{1}{9}$ times the amount of rubber coating on its surface.

10	9	4	5		11	3		7	8	6	2	1

1.2 Start Thinking

Density is the amount of matter that an object or substance has in a given unit of volume. The density of an object or substance is calculated by dividing its mass by its volume.

$$\text{Density} = \frac{\text{Mass}}{\text{Volume}}$$

Ice cubes float in a glass of water. So, how does the density of ice compare to the density of water? Which weighs more, a cubic foot of ice or a cubic foot of water? Why?

1.2 Warm Up

Convert the length using 1 inch = 2.54 centimeters.

1. Convert 4 inches to centimeters.
2. Convert 5.5 inches to centimeters.
3. Convert 27 centimeters to inches.
4. Convert 46 centimeters to inches.

1.2 Cumulative Review Warm Up

Solve the equation.

1. $4^{3x} = 4^9$
2. $6^{x-2} - 6^8$
3. $2^{x+12} = 2^{4x}$
4. $7^{3x-10} = 7^x$
5. $2^x = 16^2$
6. $3^{x+6} = 81^x$

Name ______________________________________ Date ________

1.2 Practice A

1. A cylindrical container with a radius of 12 centimeters is filled to a height of 6 centimeters with coconut oil. The density of coconut oil is 0.92 gram per cubic centimeter. What is the mass of the coconut oil to the nearest gram?

2. Describe and correct the error in finding the density of an object that has a mass of 14.5 grams and a volume of 21 cubic inches.

✗	density $= 14.5 \times 21 = 304.5$ So, the density is about 304.5 grams per cubic inch.

In Exercises 3 and 4, describe how the change affects the volume of the prism or cone.

3. multiplying the height by $\frac{3}{4}$

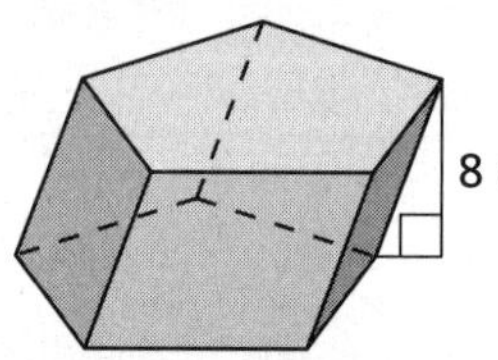

4. doubling all the linear dimensions

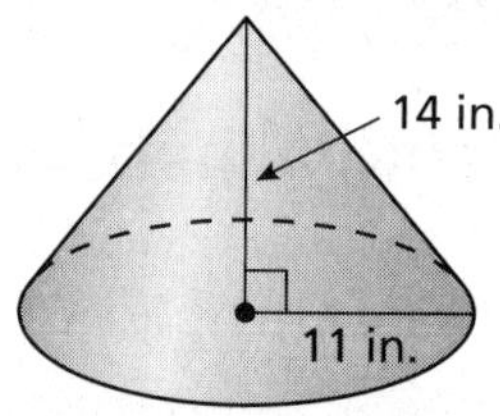

5. A cone has height h and a base with radius r. You want to change the cone so its volume is tripled. What is the new height if you only change the height? What is the new radius if you only change the radius? Explain.

6. The Pyramid Arena in Memphis, Tennessee, is about 98 meters tall and has a square base with a side length of about 180 meters. A prism-shaped building has the same square base as the Pyramid Arena. What is the height of the building if it has the same volume as the Pyramid Arena?

7. A swimming pool in the shape of a cylinder has a radius of 2 meters. The water in the pool is 1 meter deep.

   a. The density of water is about 1 gram per cubic centimeter. Find the number of kilograms of water in the pool.

   b. You add 2500 kilograms of water to the pool. What is the depth of the water in the pool?

Name______________________________ Date__________

1.2 Practice B

1. A cylindrical container with a radius of 8 centimeters is filled to a height of 10 centimeters with sulfuric acid. The density of sulfuric acid is 1.84 grams per cubic centimeter. What is the mass of the sulfuric acid to the nearest gram?

2. A block of ice is in the shape of a square with side lengths of 0.4 meter.

   a. The ice has a density of 917 kilograms per cubic meter. Find the mass of the block of ice to the nearest kilogram.

   b. The side length of the block of ice decreases by 0.05 meter this week. How many cubic centimeters of ice does the block lose? The side length decreases at a constant rate for the next 5 weeks. Does the block of ice lose the same amount of ice each week? Explain.

In Exercises 3 and 4, describe how the change affects the volume of the prism or pyramid.

3. multiplying the radius by $\frac{3}{2}$

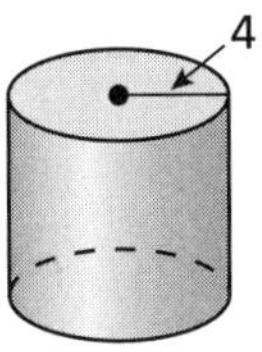

$V = 112\pi$ in.^{3}

4. tripling all the linear dimensions

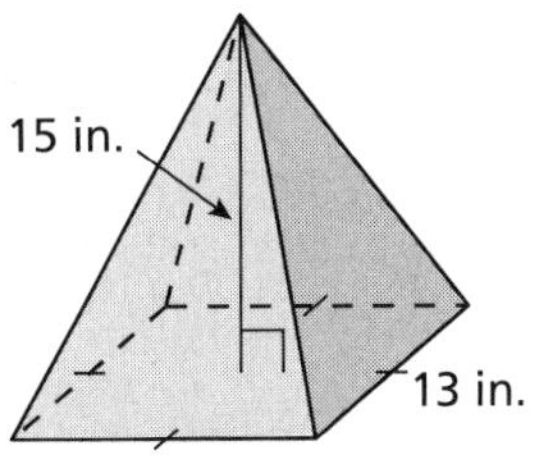

5. A cone has height h and a base with radius r. You want to change the cone so its volume is halved. What is the new height if you only change the height? What is the new radius if you only change the radius? Explain.

6. A 10 karat gold paper weight has a volume of 125 cubic centimeters. The paper weight consists of 41.7% gold and 58.3% other metal. The density of gold is 19,300 kilograms per cubic meter. The density of the other metal is 7140 kilograms per cubic meter.

   a. What is the mass of the gold in the paper weight to the nearest tenth of a kilogram?

   b. What is the mass of the other metal in the paper weight to the nearest tenth of a kilogram?

   c. What is the total mass of the paper weight?

   d. What is the total density of the paper weight in kilograms per cubic meter?

Name ______________________________ Date __________

1.2 Enrichment and Extension

Water Displacement

When an object is submerged in fluid, the object *displaces* some of the fluid. The volume of the fluid displaced is equal to the volume of the object.

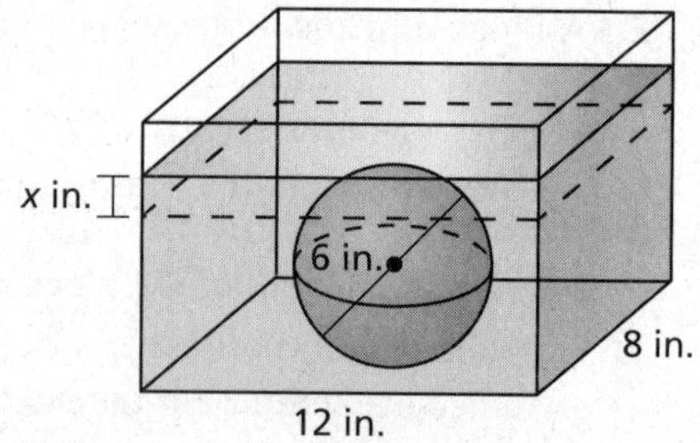

Example: The diagram shows a sphere submerged in a rectangular tank of water. How many inches does the water level rise?

First, find the volume of the sphere.

$$V = \frac{4}{3}\pi r^3 = \frac{4}{3}\pi(3)^3 = 36\pi$$

Then write an equation relating the volume of the sphere and the volume of water displaced, which is represented by a rectangular prism of width 8 inches, length 12 inches, and height x inches.

$$V = \ell wh$$

$$36\pi = (12)(8)(x)$$

$$\frac{3\pi}{8} = x$$

So, the water rises $\frac{3\pi}{8} \approx 1.2$ inches.

In Exercises 1–4, find the indicated measure.

1. A cube with a side length of 12 centimeters is submerged in a rectangular tank of water. The tank is 30 centimeters long and 18 centimeters wide. How many centimeters does the water level rise?

2. A cylinder with a radius of 6 inches and a height of 12 inches is partially submerged in a rectangular tank of water. About three-fourths of the cylinder is below the surface of the water. The tank is 30 inches long and 14 inches wide. How many inches does the water level rise?

3. To determine the density of a spherical object that weighs 2340 grams, you submerge the object in a rectangular tank of water that is 30 centimeters long and 36 centimeters wide. The water level rises about 4.5 centimeters. What is the density in grams per cubic centimeter of the object?

4. Bronze has a density of 8.7 grams per cubic centimeter. You submerge a statuette that weighs 6500 grams in a cylindrical tank of water, as shown in the diagram. Is it reasonable to conclude that the statuette is made of bronze? Explain.

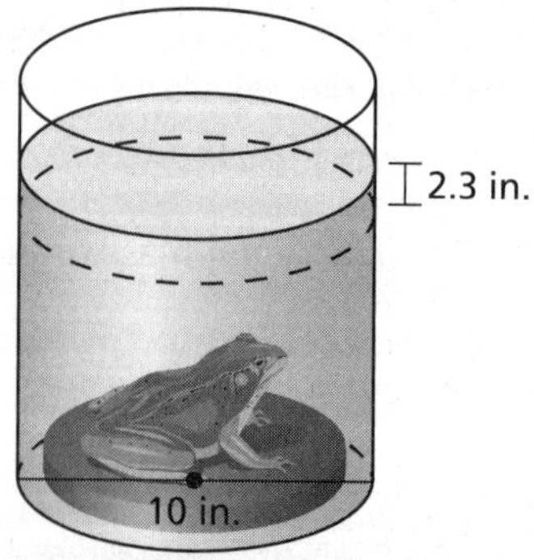

Name____ Date__________

1.2 Puzzle Time

What Must Be Added To A 400-Pound Barrel To Make It Weigh 100 Pounds?

Write the letter of each answer in the box containing the exercise number.

Find the indicated measure.

1. The diagram shows the approximate dimensions of a stick of butter. Butter has a density of 0.959 gram per cubic centimeter. Find the mass of a stick of butter to the nearest gram.

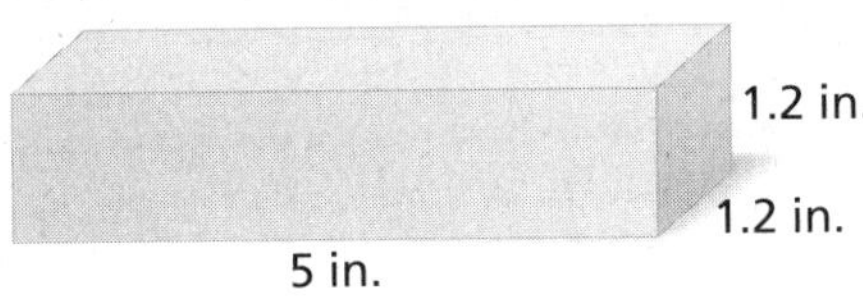

2. The diagram shows the dimensions of a lead cylindrical weight. Lead has a density of 11.342 grams per cubic centimeter. Find the mass of the cylindrical weight to the nearest gram.

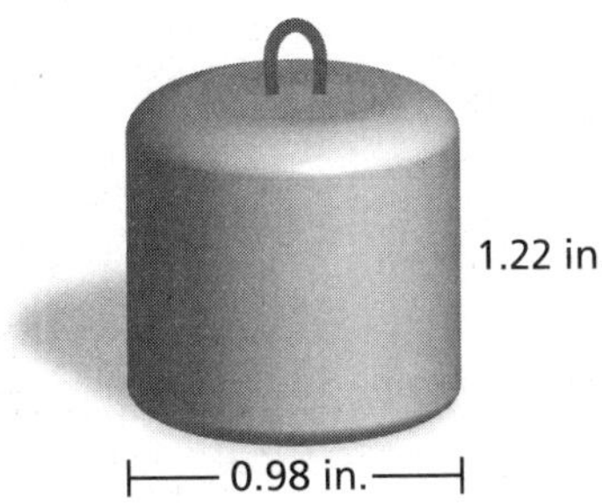

3. A piece of quartz with a volume of 64 cubic centimeters has a mass of 100 grams. What is the density of quartz to the nearest tenth of a gram per cubic centimeter?

4. A platinum cylinder has a mass of 45.14 grams, a diameter of 32.7 millimeters, and a thickness of 2.5 millimeters. What is the density of platinum to the nearest tenth of a gram per cubic centimeter?

5. A landscaper wants to haul 2 cubic yards of topsoil that has a density of about 115 pounds per cubic foot. What is the mass of the topsoil to the nearest tenth of a ton?

Answers

H. 113

A. 171

E. 21.5

L. 2.7

O. 3.1

2		1	5	3	4

1.3 Start Thinking

A round cake is pictured. Describe two different shapes that you can obtain by slicing the cake to get a cross section. How can you describe a cross section of the cake as an intersection of two geometric figures?

1.3 Warm Up

Sketch the figure described.

1. plane G and line k intersecting at all points on line k

2. $\overleftrightarrow{AB}$ and $\overleftrightarrow{BC}$ intersecting at B

3. plane A and plane B intersecting at $\overleftrightarrow{MN}$

4. lines ℓ, m, and n all intersecting at point B

5. plane D and line k intersecting at point C

6. plane C and plane D intersecting

1.3 Cumulative Review Warm Up

Find the median of the data set.

1. 1, 2, 3, 4, 5, 6, 7, 8, 9, 10, 11, 12, 13, 14, 15

2. 2, 4, 6, 8, 10, 12, 14, 16, 18, 20, 22, 24, 26, 28

3. 1, 3, 6, 6, 7, 9, 9, 14, 21, 22

4. 11, 12, 14, 17, 18, 19, 23, 23, 25, 30

5. 36, 16, 9, 47, 24, 8, 18, 58

6. 66, 51, 82, 124, 7, 91, 105, 101

Name_______________________________ Date__________

1.3 Practice A

In Exercises 1–3, describe the cross section formed by the intersection of the plane and the solid.

1.

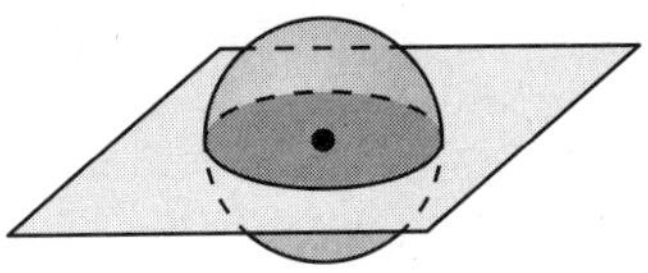

2.

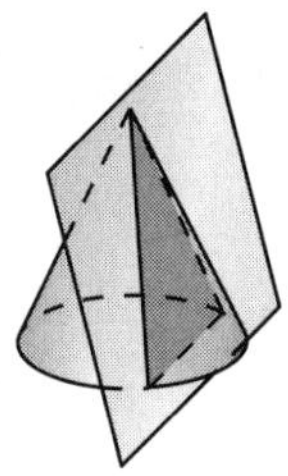

3.

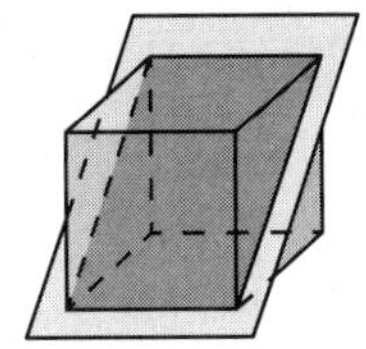

In Exercises 4–6, draw the cross section formed by the described plane that contains $\overleftrightarrow{AB}$. What is the shape of the cross section?

4. plane is perpendicular to base

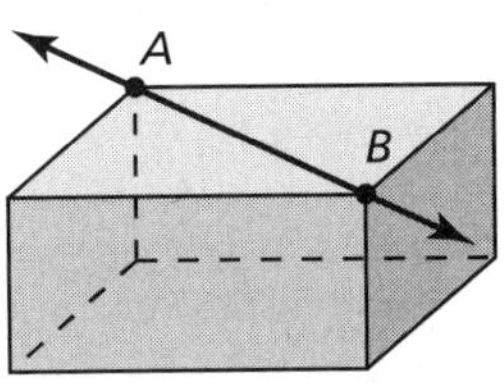

5. plane is parallel to base

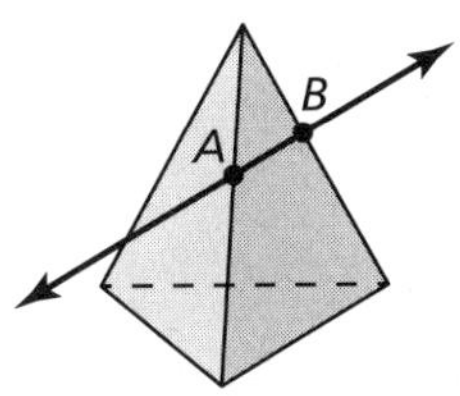

6. plane is perpendicular to base

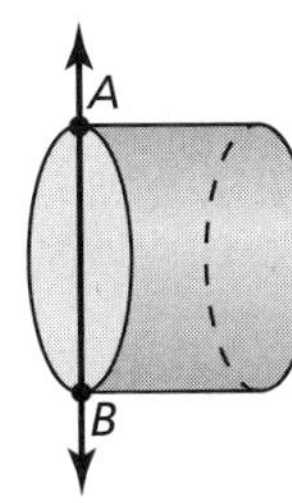

7. Draw a cross section formed by a vertical plane that divides the solid into two congruent parts. Is there more than one way to use a vertical plane to divide the figure into two congruent parts? If so, does the cross section change? Explain.

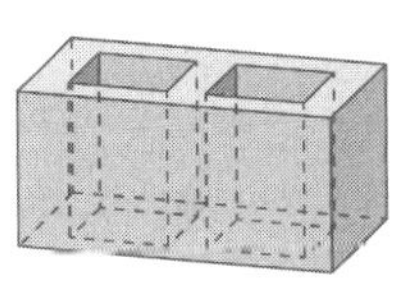

8. Sketch a cube. Is it possible for a cross section of a cube to be a square? Explain your reasoning. If so, describe or sketch two different ways in which the plane could intersect the solid.

9. Consider the rectangular prism in Exercise 4. The length of the prism is 4 inches, the width is 2 inches, and the height is 2 inches.

a. What is the perimeter of the cross section?

b. What is the area of the cross section?

Name ______________________________ Date __________

1.3 Practice B

In Exercises 1–3, describe the cross section formed by the intersection of the plane and the solid.

1.

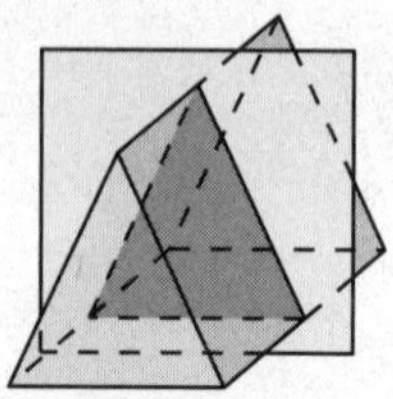

2.

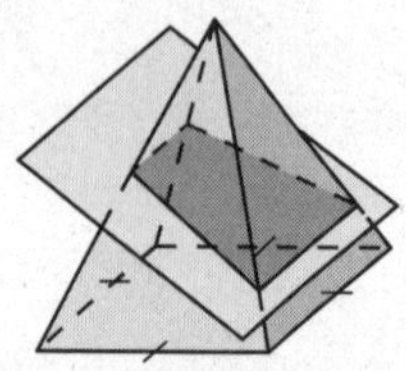

3.

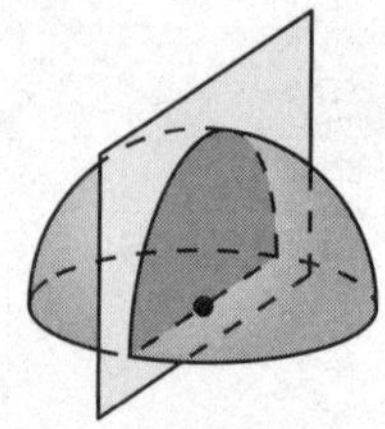

In Exercises 4–6, draw the cross section formed by the described plane that contains $\overleftrightarrow{AB}$. What is the shape of the cross section?

4. plane is perpendicular to base

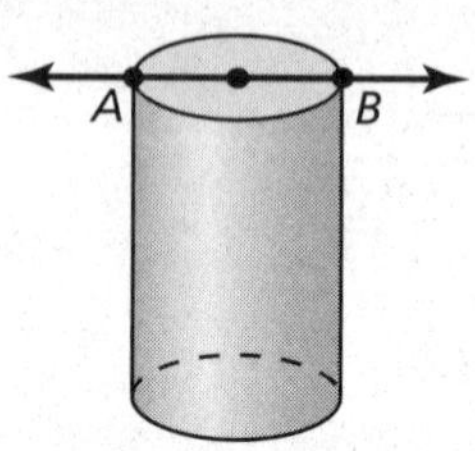

5. plane is parallel to base

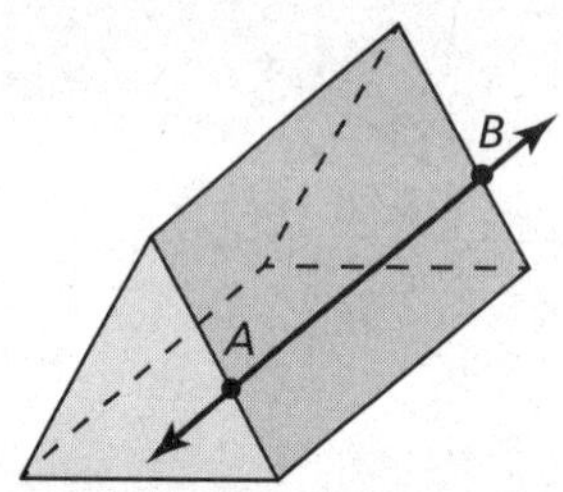

6. plane is parallel to base

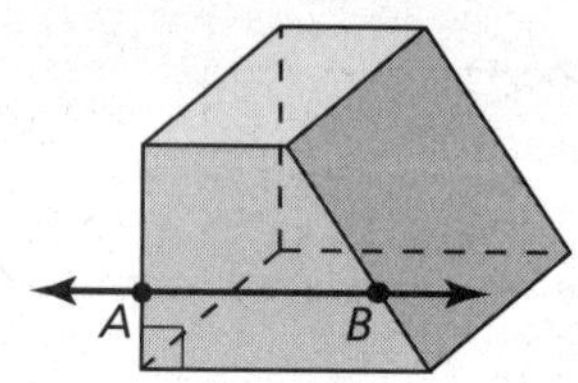

7. A cone with a height of 6 inches and radius of 4 inches is sliced in half by a horizontal plane, creating a circular cross section with a radius of 2 inches. Each piece is then sliced in half by a vertical plane, as shown.

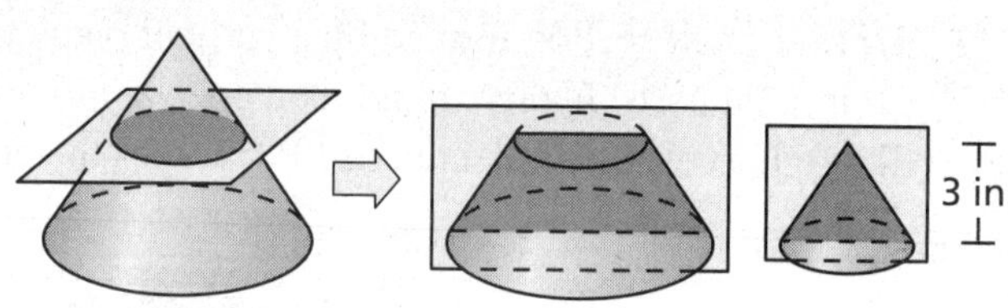

a. Describe the shape formed by each cross section.

b. What are the perimeters and areas of the cross sections?

c. Suppose the horizontal plane is tilted, slicing the original cone as shown at the right. Is the cross section a circle? If it is not, describe how it is different from a circle and sketch the cross section.

Name________________________________ Date__________

1.3 Enrichment and Extension

Comparing Volumes of Solids

Solid A is a hemisphere, as shown at the left. Solid B is formed by removing a right cone from a cylinder, as shown at the right.

Solid A

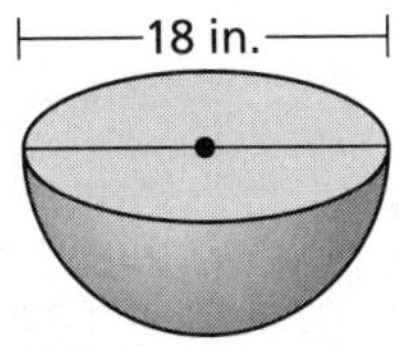

Solid B

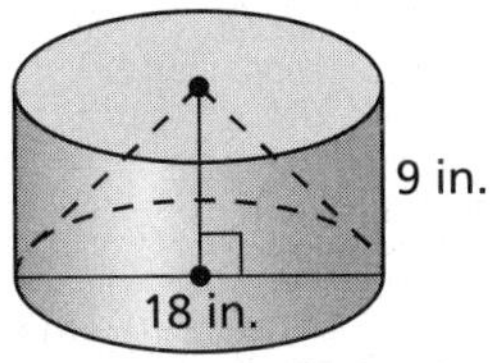

1. Consider solid A. Write an expression for the area (in square inches) of the cross section formed by a horizontal plane that intersects the solid at height h (in inches). Explain your reasoning.

2. Consider solid B. Write an expression for the area (in square inches) of the cross section formed by a horizontal plane that intersects the solid at height h (in inches). Explain your reasoning.

3. Compare your answers for Exercises 1 and 2. What do you notice?

4. Based on the heights and cross sections of the solids, what conclusion can you make about the volumes of solid A and solid B? Explain your reasoning.

5. Consider solid C, shown below. Is the volume of solid C equal to the volume of either solid A or solid B? Explain.

Solid C

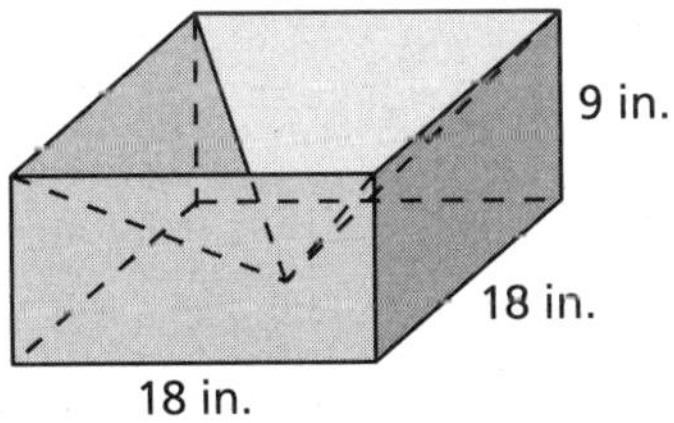

Name ______________________________ Date __________

1.3 Puzzle Time

What Happens Once In A Year, Twice In A Week, But Never In A Day?

Write the letter of each answer in the box containing the exercise number.

Complete the sentence.

1. If two planes intersect, then their intersection is a __________.

2. The intersection of a plane and a __________ is called a cross section.

3. A plane ________ intersect a cone and form an intersection that is a rectangle.

4. A plane __________ intersect a cylinder and form a cross section that is a rectangle.

Describe the shape formed by the intersection of the plane and the solid.

5. 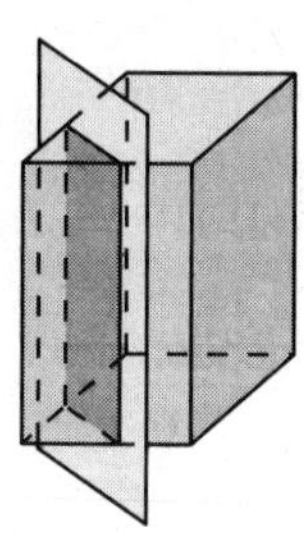

6. 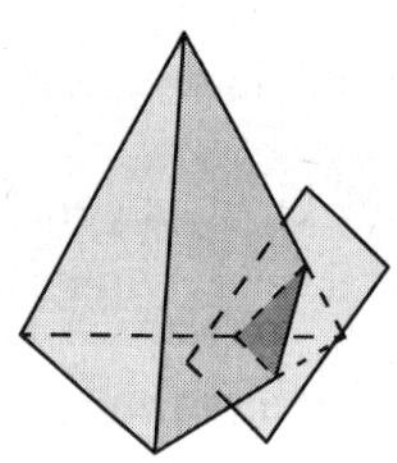

7. 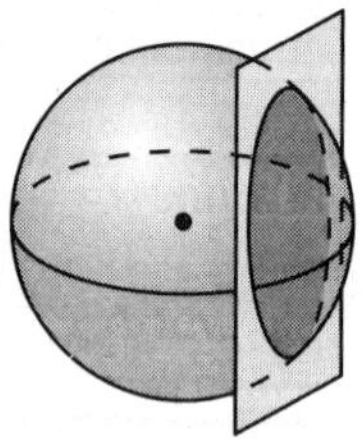

8. 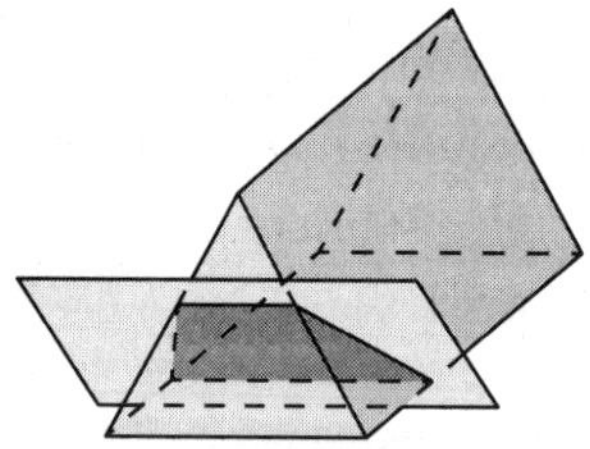

9. 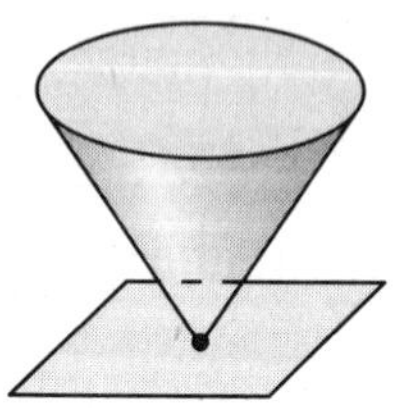

10. 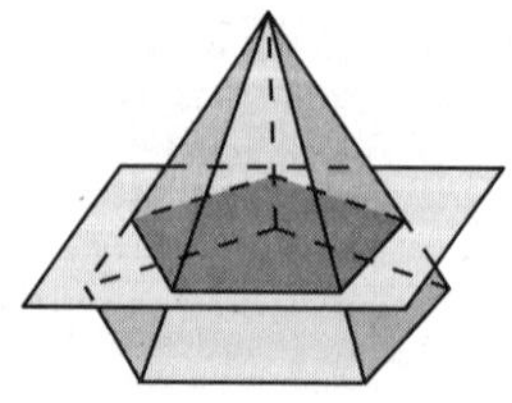

Answers

H. point
E. trapezoid
T. pentagon
L. line
E. cannot
T. circle
R. rectangle
E. triangle
E. solid
T. can

10	9	6		1	8	4	7	2	5		3

1.4 Start Thinking

Rotating a line segment $\overline{AB}$ about point A forms a circle with center A. How can you use a similar method to form a cone using the triangle shown?

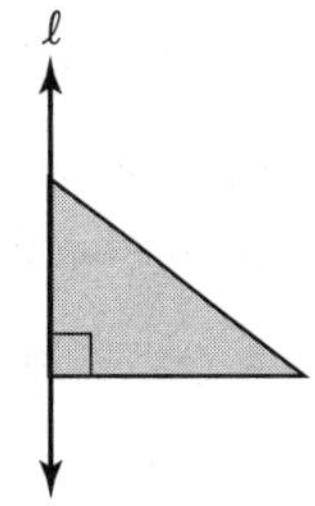

1.4 Warm Up

Find the volume of the solid.

1. 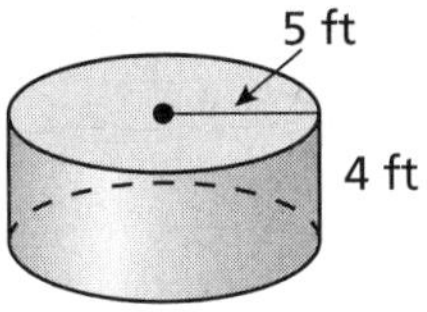

2. 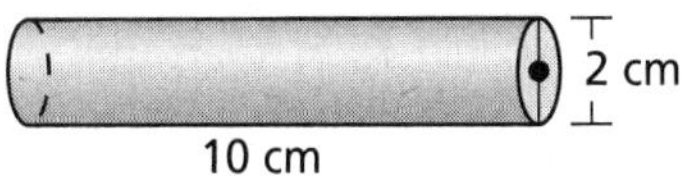

3. 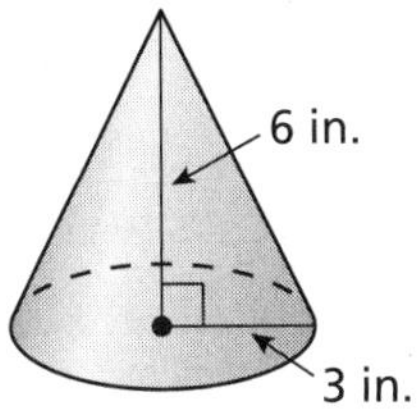

4. 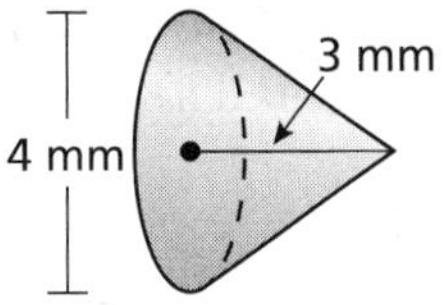

5. 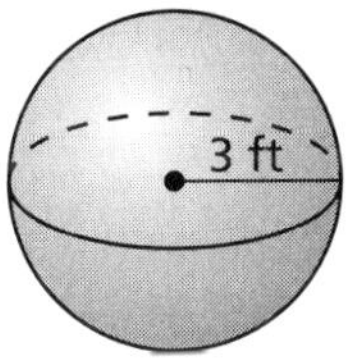

6. 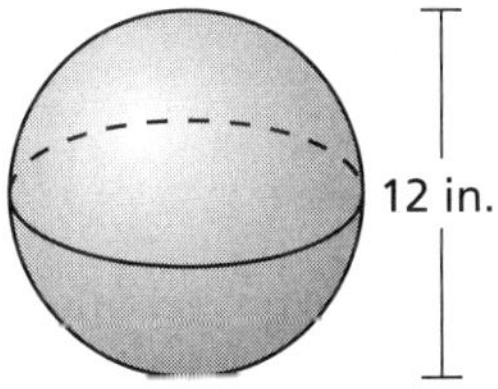

1.4 Cumulative Review Warm Up

Determine whether the sequence is arithmetic. If so, find the common difference.

1. 7, 12, 17, 22, 27, …

2. 11, 17, 22, 26, 29, …

3. −22, −12, −2, 8, 18, …

4. 11, 5, −1, −7, −13, …

5. 32, 16, 8, 4, 2, …

6. 91, 74, 57, 40, 23, …

Name ______________________________ Date __________

1.4 Practice A

In Exercises 1–3, sketch the solid produced by rotating the figure around the given axis. Then identify and describe the solid.

1.

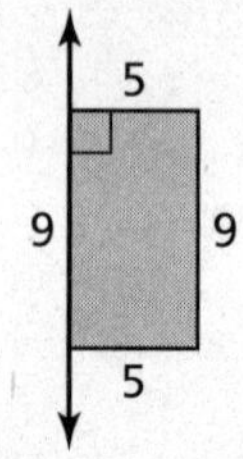

2.

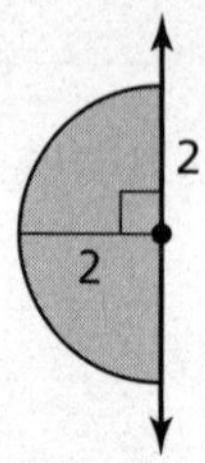

3.

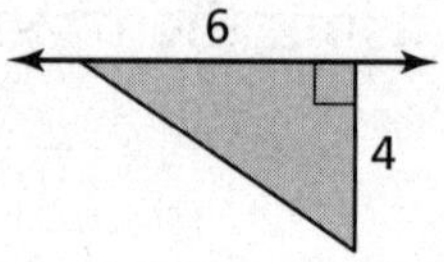

In Exercises 4–6, sketch the solid of revolution. Then identify and describe the solid.

4. a square with side length 1 rotated around one side

5. a right triangle with legs of length 3 and 7 rotated around its shorter leg

6. a semicircle with radius 8 rotated around its diameter

In Exercises 7 and 8, sketch and describe the solid produced by rotating the figure around the given axis. Then find its surface area and volume.

7.

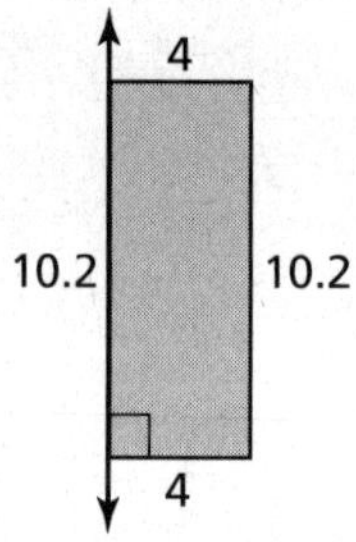

8.

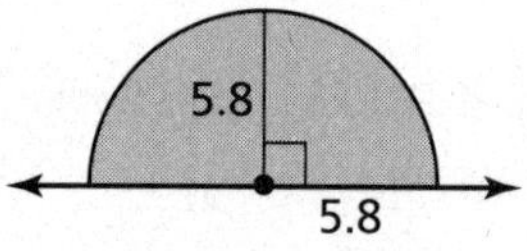

9. The figure shows the graph of a function f on an interval $[a, b]$. Sketch the solid produced when the region enclosed by the graph of f and the equations $x = a, x = b$, and $y = 0$ is rotated around the x-axis.

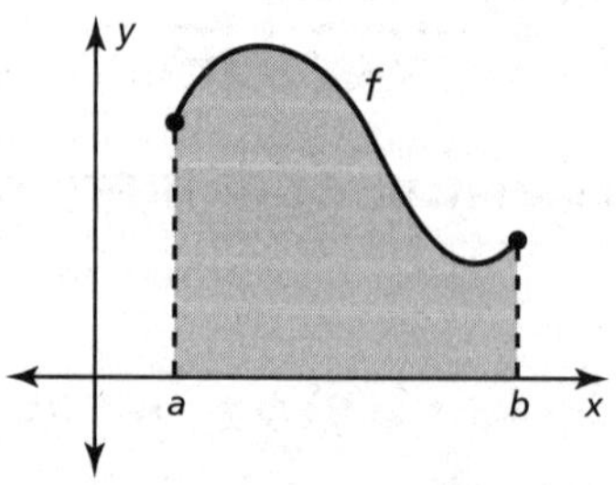

Name______________________________ Date__________

1.4 Practice B

In Exercises 1–3, sketch the solid produced by rotating the figure around the given axis. Then identify and describe the solid.

1.

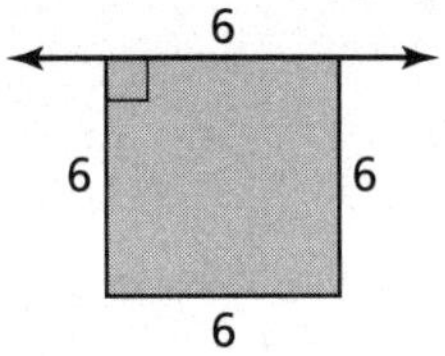

2.

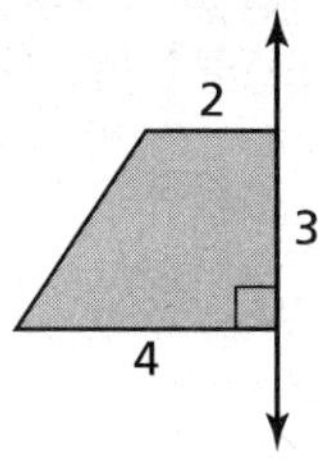

3.

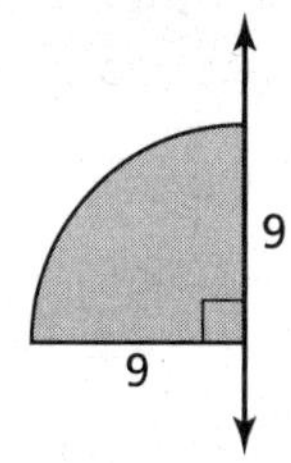

4. Sketch a two-dimensional shape and an axis of revolution that forms the object shown.

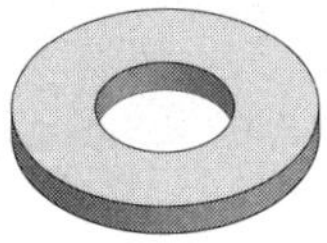

In Exercises 5–7, sketch and describe the solid that is produced when the region enclosed by the given equations is rotated around the given axis. Then find the volume of the solid.

5. $x = 0,\ y = 0,\ y = 2x + 4$; x-axis

6. $x = 0,\ y = 0,\ y = -\frac{1}{3}x + 2$; y-axis

7. $x = 5,\ y = 0,\ y = x$; x-axis

8. Sketch the composite solid produced by rotating the composite figure around the given axis. Then describe the composite solid.

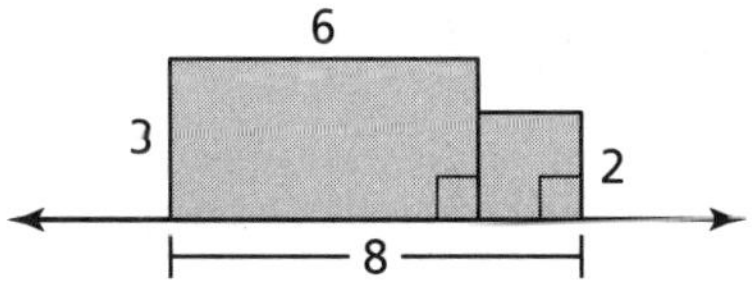

9. Your friend says when you rotate the figure shown around either the x-axis or the y-axis, the resulting solid is the same size and shape. Is your friend correct?

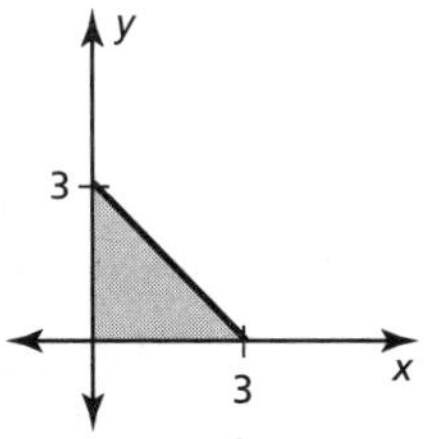

Name ______________________________ Date __________

1.4 Enrichment and Extension

Solids of Revolution

In Exercises 1–4, sketch and describe the solid that is produced when the region enclosed by the given equations is rotated around each line. Then find the volume of each solid.

1. $x = 4, y = 0, y = \frac{1}{2}x$

a. Rotate around the line $y = 2$.

b. Rotate around the line $x = 4$.

2. $x = 1, x = 6, y = 4, y = 7$

a. Rotate around the line $x = 1$.

b. Rotate around the line $x = 10$.

3. $x = 6, y = 0, y = \frac{3}{2}x - 3$

a. Rotate around the line $x = 6$.

b. Rotate around the line $y = 6$.

4. $x = -4, x = 2, y = 3, y = 4$

a. Rotate around the line $y = -3$.

b. Rotate around the line $y = 4$.

Name____ Date__________

1.4 Puzzle Time

Why Did The Clock In The Cafeteria Run Slow?

A	B	C	D	E	F

Complete each exercise. Find the answer in the answer column. Write the word under the answer in the box containing the exercise letter.

cone; $\frac{64\pi}{3}$ SECONDS	cone; 96π WENT
sphere; 519.84π ALWAYS	cylinder; 119.88π IT
cylinder; $\frac{45\pi}{4}$ FOUR	cone; 156.21π OF
cylinder; 126.54π RAN	cone; $\frac{16\pi}{3}$ BACK
cone; 64π TIME	sphere; 629.11π OUT

Identify the solid produced by rotating the figure around the given axis. Then find the surface area of the solid.

A.

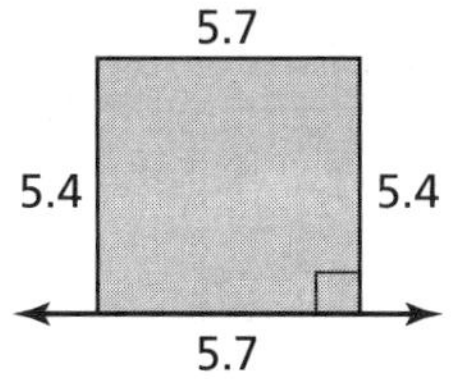

B.

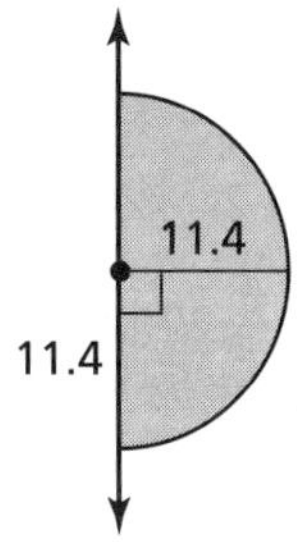

C.

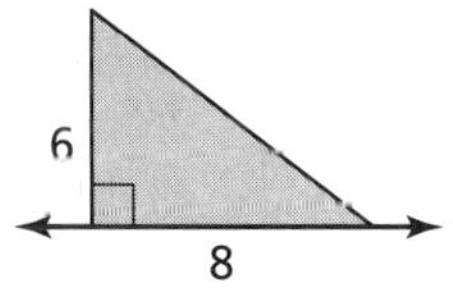

Identify the solid that is produced when the region enclosed by the given equations is rotated around the given axis. Then find the volume of the solid.

D. $x = 4, y = 0, y = \frac{1}{2}x$; x-axis

E. $x = 0, y = 0, x = \frac{3}{2}, y = 5$; y-axis

F. $x = 0, y = 0, y = x - 4$; y-axis

Name ______________________________ Date ________

Chapter 1 Cumulative Review

In Exercises 1–12, tell whether the value is a solution of the inequality.

1. $2x + 8 \geq 13;\ x = 4$

2. $7 - 3x < 9;\ x = 1$

3. $13 - 2x \leq 10;\ x = -4$

4. $1 + 5x < 2;\ x = -4$

5. $-6 < x + 9 \leq 3;\ x = 3$

6. $-7 \leq 5 - x \leq 11;\ x = -2$

7. $1 \leq 8 - 2x \leq 16;\ x = -9$

8. $3 < 4x + 3 \leq 18;\ x = 2$

9. $5x - 7 > 3$ or $-4x - 1 \leq 6;\ x = -2$

10. $4 - 2x < -1$ or $7x < 28;\ x = -2$

11. $\frac{1}{2}x + 6 \geq 8$ or $\frac{1}{4}x - 2 \leq 2;\ x = 4$

12. $\frac{1}{3}x - 1 < 2$ or $\frac{1}{6}x + 5 \geq 3;\ x = 12$

13. Your slowest time to run a mile is 12 minutes and your fastest time is 10.5 minutes. Write a compound inequality to represent your running times.

14. You plan on visiting relatives. Your vehicle gets about 21 miles per gallon in the city and 25 miles per gallon on the highway.

a. Write a compound inequality to represent the range of gas mileages for your vehicle.

b. Your relative lives in the city 31.5 miles away. How many gallons of gasoline do you use on the trip?

c. To visit another relative who lives 98 miles away, you travel 10.5 miles through the city and 87.5 miles on the highway.

In Exercises 15–28, solve the equation. Check your solution.

15. $2z + 8 = 15 - 5z$

16. $-4c + 13 = -11 + 2c$

17. $2(y - 3) - 4 = 8 + 5y$

18. $-4(2 - h) - 3 = 6 + 3(2h + 1)$

19. $\frac{1}{3}p - 5 = -2$

20. $-\frac{1}{7}b - 4 = 3$

21. $\frac{5}{2}r + \frac{3}{4} = \frac{3}{8} - \frac{1}{2}r$

22. $\frac{2}{3}s + \frac{1}{3} = \frac{3}{4}s - \frac{1}{4}$

23. $|f - 4| = 9$

24. $|-3k + 8| = 17$

25. $|2g + 2| - 1 = 6$

26. $|4v - 1| + 3 = 7$

27. $2|5 - w| - 9 = -1$

28. $4|3d - 1| + 4 = 12$

Name____ Date__________

Chapter 1 Cumulative Review (continued)

In Exercises 29–34, solve the literal equation for *y*.

29. $3x + y = 2$

30. $9x - y = 8$

31. $2y - 4x = -2$

32. $2y + x = 5$

33. $\frac{1}{3}y + 2x = 4$

34. $x - \frac{1}{5}y = 6$

In Exercises 35–38, write an expression that represents the situation.

35. You spend \$108 on x DVDs. How much does each DVD cost?

36. You have \$55 and spend x dollars. How much money do you have left?

37. You have \$32 and your friend loans you x dollars. How much money do you now have?

38. On a boat you cruise 40 miles per hour for x hours. How many miles do you cruise?

39. A taxi company charges a flat fee of \$2.50 and then \$2.25 per mile for a ride.

a. Write an expression to show the cost of traveling m miles.

b. How much does it cost to travel 9 miles?

c. How much does it cost to travel 11 miles?

d. You travel 13 miles and give the driver a \$3 tip. How much do you spend?

40. Apples cost \$2.12 per pound at the local grocery store.

a. Write an expression for the cost of buying p pounds of apples.

b. How much does it cost to buy 5 pounds of apples?

c. You buy 6 pounds of apples and pay with a \$20 bill. How much change do you receive?

d. How many pounds of apples can you buy with \$20?

In Exercises 41–48, tell which property is illustrated by the statement.

41. $4(5 + 2) = 4(5) + 4(2)$

42. $24 + (-24) = 0$

43. $3 \bullet 8 = 8 \bullet 3$

44. $(-4 \bullet 9) \bullet 7 = -4 \bullet (9 \bullet 7)$

45. $13 + 25 = 25 + 13$

46. $8 + (13 + 15) = (8 + 13) + 15$

47. $14 + 0 = 14$

48. $13 \bullet 1 = 13$

Name ____ Date __________

Chapter 1 Cumulative Review (continued)

In Exercises 49–52, draw a scatter plot of the data.

49.

x	3	4	6	9	12
f(x)	17	15	15	12	9

50.

x	10	12	14	17	18
f(x)	2	5	8	9	10

51.

x	1	1	2	3	5
f(x)	4	5	8	11	14

52.

x	5	7	8	8	11
f(x)	26	24	21	19	17

In Exercises 53–64, solve the system.

53. $x + y = 4$
$-x + 2y = -13$

54. $2x + y = 17$
$-2x + y = 1$

55. $-x + 4y = -27$
$y = 3x - 4$

56. $x = 2y + 2$
$x - 2y = -1$

57. $\frac{1}{2}x - y = -8$
$3x + 2y = 8$

58. $3x - 3y = -12$
$-2x + 2y = -8$

59. $4y = 4x + 4$
$-x + y = -1$

60. $2x + 6y = 2$
$x = 3y + 13$

61. $y = 2x + 1$
$-4x + 2y = 2$

62. $x - 2y = -13$
$4x + 2y = 18$

63. $\frac{1}{4}x + 3y = 2$
$x + 12y = 8$

64. $-\frac{1}{2}x + y = -8$
$-3x + y = -3$

In Exercises 65–72, simplify the expression.

65. $(3a + 2)^2$ **66.** $(-b + 7)^2$ **67.** $(9 + 5c)^2$ **68.** $(8 - 3d)^2$

69. $(-4m + 4)^2$ **70.** $(-2n - 1)^2$ **71.** $(6 - 5p)^2$ **72.** $(q - 2)^2$

73. You are presented with three different figures.

a. You have a cylinder with a height of 8 centimeters and a radius of 6 centimeters. Find the volume of the cylinder. Round your answer to the nearest tenth.

b. You have a sphere with a radius of 7 centimeters. Find the volume of the sphere. Round your answer to the nearest tenth.

c. You have a cone with a height of 9 centimeters and a radius of 4 centimeters. Find the volume of the cone. Round your answer to the nearest tenth.

Name______________________________ Date__________

Chapter 1 Cumulative Review (continued)

In Exercises 74–81, write an equation of the line.

74. y-intercept: 3, slope: 2

75. y-intercept: -4, slope: 1

76. y-intercept: 1, slope: $-\frac{1}{4}$

77. y-intercept: -5, slope: $-\frac{3}{4}$

78. y-intercept: 0, slope: 4

79. y-intercept: $\frac{4}{5}$, slope: 0

80. y-intercept: $-\frac{1}{2}$, slope: $-\frac{2}{7}$

81. y-intercept: 2, slope: $\frac{2}{3}$

82. A city park is triangular with a base length of 3 blocks and a height of 4 blocks. During an evening band concert, its population density is about 49 people per square block. Find the number of people in the park that evening.

83. About 125,000 people live in a circular region with a population density of about 1272 people per square mile. Find the radius of the region.

In Exercises 84 and 85, describe how the change affects the surface area of the right prism or right cylinder.

84. doubling the diameter

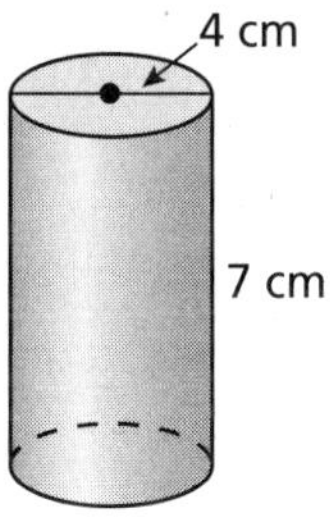

85. multiplying the base edge lengths by 3 and the height by $\frac{1}{3}$

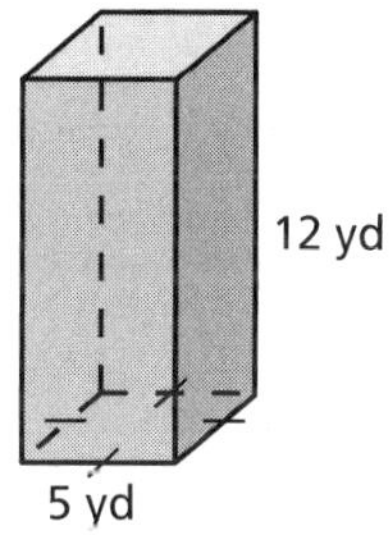

In Exercises 86–88, sketch the solid produced by rotating the figure around the given axis. Then identify and describe the solid.

86.

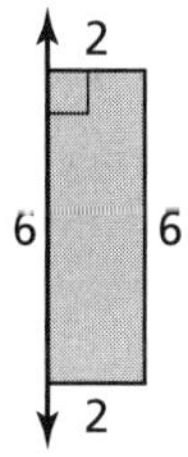

87.

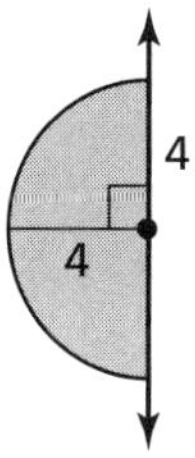

88.

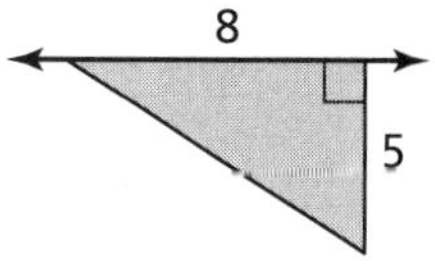

Chapter 2

Name______________________________ Date__________

Chapter 2 Linear and Quadratic Functions

Dear Family,

In this chapter, your child will be graphing based on transformations to parent functions. The transformations to be used are translations, reflections, and horizontal and vertical stretches and shrinks.

These transformations are similar to how television show animators create cartoons. Each frame is drawn with pencil and paper, transferred onto a transparency using ink and color, and then photographed. Each 20-minute cartoon is created by sequencing about 30,000 different frames. When there are slight changes in each frame, using various transformations, the objects appear to be animated.

To make your own flip book at home, follow the steps below:

1. Gather a stack of paper. The more pages or frames you have, the better your animation will look.
2. Determine what you want to illustrate. You can use stick figures and easy-to-draw objects, or you can use objects that have a lot of detail. It's your choice!
3. Draw your first picture at the bottom of the stack. It will make it easier to trace or reference part of the picture that remains the same.
4. Draw the next pages so that objects that do not move are in the same position, and the animated objects move slightly in each frame. Adding translations and rotations gives the illusion of movement. Adding horizontal stretches and shrinks may add some humor to your animation.
5. Flip the pages of your book!

You can make each page as detailed as you want. You can also use the Internet to find many free animation websites that will allow you to create your own personalized animation electronically.

Have fun creating your own animation.

Nombre ______________________________ Fecha ________

Capítulo 2 Funciones lineales y cuadráticas

Estimada familia:

En este capítulo, su hijo hará gráficas basándose en transformaciones de funciones madre. Las transformaciones que se usarán serán traslaciones, reflexiones y alargamientos y encogimientos horizontales y verticales.

Estas transformaciones se parecen a la manera en que los animadores de programas televisivos crean los dibujos animados. Cada marco se dibuja con lápiz y papel, se traslada a una transparencia usando tinta y color y luego se toma una fotografía. Cada dibujo animado de 20 minutos se crea con una secuencia de aproximadamente 30,000 marcos diferentes. Cuando hay cambios menores en cada marco, mediante el uso de varias transformaciones, los objetos parecen ser animados.

Para crear su propio foliscopio en casa, sigan los siguientes pasos:

1. Busquen una pila de hojas de papel. Cuantas más páginas o marcos tengan, mejor lucirá su animación.
2. Decidan qué quieren ilustrar. Pueden usar monigotes hechos con palitos y objetos fáciles de dibujar o pueden usar objetos que tengan muchos detalles. ¡Es su elección!
3. Dibujen el primer dibujo en la parte inferior de la pila. Será más fácil calcar o tomar como referencia a una parte del dibujo que no cambia.
4. Dibujen las siguientes páginas de manera que los objetos que no se mueven estén en la misma posición y que los objetos animados se muevan levemente en cada marco. Agregar traslaciones y rotaciones da la impresión de que hay movimiento. Agregar alargamientos y encogimientos horizontales y verticales puede sumar algo de humor a su animación.
5. ¡Pasen las páginas de su foliscopio!

En cada página, pueden incluir todos los detalles que quieran. También pueden buscar en Internet muchos sitios web de animación gratis que les permitirán crear su propia animación electrónica personalizada.

Que se diviertan creando su propia animación.

2.1 Start Thinking

Draw a coordinate plane. Tape the middle of a string to the origin to make a V with the ends of the string. Then tape the ends of the string in the coordinate plane. Trace the line formed by the string. Move the ends of the string wider and trace the new lines. What happens as the string ends are stretched wider?

2.1 Warm Up

Plot the ordered pairs from the table in a coordinate plane. Connect them with a line.

1.

x	y
-3	-3
-2	-2
-1	-1
0	0

2.

x	y
-2	-5
0	-3
3	0

3.

x	y
-2	-5
0	0
2	5

4.

x	y
-4	1
-2	3
0	5

2.1 Cumulative Review Warm Up

Find the slope of the line passing through the set of points.

1. $(-4, 4)$ and $(2, 1)$

2. $(-2, 0)$ and $(0, 2)$

3. $(5, 2)$ and $(7, 3)$

4. $(10, 300)$ and $(15, 425)$

Name ______________________________ Date ________

2.1 Practice A

In Exercises 1 and 2, identify the function family to which *f* belongs. Compare the graph of *f* to the graph of its parent function.

1.

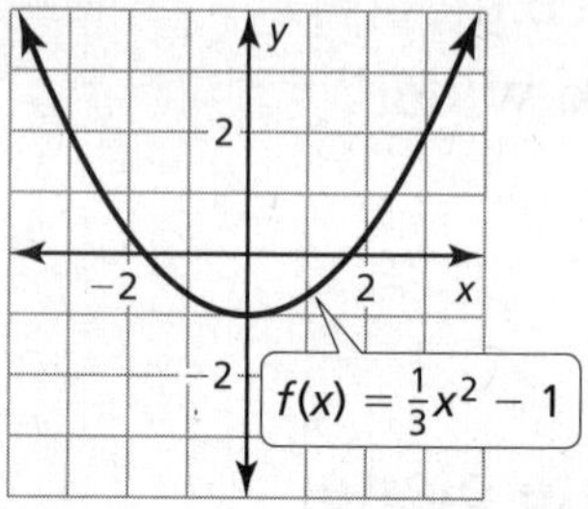

2.

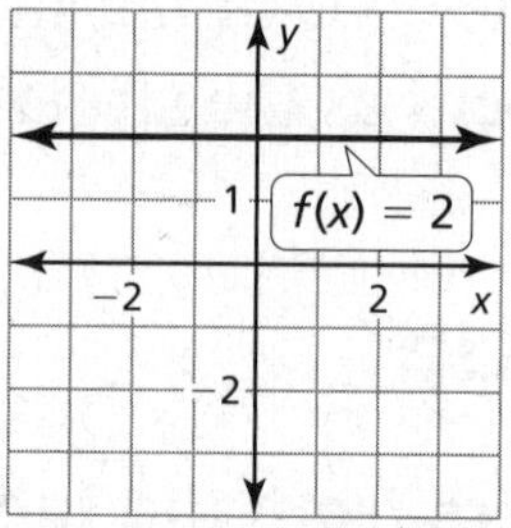

3. You purchased a computer for your business for \$800. Using straight-line depreciation, the amount of depreciation allowed for each year after the purchase is given by the function $f(x) = 800 - 114.29x$. Identify the function family to which f belongs.

In Exercises 4–9, graph the function and its parent function. Then describe the transformation.

4. $h(x) = x + 2$

5. $f(x) = x - 3$

6. $g(x) = x^2 + 2$

7. $f(x) = (x - 1)^2$

8. $h(x) = |x + 4|$

9. $f(x) = 5$

In Exercises 10–15, graph the function and its parent function. Then describe the transformation.

10. $f(x) = 3x$

11. $g(x) = \frac{1}{2}x$

12. $h(x) = 3x^2$

13. $g(x) = \frac{1}{4}x^2$

14. $h(x) = 2|x|$

15. $f(x) = \frac{5}{2}x$

In Exercises 16–18, use a graphing calculator to graph the function and its parent function. Then describe the transformations.

16. $f(x) = \frac{1}{3}x - 1$

17. $h(x) = 2|x| - 3$

18. $g(x) = \frac{5}{3}x^2 + 2$

19. In the same coordinate plane, sketch the graph of a parent absolute-value function and the graph of an absolute-value function that has no x-intercepts. Describe the transformation(s) of the parent function.

Name_______________________________________ Date__________

2.1 Practice B

In Exercises 1 and 2, identify the function family to which *f* belongs. Compare the graph of *f* with the graph of its parent function.

1.

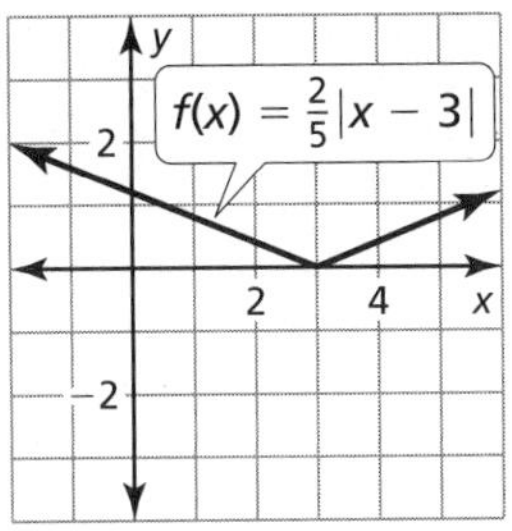

2.

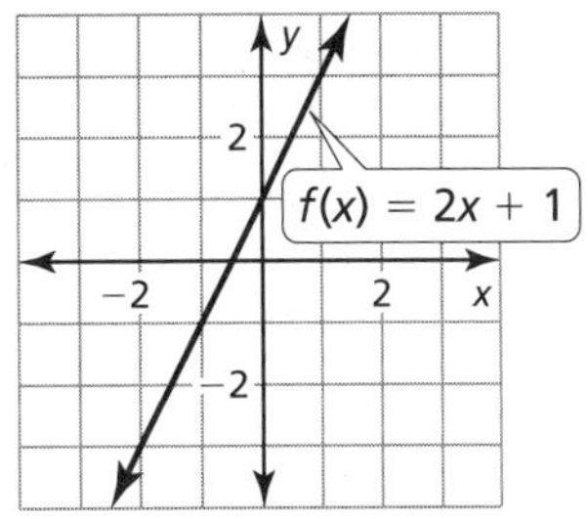

In Exercises 3–8, graph the function and its parent function. Then describe the transformation.

3. $h(x) = x + 2$ **4.** $f(x) = -x$ **5.** $g(x) = -x^2$

6. $f(x) = (x + 2)^2$ **7.** $h(x) = |x| - 2$ **8.** $f(x) = -3$

In Exercises 9–11, graph the function and its parent function. Then describe the transformation.

9. $f(x) = \frac{3}{5}x$ **10.** $h(x) = \frac{3}{2}|x|$ **11.** $h(x) = \frac{4}{3}x^2$

In Exercises 12–14, use a graphing calculator to graph the function and its parent function. Then describe the transformations.

12. $g(x) = \frac{1}{10}x^2 + 5$ **13.** $h(x) = (x - 5)^2 + \frac{4}{9}$ **14.** $f(x) = -|x + 2| - \frac{1}{3}$

In Exercises 15–18, identify the function family and describe the domain and range. Use a graphing calculator to verify your answer.

15. $h(x) = |x + 5| + 3$ **16.** $g(x) = -2x - 10$ **17.** $g(x) = 7x^2 - 3$

18. You are throwing a football with your friends. The height (in feet) of the ball above the ground *t* seconds after it is released from your hand is modeled by the function $f(t) = -16t^2 + 45t + 6$.

a. Without graphing, identify the type of function modeled by the equation.

b. What is the value of *t* when the ball is released from your hand? Explain.

c. How many feet above the ground is the ball when it is released from your hand? Explain.

Name ______________________ Date ________

2.1 Enrichment and Extension

Parent Functions and Transformations

1. Graph two trapezoids, one with vertices $A(0, 1)$, $B(1, 0)$, $C(4, 0)$, and $D(5, 1)$, and the other with vertices $A'(-6, -2)$, $B'(-5, -1)$, $C'(-2, -1)$, and $D'(-1, -2)$. Then describe a possible sequence of transformations from trapezoid $ABCD$ to trapezoid $A'B'C'D'$.

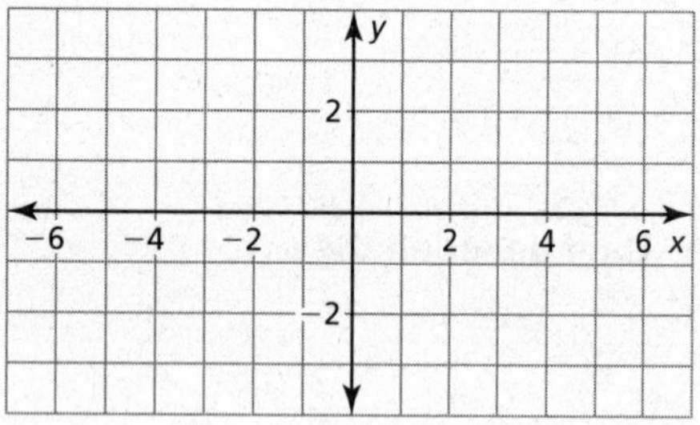

2. You purchase a car from a dealership for \$15,000. The trade-in value of the car x years after the purchase is given by the function $f(x) = 15{,}000 - 300x^2$.

   a. What type of function is used to model the data?

   b. After how many years will the trade-in value be \$0?

   c. In this situation, what are the domain and range?

In Exercises 3–5, use the table to state the domain and range of the function and describe the transformation of the function from its parent function. Use a graphing calculator to check your answers.

	Domain	Range	Vertical stretch or shrink	Horizontal stretch or shrink	Reflection(s)	Translation(s)
3. $y = -\frac{1}{2}x + 5$						
4. $y = -4x^2 - 3$						
5. $y = \left\lvert -\frac{1}{3}x - 1 \right\rvert + 5$						

Name____________________ Date__________

2.1 Puzzle Time

Why Didn't The Man Fix Dinner?

A	B	C	D	E	F
G	H	I	J	K	L
M					

Complete each exercise. Find the answer in the answer column. Write the word under the answer in the box containing the exercise letter.

$f(x) = \frac{2}{3}x$	BECAUSE
$f(x) = \frac{1}{2}x - 1$	PEOPLE
$f(x) = -x^2$	DO
$f(x) = \lvert x - 5 \rvert + 3$	IT
$f(x) = x^2 - 1$	IF
$f(x) = 5x - 3$	BROKEN
$f(x) = (x + 1)^2 + 2$	FIX

Identify the function that matches the given description of the graph.

A. vertical shrink of the parent linear function

B. vertical shrink followed by a translation 1 unit down of the parent linear function

C. reflection in the x-axis followed by a vertical stretch and a translation 3 units right of the parent absolute value function

D. vertical stretch followed by a translation 7 units down of the parent linear function

E. translation 1 unit down of the parent quadratic function

F. vertical stretch of the parent linear function

G. reflection in the x -axis followed by a translation 1 unit up of the parent quadratic function

H. reflection in the x -axis followed by a vertical stretch and a translation 4 units up of the parent absolute value function

I. vertical stretch followed by a translation 3 units down of the parent linear function

J. reflection in the x-axis of the parent quadratic function

K. vertical stretch of the parent absolute value function

L. translation 1 unit left and 2 units up of the parent quadratic function

M. translation 5 units right and 3 units up of the parent absolute value function

$f(x) = -2\lvert x \rvert + 4$	NOT
$f(x) = 7x$	IT
$f(x) = -4\lvert x - 3 \rvert$	ALWAYS
$f(x) = 9\lvert x \rvert$	NOT
$f(x) = -x^2 + 1$	IS
$f(x) = 2x - 7$	SAY
$f(x) = (x - 1)^2 + 2$	HUNGRY

2.2 Start Thinking

Graph the line $h(x) = \frac{2}{3}x + 2$ in a coordinate plane. What happens if 1 is added to the right side of the equation? What happens if -1 is added? Explain what happens to each point on the line when a number is added to one side of the equation of the line.

2.2 Warm Up

Graph the point and its reflection.

1. point $(5, 2)$ reflected in the x-axis
2. point $(-1, 0)$ reflected in the y-axis
3. point $(1, 2)$ reflected in the y-axis
4. point $(3, -3)$ reflected in the x-axis
5. point $(-3, 3)$ reflected in the line through $(-5, 1)$ and $(-2, 1)$
6. point $(-4, 3)$ reflected in the line through $(2, 2)$ and $(4, 2)$

2.2 Cumulative Review Warm Up

Determine the number of lines of symmetry, if any, for the letter. Draw each line of symmetry.

1. T
2. E
3. L
4. W
5. H
6. R

Name______________________________ Date__________

2.2 Practice A

In Exercises 1–4, write a function *g* whose graph represents the indicated transformation of the graph of *f*. Use a graphing calculator to check your answer.

1. $f(x) = x - 2$; translation 5 units left

2. $f(x) = x + 1$; translation 4 units right

3. $f(x) = |3x + 2| + 4$; translation 3 units down

4. $f(x) = 4x - 5$; translation 3 units up

In Exercises 5–8, write a function *g* whose graph represents the indicated transformation of the graph of *f*. Use a graphing calculator to check your answer.

5. $f(x) = -3x + 7$; reflection in the x-axis

6. $f(x) = \frac{1}{3}x - 2$; reflection in the x-axis

7. $f(x) = |4x| - 6$; reflection in the y-axis

8. $f(x) = |3x - 5| + 3$; reflection in the y-axis

In Exercises 9–12, write a function *g* whose graph represents the indicated transformation of the graph of *f*. Use a graphing calculator to check your answer.

9. $f(x) = x + 3$; vertical stretch by a factor of 4

10. $f(x) = 4x + 3$; vertical shrink by a factor of $\frac{1}{3}$

11. $f(x) = |3x| + 2$; horizontal shrink by a factor of $\frac{1}{3}$

12. $f(x) = |x + 1|$; horizontal stretch by a factor of 3

In Exercises 13 and 14, write a function *g* whose graph represents the indicated transformation of the graph of *f*.

13. $f(x) = x$; vertical shrink by a factor of $\frac{1}{3}$ followed by a translation 4 units down

14. $f(x) = |x|$; translation 3 units left followed by a horizontal shrink by a factor of $\frac{1}{2}$

Name ______________________________ Date __________

2.2 Practice B

In Exercises 1–4, write a function *g* whose graph represents the indicated transformation of the graph of *f*. Use a graphing calculator to check your answer.

1. $f(x) = 5x - 2$; translation 5 units right

2. $f(x) = 3x + 6$; translation 4 units up

3. $f(x) = 3 - |x - 2|$; translation 2 units left

4. $f(x) = |2x| + 3$; translation 2 units down

In Exercises 5–8, write a function *g* whose graph represents the indicated transformation of the graph of *f*. Use a graphing calculator to check your answer.

5. $f(x) = -x + 3$; reflection in the y-axis

6. $f(x) = \frac{2}{3}x - 4$; reflection in the x-axis

7. $f(x) = -5 + |x - 8|$; reflection in the y-axis

8. $f(x) = |4x - 1| + 2$; reflection in the y-axis

In Exercises 9–12, write a function *g* whose graph represents the indicated transformation of the graph of *f*. Use a graphing calculator to check your answer.

9. $f(x) = 3 - x$; horizontal stretch by a factor of 2

10. $f(x) = 3x + 5$; vertical shrink by a factor of $\frac{1}{3}$

11. $f(x) = |3x| + 2$; horizontal shrink by a factor of $\frac{1}{3}$

12. $f(x) = -2|x - 2| + 4$; vertical stretch by a factor of 2

In Exercises 13 and 14, write a function *g* whose graph represents the indicated transformation of the graph of *f*.

13. $f(x) = x$; translation 5 units up followed by a vertical shrink by a factor of $\frac{1}{4}$

14. $f(x) = |x|$; reflection in the x-axis followed by a translation 2 units left

Name______________________________ Date__________

2.2 Enrichment and Extension

Transformations of Linear and Absolute Value Functions

In Exercises 1–6, write a function *g* whose graph represents the indicated transformations of the graph of *f*. Then find the *x*-intercept of the graph of *g*. Use a graphing calculator to check your answers.

$$f(x) = 2x - 1$$

1. translation 3 units to the right followed by a translation 1 unit down
2. translation 1 unit to the left followed by a reflection in the x-axis
3. vertical stretch by a factor of 3 followed by a translation 3 units down
4. horizontal shrink by a factor of $\frac{1}{3}$ followed by a translation 5 units up
5. translation 3 units to the right followed by a vertical stretch by a factor of 2
6. translation 1 unit up followed by a reflection in the x-axis and a translation 3 units to the left

In Exercises 7–12, write a function *g* whose graph represents the indicated transformations of the graph of *f*. Then find all *x*-intercept(s) of the graph of *g*. Use a graphing calculator to check your answers.

$$f(x) = |x + 2| - 1$$

7. translation 3 units to the right followed by a translation 1 unit down
8. translation 1 unit to the left followed by a translation 2 units up
9. translation 1 unit up followed by a reflection in the x-axis and a translation 3 units to the left
10. translation 1 unit to the right followed by a vertical stretch by a factor of 2 and a translation 4 units down
11. horizontal shrink by a factor of $\frac{1}{4}$ followed by a translation 10 units to the right and 1 unit up, and a reflection in the x-axis
12. translation 5 units to the right followed by a translation 3 units down, a vertical shrink by a factor of $\frac{1}{2}$, and a reflection in the x-axis

Name ______________________________ Date __________

2.2 Puzzle Time

What U.S. President Died July 4, 1831?

Write the letter of each answer in the box containing the exercise number.

Write a function *g* whose graph represents the indicated transformation of the graph of *f*.

1. $f(x) = x + 4$; translation 3 units left
2. $f(x) = x - 7$; translation 5 units right
3. $f(x) = |2x - 5| + 3$; translation 2 units up
4. $f(x) = -4x - 8$; reflection in the x-axis
5. $f(x) = |2x + 1| - 6$; reflection in the y-axis
6. $f(x) = -x + 5$; horizontal shrink by a factor of $\frac{1}{2}$
7. $f(x) = |2x - 4|$; vertical stretch by a factor of 4

Write a function *g* whose graph represents the indicated transformations of the graph of *f*.

8. $f(x) = x$; vertical stretch by a factor of 3 followed by a translation 2 units down
9. $f(x) = x$; translation 1 unit up followed by a vertical shrink by a factor of $\frac{1}{4}$
10. $f(x) = |x|$; reflection in the x-axis followed by a translation 2 units right
11. $f(x) = |x|$; vertical shrink by a factor of $\frac{1}{2}$ followed by a translation 5 units up and 1 unit left

Answers

R. $g(x) = \frac{1}{4}x + \frac{1}{4}$

O. $g(x) = -|x - 2|$

A. $g(x) = x - 12$

M. $g(x) = |2x - 5| + 5$

M. $g(x) = -2x + 5$

E. $g(x) = 4x + 8$

E. $g(x) = \frac{1}{2}|x + 1| + 5$

S. $g(x) = |-2x + 1| - 6$

N. $g(x) = 3x - 2$

J. $g(x) = x + 7$

O. $g(x) = 4|2x - 4|$

1	2	3	4	5		6	7	8	9	10	11

2.3 Start Thinking

Your business is debuting a new product to be sold at $14.99 per unit. Your business produces this product at a cost of $8.50 per unit. Your goal is to make $150,000 in profit from this product in the next year. How could you model this situation with an equation and a graph to "sell" this idea to your employees?

2.3 Warm Up

Determine a reasonable domain and range for the situation. Write your answer in set notation.

1. A professional basketball player earns $150,000 for each game played, and there are 82 games in a season.
2. You eat up to 5 meals a day with an average of 844 milligrams of potassium at each meal.
3. The average amount of money spent on food per person at an amusement park that can accommodate 2500 people is $5.25.

2.3 Cumulative Review Warm Up

Find the mean of the data set. Round to the nearest tenth, if necessary.

1. 5, 2, 7, 4, 6, 6, 6
2. 11, 6, 8, 8, 11, 13, 9, 7, 11
3. 10, 9, 6, 10.1, 10.9, 9.6, 9.8, 16.9, 10.1
4. Salaries at a company: $216,000 $95,000 $80,600

Name ______________________ Date ________

2.3 Practice A

In Exercises 1 and 2, use the graph to write an equation of the line and interpret the slope.

1. 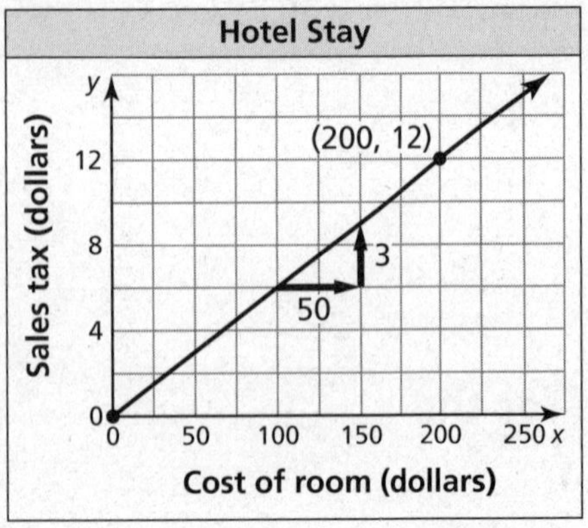

2. 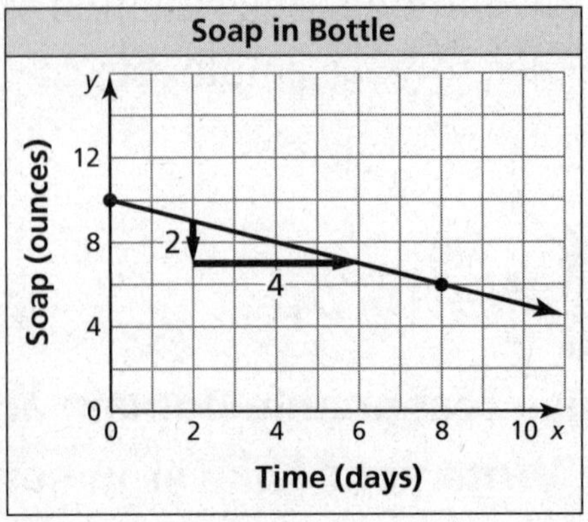

3. Two car washes charge a basic fee plus a fee based on the number of extras that are chosen. The table below shows the total costs for different car washes at Bubbles Car Wash. The total cost y (in dollars) for a car wash with x extras at Soapy Car Wash is represented by the equation $y = x + 9$. Which car wash charges more for the basic fee? How many extras must be chosen for the total costs to be the same?

Number of extras, x	2	4	6	8
Total cost, y	9	12	15	18

In Exercises 4 and 5, determine whether the data show a linear relationship. If so, write an equation of a line of fit. Estimate y when $x = 15$ and explain its meaning in the context of the situation.

4.

Weeks, x	3	6	10	12	16
Height of basil plant (inches), y	1	2	5	9	15

5.

Minutes, x	6	10	14	20	24
Cars washed, y	3	5	7	10	12

6. A set of data points has a correlation coefficient $r = -0.86$. Your friend claims that because the correlation coefficient is close to -1, it is reasonable to use the line of best fit to make predictions. Is your friend correct? Explain your reasoning.

Name____ Date__________

2.3 Practice B

In Exercises 1 and 2, use the graph to write an equation of the line and interpret the slope.

1.

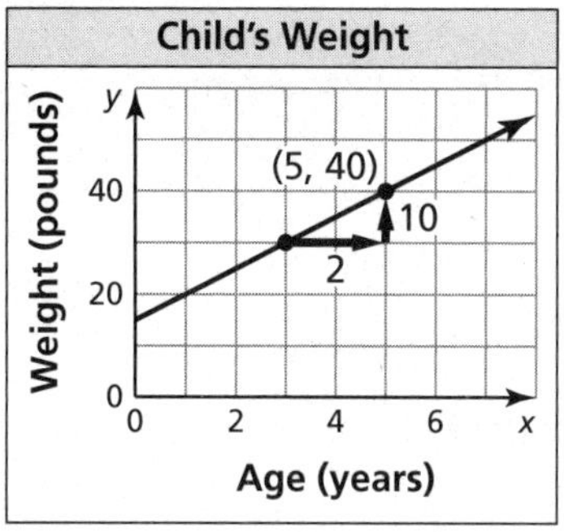

2.

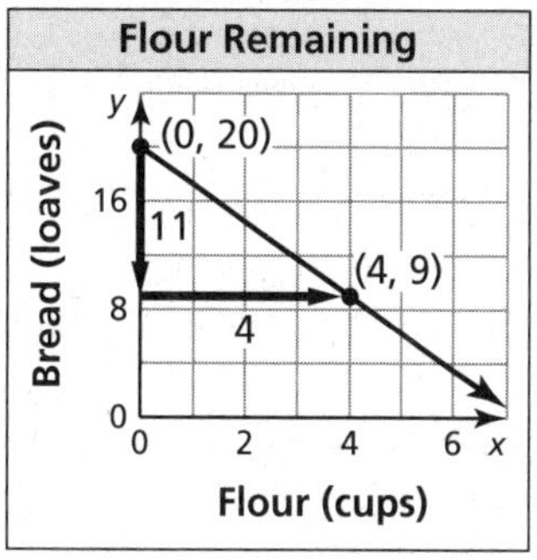

In Exercises 3 and 4, determine whether the data show a linear relationship. If so, write an equation of a line of fit. Estimate *y* when $x = 15$ and explain its meaning in the context of the situation.

3.

Days, *x*	3	7	11	14	20
Number of tickets sold, *y*	76	164	252	318	450

4.

Minutes running, *x*	6	10	17	25	40
Calories burned, *y*	70	118	200	295	472

In Exercises 5 and 6, use the *linear regression* feature on a graphing calculator to find an equation of the line of best fit for the data. Find and interpret the correlation coefficient.

5.

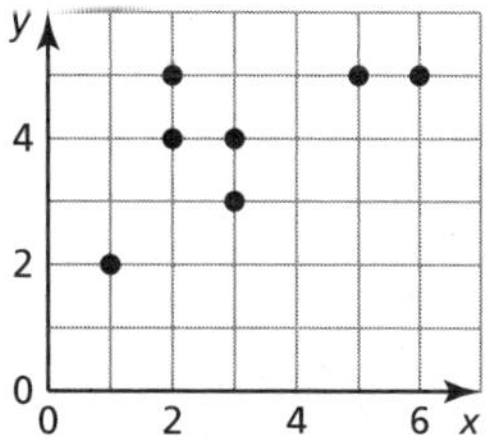

6.

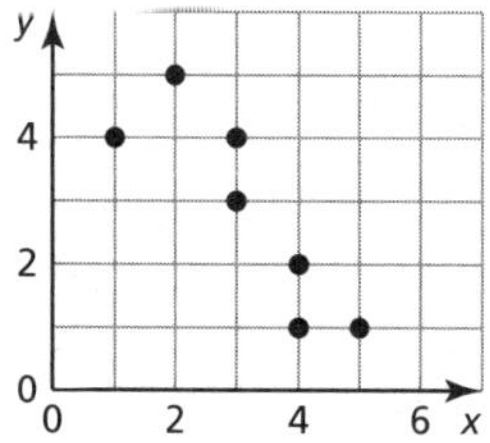

Name ______________________________ Date __________

2.3 Enrichment and Extension

Modeling with Linear Functions

In Exercises 1–4, write a linear equation in slope-intercept form.

1. a line that passes through the points $(-1, 4)$ and $(1, 4)$

2. a line whose x-intercept is 3 and y-intercept is -2

3. a line that has a slope of 0 and passes through the point $(2, 7)$

4. a line that has the same slope as $6y + 10 = 3x$ and the same y-intercept as $4x - 3y = 9$

In Exercises 5–10, find the value of *A* that makes the statement true.

5. The lines $3x - 4y = 6$ and $Ax + 12y = 20$ are parallel.

6. The lines $7x + 2y = 8$ and $Ax + 4y = 11$ are parallel.

7. The lines $6x - 8y = 11$ and $Ax + 2y = 15$ are parallel.

8. The lines $5x + 10y = 40$ and $Ax + 9y = 27$ are perpendicular.

9. The lines $4x - 3y = 10$ and $Ax + 4y = 7$ are perpendicular.

10. The lines $8x + 32y = 14$ and $Ax + \frac{1}{3}y = 12$ are perpendicular.

In Exercises 11 and 12, use the table to answer the question.

11. The table shows the costs of tuition (in thousands of dollars) for different years. Write an equation of a line of fit for the data. Estimate the cost of tuition in 2015.

12. Suppose the tuition in 2008 was \$24,000. Write an equation of a line of fit to estimate the cost of tuition in 2015. How much more would the tuition be in 2015 in this situation in comparison to the situation in Exercise 11?

Yearly college tuition	
Year, *x*	**Cost, *y***
2004	17
2005	18
2008	22

Name____ Date__________

2.3 Puzzle Time

What Do You Call It When A Chicken Stumbles As It Crosses The Street?

Write the letter next to each answer in the box containing the exercise number.

Use the graph to write an equation of the line.

1.

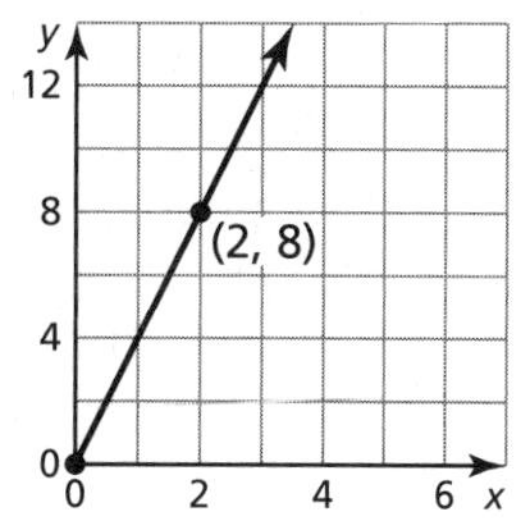

2.

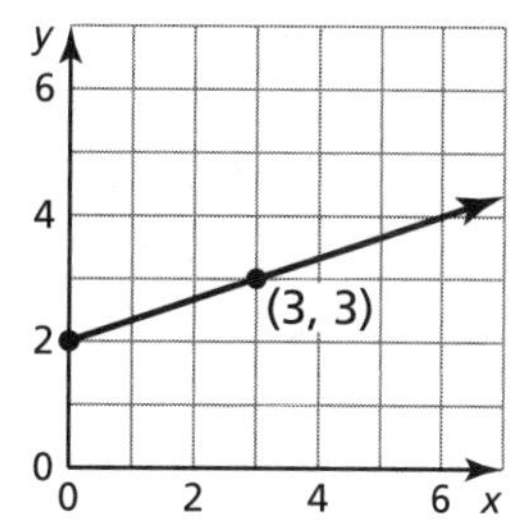

3.

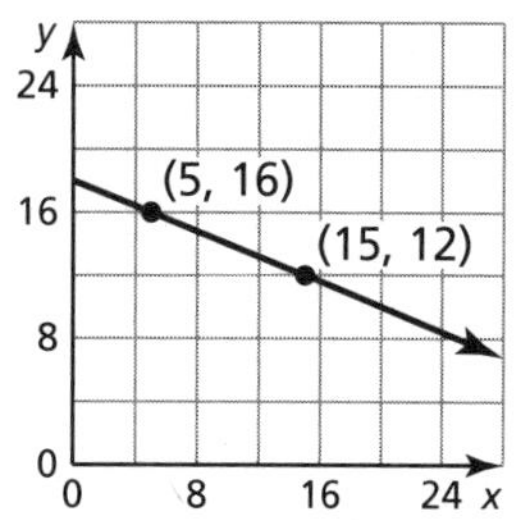

4.

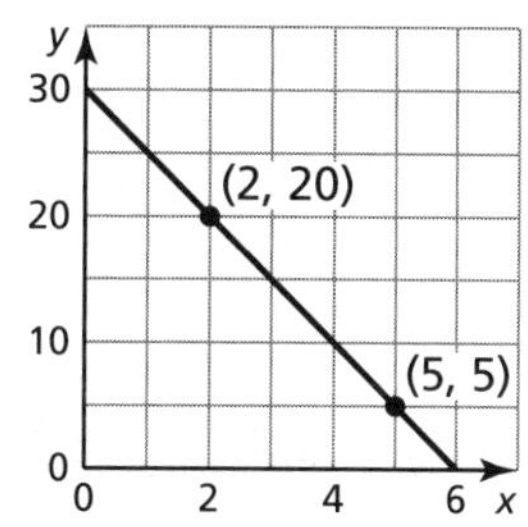

Answers

O. $y = \frac{1}{3}x + 2$

A. $y = -\frac{2}{5}x + 18$

P. $y = \frac{1}{2}x + 3$

R. $y = -\frac{4}{5}x + 8$

D. $y = -5x + 30$

T. $y = -\frac{2}{3}x + 5\frac{1}{3}$

R. $y = 4x$

I. $y = \frac{1}{2}x + 2\frac{1}{2}$

The data in the table show a linear relationship. Write an equation of the line of fit.

5.

x	y
−7	10
−4	8
−1	6
2	4

6.

x	y
5	4
2.5	6
0	8
−2.5	10

7.

x	y
−13	−4
−17	−6
−21	−8
−25	−10

8.

x	y
−6	0
−4	1
−2	2
0	3

1	2	3	4		5	6	7	8

2.4 Start Thinking

Companies often try out different business models for the same product to determine which generates the most revenue. A graph depicts three business models that intersect at a point. What does the point of intersection represent?

2.4 Warm Up

Rewrite the equation in slope-intercept form.

1. $3x - 2y = 3$

2. $2x + 6y = 21$

3. $3x + 2y = 0$

4. $7x + 2y = 13$

5. $x - 4y = 5$

6. $7x - 8y = 19$

2.4 Cumulative Review Warm Up

Determine whether the ordered pair is a solution to the equation.

1. $(3, 2)$; $4x - 3y = 6$

2. $(0, -1)$; $x - 6y = -9$

3. $(5, 2)$; $6x + y = 15$

4. $(-2, 4)$; $x - 2y = -10$

5. $(3, -4)$; $2y = x + 8$

6. $(4, -2)$; $y = 3x - 14$

Name______________________________ Date__________

2.4 Practice A

In Exercises 1 and 2, solve the system using the elimination method.

1. $x - 6y + 2z = 5$
   $2x - 3y + z = 4$
   $3x + 4y - z = -2$

2. $x + y - z = -2$
   $2x - y + z = 8$
   $-x + 2y + 2z = 10$

3. Describe and correct the error in the first step of solving the system of linear equations.

   $5x + 3y - z = 15$
   $-x + 2y + 3z = 10$
   $3x - 4y + 3z = 8$

   ✗
   $$\begin{aligned} 5x + 3y - z &= 15 \\ -5x + 10y + 15z &= 10 \\ \hline 13y + 14z &= 25 \end{aligned}$$

In Exercises 4 and 5, solve the system using the elimination method.

4. $x + 4y - 3z = 1$
   $3x + 12y - 9z = 8$
   $2x + 4y - 4z = -12$

5. $x + y - z = 2$
   $x - y - z = 2$
   $3x + y - 3z = 6$

6. Three bouquets of flowers are ordered at a florist. Three roses, 2 carnations, and 1 tulip cost \$14, 6 roses, 2 carnations, and 6 tulips cost \$38, and 1 rose, 12 carnations, and 1 tulip cost \$18. How much does each item cost?

In Exercises 7 and 8, solve the system of linear equations using the substitution method.

7. $y = -3$
   $2x + y = 5$
   $x - 2y + z = 6$

8. $x - y = 5$
   $-x + 4y + 2z = 3$
   $-x + 3y - 5z = -6$

9. A triangle has a perimeter of 90 centimeters.

   a. Write and use a linear system to determine the lengths of sides ℓ, m, and n.

   b. Is the triangle a right triangle? Explain.

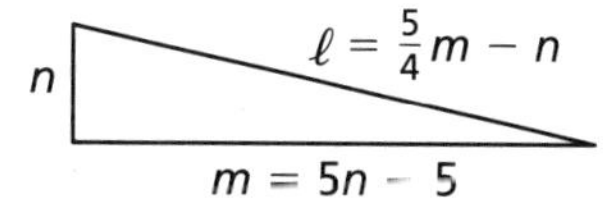

Name ______________________________ Date __________

2.4 Practice B

In Exercises 1 and 2, solve the system using the elimination method.

1. $3x - y + z = -1$

$3x + 2y - 5z = -16$

$3x + 3y + 2z = 6$

2. $4x + 3y - 5z = -9$

$6x + 6y - 3z = 6$

$3x - 3y + 4z = 19$

3. Describe and correct the error in the first step of solving the system of linear equations.

$5x + 3y - z = 15$

$-x + 2y + 3z = 10$

$3x - 4y + 3z = 8$

$$\begin{array}{r} -15x - 9y - 3z = 45 \\ 3x - 4y + 3z = 8 \\ \hline -12 - 13y = 53 \end{array}$$

In Exercises 4 and 5, solve the system using the elimination method.

4. $x - y - z = 5$

$4x - 4y - 4z = 15$

$3x - y - 4z = -2$

5. $-x + y + z = 3$

$x + y + 3z = 5$

$3y + 6z = 12$

In Exercises 6 and 7, solve the system of linear equations using the substitution method.

6. $2x - y = 6$

$4x - 3y - 2z = 14$

$-x + 2y - 3z = 12$

7. $6x + 3y - 9z = 10$

$-2x - y + 3z = 3$

$x - 2y - z = 1$

8. Your friend claims that she has a bag of 30 coins containing nickels, dimes, and quarters. The total value of the 30 coins is \$3. There are twice as many nickels as there are dimes. Is your friend correct? Explain your reasoning.

9. Each equation in this system represents a line.

$x - 2y - 3 = 0$

$2x + y + 1 = 0$

$3x + 4y + 5 = 0$

a. Solve the system of linear equations using either the elimination method or the substitution method.

b. Do the lines intersect at a point? Explain.

Name____________________ Date__________

2.4 Enrichment and Extension

Solving Linear Systems

Example: Solve the four-variable system using the elimination method.

$3w + 3x + 9y - 6z = 33$	**Equation 1**
$2w + 5x + 4y - 2z = 20$	**Equation 2**
$-w + 2x - 3y + z = -1$	**Equation 3**
$w - 3x + 2y + 3z = -17$	**Equation 4**

Solution:

Step 1 Rewrite the system as a linear system in *two* variables.

$$\begin{array}{r} 3w + 3x + 9y - 6z = 33 \\ -3w + 6x - 9y + 3z = -3 \\ \hline 9x \qquad\quad - 3z = 30 \end{array}$$

Add 3 times Equation 3 to Equation 1 (to eliminate w and y).

New Equation 1

$$\begin{array}{r} 2w + 5x + 4y - 2z = 20 \\ -2w + 6x - 4y - 6z = 34 \\ \hline 11x \qquad\quad - 8z = 54 \end{array}$$

Add -2 times Equation 4 to Equation 2 (to eliminate w and y).

New Equation 2

Step 2 Solve the new linear system for both of its variables.

$$\begin{array}{r} 72x - 24z = 240 \\ -33x + 24z = -162 \\ \hline 39x \qquad = 78 \end{array}$$

$$x = 2$$

$$z = -4$$

Add -3 times the new Equation 2 to 8 times the new Equation 1 (to eliminate z).

Step 3 Substitute $x = 2$ and $z = -4$ into any of the two original equations. Solve for the remaining variables using elimination.

The solution is $w = 1$, $x = 2$, $y = 0$, and $z = -4$, or the ordered quadruple $(1, 2, 0, -4)$.

In Exercises 1–3, solve the system using the elimination method.

1. $2w + 4x + 2y - z = 1$
$4w + 3x + 8y - 3z = 19$
$-w - 5x - y + 3z = 8$
$w + 2x + 2y - z = 2$

2. $5w + x + 2y - 2z = -13$
$3w + 2x - 3y - z = -7$
$3w - 2x - 3y + 4z = 25$
$w + 4x - y - 2z = -19$

3. $6w + x + 6y - 9z = 10$
$3w + 7x - y + 9z = 34$
$w + 5x + 4y - 3z = -5$
$2w - x - 3y - 3z = -9$

Name ______________________ Date ________

2.4 Puzzle Time

Did You Hear About...

A	B	C	D	E	F
G	H	I	J		

Complete each exercise. Find the answer in the answer column. Write the word under the answer in the box containing the exercise letter.

Answer	Word
$(-2, -3, 3)$	TALLEST
$(-5, 4, -3)$	MEASURED
$(1, 3, -3)$	FEET
$(1, 3, 1)$	LIVING
$(5, -1, -4)$	TEN

Solve the system using the elimination method.

A. $-2x - 6y - 2z = -2$
$3x + 2y + 5z = 7$
$-3x - 3y + 3z = 9$

B. $-x - 5y - 5z = 2$
$4x - 5y + 4z = 19$
$x + 5y - z = -20$

C. $4x - 4y + 4z = -4$
$4x + y - 2z = 5$
$-3x - 3y - 4z = -16$

D. $x - 6y + 4z = -12$
$x + y - 4z = 12$
$2x + 2y + 5z = -15$

Solve the system using the substitution method.

E. $5x + 4y - 6z = -24$
$-2y + 2z = 0$
$y - z = 2$

F. $-4x + 2z = 14$
$-x + y - z = 12$
$-2x - 4z = 22$

G. $6x - 9y + z = -12$
$2x - 3y - z = -4$
$-8x + 12y = 16$

H. $3x - 3y = -6$
$3x + 3y + z = 9$
$-4x + 5y + z = 8$

Solve the system using the method of your choice.

I. $5x - 4y + 2z = 21$
$-x - 5y + 6z = -24$
$-x - 4y + 5z = -21$

J. $4x - 3y + z = -10$
$2x + y + 3z = 0$
$-x + 2y - 5z = 17$

Answer	Word
$(-2 + 1.5y, y, 0)$	SIX
$(0, 0, -3)$	HORSE
no solution	WHO
$(-1, 0, 2)$	THE
$(1, 4, -2)$	INCHES

2.5 Start Thinking

Make a table of values and use it to graph the following functions on the same coordinate plane. Use the same x-values for each function.

$f(x) = x^2$

$f(x) = (x - 1)^2$

$f(x) = 2x^2$

$f(x) = x^2 + 1$

Describe how the graphs of the last three functions differ from the graph of $f(x) = x^2$.

2.5 Warm Up

Multiply.

1. $(3x - 2)(2x - 4)$

2. $(5x + 2)(4x + 1)$

3. $(4x + y)(2x - 3y)$

4. $3a(4a + 1)$

5. $(4x + 1)(5x - 2)$

6. $(5y + 4)(3y + 2)$

2.5 Cumulative Review Warm Up

Write a function *g* whose graph represents the indicated transformation of the graph of *f*.

1. $f(x) = x + 6$; translation 3 units right

2. $f(x) = x - 3$; translation 1 unit left

3. $f(x) = |5x - 2| - 3$; translation 1 unit down

Name ________________________________ Date __________

2.5 Practice A

In Exercises 1–6, describe the transformation of $f(x) = x^2$ represented by g. Then graph each function.

1. $g(x) = x^2 - 2$

2. $g(x) = x^2 + 1$

3. $g(x) = (x + 1)^2$

4. $g(x) = (x - 2)^2$

5. $g(x) = (x - 5)^2$

6. $g(x) = (x + 2)^2 - 1$

In Exercises 7–9, describe the transformation of $f(x) = x^2$ represented by g. Then graph each function.

7. $g(x) = -2x^2$

8. $g(x) = (-2x)^2$

9. $g(x) = \frac{1}{4}x^2$

10. Describe and correct the error in analyzing the graph of $f(x) = -\frac{1}{3}x^2$.

X 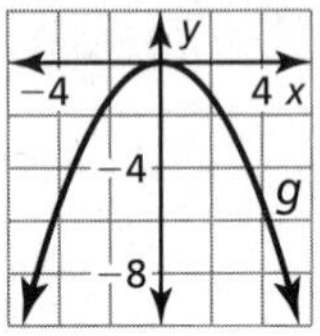

The graph of g is a reflection in the x-axis, followed by a vertical stretch by a factor of $\frac{1}{3}$ of the graph of the parent quadratic function.

In Exercises 11 and 12, describe the transformation of the graph of the parent quadratic function. Then identify the vertex.

11. $f(x) = 2(x + 3)^2 + 2$

12. $f(x) = -5x^2 - 1$

In Exercises 13 and 14, write a rule for g described by the transformations of the graph of f. Then identify the vertex.

13. $f(x) = x^2$; vertical stretch by a factor of 3 and a reflection in the x-axis, followed by a translation 3 units down

14. $f(x) = 4x^2 + 5$; horizontal stretch by a factor of 2 and a translation 2 units up, followed by a reflection in the x-axis

15. Let the graph of g be a translation 4 units down and 3 units right, followed by a horizontal shrink by a factor of $\frac{1}{2}$ of the graph of $f(x) = x^2$.

a. Identify the values of a, h, and k. Write the transformed function in vertex form.

b. Suppose the horizontal shrink was performed first, followed by the translations. Identify the values of a, h, and k, and write the transformed function in vertex form.

Name________________________________ Date__________

2.5 Practice B

In Exercises 1–6, describe the transformation of $f(x) = x^2$ represented by g. Then graph each function.

1. $g(x) = x^2 + 3$
2. $g(x) = (x + 5)^2$
3. $g(x) = (x + 6)^2 - 4$
4. $g(x) = (x - 1)^2 + 5$
5. $g(x) = (x - 4)^2 + 3$
6. $g(x) = (x + 8)^2 - 2$

In Exercises 7–9, describe the transformation of $f(x) = x^2$ represented by g. Then graph each function.

7. $g(x) = -\left(\frac{1}{2}x\right)^2$
8. $g(x) = \frac{1}{3}x^2 + 2$
9. $g(x) = \frac{1}{3}(x + 1)^2$

In Exercises 10 and 11, describe the transformation of the graph of the parent quadratic function. Then identify the vertex.

10. $f(x) = -3(x + 6)^2 - 4$
11. $f(x) = \frac{1}{3}(x - 2)^2 + 1$

In Exercises 12 and 13, write a rule for g described by the transformations of the graph of f. Then identify the vertex.

12. $f(x) = x^2$; vertical shrink by a factor of $\frac{1}{2}$ and a reflection in the y-axis, followed by a translation 2 units left

13. $f(x) = (x + 4)^2 + 2$; horizontal shrink by a factor of $\frac{1}{3}$ and a translation 2 units up, followed by a reflection in the x-axis

14. Justify each step in writing a function g based on the transformations of $f(x) = 4x^2 - 3x$.

    translation 3 units up followed by a reflection in the y-axis

$h(x) = f(x) + 3$	
$= 4x^2 - 3x + 3$	
$g(x) = h(-x)$	
$= 4x^2 + 3x + 3$	

Name ______________________________ Date ________

2.5 Enrichment and Extension

Transformations of Quadratic Functions

Ten parabolas are shown below.

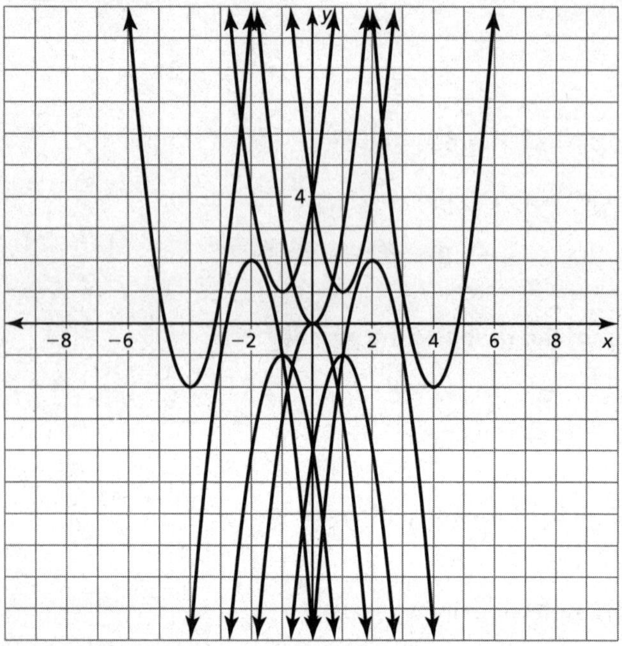

The equations for three of the graphs are

$y = -3(x - 1)^2 - 1$

$y = 3x^2$, and

$y = 3(x + 4)^2 - 2.$

Write equations of the form $y = a(x - h)^2 + k$ for the other seven parabolas with the help of a graphing calculator.

Name____ Date__________

2.5 Puzzle Time

What Is The Most Densely Populated Country On The Mainland Of The Americas?

Write the letter of each answer in the box containing the exercise number.

Describe the transformation of $f(x) = x^2$ represented by g.

1. $g(x) = -2x^2$
2. $g(x) = (x - 1)^2$
3. $g(x) = x^2 - 1$
4. $g(x) = (x + 1)^2$
5. $g(x) = \frac{1}{2}x^2 - 2$
6. $g(x) = (x - 2)^2 - 1$

Write a rule for g described by the transformations of the graph of f.

7. $f(x) = x^2$; vertical stretch by a factor of 2 and a reflection in the x-axis, followed by a translation 3 units down
8. $f(x) = x^2$; vertical shrink by a factor of $\frac{1}{2}$ followed by a translation 3 units left
9. $f(x) = 4x^2 + 10$; horizontal stretch by a factor of 2 followed by a translation 3 units up
10. $f(x) = (x - 2)^2 - 8$; horizontal shrink by a factor of $\frac{1}{2}$ and a translation 5 units down, followed by a reflection in the x-axis

Answers

O. $g(x) = x^2 + 13$

L. translation 1 unit right

R. $g(x) = -(2x - 2)^2 + 13$

S. translation 1 unit down

V. translation 2 units right followed by a translation 1 unit down

L. vertical shrink by a factor of $\frac{1}{2}$ followed by a translation 2 units down

E. reflection in the x-axis and a vertical stretch by a factor of 2

A. $g(x) = -2x^2 - 3$

D. $g(x) = \frac{1}{2}(x + 3)^2$

A. translation 1 unit left

1	2		3	4	5	6	7	8	9	10

2.6 Start Thinking

Complete the table for the function $f(x) = |x|$. Graph the values from the table on a piece of graph paper.

x	-2	-1	0	1	2
$f(x)$					

What is the shape of the graph? Are opposite integers the same distance from the y-axis on the graph? Is this graph symmetric? Why or why not?

2.6 Warm Up

Give the coordinates of the image of point $P(-5, 3)$ after each reflection.

1. reflection in the y-axis
2. reflection in the x-axis
3. reflection in the line through $(5, -6)$ and $(8, -6)$
4. reflection in the line through $(-1, -1)$ and $(-1, -2)$

2.6 Cumulative Review Warm Up

Determine whether the data show a linear relationship. If so, write an equation of a line of fit, estimate y when $x = 20$, and explain its context in the situation.

1.

Minutes jogging, x	2	5	10	15
Calories burned, y	22	55	110	165

2.

Years, x	10	12	17	21
Height (feet), y	4.2	5.0	6.0	6.1

Name______________________________ Date__________

2.6 Practice A

In Exercises 1–12, graph the function. Label the vertex and axis of symmetry.

1. $f(x) = (x - 2)^2$

2. $f(x) = (x + 1)^2$

3. $g(x) = (x + 2)^2 + 4$

4. $h(x) = (x - 3)^2 - 2$

5. $y = -3(x - 1)^2 + 3$

6. $f(x) = 4(x + 2)^2 - 1$

7. $y = x^2 - 2x + 1$

8. $y = 3x^2 + 6x + 1$

9. $y = -3x^2 + 6x + 4$

10. $f(x) = -x^2 + 6x - 3$

11. $g(x) = -x^2 + 2$

12. $f(x) = 5x^2 - 4$

13. Explain why you cannot use the axes of symmetry to distinguish between the quadratic functions $y = 3x^2 + 12x + 1$ and $y = x^2 + 4x + 5$.

14. Which function represents the narrowest parabola? Explain your reasoning.

A. $y = x^2 + 3$

B. $y = 0.5x^2 - 2$

C. $y = 3(x + 2)^2$

D. $y = -2x^2 + 1$

In Exercises 15–18, find the minimum or maximum value of the function. Describe the domain and range of the function, and where the function is increasing and decreasing.

15. $y = 5x^2 + 2$

16. $y = 4x^2 - 3$

17. $y = -x^2 + 4x - 1$

18. $f(x) = -2x^2 + 4x + 9$

19. The number of customers in a grocery store is modeled by the function $y = -x^2 + 10x + 50$, where y is the number of customers in the store and x is the number of hours after 7:00 A.M.

a. At what time is the maximum number of customers in the store?

b. How many customers are in the store at the time in part (a)?

In Exercises 20 and 21, use completing the square to find the vertex of the parabola or the center and radius of the circle. Then graph the equation.

20. $-4x^2 + 16x - y + 2 = 0$

21. $x^2 + y^2 - 4x + 12y - 8 = 0$

Name ______________________________ Date __________

2.6 Practice B

In Exercises 1–12, graph the function. Label the vertex and axis of symmetry.

1. $f(x) = -3(x - 2)^2 - 4$

2. $f(x) = 3(x + 1)^2 + 5$

3. $g(x) = -\frac{1}{2}(x + 3)^2 + 2$

4. $h(x) = \frac{1}{2}(x - 2)^2 - 1$

5. $y = 0.6(x - 2)^2$

6. $f(x) = 0.25x^2 - 1$

7. $y = -x^2 + 8$

8. $y = 7x^2 + 2$

9. $y = 1.5x^2 - 6x + 3$

10. $f(x) = 0.5x^2 + 3x - 1$

11. $y = \frac{5}{2}x^2 - 5x + 1$

12. $f(x) = -\frac{3}{2}x^2 - 6x - 4$

13. A quadratic function is decreasing to the left of $x = 3$ and increasing to the right of $x = 3$. Will the vertex be the highest or lowest point on the graph of the parabola? Explain.

14. The graph of which function has the same axis of symmetry as the graph of $y = 2x^2 - 8x + 3$? Explain your reasoning.

A. $y = -4x^2 + 16x - 5$

B. $y = 2x^2 + 8x + 7$

C. $y = 3x^2 - 6x + 7$

D. $y = -6x^2 + 10x - 1$

In Exercises 15–18, find the minimum or maximum value of the function. Describe the domain and range of the function, and where the function is increasing and decreasing.

15. $y = 3x^2 + 12$

16. $y = -x^2 - 6x$

17. $y = -\frac{1}{3}x^2 - 2x + 3$

18. $f(x) = \frac{1}{2}x^2 + 3x + 7$

19. The height of a bridge is given by $y = -3x^2 + x$, where y is the height of the bridge (in miles) and x is the number of miles from the base of the bridge.

a. How far from the base of the bridge does the maximum height occur?

b. What is the maximum height of the bridge?

Name______________________________ Date__________

2.6 Enrichment and Extension

Characteristics of Quadratic Functions

In Exercises 1–4, graph the quadratic function. Label the vertex and axis of symmetry. Then solve the related linear equation.

	Quadratic Function	*Related Linear Equation*
1.	$f(x) = -5.1x^2 - 15.3x + 2.4$	$-10.2x - 15.3 = 0$
2.	$f(x) = 1.9x^2 + 8x - 4$	$3.8x + 8 = 0$
3.	$f(x) = \frac{3}{8}x^2 - \frac{1}{2}x + \frac{1}{4}$	$\frac{3}{4}x - \frac{1}{2} = 0$
4.	$f(x) = -\frac{1}{10}x^2 + \frac{7}{2}x - \frac{5}{6}$	$-\frac{1}{5}x + \frac{7}{2} = 0$

5. Consider the functions and related linear equations in Exercises 1–4. Compare the vertex of each quadratic function with the solution of its related linear equation. Compare the coefficient of the x^2-term of each quadratic function with the coefficient of the x-term of its related linear equation. Compare the coefficient of the x-term of each quadratic function with the constant term of its related linear equation.

6. Use your answers from Exercise 5 to write a related linear equation for each of the quadratic functions below. Then solve the linear equation and use the result to find the vertex of the quadratic function.

a. $f(x) = 4x^2 - 5x + 12$

b. $f(x) = -12x^2 + 6x - 5$

c. $f(x) = -3x^2 - x + 7$

7. For the graph of the quadratic function $f(x) = ax^2 + bx + c$, show that the vertex is given by $\left(-\frac{b}{2a}, -\frac{b^2}{4a} + c\right)$.

8. Lifeguards at a beach rope off a rectangular swimming area (see figure) with P feet of rope. What is the maximum possible swimming area in terms of P?

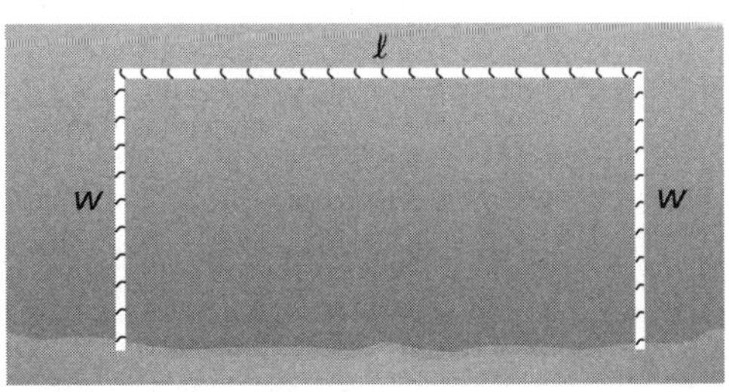

Name ______________________________ Date __________

2.6 Puzzle Time

What Is Roz Savage Famous For?

1	2	3	4	5	6
7	8	9	10	11	12

Complete each exercise. Find the answer in the answer column. Write the word under the answer in the box containing the exercise number.

minimum, 0	WOMAN
maximum, 2	THE
(−1, −9.5) $x = -1$	ROWER
(4, 1) $r = 4$	OCEANS
(1, 0) $r = 5$	ROW
(1, −4) $x = 1$	BRITISH

Find the vertex and axis of symmetry of the graph of the function.

1. $f(x) = 9x^2 - 3$

2. $y = -x^2 + 2x - 5$

3. $g(x) = -0.5x^2 - x - 10$

4. $f(x) = -2x^2 + 8x - 1$

Find the minimum or maximum value of the function.

5. $f(x) = -3x^2 + 12x - 10$

6. $y = -x^2 + 8$

7. $g(x) = x^2 - 2x + 1$

8. $y = 2x^2 - 20x$

Use completing the square to find the center and radius of the circle.

9. $x^2 + y^2 - 2x - 24 = 0$

10. $x^2 + y^2 + 4y - 5 = 0$

11. $x^2 + y^2 + 6x - 4y + 9 = 0$

12. $x^2 + y^2 - 8x - 2y + 1 = 0$

maximum, 8	FIRST
(−3, 2) $r = 2$	THREE
minimum, −50	TO
(0, −2) $r = 3$	ACROSS
(0, −3) $x = 0$	THIS
(2, 7) $x = 2$	WAS

2.7 Start Thinking

Use the graph of the parabola $y = \frac{1}{4}(x - 4)^2$ to label the x- and y-axis and give the graph an appropriate title for a real-life situation. Then find y when $x = 25$ and describe what this signifies in terms of the labels you created.

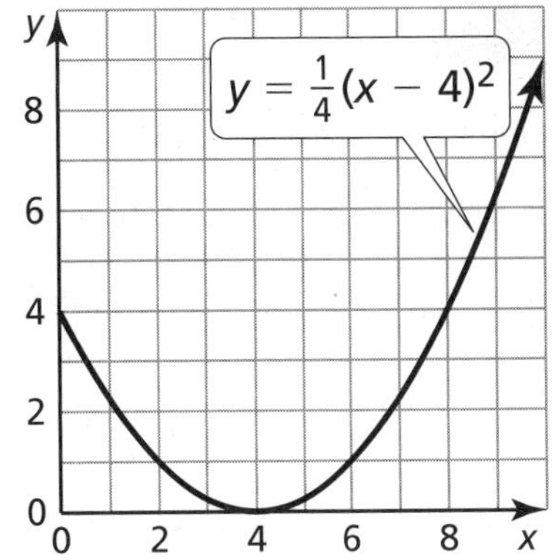

2.7 Warm Up

Write an equation of a line in point-slope form with the information given.

1. passes through: $(6, 0)$; slope: $-\frac{1}{6}$
2. passes through: $(1, 3)$; slope: $\frac{1}{2}$
3. passes through: $(4, -1)$; slope: -2
4. passes through: $(3, -3)$; slope: 3
5. passes through: $(4, -18)$; slope: $-\frac{1}{4}$
6. passes through: $(6, -1)$; slope: -3

2.7 Cumulative Review Warm Up

Use a graphing calculator to graph the function and its parent function. Then describe the transformation.

1. $f(x) = 4x - 1$
2. $h(x) = -2|x|$
3. $g(x) = 2x^2 + 7$
4. $f(x) = -(x - 2)^2 - \frac{2}{3}$

Name ______________________________ Date __________

2.7 Practice A

In Exercises 1–3, write an equation of the parabola in vertex form.

1. passes through $(6, 4)$ and has vertex $(2, -3)$

2. passes through $(-3, -10)$ and has vertex $(3, -8)$

3. passes through $(0, -5)$ and has vertex $(-1, 4)$

In Exercises 4–6, write an equation of the parabola in intercept form.

4. x-intercepts of 10 and 6; passes through $(11, 8)$

5. x-intercepts of 2 and 8; passes through $(0, 3)$

6. x-intercepts of -14 and -2; passes through $(-16, -8)$

7. Use the parabola shown.

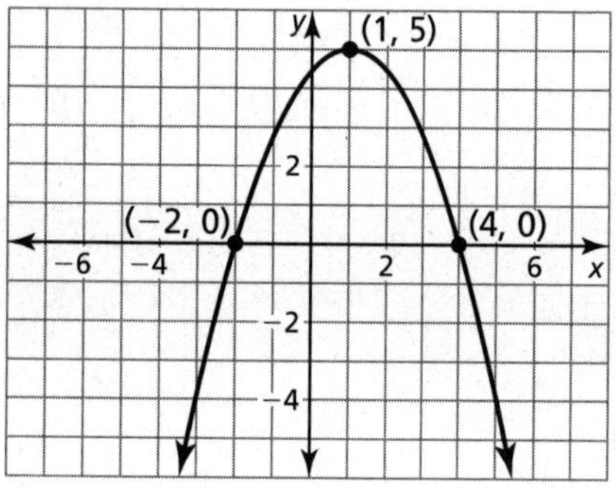

   a. Write an equation of the parabola in vertex form.

   b. Expand the equation in part (a) to the form $y = ax^2 + bx + c$.

   c. Write an equation of the parabola in intercept form.

   d. Expand the equation in part (c) to the form $y = ax^2 + bx + c$.

   e. Do both methods give an equation that represents the parabola? Which method did you find easier? Explain.

8. A basketball is thrown up in the air. The table shows the heights y (in feet) of the basketball after x seconds. Write and solve an equation to determine how long the ball is above 6 feet. How long is the ball in the air?

Time, x	0	6	12	18
Basketball height, y	5	10	10	5

Name____ Date__________

2.7 Practice B

In Exercises 1–3, write an equation of the parabola in vertex form.

1. passes through $(4, -7)$ and has vertex $(1, -6)$

2. passes through $(5, -4)$ and has vertex $(-2, 5)$

3. passes through $(2, 2)$ and has vertex $(-1, -1)$

In Exercises 4–6, write an equation of the parabola in intercept form.

4. x-intercepts of 12 and 8; passes through $(9, 5)$

5. x-intercepts of -7 and -1; passes through $(1, 1)$

6. x-intercepts of -9 and 9; passes through $(0, 4)$

7. Describe and correct the error in writing an equation of the parabola.

> ✗ Vertex: $(3, -5)$
>
> Passes through $(1, -7)$
>
> $$y = a(x - h)^2 + k$$
> $$-5 = a(3 - 1)^2 + (-7)$$
> $$-5 = 4a - 7$$
> $$2 = 4a$$
> $$\frac{1}{2} = a$$
>
> The equation is $y = \frac{1}{2}(x - 1) - 7$.

8. The graph shows the area y (in square feet) of rectangles that have a perimeter of 200 feet and a length of x feet.

a. Interpret the meaning of the vertex in this situation.

b. Write an equation for the parabola to predict the area of the rectangle when the length is 2 feet.

c. Compare the average rates of change in the area from 0 to 50 feet and 50 to 100 feet.

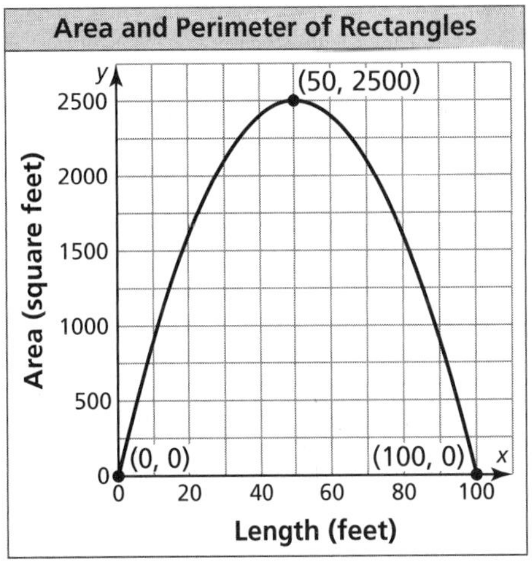

Name ______________________________ Date __________

2.7 Enrichment and Extension

Modeling with Quadratic Functions

In Exercises 1–5, analyze the differences in the outputs to determine whether the data are *linear* or *quadratic*. Write an equation that fits the data. If quadratic, write the equation in (a) standard form and (b) vertex form, and (c) state the transformation from the parent function $f(x) = x^2$.

1.

Altitude (1000 feet), x	1	1.5	2	2.5	3
Boiling water temperature (°F), y	210.3	209.4	208.5	207.6	206.7

2.

Time (seconds), x	1	2	3	4	5
Height (feet), y	73.5	78.4	73.5	58.8	34.3

3.

Units sold, x	1	2	3	4	5
Profit (thousands of dollars), y	39	60	75	84	87

4.

Depth (feet), x	0	10	20	30	40
Pressure (pounds per square inch), y	14.7	19.03	23.36	27.69	32.02

5.

Time (seconds), x	1	1.5	2	2.5	3
Height (feet), y	12	12.75	11	6.75	0

Name____________________ Date__________

2.7 Puzzle Time

Don Featherstone Is Famous For Being The First To Make This Plastic Lawn Ornament In 1957.

Write the letter of each answer in the box containing the exercise number.

Write an equation of the parabola in vertex form.

1. passes through $(-5, 0)$ and has vertex $(-2, 1)$
2. passes through $(4, 10)$ and has vertex $(0, 0)$
3. passes through $(6, -1)$ and has vertex $(14, -20)$
4. passes through $(-5, -3)$ and has vertex $(7, 12)$
5. passes through $(9, 15)$ and has vertex $(-6, 21)$
6. passes through $(0, 0)$ and has vertex $(-10, -4)$

Write an equation of the parabola in intercept form.

7. x-intercepts of 10 and -2; passes through $(1, 4)$
8. x-intercepts of 9 and -3; passes through $(3, 4)$
9. x-intercepts of -5 and -1; passes through $(-10, 2)$
10. x-intercepts of 4 and 6; passes through $(2, 3)$
11. x-intercepts of -10 and 8; passes through $(-6, -4)$
12. x-intercepts of 0 and 4; passes through $(12, -8)$

Answers

I. $y = \frac{5}{8}x^2$

I. $y = \frac{2}{45}(x + 5)(x + 1)$

L. $y = \frac{1}{25}(x + 10)^2 - 4$

N. $y = \frac{19}{64}(x - 14)^2 - 20$

M. $y = -\frac{1}{9}(x - 9)(x + 3)$

F. $y = -\frac{2}{75}(x + 6)^2 + 21$

A. $y = -\frac{4}{27}(x - 10)(x + 2)$

N. $y = \frac{3}{8}(x - 4)(x - 6)$

G. $y = \frac{1}{14}(x + 10)(x - 8)$

K. $y = -\frac{5}{48}(x - 7)^2 + 12$

O. $y = -\frac{1}{12}x(x - 4)$

P. $y = -\frac{1}{9}(x + 2)^2 + 1$

1	2	3	4		5	6	7	8	9	10	11	12

Name ______________________________ Date ________

Chapter 2 Cumulative Review

In Exercises 1–16, solve the proportion.

1. $\frac{-6}{4} = \frac{x}{12}$ **2.** $\frac{5}{-3} = \frac{x}{9}$ **3.** $\frac{-19}{6} = \frac{38}{x}$ **4.** $\frac{7}{13} = \frac{-14}{x}$

5. $\frac{-7}{15} = \frac{-35}{x}$ **6.** $\frac{x}{4} = \frac{-1}{2}$ **7.** $\frac{-2}{x} = \frac{6}{-9}$ **8.** $\frac{-24}{36} = \frac{-8}{x}$

9. $\frac{3}{-5} = \frac{x}{-35}$ **10.** $\frac{12}{x} = \frac{60}{45}$ **11.** $\frac{x}{-2} = \frac{-80}{20}$ **12.** $\frac{4}{10} = \frac{x}{15}$

13. $\frac{13}{-17} = \frac{x}{-34}$ **14.** $\frac{5}{x} = \frac{-40}{56}$ **15.** $\frac{-2}{15} = \frac{x}{-60}$ **16.** $\frac{81}{x} = \frac{-9}{5}$

17. Your chemistry test has 64 questions. Your teacher rounds to the nearest whole percent.

a. You have 57 correct answers. What percent of your answers are correct?

b. You have 61 correct answers. What percent of your answers are correct?

c. You want to earn at least an 85%. How many correct answers must you have?

d. You want to earn at least a 93%. How many correct answers must you have?

18. Your English literature test has 28 questions. Your teacher rounds to the nearest whole percent.

a. You have 25 correct answers. What percent of your answers are correct?

b. You have 21 correct answers. What percent of your answers are correct?

c. You want to earn at least an 85%. How many correct answers must you have?

d. You want to earn at least a 93%. How many correct answers must you have?

e. You want to earn at least a 77%. How many correct answers must you have?

19. You want to mix leftover yellow and blue paint to make green paint. The ratio to make the green paint is 2 parts yellow paint to 1 part blue paint. You have 6 cups of yellow paint. How many cups of blue paint do you need to make the green paint?

20. You want to make an environmentally friendly carpet cleaner using salt and white vinegar. The recipe ratio of white vinegar to salt is 8:2. You have one cup of white vinegar. How many tablespoons of salt are needed to make the carpet cleaner?

Name______________________________ Date__________

Chapter 2 Cumulative Review (continued)

In Exercises 21–35, find the distance between the two points. Round your answer to the nearest hundredth.

21. $(-4, 3)$ and $(-1, -2)$
22. $(3, 7)$ and $(-4, 6)$
23. $(-9, 8)$ and $(-11, 10)$
24. $(2, 5)$ and $(-1, -5)$
25. $(-9, -10)$ and $(1, 6)$
26. $(2, 3)$ and $(1, 10)$
27. $(8, 6)$ and $(-4, 3)$
28. $(-2, -9)$ and $(4, -5)$
29. $(12, -7)$ and $(9, 3)$
30. $(5, -8)$ and $(2, 11)$
31. $(5, 10)$ and $(-1, 2)$
32. $(-3, -2)$ and $(5, 4)$
33. $(7, -3)$ and $(-4, -8)$
34. $(8, 1)$ and $(-9, 2)$
35. $(-2, -1)$ and $(7, 7)$

In Exercises 36–44, find the *x*-intercept of the graph of the linear equation.

36. $y = \frac{3}{7}x - 5$
37. $y = 12x + 27$
38. $y = -7x + 39$
39. $y = 5(x + 2)$
40. $y = -7(x + 10)$
41. $y = 4(x - 8)$
42. $-6x + 5y = -30$
43. $11x + 7y = -33$
44. $14x - 8y = 28$

In Exercises 45–53, solve the equation for *x*.

45. $y = 8x + 24$
46. $y = \frac{1}{2}x + 3$
47. $y = -5x + 35$
48. $y = \dfrac{x - 7}{4}$
49. $y = \dfrac{-3x + 8}{-6}$
50. $y = \dfrac{4x + 5}{6}$
51. $3x - 6y = 30$
52. $-8x - 2y = 40$
53. $5x - 15y = -40$

In Exercises 54–65, solve the equation. Check for extraneous solutions.

54. $4\sqrt{x} - 2 = 0$
55. $-7\sqrt{x} - 5 = 0$
56. $6\sqrt{x + 10} + 3 = 0$
57. $-\sqrt{x - 9} - 8 = 0$
58. $3\sqrt{x + 8} - 6 = 4$
59. $7\sqrt{x - 5} - 4 = 3$
60. $\sqrt{2x} - 9 = 0$
61. $\sqrt{3x} - 10 = 0$
62. $\sqrt{4x} - 1 = 1$
63. $\sqrt{9x} - 7 = 3$
64. $\sqrt{2x - 4} = \sqrt{x + 4}$
65. $\sqrt{5x + 10} = \sqrt{x - 6}$

66. You are riding in a car and traveling at an average speed of 65 miles per hour. The destination is 325 miles away. How long does it take you to get there?

67. You are riding in a car and traveling at an average speed of 48 miles per hour. The destination is 108 miles away. How long does it take you to get there?

68. You are on a cruise ship that travels at an average speed of 24 knots per hour. The first port of call is 1449 miles away from where you set sail (1 knot is about 1.15 miles). How long does it take you to get there?

Chapter 2 Cumulative Review (continued)

In Exercises 69–83, write an equation for the line that passes through the points.

69. $(11, -2)$ and $(3, -1)$ **70.** $(12, 7)$ and $(11, 3)$ **71.** $(4, 6)$ and $(-4, 2)$

72. $(-5, 8)$ and $(-1, -6)$ **73.** $(8, -6)$ and $(-2, 7)$ **74.** $(-12, 8)$ and $(2, 2)$

75. $(1, 8)$ and $(-9, 2)$ **76.** $(-2, -10)$ and $(6, 9)$ **77.** $(3, 5)$ and $(6, -7)$

78. $(-4, 3)$ and $(8, -12)$ **79.** $(12, 5)$ and $(4, -6)$ **80.** $(11, -3)$ and $(3, 3)$

81. $(-5, 4)$ and $(9, 1)$ **82.** $(6, -3)$ and $(-1, -4)$ **83.** $(9, -6)$ and $(5, -1)$

In Exercises 84–98, factor the trinomial.

84. $x^2 + x - 6$ **85.** $x^2 - x - 42$ **86.** $x^2 - 4x - 77$

87. $x^2 + 13x + 12$ **88.** $x^2 - 2x - 24$ **89.** $x^2 + 4x - 5$

90. $x^2 + 11x + 30$ **91.** $x^2 - 20x + 96$ **92.** $2x^2 - 3x - 35$

93. $3x^2 - 2x - 40$ **94.** $2x^2 + 19x - 10$ **95.** $3x^2 - 32x + 20$

96. $4x^2 + 36x - 144$ **97.** $5x^2 + 7x - 6$ **98.** $6x^2 + 10x - 44$

In Exercises 99–113, write the quadratic function in standard form.

99. $y = -(x + 8)(x - 4)$ **100.** $y = 4(x + 4)(x - 3)$ **101.** $y = -2(x - 1)(x + 2)$

102. $y = -6(x + 3)(x - 4)$ **103.** $y = -5(x - 3)(x - 8)$ **104.** $y = 3(x + 4)(x - 7)$

105. $y = 6(x - 3)^2$ **106.** $y = -7(x + 1)^2$ **107.** $y = 2(x - 4)^2$

108. $y = (x - 4)^2 + 1$ **109.** $y = (x + 5)^2 - 7$ **110.** $y = (x - 8)^2 + 9$

111. $y = 3(x + 1)^2 + 4$ **112.** $y = -7(x + 5)^2 - 3$ **113.** $y = -9(x - 3)^2 + 7$

114. You painted a picture that is 18 inches by 24 inches. You want to put wooden pieces around the outside as a frame. There are 648 square inches of wooden pieces. You are using all the material, and you want an even boarder around the entire picture. What should be the width of the border?

115. You want to create a decorative border for your garden that measures 12 feet by 7 feet. You decide to put a 1-foot border around the garden. How many square feet of rocks do you need?

116. You have a piece of canvas that you want to paint. You have enough paint to cover 252 square inches. You know you want one side to be 14 inches. How long should the other side be if you want to use up all the paint?

117. You have a square table and want to create a tablecloth for it using a piece of fabric. The side length of the table is 72 inches. How much fabric do you need?

Name______________________________ Date__________

Chapter 2 Cumulative Review (continued)

118. A part for a toy train is made by drilling a hole that has a diameter of 0.5 centimeter through a wooden ball that has a diameter of 5 centimeters.

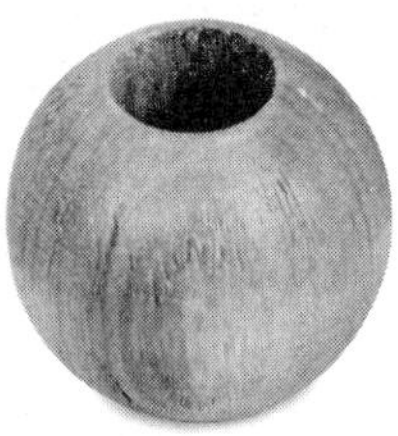

a. Estimate the volume of the wooden ball after the hole is made. Explain your reasoning.

b. Do the surface area and volume of the wooden ball decrease after the hole is made? Explain.

In Exercises 119–132, use the Quadratic Formula to solve the equation.

119. $x^2 - 4x + 58 = 0$

120. $x^2 + 2x - 14 = 0$

121. $2x^2 + 14x - 45 = 0$

122. $3x^2 - 5x + 6 = 0$

123. $-4x^2 - 7x + 5 = 0$

124. $5x^2 - 4x + 32 = 0$

125. $6x^2 - 7x + 4 = 0$

126. $-4x^2 + 13x - 21 = 0$

127. $2x^2 + 3x = -8$

128. $-2x^2 + 5x = 18$

129. $-3x^2 + 16x = 34$

130. $-4x^2 - 15x = 28$

131. $5x^2 - 4x + 23 = 3x^2 + 14x + 34$

132. $3x^2 - 4x + 14 = 9x^2 - 7x + 19$

In Exercises 133–138, graph the function and its parent function. Then describe the transformation.

133. $g(x) = 2x$

134. $h(x) = -\frac{1}{2}x$

135. $c(x) = x - 4$

136. $d(x) = 2x^2$

137. $k(x) = |x + 4|$

138. $m(x) = |x - 3| + 2$

In Exercises 139–146, write a function *g* whose graph represents the indicated transformation of the graph of *f*.

139. $f(x) = -2x$; translation 3 units down

140. $f(x) = \frac{4}{7}x$; translation 1 unit up

141. $f(x) = |x| + 4$; translation 2 units left

142. $f(x) = \frac{1}{2}|x| - 7$; translation 6 units right

143. $f(x) = |x - 3| + 5$; reflection in the x-axis

144. $f(x) = |x + 1| - 2$; reflection in the x-axis

145. $f(x) = 3x - 1$; reflection in the y-axis

146. $f(x) = \frac{3}{4}x + 5$; reflection in the y-axis

In Exercises 147–149, describe the transformation of the graph of the parent quadratic function. Then identify the vertex.

147. $f(x) = 2(x + 4)^2 + 3$

148. $f(x) = -4x^2 + 6$

149. $f(x) = \frac{1}{4}(x - 2)^2$

Chapter 3

Name________________________________ Date__________

Chapter 3 Polynomial Functions

Dear Family,

There are many patterns in the world that can be explained using math. Once a pattern is determined, the information can be used to make predictions and simplify problems. One of the most commonly used patterns in mathematics is Pascal's Triangle, named after the French mathematician Blaise Pascal.

Pascal's Triangle	
1	← Row 0
1 1	← Row 1
1 2 1	← Row 2
1 3 3 1	← Row 3
1 4 6 4 1	← Row 4

To find the next row of Pascal's Triangle, the first and last numbers will be 1. Then each number between each 1 is the sum of the two numbers above it.

Work together to copy the first 5 rows of Pascal's Triangle onto a sheet of paper, and then find the next 10 rows of the pattern. Use the first 15 rows of Pascal's Triangle to answer the following questions.

- What patterns do you notice within the triangle?
- What pattern do you notice when you find the sum of each row?
- What is the relationship between Pascal's Triangle and the powers of 11?

You can use the Internet to verify the rows of the Triangle and the answers to the questions above.

In this chapter, your student will work with polynomial functions. To find the binomial expansion of $(x+y)^n$, your student can use Pascal's Triangle to determine the coefficients of the variables.

There is a lot of interesting information about Pascal's Triangle, including its history and applications in the mathematical and real world, such as in probability and statistics.

Have fun looking for patterns in your daily life!

Nombre ______________________________ Fecha ________

Capítulo 3 Funciones polinomiales

Estimada familia:

Hay muchos patrones en el mundo que pueden explicarse con matemáticas. Luego de que se determina un patrón, la información puede usarse para hacer predicciones y simplificar problemas. Uno de los patrones más usados en matemáticas es el triángulo de Pascal, que lleva su nombre por el matemático francés Blaise Pascal.

Triángulo de Pascal

1	← Fila 0
1 1	← Fila 1
1 2 1	← Fila 2
1 3 3 1	← Fila 3
1 4 6 4 1	← Fila 4

Para hallar la siguiente fila en el triángulo de Pascal, el primer y el último número serán 1. Luego, cada número entre cada 1 es la suma de los dos números que están arriba de el.

Trabajen juntos para copiar las primeras 5 filas del Triángulo de Pascal en una hoja de papel y luego, hallen las siguientes 10 filas del patrón. Usen las primeras 15 filas del Triángulo de Pascal para responder las siguientes preguntas.

- ¿Qué patrones observan dentro del triángulo?
- ¿Qué patrón observan cuando hallan la suma de cada fila?
- ¿Cuál es la relación entre el Triángulo de Pascal y las potencias de 11?

Pueden consultar en Internet para verificar las filas del triángulo y las respuestas a las preguntas mencionadas.

En este capítulo, su hijo trabajará con funciones polinomiales. Para hallar la expansión binomial de $(x + y)^n$, su hijo puede usar el triángulo de Pascal para determinar los coeficientes de las variables.

Hay mucha información interesante sobre el triángulo de Pascal, incluyendo su historia y usos en el mundo matemático y en el mundo real, tal como en probabilidad y estadística.

3.1 Start Thinking

Use a graphing calculator to graph the functions $f(x) = x^3$ and $g(x) = x^4$. Compare and contrast the graphs of the two functions.

Explain why the graph of $g(x) = x^4$ is always positive. Are there any points the two graphs have in common? If so, what are they?

3.1 Warm Up

Evaluate the function for the given value of *x*.

1. $f(x) = 7x - 6; x = -2$
2. $g(x) = x^2 + 3; x = 6$
3. $f(x) = -3x + 4; x = -2$
4. $g(x) = x^2 - 6x; x = -4$
5. $f(x) = 1.7x - 73; x = 16$
6. $h(x) = 8.49x; x = 4$

3.1 Cumulative Review Warm Up

Determine whether the given characteristics describe a parabola that opens up or down.

1. Focus: $(0, -5)$

   Directrix: $y = 5$
2. Focus: $(0, 5)$

   Directrix: $y = -5$
3. Focus: $(0, -1)$

   Directrix: $y = 1$
4. Focus: $(0, 1)$

   Directrix: $y = -1$

Name ______________________________ Date __________

3.1 Practice A

In Exercises 1–4, decide whether the function is a polynomial function. If so, write it in standard form and state its degree, type, and leading coefficient.

1. $f(x) = 4x^2 - 3x + 5x^3 - 7$

2. $h(x) = 5x^3 - 7x^{-2} + x - 1$

3. $g(x) = x^4 - \frac{1}{3}x^2 + 10 - 4x^3 + 2x$

4. $f(x) = 8x^2 - \sqrt{3}x + 2$

In Exercises 5–7, evaluate the function for the given value of *x*.

5. $f(x) = -2x^4 + x^3 + 5x^2 - 3x - 7;\ x = -1$

6. $g(x) = 5x^4 - 2x^3 + 9x - 10;\ x = -6$

7. $h(x) = x^5 - 4x^3 + 3x^2 + 11x - 8;\ x = 7$

In Exercises 8 and 9, describe the end behavior of the graph of the function.

8. $g(x) = 6x^4 - 3x^3 + 12x^2 + 8x + 2$

9. $h(x) = -5x^9 + 6x^7 - 5x^4 + x^2 - 1$

In Exercises 10–13, graph the polynomial function.

10. $q(x) = x^4 - 2$

11. $h(x) = x^3 - 2x + 3$

12. $k(x) = 2x^2 + 3 - x^3$

13. $f(x) = x^5 - 2x^3 + 1$

In Exercises 14 and 15, describe the *x*-values for which (a) *f* is increasing or decreasing, (b) $f(x) > 0$, and (c) $f(x) < 0$.

14.

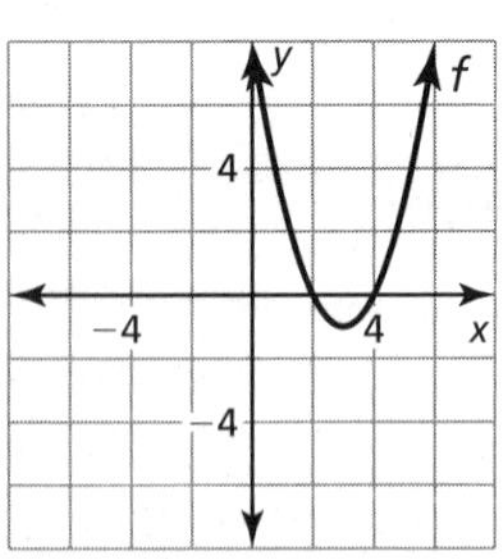

15.

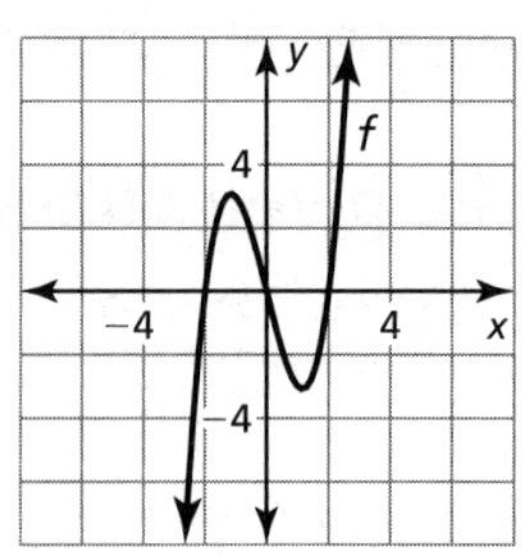

16. Suppose $f(x) \to -\infty$ as $x \to -\infty$ and $f(x) \to -\infty$ as $x \to +\infty$. Describe the degree and leading coefficient of the function.

Name____ Date__________

3.1 Practice B

In Exercises 1–4, decide whether the function is a polynomial function. If so, write it in standard form and state its degree, type, and leading coefficient.

1. $h(x) = 6x^3 - 9x^{-3} + x^2 - 5x - 1$

2. $f(x) = 11x^2 - \sqrt{7} + 12x$

3. $g(x) = 2x^4 - \frac{1}{3}x^2 - \sqrt{14}x^3 + 2x - \frac{5}{3}$

4. $f(x) = 2x^3 + 9x^2 - 5x + \frac{4}{x} - 1$

In Exercises 5–7, evaluate the function for the given value of *x*.

5. $f(x) = -x^3 + 5x^2 + 9x + 4;\ x = -11$

6. $g(x) = 3x^3 + 6x^2 + 12x - 10;\ x = \frac{1}{3}$

7. $h(x) = 9x^3 - 8x^2 + 11x + 8;\ x = -\frac{1}{2}$

In Exercises 8 and 9, describe the end behavior of the graph of the function.

8. $g(x) = -5x^4 + 7x^3 - 7x^6 + x^2 - 9x + 2$

9. $h(x) = -2x^3 + 5x^2 + 4x^5 - 3x^4 + 12x^2 - 4$

In Exercises 10–13, graph the polynomial function.

10. $q(x) = x^4 - x^3 - 5x^2$

11. $h(x) = 4 - 2x^2 - x^4$

12. $k(x) = x^5 - 2x^4 + x - 2$

13. $f(x) = x^6 - 3x^5 + 2x^3 + x + 1$

In Exercises 14 and 15, sketch a graph of the polynomial function *f* having the given characteristics. Use the graph to describe the degree and leading coefficient of the function *f*.

14. f is increasing when $x < 1$; f is decreasing when $x > 1$.

$f(x) > 0$ when $-1 < x < 3$; $f(x) < 0$ when $x < -1$ and $x > 3$.

15. f is increasing when $x < -1.1$ and $x > 2.4$; f is decreasing when $-1.1 < x < 2.4$.

$f(x) > 0$ when $-2 < x < 0$ and $x > 4$; $f(x) < 0$ when $x < -2$ and $0 < x < 4$.

16. The function $h(t) = -4.9t^2 + 28.62t + 2.4$ models the height h of a high pop-up hit by a baseball player after t seconds. Use a graphing calculator to graph the function. State an appropriate window to view the maximum height of the ball and when the ball hits the ground.

Name ______________________________ Date __________

3.1 Enrichment and Extension

Quartic Regression

Quartic regression is a process by which the equation of "best fit" is found for a set of data in the form $y = ax^4 + bx^3 + cx^2 + dx + e$. Consider the data set shown below.

The median monthly rent (in dollars) in the United States for different years between 1950 and 2000 is given in the table below. Find a quartic regression model for the data by using $x = 0$ for 1950.

x	1950	1960	1970	1980	1990	2000
y	257	350	415	481	571	602

To input data, press **STAT** and then **EDIT**. Input x-values starting with 0 and y-values.

```
EDIT CALC TESTS
1:1-Var Stats
2:2-Var Stats
3:Med-Med
4:LinReg(ax+b)
5:QuadReg
6:CubicReg
7↓QuartReg
```

To calculate the **Quartic Regression**, press **STAT** and then **RIGHT ARROW** to **CALC**.

Select **7:QuartReg**. After **QuartReg** appears on the screen, press **ENTER**.

The equation $y \approx -0.0002x^4 + 0.0225x^3 - 0.694x^2 + 14.46x + 256.6$ should appear.

In Exercises 1–4, find a quartic regression equation for the data given in the table.

1.

x	1	2	3	4	5
y	1	−2	3	0	1

2.

x	2	4	6	8	10
y	3	−1	4	−3	4

3.

x	−1	0	1	2	3	4
y	6	−1	−3	−1.5	5	1

4.

x	23	42	112	32	98	112
y	32	87	423	23	19	233

Name______________________________ Date__________

3.1 Puzzle Time

What Do You Get When You Cross An Ear Of Corn With A Spider?

Write the letter of each answer in the box containing the exercise number.

Write the polynomial function in standard form and state its degree, type, and leading coefficient.

1. $f(x) = \frac{3}{4}x^3 - 2x + x^4$

2. $f(x) = 3x^2 - x^3 + 7x - 3$

Evaluate the function for the given value of *x*.

3. $f(x) = 9x^4 + 8x^3 - 6x^2 + 2x - 5; x = -2$

4. $f(x) = 5x^3 - 6x^2 - 3x + 2; x = 3$

Describe the end behavior of the graph of the functions.

5. $f(x) = 2x^3 - 12x + 3x^4 - 6$

6. $f(x) = 8x^3 + 5x^2 + 2x$

7. $f(x) = 2x - 5x^7$

Answers

B. $f(x) = -x^3 + 3x^2 + 7x - 3$
degree 3 (cubic)
leading coefficient of -1

W. 74

O. $f(x) \to +\infty$ as $x \to -\infty$ and
$f(x) \to -\infty$ as $x \to +\infty$

B. $f(x) \to +\infty$ as $x \to -\infty$ and
$f(x) \to +\infty$ as $x \to +\infty$

E. 47

P. -193

C. $f(x) = x^4 + \frac{3}{4}x^3 - 2x$
degree 4 (quartic)
leading coefficient of 1

S. $f(x) \to -\infty$ as $x \to -\infty$ and
$f(x) \to +\infty$ as $x \to +\infty$

T. $f(x) \to -\infty$ as $x \to -\infty$ and
$f(x) \to -\infty$ as $x \to +\infty$

1	7	5	4	3	2	6

3.2 Start Thinking

Use FOIL to multiply $(x + 1)(x - 1)$. Show your steps. Then choose another number to replace the constant 1 in the binomials and multiply. Show your steps. Is there a pattern? If so, explain the pattern.

Change the signs in the binomials so they are both the same, and multiply again. Show your steps. Does the pattern still hold? Explain why or why not.

3.2 Warm Up

Simplify.

1. $(5 - 4)7$

2. $-1(x - 7)$

3. $(3 + 4m)7$

4. $14r - 4r$

5. $6z^2 - 2z - 9z^2$

6. $6m - 3m - 4p - 5m$

3.2 Cumulative Review Warm Up

Find the value of *c* that makes the expression a perfect square trinomial. Then write the expression as the square of a binomial.

1. $x^2 + 5x + c$

2. $z^2 + 6z + c$

3. $w^2 - 12w + c$

4. $x^2 - 25x + c$

5. $x^2 - 8x + c$

6. $s^2 + 27x + c$

Name____ Date__________

3.2 Practice A

In Exercises 1–3, find the sum.

1. $(-6x^2 + 3x - 7) + (10x^2 + 4x - 2)$

2. $(10x^4 + 3x^2 - 5x + 4) + (7x^5 - 5x^4 + 2x - 9)$

3. $(5x^4 + 3x^2 - 6x - 10) + (2x^3 - 7x^2 + 6x + 1)$

In Exercises 4–6, find the difference.

4. $(4x^3 + 6x^2 - 9x + 1) - (8x^3 + 2x^2 - 5x - 1)$

5. $(10x^4 - 4x^3 - 7x^2 + 5x + 9) - (2x^4 - 5x^3 - 4x^2 + 9x + 3)$

6. $(7x^5 + 4x^3 - 2x^2 + 12x + 5) - (6x^4 - 9x^3 + x^2 - 3)$

7. A city is planning a new sports park. The total area (in square feet) of the park is modeled by the expression $9x^2 + 4x - 5$. The area of the park designated for soccer fields is modeled by the expression $2x^2 - 5x + 3$. Write an expression that models the area of the park that is not designated for soccer fields.

In Exercises 8–11, find the product.

8. $5x^2(3x^2 + 7x + 6)$

9. $-2x^4(10x^3 - 9x^2 - 7x + 4)$

10. $(8x^2 - 3x + 1)(-3x + 2)$

11. $(-x - 6)(3x^2 + 2x + 9)$

12. Describe and correct the error in performing the operation.

✗ $-3x^2(4x^2 - 5x + 7) = -12x^4 - 15x^3 + 21x^2$

In Exercises 13–16, find the product of the binomials.

13. $(x - 1)(x + 4)(x - 3)$

14. $(x - 6)(x - 9)(x + 2)$

15. $(x + 3)(2x + 1)(2x - 3)$

16. $(3x + 5)(x - 4)(4x + 1)$

In Exercises 17–19, find the product.

17. $(x + 8)(x - 8)$

18. $(y + 4)^2$

19. $(2p - 3)^2$

20. Use Pascal's Triangle to expand $(k + 3)^4$.

Name ____ Date __________

3.2 Practice B

In Exercises 1 and 2, find the sum.

1. $(8x^7 - 6x^5 + 4x^3 - 6x) + (15x^6 + 4x^5 - 3x^3 + 2)$

2. $(8x^4 - 2x^3 + 9x^2 + 7x + 14) + (6x^4 - 5x^3 - 9x^2 - 11x - 9)$

In Exercises 3 and 4, find the difference.

3. $(9x^5 + 5x^4 - 9x^2 + 10x) - (12x^5 + 2x^4 - x^2 - 9)$

4. $(12x^4 - 6x^2 + 2x + 14) - (3x^4 - 5x^3 + 9x + 3)$

In Exercises 5–8, find the product.

5. $(x^2 - 7x - 2)(x^2 - 3x - 6)$

6. $(2x^2 + 3x - 1)(-5x^2 - 2x + 4)$

7. $(4x^2 - 3x + 6)(x^2 - 2x + 2)$

8. $(3x^2 - 6x - 5)(x^4 + 2x^2 + 5x)$

9. Describe and correct the error in performing the operation.

> ✗ $4x^2(3x^4 - 2x^3 + 7) = 12x^8 - 8x^6 + 28x^2$

In Exercises 10–13, find the product of the binomials.

10. $(x - 3)(2x + 2)(3x - 1)$

11. $(2x + 3)(x - 5)(4x + 1)$

12. $(2x - 1)(3 - 2x)(4x + 5)$

13. $(5 - 2x)(2 - x)(4x + 3)$

In Exercises 14–16, find the product.

14. $(3x + 5)(3x - 5)$

15. $(6t + 7)^2$

16. $(pq + 2)^2$

17. A rectangular pool has a level floor. The length of the pool is $(3x - 1)$ feet, the width of the pool is $(x + 6)$ feet, and the depth of the pool is $(x + 6)$ feet.

a. Write an expression for the volume of the pool as a product of binomials.

b. Write an expression for the volume of the pool as a polynomial in standard form.

18. Use Pascal's Triangle to expand $(2m - 5)^5$.

19. Use the Binomial Theorem to write the binomial expansion of $(3s + t)^3$.

Name____ Date__________

3.2 Enrichment and Extension

Galton Boards

A *Galton board* is a triangular array of pegs named after English statistician Francis Galton (1822–1911).

When a chip is dropped onto the top peg, it is equally likely to fall to the left or to the right of the peg. The chip eventually falls into one of the bins at the bottom, as shown in the diagram.

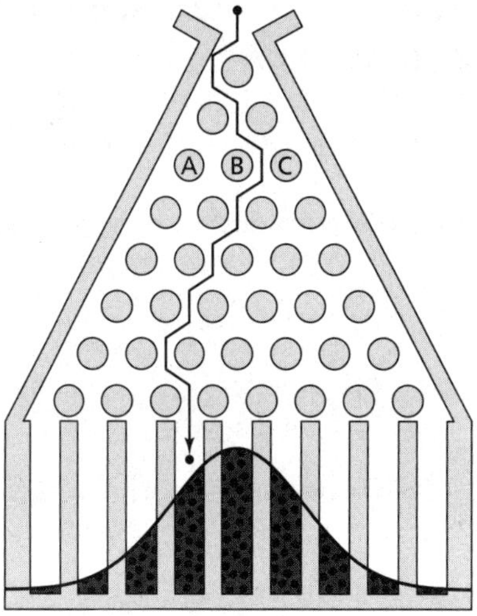

Example: How many different paths can a chip travel to Peg A? Peg B? Peg C?

Solution: The chip can strike the first peg, fall to the left, strike the left peg in the next row, and fall to the left again to strike Peg A. So, there is 1 path to Peg A.

The chip can strike the first peg, fall to the left, strike the left peg in the next row, and fall to the right to strike Peg B. Or, the chip can strike the first peg, fall to the right, strike the right peg in the next row, and fall to the left to strike Peg B. So, there are 2 paths to Peg B.

The chip can strike the first peg, fall to the right, strike the right peg in the next row, and fall to the right again to strike Peg C. So, there is 1 path to Peg C.

In Exercises 1–5, use the Galton board at the right to answer the following questions.

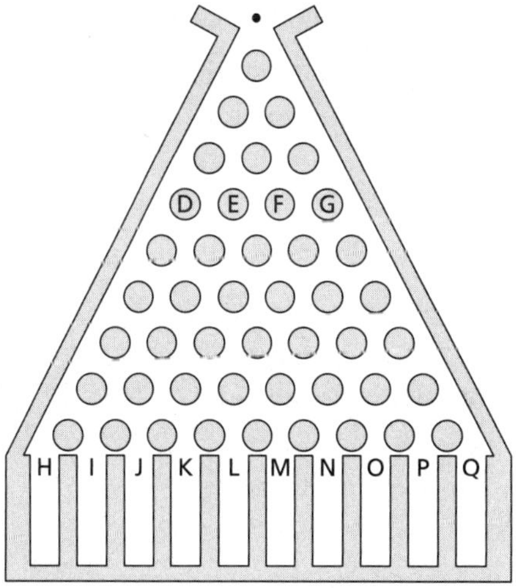

1. How many different paths can a chip travel to Peg D? Peg E? Peg F? Peg G?

2. Explain why the number of paths to Peg E is the same as the number of paths to Peg F.

3. Do you notice a pattern in the numbers of different paths a chip can travel to the pegs in each row? If so, describe the pattern.

4. How many different paths can a chip travel to each bin, labeled H through Q?

5. Why do you think the majority of chips tend to fall into the center bins?

Name ______________________________ Date __________

3.2 Puzzle Time

What Do You Get When You Cross A Centipede With A Parrot?

Write the letter of each answer in the box containing the exercise number.

Find the sum, difference, or product.

1. $(10x^2 + 5) - (6x^2 - x + 4)$
2. $(x^5 + 3x^4 - x + 7) + (3x^5 - 8x^4 - x^3 + 12)$
3. $(-3x^3 + 8x^2 - 4) - (-2x^3 - x^2)$
4. $-x^2(4x^3 - 2x^2 + 5x - 7)$
5. $(4x^3 - 2) - (3x^2 - 2)$
6. $(x - 4)(x + 3)$
7. $(5x^2 - x + 2)(-4x + 1)$
8. $(2 - x)(10x^2 - 7x + 9)$
9. $(-2x^2 - 5x + 1)(x^2 + 3x - 2)$
10. $(8x^3 - 2x^2 + 6x - 18) + (4x^3 - x^2 - 5x + 7)$
11. $(15x - 8) - (20x + 8)$
12. $(x - 2)(x + 4)(x - 10)$
13. $(2x - 1)(3 - x)(4 + 2x)$

Answers

A. $-x^3 + 9x^2 - 4$

A. $4x^2 + x + 1$

T. $-10x^3 + 27x^2 - 23x + 18$

W. $4x^5 - 5x^4 - x^3 - x + 19$

K. $-5x - 16$

K. $4x^3 - 3x^2$

I. $x^2 - x - 12$

E. $-20x^3 + 9x^2 - 9x + 2$

E. $-4x^3 + 6x^2 + 22x - 12$

A. $-2x^4 - 11x^3 - 10x^2 + 13x - 2$

L. $12x^3 - 3x^2 + x - 11$

L. $-4x^5 + 2x^4 - 5x^3 + 7x^2$

I. $x^3 - 8x^2 - 28x + 80$

1		2	3	4	5	6	7		8	9	10	11	12	13

3.3 Start Thinking

Factor the trinomial $x^2 + x - 2$ into the product of two binomials. Use your knowledge of inverse operations to explain why the equation $\dfrac{x^2 + x - 2}{x - 1} = x + 2$ must be true.

Using this example as evidence, can you conclude that factoring and division are the same? Explain why or why not.

3.3 Warm Up

Factor the expression completely.

1. $13t + 39y$

2. $3k^2 - 3k$

3. $5a^2b^2 - a^2b + 11ab^2$

4. $x^2 - 25$

5. $n^2 - 13n + 22$

6. $3x^2 + 30x + 63$

3.3 Cumulative Review Warm Up

Write a function *g* whose graph represents the indicated transformations of the graph of *f*.

1. $f(x) = x$; vertical stretch by a factor of 3, followed by a translation 2 units down

2. $f(x) = x$; vertical shrink by a factor of $\frac{1}{4}$, followed by a translation 1 unit up

3. $f(x) = |x|$; horizontal stretch by a factor of 3, followed by a translation 1 unit left

Name ______________________________ Date __________

3.3 Practice A

In Exercises 1–4, divide using polynomial long division.

1. $\left(x^2 + x + 12\right) \div (x - 5)$

2. $\left(2x^2 - x - 1\right) \div (x - 2)$

3. $\left(x^3 + x^2 - 9x - 6\right) \div \left(x^2 - 9\right)$

4. $\left(6x^3 - x^2 + 12x\right) \div \left(x^2 + 2\right)$

In Exercises 5–10, divide using synthetic division.

5. $\left(x^2 + 6x + 1\right) \div (x - 3)$

6. $\left(3x^2 - 11x - 4\right) \div (x - 1)$

7. $\left(2x^2 - x + 5\right) \div (x + 2)$

8. $\left(x^3 - 2x + 6\right) \div (x + 3)$

9. $\left(x^2 + 25\right) \div (x - 5)$

10. $\left(5x^2 - 3x + 2\right) \div (x - 1)$

11. Describe and correct the error in using synthetic division to divide $x^3 + 2x^2 + 7$ by $x + 3$.

$$\text{✗} \quad \begin{array}{r|rrrr} 3 & 1 & 2 & 0 & 7 \\ & & 3 & 15 & 45 \\ \hline & 1 & 5 & 15 & 52 \end{array}$$

$$\frac{x^3 + 2x^2 + 7}{x + 3} = x^2 + 5x + 15 + \frac{52}{x + 3}$$

In Exercises 12–15, use synthetic division to evaluate the function for the indicated value of *x*.

12. $f(x) = -x^2 - 7x + 18;\ x = -2$

13. $f(x) = 2x^2 - 3x + 6;\ x = 5$

14. $f(x) = x^3 + 2x^2 - 3x + 4;\ x = -1$

15. $f(x) = x^3 + 2x^2 - 5x + 12;\ x = -3$

16. You divide two polynomials and obtain the result $x^2 - 3 + \dfrac{6}{x + 1}$. What is the dividend? How did you find it?

Name____ Date__________

3.3 Practice B

In Exercises 1–3, divide using polynomial long division.

1. $(x^3 + 3x^2 - 4x - 6) \div (x^2 - 4)$

2. $(4x^4 + 2x^3 - 9x^2 - 36) \div (x^2 + x - 4)$

3. $(2x^4 - 40x^2 - 28) \div (x^2 - 5x - 2)$

In Exercises 4–9, divide using synthetic division.

4. $(4x^2 - 15x + 7) \div (x - 2)$

5. $(x^3 - 9x + 12) \div (x + 3)$

6. $(x^2 + 16) \div (x - 4)$

7. $(2x^3 - 5x^2 + 3) \div (x + 1)$

8. $(x^4 + 5x^3 - 6x^2 - 11x + 14) \div (x + 4)$

9. $(x^4 + 2x^3 + 4x - 20) \div (x + 6)$

10. Describe and correct the error in using synthetic division to divide $x^3 + 2x^2 + 7$ by $x + 3$.

X				
−3	1	2	0	7
		−3	3	−9
	1	−1	3	−2

$$\frac{x^3 + 2x^2 + 7}{x + 3} = x^3 - x^2 + 3x - 2$$

In Exercises 11–14, use synthetic division to evaluate the function for the indicated value of *x*.

11. $f(x) = x^3 + x^2 - 4x + 3;\ x = -1$

12. $f(x) = -x^3 - 6x^2 + 6;\ x = -2$

13. $f(x) = x^4 + 5x^2 - 8x + 1;\ x = 4$

14. $f(x) = -x^4 - x^2 - 5;\ x = 3$

15. What is the value of *k* such that $(x^3 + kx^2 - 9x - 36) \div (x + 4)$ has a remainder of zero?

Name ____________________ Date __________

3.3 Enrichment and Extension

Finding Oblique Asymptotes

A line is an *asymptote* when a graph approaches it more and more closely. An *oblique asymptote*, or *slant asymptote*, exists if the following statement is true.

> A line $y = ax + b$ is a slant asymptote of the graph of f when $f(x) - (ax + b)$ approaches 0 as x approaches positive or negative infinity.

Example: Find the oblique asymptote of the graph of $f(x) = \dfrac{x^2 - 6}{x + 4}$.

Solution: Use synthetic or long division.

$f(x) = x - 4 + \dfrac{10}{x + 4}$, which can be rewritten as $f(x) - (x - 4) = \dfrac{10}{x + 4}$.

As x approaches positive or negative infinity, the remainder term $\dfrac{10}{x + 4}$ approaches 0.

So, the oblique asymptote is $y = x - 4$.

In Exercises 1–6, find the oblique asymptote of the graph of the function.

1. $f(x) = \dfrac{-3x^2 + 2}{x - 1}$

2. $f(x) = \dfrac{x^2 + 3x + 2}{x - 2}$

3. $f(x) = \dfrac{6x^2 + 13x - 5}{3x + 5}$

4. $f(x) = \dfrac{x^2 + 8x + 15}{x + 2}$

5. $f(x) = \dfrac{2x^3 + 4x^2 - 9}{3 - x^2}$

6. $f(x) = \dfrac{ax^2 + bx + c}{x + d}$

Name____ Date__________

3.3 Puzzle Time

Why Did The Other Vegetables Like The Corn?

A	B	C	D	E	F
G	H				

Complete each exercise. Find the answer in the answer column. Write the word under the answer in the box containing the exercise letter.

8 LEND
3 WAS
5 AN
6 WILLING

Divide using polynomial long division. Match the expression with the equivalent expression at the bottom.

A. $(x^2 + x - 10) \div (x - 2)$

B. $(x^3 + x^2 + x + 4) \div (x^2 + 2)$

C. $(8x^3 + 2x^2) \div (x^2 - 1)$

D. $(4x^4 - 36x^2 - 30x - 12) \div (x^2 - 3x)$

Divide using synthetic division. Match the expression with the equivalent expression at the bottom.

E. $(x^2 + 3x - 1) \div (x + 1)$ **F.** $(5x^2 - 2x + 8) \div (x - 4)$

G. $(x^3 - 2x^2 + x - 6) \div (x + 3)$

H. $(x^2 + 16) \div (x - 4)$

4 EAR
2 IT
1 ALWAYS
7 TO

Answers

1. $8x + 2 + \dfrac{8x + 2}{x^2 - 1}$ **2.** $x + 3 - \dfrac{4}{x - 2}$

3. $x + 1 + \dfrac{-x + 2}{x^2 + 2}$ **4.** $x + 4 + \dfrac{32}{x - 4}$

5. $x^2 - 5x + 16 + \dfrac{-54}{x + 3}$ **6.** $4x^2 + 12x + \dfrac{-30x - 12}{x^2 - 3x}$

7. $x + 2 + \dfrac{-3}{x + 1}$ **8.** $5x + 18 + \dfrac{80}{x - 4}$

3.4 Start Thinking

Consider the polynomial $21xy + 3x + 40y - 5$. To factor a polynomial with four terms, consider each term separately, focusing first on the coefficients.

Is there any way you can group two of the terms together so that their coefficients have a common factor greater than 1? If so, what groups can you make? Can you group the polynomial together in such a way that it makes two groups, each with coefficients that have a common factor larger than 1? If so, state the groups.

3.4 Warm Up

Find the greatest common factor of the polynomial.

1. $9y + 3xy^3$

2. $6r^3s - 8rs$

3. $3x + 3xy - 3xz$

4. $4y^2z + 4yz - 5y^3z$

5. $5ab^3 - 5a^3b + 11a^3b$

6. $-x^2y + xy - xy$

3.4 Cumulative Review Warm Up

Graph the function. Label the vertex and the axis of symmetry.

1. $y = 4x^2 + 4x - 5$

2. $y = -4x^2 + 5x$

3. $y = 5x^2 - 23x + 8$

4. $y = x^2 + 2x - 1$

5. $y = x^2 - x + 3$

6. $y = 2x^2 + 3x + 1$

Name____ Date__________

3.4 Practice A

In Exercises 1–6, factor the polynomial completely.

1. $x^3 - x^2 - 12x$

2. $9p^7 - 36p^5$

3. $3n^6 - 33n^5 + 72n^4$

4. $2k^4 - 242k^2$

5. $2w^4 - 7w^3 - 15w^2$

6. $3q^6 - 17q^5 - 28q^4$

In Exercises 7–9, factor the polynomial completely.

7. $x^3 + 27$

8. $y^3 + 1000$

9. $w^3 - 125$

In Exercises 10–13, factor the polynomial completely.

10. $y^3 - 3y^2 + 4y - 12$

11. $q^3 - 2q^2 + 9q - 18$

12. $2d^3 + 10d^2 + 3d + 15$

13. $x^3 - 6x^2 - 9x + 54$

In Exercises 14–16, factor the polynomial completely.

14. $36p^4 - 25$

15. $n^4 + 11n^2 + 28$

16. $y^4 - 16$

In Exercises 17–20, determine whether the binomial is a factor of $f(x)$.

17. $f(x) = 3x^3 + 7x^2 - 8x - 5;\ x + 5$

18. $f(x) = 2x^3 + 15x^2 - 23x + 36;\ x + 9$

19. $f(x) = 6x^5 - 8x^4 - 6x^3 - 4x^2;\ x - 2$

20. $f(x) = 12x^3 - 69x^2 + 39x + 30;\ x - 6$

21. Fill in the blank of the divisor so that the remainder is 0. Justify your answer.

$f(x) = x^3 + 5x^2 - 6x;\ (x - \_\_\_)$

22. What is the value of k such that $x - 6$ is a factor of $f(x) = 3x^3 - 17x - kx + 18$? Justify your answer.

23. Factor each polynomial completely.

a. $5a^2c - 3a^2d + 5b^2c - 3b^2d$

b. $x^{2n} + 6x^n + 9$

Name ______________________ Date ________

3.4 Practice B

In Exercises 1–6, factor the polynomial completely.

1. $5t^5 - 320t^3$
2. $2p^6 - 26p^5 + 84p^4$
3. $3x^4 - 432x^2$
4. $5a^6 - 16a^5 - 45a^4$
5. $12j^9 - 28j^8 + 15j^7$
6. $15q^{10} + 38q^9 + 24q^8$

In Exercises 7–9, factor the polynomial completely.

7. $2p^9 - 16p^6$
8. $25k^8 + 1600k^5$
9. $54w^7 - 16w^4$

In Exercises 10–13, factor the polynomial completely.

10. $x^3 - 7x^2 + 5x - 35$
11. $m^3 - 2m^2 - 16m + 32$
12. $9w^3 - 27w^2 - 4w + 12$
13. $25s^3 + 100s^2 - s - 4$

In Exercises 14–16, factor the polynomial completely.

14. $81g^4 - 625$
15. $2t^8 + 6t^5 - 20t^2$
16. $5v^{10} - 25v^6 + 30v^2$

In Exercises 17–20, determine whether the binomial is a factor of $f(x)$.

17. $f(x) = 4x^3 - 15x^2 - 30x + 25;\ x - 5$
18. $f(x) = 2x^3 + 16x^2 - 4x - 50;\ x + 7$
19. $f(x) = 8x^5 + 43x^4 - 58x^3 + 60x^2 - 70;\ x - 4$
20. $f(x) = 42x^4 + 143x^3 + 37x^2 - 27x + 45;\ x - 2$
21. Fill in the blank of the divisor so that the remainder is 0. Justify your answer.

    $f(x) = 2x^3 + 7x^2 - 4x;\ (x + \_\_\_)$

22. The standard equation of a circle with radius r and center (h, k) is $(x - h)^2 + (y - k)^2 = r^2$. Rewrite the equation of each circle in standard form. Identify the center and radius of the circle. Then graph the circle.

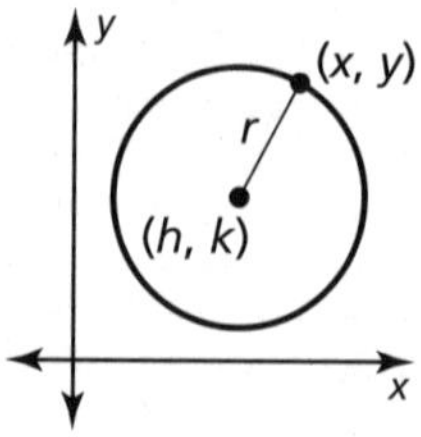

a. $x^2 + 8x + 16 + y^2 = 9$

b. $x^2 - 10x + 25 + y^2 = 4$

c. $x^2 - 4x + 4 + y^2 + 6y + 9 = 16$

Name ________________________________ Date ________

3.4 Enrichment and Extension

Factoring Polynomials

In Exercises 1–6, find the value of *k* that makes the linear expression a factor of the nonlinear expression.

1. $x^3 - k; x - 3$

2. $x^3 + 3x^2 - x + k; x - 2$

3. $x^2 - kx + 2; x - 1$

4. $x^2 - kx + 4; x + 2$

5. $kx^3 - 2x^2 + x - 6; x + 3$

6. $x^5 + kx^3 - 5x^4 + 5x - 15; x - 3$

The pattern for the sum or difference of two cubes can be applied to the sum or difference of higher powers.

Recall:

$$\left(a^3 + b^3\right) = (a + b)\left(a^2 - ab + b^2\right)$$

$$\left(a^3 - b^3\right) = (a - b)\left(a^2 + ab + b^2\right)$$

When given $a^n \pm b^n$, where n is a whole number and $n \geq 3$:

- The first factor will be $a + b$, given an original sign of addition.
- The first factor will be $a - b$, given an original sign of subtraction.
- The first term of the second factor will be a^{n-1}, the following terms have alternating signs, and the last term will be b^{n-1}, given an original sign of addition.
- The first term of the second factor will be a^{n-1}, the following terms will be positive, and the last term will be b^{n-1}, given an original sign of subtraction.
- In the second factor, the exponents of a will decrease by 1 and the exponents of b will increase by 1 from left to right.

Example: Factor $x^9 + y^9$.

Solution:
$$x^9 + y^9 = (x + y)\left(x^8 - x^7y + x^6y^2 - x^5y^3 + x^4y^4 - x^3y^5 + x^2y^6 - xy^7 + y^8\right)$$

In Exercises 7–10, factor the expression.

7. $x^5 - y^5$

8. $a^7 + b^7$

9. $a^{14} - b^{14}$

10. $x^{10} - y^{10}$

Name ______________________________ Date __________

3.4 Puzzle Time

What Type Of Player Gives Refunds?

Write the letter of each answer in the box containing the exercise number.

Factor the polynomial completely.

1. $x^3 + 125$
2. $1 - x^3$
3. $x^3 + 8$
4. $x^3 + x^2 - 6x$
5. $2x^4 - 162x^2$
6. $x^3 + 6x^2 + 3x + 18$
7. $x^3 - x^2 + 2x - 2$
8. $x^3 + 2x^2 - 4x - 8$
9. $9x^4 - 36$
10. $2x^4 + 4x^2$
11. $x^4 - 4x^2 - 5$

Answers

K. $(x^2 - 5)(x^2 + 1)$

A. $9(x^2 + 2)(x^2 - 2)$

A. $(x + 2)(x^2 - 2x + 4)$

R. $x(x - 2)(x + 3)$

T. $2x^2(x - 9)(x + 9)$

Q. $(x + 5)(x^2 - 5x + 25)$

E. $(x^2 + 3)(x + 6)$

R. $(x^2 + 2)(x - 1)$

U. $(1 - x)(1 + x + x^2)$

B. $(x + 2)^2(x - 2)$

C. $2x^2(x^2 + 2)$

1	2	3	4	5	6	7	8	9	10	11

3.5 Start Thinking

Using a graphing calculator, graph the function $f(x) = x^3 - 6x^2$. What are the two ordered pairs that represent where the graph of the function appears to cross the x-axis? Substitute the x- and y-values into the function $\left(\text{note that } f(x) \text{ and } y \text{ are equivalent}\right)$ and simplify.

Describe what happens when a point where the graph crosses the x-axis is inserted into the function.

3.5 Warm Up

Solve.

1. $4t - 6 = -7$

2. $8x - 28 = -68$

3. $-3 - \dfrac{r}{10} = 3$

4. $15 = \dfrac{5 - z}{-1}$

5. $-7m + 9 = 23$

6. $\dfrac{3b}{2} - 7 = 23$

3.5 Cumulative Review Warm Up

Solve the inequality algebraically.

1. $5x^2 > 25$

2. $x^2 + 12x \leq -27$

3. $x^2 + 6x + 6 \geq 1$

4. $x^2 < 5$

5. $x^2 + 2x - 4 > -1$

6. $x^2 - 9x < -8$

Name ______________________________ Date __________

3.5 Practice A

In Exercises 1–6, solve the equation.

1. $q^3 - q^2 - 30q = 0$

2. $k^3 + 6k^2 + 9k = 0$

3. $3y^4 - 6y^3 = -3y^2$

4. $n^3 + 2n^2 - 9n - 18 = 0$

5. $3p^3 = 21p$

6. $8u^6 = 16u^4$

In Exercises 7–10, find the zeros of the function. Then sketch a graph of the function.

7. $f(x) = x^4 + x^3 - 12x^2$

8. $g(x) = x^4 - 8x^2 + 16$

9. $h(x) = x^5 - 2x^4 - 15x^3$

10. $f(x) = -3x^3 - 15x^2 - 12x$

11. According to the Rational Root Theorem, which is *not* a possible solution of the equation $3x^4 - 6x^3 + 2x + 4 = 0$?

A. 4 **B.** $\frac{1}{3}$ **C.** -3 **D.** $-\frac{2}{3}$

12. Describe and correct the error in listing the possible rational zeros of the function.

> ✗ $f(x) = x^3 + 3x^2 - 8x - 18$
>
> Possible zeros: $\pm 2, \pm 3, \pm 6, \pm 9$

In Exercises 13 and 14, find all the real solutions of the equation.

13. $x^4 - 8x^2 - 9 = 0$

14. $x^3 + 2x^2 - 5x - 6 = 0$

15. Write a third or fourth degree polynomial function that has zeros at $\pm\frac{3}{2}$. Justify your answer.

16. Determine the value of k for each equation so that the given x-value is a solution.

a. $x^3 + 2x^2 - 9x + k = 0;\ x = 3$

b. $x^3 - 3x^2 + kx - 12 = 0;\ x = -4$

Name______________________________ Date__________

3.5 Practice B

In Exercises 1–6, solve the equation.

1. $4x^4 + 12x^3 + 9x^2 = 0$

2. $6h^5 = 12h^3$

3. $16q^4 - 8q^2 + 1 = 0$

4. $w^4 + 81 = 18w^2$

5. $p^3 - 25p = 50 - 2p^2$

6. $y^3 - 8y^2 = 9y - 72$

In Exercises 7–10, find the zeros of the function. Then sketch a graph of the function.

7. $f(x) = -5x^4 + 20x^3 + 60x^2$

8. $g(x) = -x^3 - x^2 + 30x$

9. $h(x) = x^3 + x^2 - 4x - 4$

10. $f(x) = x^3 - 4x^2 - 9x + 36$

11. According to the Rational Root Theorem, which is *not* a possible zero of the function $f(x) = 24x^4 - 16x^3 + 21x - 27$?

A. $-\frac{3}{8}$ **B.** -2 **C.** $-\frac{1}{3}$ **D.** $-\frac{9}{4}$

12. Describe and correct the error in listing the possible rational zeros of the function.

> ✗ $f(x) = 2x^3 + 5x^2 - 2x - 6$
>
> Possible zeros: $\pm1, \pm2, \pm3, \pm6$

In Exercises 13 and 14, find all the real solutions of the equation.

13. $2x^3 - 3x^2 + 18x - 27 = 0$

14. $x^3 - 5x^2 - 2x + 24 = 0$

15. Write a third or fourth degree polynomial function that has zeros at $\pm\frac{7}{5}$. Justify your answer.

16. The sidewalk hazard marker is shaped like a pyramid, with a height 2 centimeters greater than the length of each side of its square base. The volume of the marker is 297 cubic centimeters. What are the dimensions of the sidewalk hazard marker?

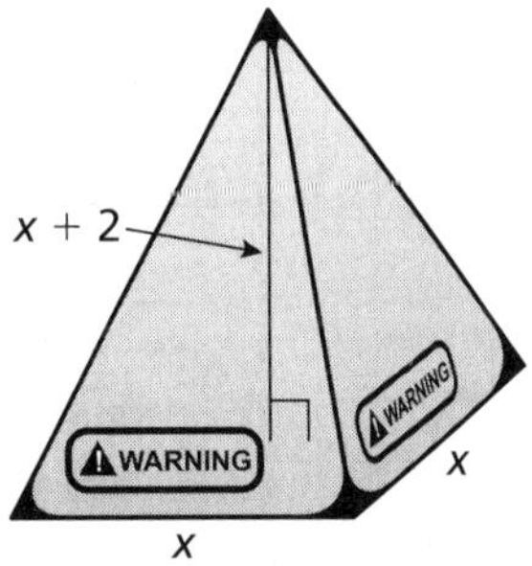

Name ______________________ Date ________

3.5 Enrichment and Extension

The Intermediate Value Theorem

You can use the Intermediate Value Theorem, stated below, to help you locate the real zeros of a polynomial function.

Intermediate Value Theorem

Let a and b be real numbers such that $a < b$, and let f be a polynomial function for which $f(a) \neq f(b)$. Then, in the interval $[a, b]$, f takes on each value between $f(a)$ and $f(b)$.

So, when the values of $f(a)$ and $f(b)$ are of opposite signs, there must be at least one real zero between $x = a$ and $x = b$.

In Exercises 1 and 2, use the above information.

1. For each polynomial function, make a table that shows the values of the function for integer values of x from –3 to 3. Use the Intermediate Value Theorem to find a one-unit-long interval (or intervals, if possible) in which the function has a zero.

a. $f(x) = x^3 + x^2 - 1$

b. $f(x) = -2x^4 - 3x^2 + x + 12$

c. $f(x) = 4x^4 - 15x^2 + 4$

d. $f(x) = 10x^3 + 28x^2 - 6x - 3$

e. $f(x) = x^5 - x^4 - 3x^3 + 2x^2 - 2$

f. $f(x) = x^6 + 6x^4 - 12x^2 + 3$

2. Consider the polynomial function $f(x) = x^3 + 2x^2 + x + 1$.

a. List the possible rational zeros of f. Are these actual zeros? Explain.

b. Copy and complete the table. Use the Intermediate Value Theorem to find a one-unit-long interval in which the function has a zero.

x	–3	–2	–1	0	1	2	3
f(x)							

c. Divide the interval you found in part (b) into tenths. Evaluate the function at each of these values. Use the Intermediate Value Theorem to find an interval in which the function has a zero.

d. Repeat the process until you are able to approximate the zero to five decimal places.

Name____ Date__________

3.5 Puzzle Time

What Did The Weatherman Say About His Meteorology Test?

A	B	C	D	E	F
G	H	I	J		

Complete each exercise. Find the answer in the answer column. Write the word under the answer in the box containing the exercise letter.

−14, 0, 2	IT
−1, 2	A
−1, 0, 5	FOGGY
−2, 0, 2	BREEZE
−4, 0, 4	WITH

Solve the equation.

A. $2x^3 + 24x^2 = 56x$

B. $2x^5 - 6x^4 = 56x^3$

C. $x^3 - x = 0$

D. $34x^3 - 48x = 22x^3$

E. $4x^4 - 64x^2 = 0$

F. $2x^3 - 4x^2 - 96x = -8x^2$

Find all the real zeros of the function.

G. $f(x) = x^4 - x^3 + x^2 - 3x - 6$

H. $f(x) = 3x^3 - 5x^2 - 47x - 15$

I. $f(x) = x^3 - 4x^2 - 5x$

J. $f(x) = x^3 - 3x^2 - 4x + 12$

$-3, -\frac{1}{3}, 5$	FEW
−4, 0, 7	WAS
−2, 2, 3	PATCHES
−8, 0, 6	ONLY
−1, 0, 1	A

3.6 Start Thinking

Use a graphing calculator to complete the table.

Function	Number of x-intercepts
$f(x) = x + 4$	
$g(x) = x^2 - 5$	
$h(x) = x^3 + 3x^2 - x - 1$	
$j(x) = x^4 - x^3 - 4x^2 + 1$	

What is the pattern? Do you think the pattern will continue infinitely? Why or why not?

3.6 Warm Up

Identify the degree of the polynomial.

1. $8ab - 5b^3 + a^4$

2. $2x^3$

3. $4m^3p^2 - 7$

4. $x^6 + 11 - 4x$

5. $3z + 3z + 6z^3$

6. $6x^3y^2 - 3x$

3.6 Cumulative Review Warm Up

Graph the function and its parent function. Then describe the transformation, if any.

1. $g(x) = x - 5$

2. $f(x) = x^2$

3. $g(x) = |x + 6|$

4. $h(x) = -x$

Name ______________________________ Date __________

3.6 Practice A

In Exercises 1–4, identify the number of solutions of the polynomial equation. Then find all solutions of the equation.

1. $x^3 - 125 = 0$

2. $27g^5 + g^2 = 0$

3. $t^5 - t^3 + t^2 - 1 = 0$

4. $x^3 - 2x^2 + 5x - 10 = 0$

In Exercises 5–8, find all zeros of the polynomial function.

5. $f(x) = x^4 - 5x^2 - 36$

6. $f(x) = x^4 + x^3 - 7x^2 - x + 6$

7. $g(x) = x^4 - x^3 + 9x^2 - 9x$

8. $h(x) = x^4 - 11x^2 + 18$

In Exercises 9 and 10, determine the number of imaginary zeros for the function with the given degree and graph. Explain your reasoning.

9. Degree: 5

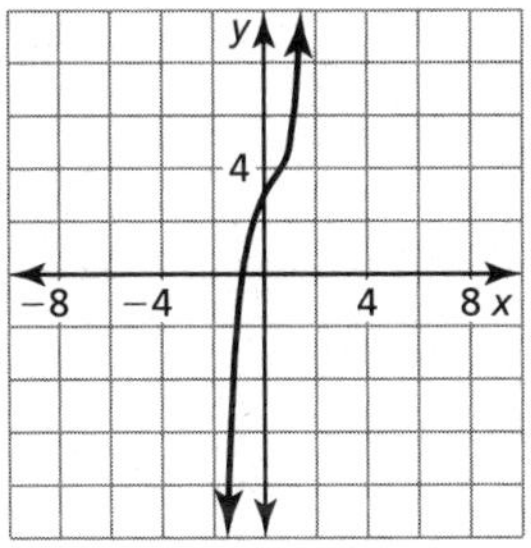

10. Degree: 4

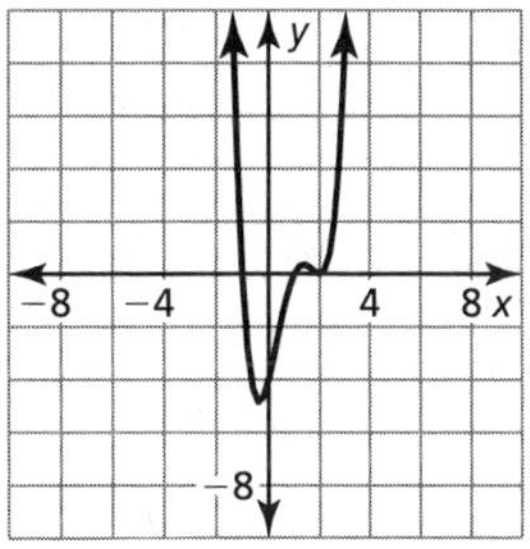

In Exercises 11–13, write a polynomial function *f* of least degree that has rational coefficients, a leading coefficient of 1, and the given zeros.

11. $-4, 1, 2$

12. $2, 3, -1$

13. $2, \sqrt{3}$

14. Write a polynomial function of degree 5 with zeros -1, 2, and i. Justify your answer.

In Exercises 15–17, determine the possible numbers of positive real zeros, negative real zeros, and imaginary zeros for the function.

15. $g(x) = x^3 - 2x^2 - 3x + 1$

16. $g(x) = x^4 + x^2 - 10$

17. $g(x) = x^5 - 2x^4 + x^2 - 3x - 2$

Name ______________________________ Date __________

3.6 Practice B

In Exercises 1–4, identify the number of solutions of the polynomial equation. Then find all solutions of the equation.

1. $8x^3 + 27 = 0$

2. $4p^5 - 32p^2 = 0$

3. $t^8 - t^4 - t^2 + 1 = 0$

4. $x^5 - 9x^3 + 8x^2 - 72 = 0$

In Exercises 5–8, find all zeros of the polynomial function.

5. $h(x) = x^4 - 4x^3 + 3x^2 + 4x - 4$

6. $f(x) = x^4 - 12x^3 + 54x^2 - 108x + 81$

7. $g(x) = x^5 + 4x^4 + x^3 - 14x^2 - 20x - 8$

8. $f(x) = x^5 + 2x^4 - 13x^3 - 26x^2 + 36x + 72$

In Exercises 9 and 10, determine the number of imaginary zeros for the function with the given degree and graph. Explain your reasoning.

9. Degree: 4

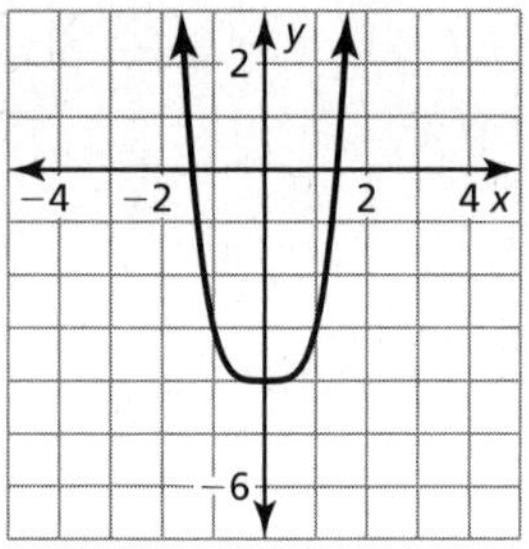

10. Degree: 3

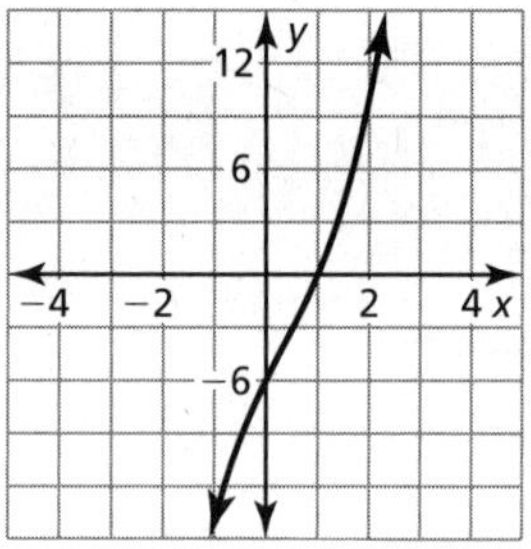

In Exercises 11–13, write a polynomial function *f* of least degree that has rational coefficients, a leading coefficient of 1, and the given zeros.

11. $2, 3 + i$

12. $2i, 1 - i$

13. $3, -\sqrt{7}$

14. Two zeros of $f(x) = x^3 - 2x^2 + 9x - 18$ are $3i$ and $-3i$. Explain why the third zero must be a real number.

15. Use Descartes' Rule of Signs to determine which function has no positive real zeros.

A. $f(x) = x^4 - 3x^2 + 6x - 7$

B. $f(x) = x^4 + 2x^2 + 4x - 3$

C. $f(x) = x^4 + x^2 + 10$

D. $f(x) = x^4 + 5x^3 - 9x - 7$

Name ______________________________ Date __________

3.6 Enrichment and Extension

Upper and Lower Bound Rules

The sign pattern in the last row of the synthetic division process can be used to place upper and lower bounds for the real zeros of f.

Upper and Lower Bound Rules

Let f be a polynomial function with real coefficients and a positive leading coefficient. Suppose $f(x)$ is divided by $x - k$ using synthetic division.

1. If $k > 0$ and each number in the last row is either positive or zero, then f has no real zero greater than k.
2. If $k < 0$ and the numbers in the last row are alternatively positive and negative (zero counts as positive or negative), then f has no real zero less than k.

In Exercises 1–10, use synthetic division to show that the real zeros of the polynomial function satisfy the given condition.

1. $f(x) = 6x^3 - 4x^2 + 3x - 5$; no real zero greater than 1

2. $f(x) = x^3 + 3x^2 - 2x + 8$; no real zero less than -4

3. $f(x) = 2x^5 - x^4 + 2x^3 - 2x^2 + 4x - 4$; no real zero greater than 1

4. $f(x) = x^4 - x^3 + 3x^2 - 8x + 8$; no real zero greater than 2

5. $f(x) = x^5 + 2x^3 - 2x^2 + 5x + 5$; no real zero less than -1

6. $f(x) = x^5 - 3x^3 + x + 2$; no real zero less than -3

7. $f(x) = x^3 - 4x^2 + 1$; no real zero less than -1

8. $f(x) = x^4 - 4x^3 - 16x - 15$; no real zero greater than 5

9. $f(x) = x^4 + x^3 - x^2 + 3$; no real zero less than -2

10. $f(x) = 2x^4 - 8x + 3$; no real zero greater than 3

Name ____ Date __________

3.6 Puzzle Time

What Do You Get When You Cross A Dog With Canvas?

Write the letter of each answer in the box containing the exercise number.

Find all solutions of the equation.

1. $x^3 + 8 = 0$
2. $x^3 - 8x^2 + 15x = 0$
3. $x^3 - 3x^2 - 2x + 6 = 0$
4. $x^4 - 5x^2 - 36 = 0$

Write a polynomial function *f* of least degree that has rational coefficients, a leading coefficient of 1, and the given zeros.

5. $-4, 1, 2$
6. $-6, -1, 3$
7. $7, 1 + i$
8. $2i, 1 - i$

Answers

N. $f(x) = x^3 - 9x^2 + 16x - 14$

T. $f(x) = x^4 - 2x^3 + 6x^2 - 8x + 8$

P. $0, 3, 5$

A. $2, 1 - i\sqrt{3}, 1 + i\sqrt{3}$

E. $f(x) = x^3 + 4x^2 - 15x - 18$

U. $-\sqrt{2}, \sqrt{2}, 3$

P. $-3, 3, 2i, 2i$

T. $f(x) = x^3 + x^2 - 10x + 8$

1		2	3	4		5	6	7	8

3.7 Start Thinking

Use the graph of the function $g(x) = x^5$ to fill in the table.

Function	Transformation	Function	Transformation
$g(x) = (x - 2)^5$		$g(x) = -x^5$	
$g(x) = (x + 2)^5$		$g(x) = (2x)^5$	
$g(x) = x^5 - 2$		$g(x) = \left(\frac{1}{2}x\right)^5$	
$g(x) = x^5 + 2$		$g(x) = 2x^5$	

How are the transformations related to other parent functions?

3.7 Warm Up

Describe the transformation.

1. $g(x) = 5x$

2. $h(x) = -\frac{1}{3}x^2$

3. $g(x) = 2x$

4. $f(x) = -\frac{1}{2}|x|$

3.7 Cumulative Review Warm Up

Describe the transformation of $f(x) = x^2$ represented by g. Then graph each function.

1. $g(x) = x^2 + 5$

2. $g(x) = (x - 1)^2$

3. $g(x) = (x + 2)^2$

4. $g(x) = (x - 5)^2 + 3$

Name ______________________________ Date ________

3.7 Practice A

In Exercises 1 and 2, describe the transformation of *f* represented by *g*. Then graph each function.

1. $f(x) = x^3,\ g(x) = x^3 - 2$

2. $f(x) = x^3,\ g(x) = (x + 3)^3$

In Exercises 3–6, describe the transformation of *f* represented by *g*. Then graph each function.

3. $f(x) = x^4,\ g(x) = -5x^4$

4. $f(x) = x^3,\ g(x) = 4x^3 - 3$

5. $f(x) = x^5,\ g(x) = \frac{2}{3}x^5 - 5$

6. $f(x) = x^4,\ g(x) = \frac{1}{2}(x - 2)^4$

In Exercises 7 and 8, write a rule for *g* and then graph each function. Describe the graph of *g* as a transformation of the graph of *f*.

7. $f(x) = x^3 + 2,\ g(x) = f(x - 1)$

8. $f(x) = x^4 - 3x + 1,\ g(x) = 2f(x)$

9. Describe and correct the error in graphing the function $g(x) = (x - 3)^2 + 2$.

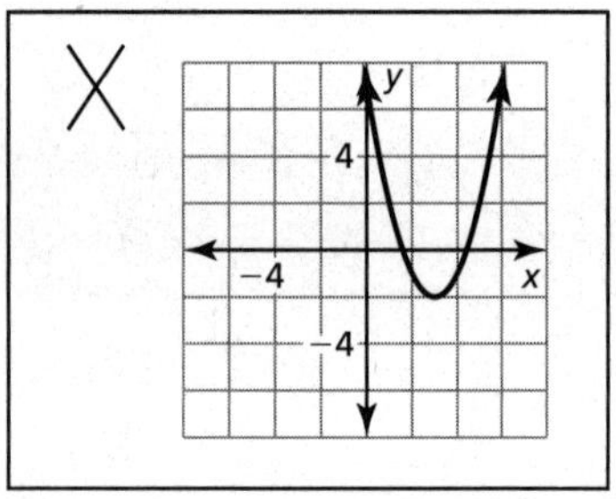

In Exercises 10 and 11, write a rule for *g* that represents the indicated transformations of the graph of *f*.

10. $f(x) = x^3 + 5$; translation 2 units right, followed by a reflection in the y-axis

11. $f(x) = x^4 - 3x + 1$; vertical shrink by a factor of $\frac{1}{3}$, followed by a translation 2 units down

12. The volume V (in cubic yards) of a rectangular box is given by $V = x^3 + 4x + 3$.

a. The function $W(x) = V(3x)$ gives the volume (in cubic feet) of the box when x is measured in yards. Write a rule for W.

b. The function $Z(x) = W(12x)$ gives the volume (in cubic inches) of the box when x is measured in yards. Write a rule for Z.

Name____________________ Date__________

3.7 Practice B

In Exercises 1 and 2, describe the transformation of *f* represented by *g*. Then graph each function.

1. $f(x) = x^4,\ g(x) = (x - 3)^4 - 2$

2. $f(x) = x^5,\ g(x) = (x - 1)^5 + 4$

In Exercises 3–6, describe the transformation of *f* represented by *g*. Then graph each function.

3. $f(x) = x^5,\ g(x) = -3x^5$

4. $f(x) = x^4,\ g(x) = 3x^4 + 2$

5. $f(x) = x^4,\ g(x) = \frac{1}{3}x^4 - 3$

6. $f(x) = x^4,\ g(x) = \frac{2}{3}(x + 3)^4$

In Exercises 7 and 8, write a rule for *g* and then graph each function. Describe the graph of *g* as a transformation of the graph of *f*.

7. $f(x) = x^3 - 4x^2 + 2,\ g(x) = -\frac{1}{4}f(x)$

8. $f(x) = x^4 + x + 1,\ g(x) = f(-x) + 2$

9. Describe and correct the error in describing the transformation of the graph of $f(x) = x^4$ represented by the graph of $g(x) = 4x^4 + 3$.

> ✗ The graph of *g* is a vertical shrink by a factor of $\frac{1}{4}$, followed by a translation 3 units up of the graph of *f*.

In Exercises 10 and 11, write a rule for *g* that represents the indicated transformations of the graph of *f*.

10. $f(x) = x^3 - 3x^2 + 2$; horizontal stretch by a factor of 3 and a translation 3 units up, followed by a reflection in the *x*-axis

11. $f(x) = 3x^5 \quad x^3 + 5x^2 + 1$; reflection in the *y*-axis and a vertical shrink by a factor of $\frac{1}{2}$, followed by a translation 1 unit up

12. The volume *V* (in cubic inches) of a rectangular box is given by $V = 2x^3 + 9$.

    a. The function $W(x) = V\left(\frac{x}{12}\right)$ gives the volume (in cubic feet) of the box when *x* is measured in inches. Write a rule for *W*. Find and interpret $W(6)$.

    b. The function $Z(x) = W\left(\frac{x}{3}\right)$ gives the volume (in cubic yards) of the box when *x* is measured in inches. Write a rule for *Z*.

Name ______________________________ Date __________

3.7 Enrichment and Extension

Transformations of Polynomial Functions

The graphs of 10 cubic functions are shown below.

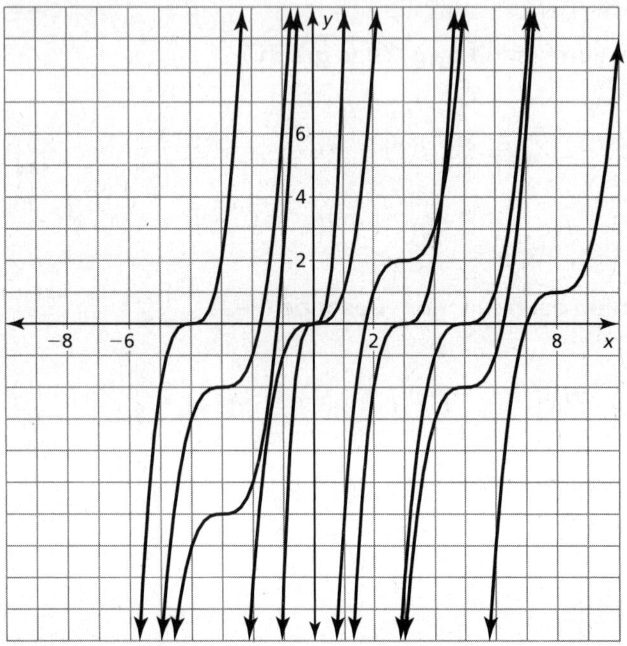

The equations for three of the graphs are

$y = x^3$,

$y = 2(x + 4)^3$, and

$y = (x + 3)^3 - 2$.

Write the equations for the other seven graphs in the form $y = a(x - h)^3 + k$.

Name ______________________ Date __________

3.7 Puzzle Time

What Day Does A Fish Dislike The Most?

Write the letter of each answer in the box containing the exercise number.

Match the function with the correct transformation of the graph of $f(x) = x^4$.

1. $g(x) = (x - 2)^4 - 1$

2. $g(x) = (x + 1)^4 - 2$

3. $g(x) = (x - 1)^4 - 2$

4. $g(x) = (x + 2)^4 - 1$

5. $g(x) = (x - 2)^4 + 1$

6. $g(x) = (x + 2)^4 + 1$

A.

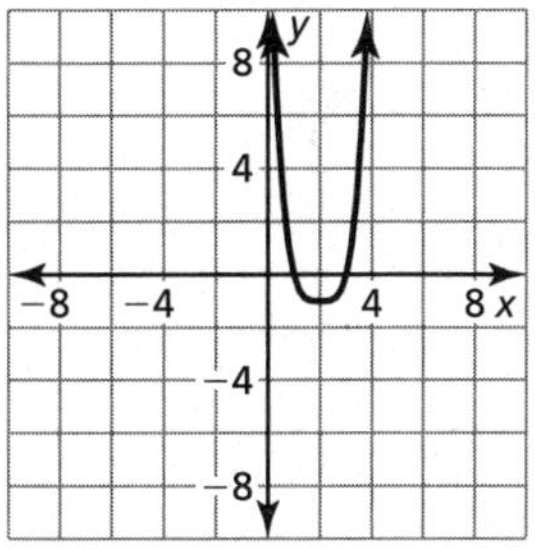

B.

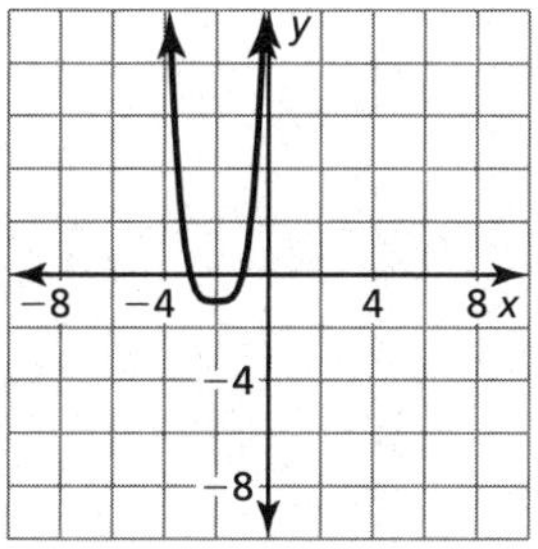

C.

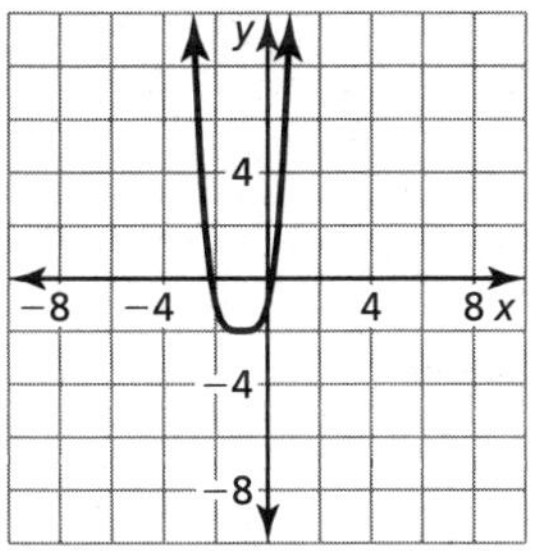

D.

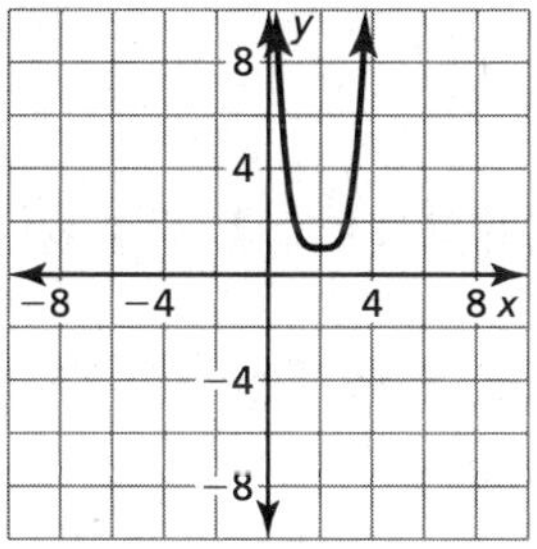

E.

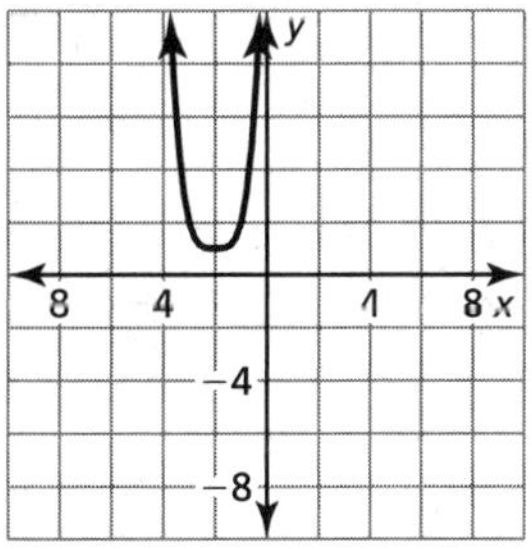

F.

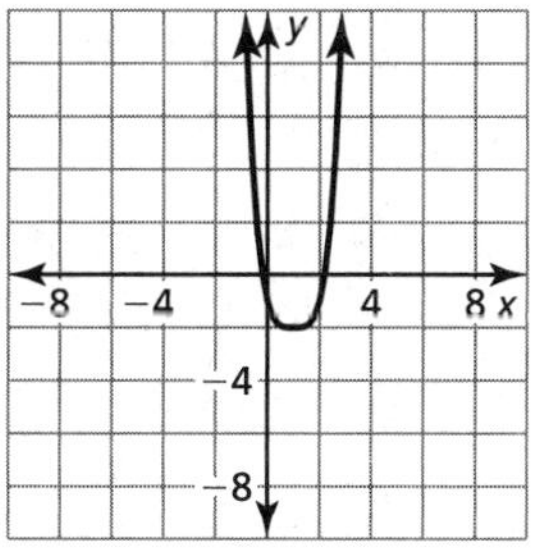

Answers

Y. E

R. C

Y. F

D. B

A. D

F. A

1	2	3	4	5	6

3.8 Start Thinking

Use a graphing calculator to graph the function $f(x) = 7x^3 + 4x^2 - 18x - 5$. Describe the shape of the function. Use the *table* feature to look at the y-values where the sign changes from positive to negative or vice versa.

What can you conclude from this data as it relates to the zeros of the function? Does this view show the exact values that are solutions? Explain.

3.8 Warm Up

Find the vertex of the function.

1. $y = x^2 - 10x - 8$

2. $y = -x^2 + 6x + 3$

3. $y = 3x^2 + 12x - 7$

4. $y = -3x^2 + 8x$

5. $y = -x^2 + 16x - 19$

6. $y = -x^2 + 4x + 6$

3.8 Cumulative Review Warm Up

Write the quadratic function in vertex form. Then identify the vertex.

1. $f(x) = x^2 + 6x - 20$

2. $g(x) = x^2 + 4x + 2$

3. $g(x) = x^2 - 10x - 38$

4. $h(x) = x^2 - 20x - 91$

5. $h(x) = x^2 - 2x + 49$

6. $f(x) = x^2 - 6x + 3$

Name______________________________ Date__________

3.8 Practice A

In Exercises 1–4, graph the function.

1. $f(x) = (x + 2)^2(x - 3)$
2. $g(x) = (x - 1)^2(x + 1)(x + 3)$
3. $h(x) = 2(x - 1)(x - 2)(x + 2)$
4. $f(x) = 3(x - 1)^2(x + 1)^2$
5. Describe and correct the error in using factors to graph $f(x) = (x - 1)^2(x + 3)$.

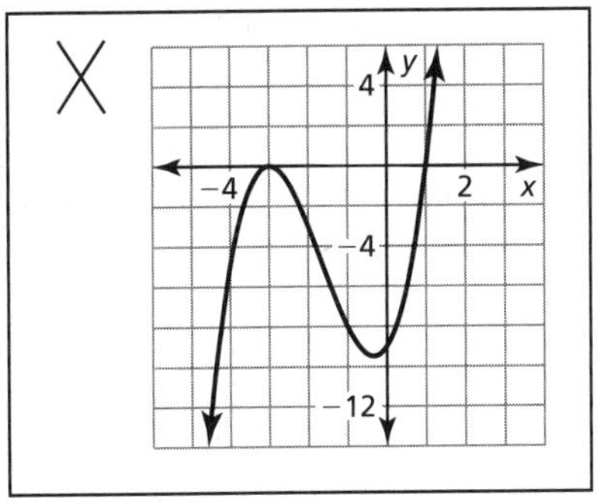

In Exercises 6–9, find all real zeros of the function.

6. $f(x) = x^3 + 3x^2 - 4x - 12$
7. $f(x) = x^3 + 7x^2 - x - 7$
8. $f(x) = x^3 - 5x^2 + x - 5$
9. $f(x) = 2x^3 - 3x^2 - 18x + 27$

In Exercises 10–13, graph the function. Identify the *x*-intercepts and the points where the local maximums and local minimums occur. Determine the intervals for which the function is increasing and decreasing.

10. $f(x) = 2x^3 - 5x^2 + 3$
11. $g(x) = -x^4 + 2x$
12. $h(x) = x^4 - 2x^2 + 3x$
13. $f(x) = x^4 - 4x^3 + 5x - 2$

In Exercises 14 and 15, estimate the coordinates of each turning point. State whether each corresponds to a local maximum or a local minimum. Then estimate the real zeros and find the least possible degree of the function.

14.

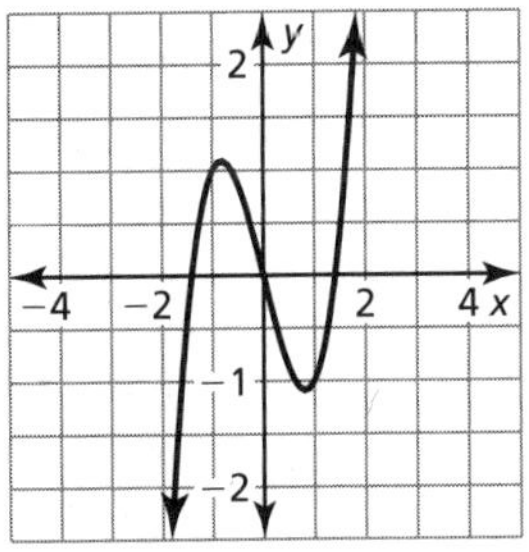

15.

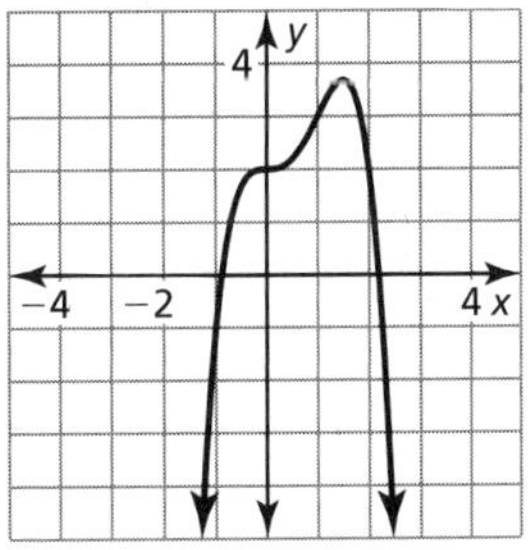

Name ______________________________ Date __________

3.8 Practice B

In Exercises 1–4, graph the function.

1. $f(x) = 4(x + 3)^2(x - 2)^2$

2. $g(x) = \frac{1}{2}(x - 4)(x + 3)(x - 6)$

3. $h(x) = \frac{1}{5}(x - 3)(x - 4)(x + 8)$

4. $f(x) = (x - 2)(x^2 + x + 2)$

5. Describe and correct the error in using factors to graph $f(x) = x^2(x + 2)^3$.

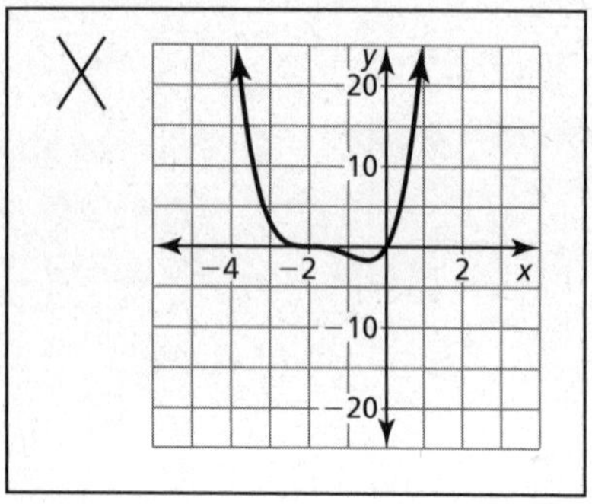

In Exercises 6–9, find all real zeros of the function.

6. $f(x) = 2x^3 - x^2 + 8x - 4$

7. $f(x) = 2x^3 + 7x^2 + x - 10$

8. $f(x) = 4x^3 - 3x^2 - 36x + 27$

9. $f(x) = 2x^3 + 3x^2 + 10x + 15$

In Exercises 10–13, graph the function. Identify the *x*-intercepts and the points where the local maximums and local minimums occur. Determine the intervals for which the function is increasing and decreasing.

10. $f(x) = 0.5x^3 - 3x^2 + 1.5$

11. $g(x) = 0.4x^3 - 3x$

12. $h(x) = x^5 - 3x^2 - 9x - 2$

13. $f(x) = x^4 - 3x^3 + 3x^2 + x - 2$

14. You are making a rectangular box out of a 12-inch by 8-inch piece of cardboard. The box will be formed by making the cuts shown in the diagram and folding up the sides. You want the box to have the greatest volume possible.

a. How long should you make the cuts?

b. What is the maximum volume?

c. What are the dimensions of the finished box?

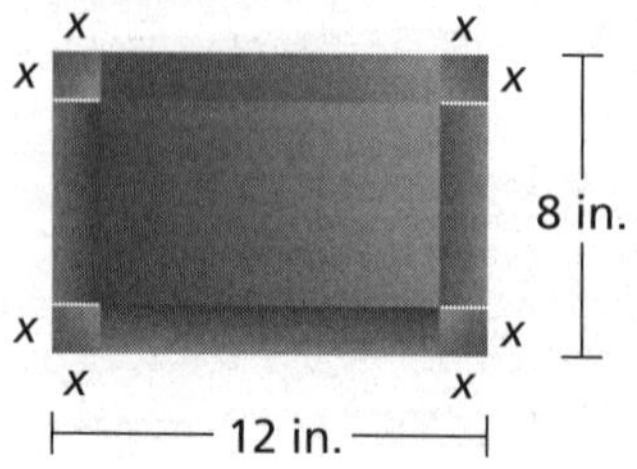

Name____ Date__________

3.8 Enrichment and Extension

Analyzing Graphs of Polynomial Functions

The *bisection method*, also known as the *interval method*, is a root-finding method that repeatedly bisects an interval, to obtain increasingly narrow subintervals in which a root must lie. This simple but slow method can be used to obtain a rough approximation of a real zero of a polynomial function.

Example: Use the bisection method to approximate the real zero of the polynomial function $f(x) = x^3 - 3$.

Solution: Find two real numbers a and b such that $f(a)$ and $f(b)$ have opposite signs. Because $f(1) = -2$ and $f(2) = 5$, $a = 1$ and $b = 2$ satisfy the condition.

The function is continuous, so the zero must lie in the interval $[1, 2]$. Evaluate the function at the average of the two endpoints of the interval: $f(1.5) = 0.375$.

Because 1.5 produces a positive result, replace the endpoint 2 with the average 1.5. The interval is now $[1, 1.5]$. Continue this process until the function value approaches 0.

The zero of the function is about 1.44.

In Exercises 1–5, use the bisection method to approximate (to the nearest hundredth) the zero of the function between the two values of *x*.

1. $f(x) = x^2 - 10,\ f(3) = -1,\ f(4) = 6$

2. $f(x) = x^3 - x - 2,\ f(1) = -2,\ f(2) = 4$

3. $f(x) = 4x^3 - 3x + 5,\ f(-2) = -21,\ f(-1) = 4$

4. $f(x) = 3x^3 - 10x^2 - 20x + 110,\ f(-3) = -1,\ f(-2) = 86$

5. $f(x) = x^5 - 2x^3 - 12,\ f(1) = -13,\ f(2) = 4$

Name ______________________________ Date __________

3.8 Puzzle Time

What Do You Call A Computer That Only Types In Uppercase?

Write the letter of each answer in the box containing the exercise number.

Match the function with its graph.

Answers	
H. A	**S.** B
Y. C	**F.** D
I. E	**T.** F

1. $f(x) = (x - 2)(x - 3)(x + 3)$

2. $f(x) = (x - 3)^2(x - 2)$

3. $f(x) = (x + 2)(x + 3)(x - 3)$

4. $f(x) = (x + 2)^2(x - 3)$

5. $f(x) = (x + 3)^2(x - 2)$

6. $f(x) = (x - 2)^2(x + 3)$

A.

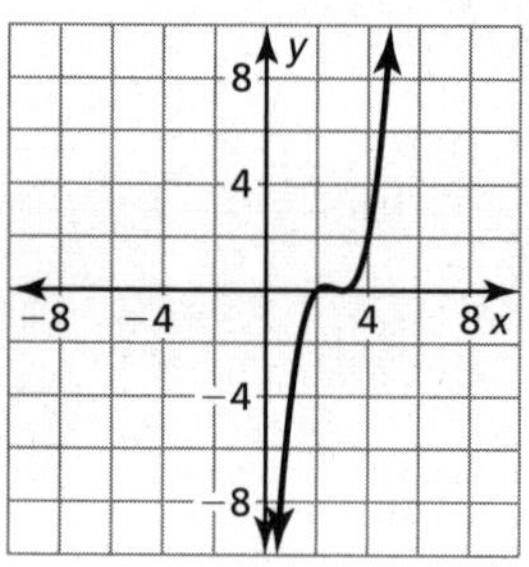

B.

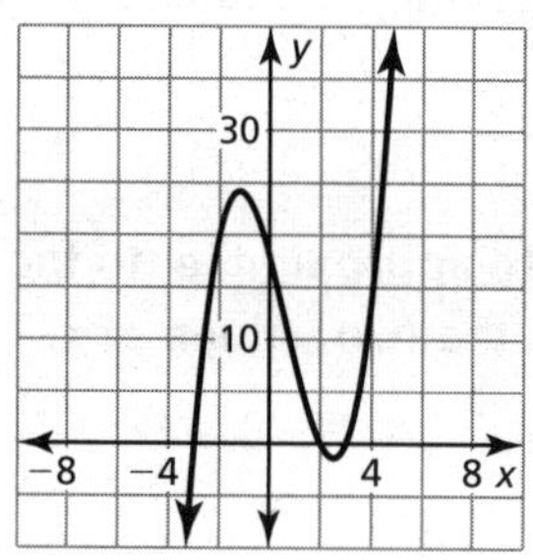

C.

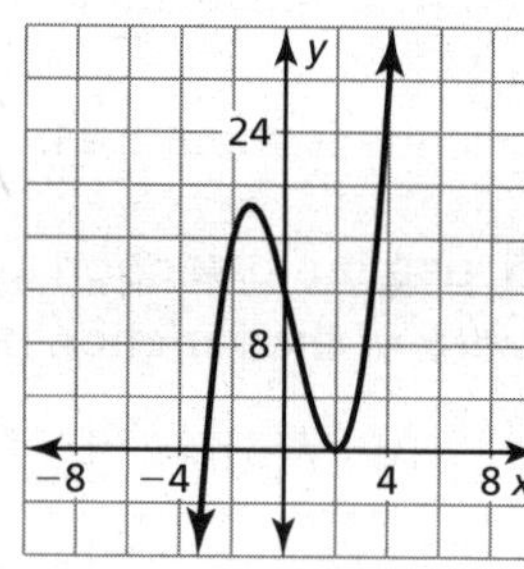

D.

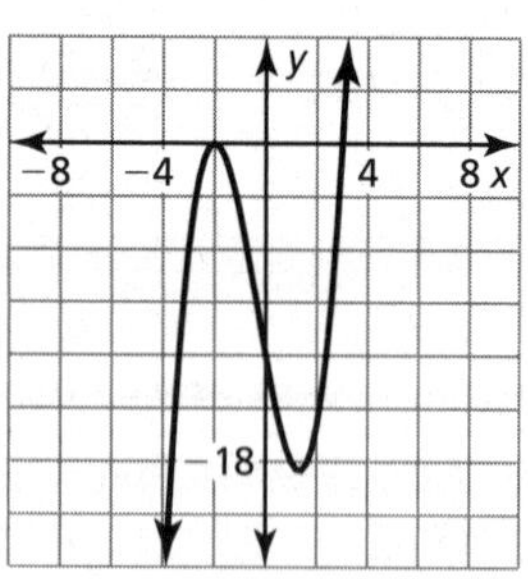

E.

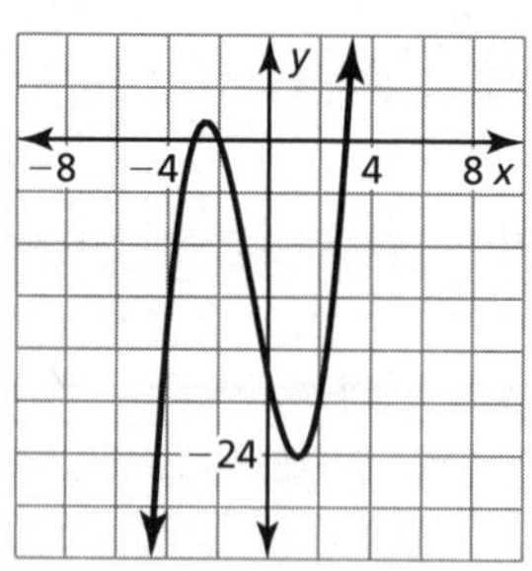

F.

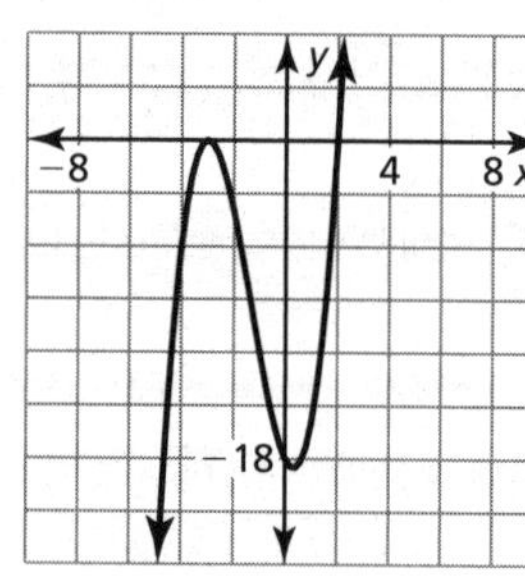

1	2	3	4	5	6

3.9 Start Thinking

Plot any two points in a coordinate plane. Connect them with a line. Plot a third point that is not on the line defined by the first two points. Connect them with a smooth curve, negating the need for the line connecting only the first two points.

What type of an equation could fit the three points? What familiar figure does the smooth curve represent? Plot a fourth point that is not on the line or smooth curve. Connect the four points with a smooth curve. What function does this curve represent?

3.9 Warm Up

Find the rate of change.

1.

x	4	6	8	10	12
y	−7	1	9	17	25

2.

x	−6	−3	0	3	6
y	4	3	2	1	0

3.

x	4	6	8	10	12
y	−7	1	9	17	25

4.

x	−5	−3	−1	1	3
y	−2	0	2	4	6

3.9 Cumulative Review Warm Up

Solve the system of linear equations using the substitution method.

1. $y = 5x + 1$

$y = -4x + 10$

2. $y = 3x - 2$

$-2x + y = -5$

Name ______________________________ Date ________

3.9 Practice A

In Exercises 1 and 2, write a cubic function whose graph is shown.

1.

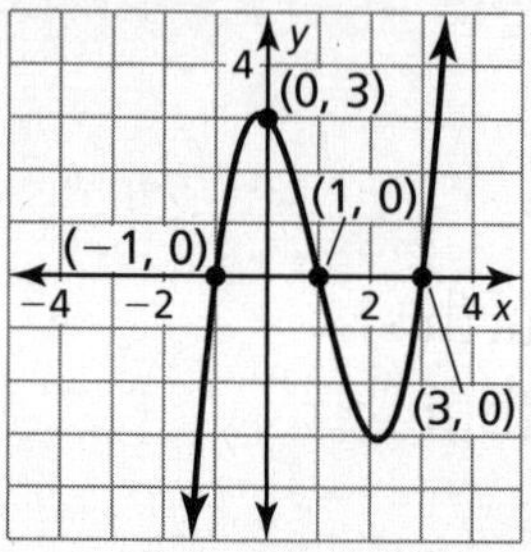

2.

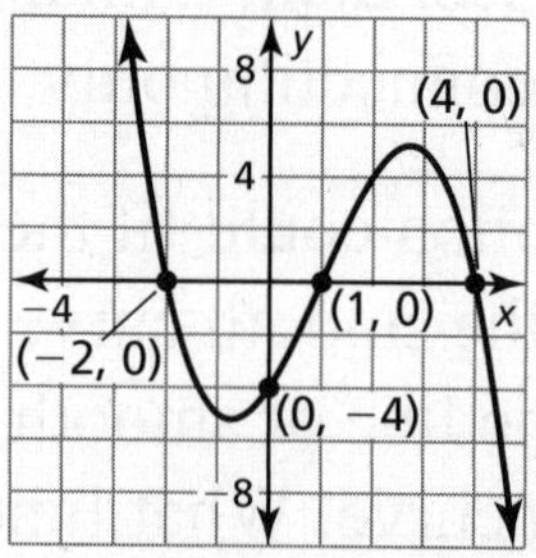

In Exercises 3–5, use finite differences to determine the degree of the polynomial function that fits the data. Then use technology to find the polynomial function.

3.

x	1	2	3	4	5	6	7
$f(x)$	1	3	7	14	25	41	63

4.

x	−4	−2	0	2	4	6
$f(x)$	−3	2	8	15	23	32

5.

x	1	2	3	4	5	6	7
$f(x)$	30	20	4	−16	−38	−60	−80

6. The data in the table show the cumulative number of customers during a 6-hour period.

x	1	2	3	4	5	6
$f(x)$	2	7	13	20	28	37

a. Find a polynomial model for the data.

b. The store is open 24 hours each day. Does this model seem reasonable for the next 6-hour period? Explain.

Name____ Date__________

3.9 Practice B

In Exercises 1 and 2, write a cubic function whose graph is shown.

1. 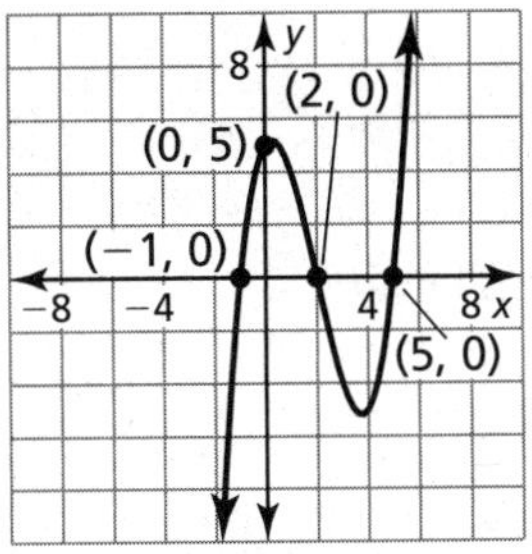

2. 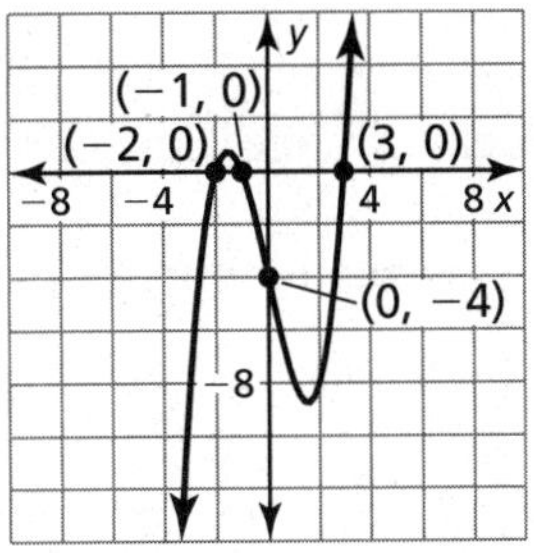

In Exercises 3–5, use finite differences to determine the degree of the polynomial function that fits the data. Then use technology to find the polynomial function.

3.

x	1	2	3	4	5	6
$f(x)$	−10	−14	−13	−7	4	20

4.

x	1	2	3	4	5	6	7	8
$f(x)$	24	12	6	9	24	54	102	171

5.

x	1	2	3	4	5	6	7
$f(x)$	20	4	0	4	16	40	84

6. The data in the table show the wave height (in inches) over a 7-second period.

x	1	2	3	4	5	6	7
$f(x)$	0.5	3	6.5	12	20.5	33	50.5

a. Find a polynomial model for the data.

b. Does this model seem reasonable for the next 7-second interval? Explain.

Name ____ Date __________

3.9 Enrichment and Extension

Polynomial Curve Fitting Using Linear Systems

Given a distinct set of n points, precisely one polynomial function f of degree $n - 1$ (or less) fits the points. You can use a system of linear equations to determine the coefficients of f.

Example: Find the cubic polynomial function $f(x) = ax^3 + bx^2 + cx + d$ whose graph passes through the points $(-3, -5)$, $(-1, 2)$, $(1, 2)$, and $(3, 5)$.

Solution: Substituting the given points into f produces the system of linear equations below.

$$\begin{aligned} -27a + 9b - 3c + d &= -5 \\ -a + b - c + d &= 2 \\ a + b + c + d &= 2 \\ 27a + 9b + 3c + d &= 5 \end{aligned}$$

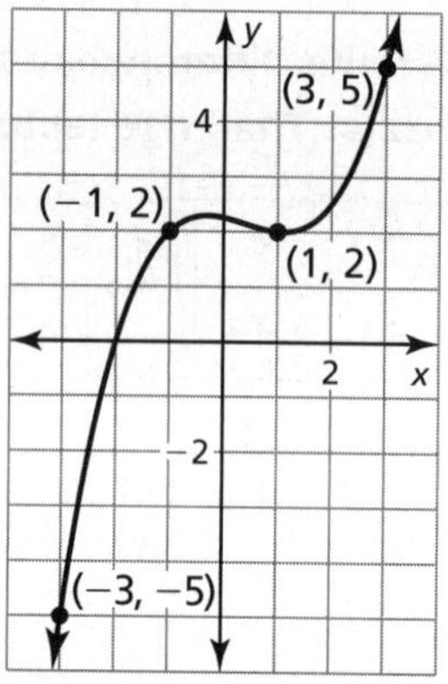

The solution of this system is

$$a = \frac{5}{24}, b = -\frac{1}{4}, c = -\frac{5}{24}, \text{ and } d = \frac{9}{4}.$$

So, the cubic polynomial function is

$$f(x) = \frac{5}{24}x^3 - \frac{1}{4}x^2 - \frac{5}{24}x + \frac{9}{4}.$$

Check this by substituting each x-coordinate of the given points into the function and verifying that you obtain the corresponding y-coordinate.

In Exercises 1–7, use a system of linear equations and a graphing calculator to find the polynomial function of degree $(n - 1)$ whose graph passes through the n given points.

1. $(-1, 1), (0, 2), (1, 1), (2, 5)$
2. $(1, -2), (2, -10), (3, -14), (4, -32)$
3. $(-2, -29), (-1, 3), (1, 7), (2, -9), (3, -149)$
4. $(-2, 12), (-1, 0), (0, 4), (1, 2), (2, 6)$
5. $(-3, -500), (-1, 0), (0, -5), (1, 4), (2, 15), (3, 184)$
6. $(-2, 35), (-1, 4), (0, 3), (1, 2), (2, -29), (3, -240)$
7. $(-3, 2183), (-2, 188), (-1, -1), (0, -4), (1, -1), (2, 188), (3, 2183)$

Name____ Date__________

3.9 Puzzle Time

What Do You Throw Out When You Need It And Take In When You Don't Need It?

Write the letter of each answer in the box containing the exercise number.

Write a cubic function whose graph passes through the given points.

1. $(-2, 0), (-1, 0), (0, -8), (2, 0)$

2. $(-1, 0), (1, 0), (2, 0), (3, 16)$

3. $(-10, 0), (-5, 0), (4, 0), (-8, -2)$

4. $(2, 0), (1, 0), (4, 0), (5, -3)$

5.

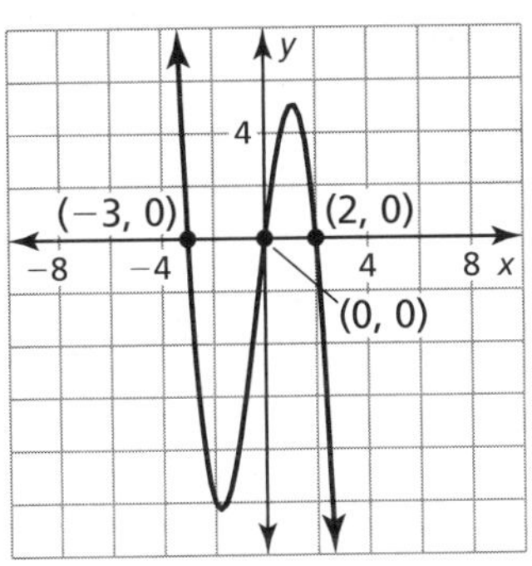

6.

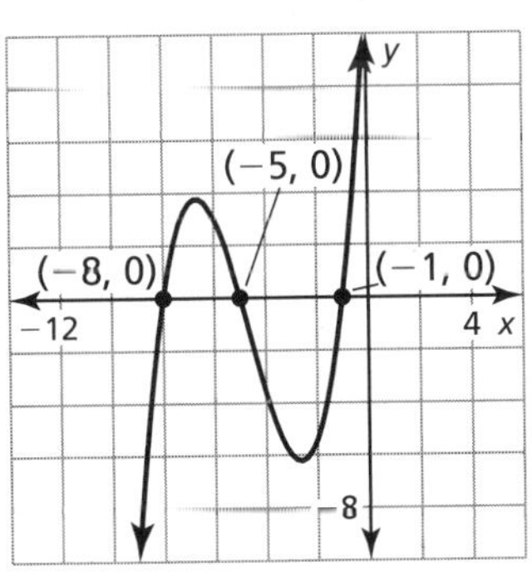

Answers

R. $f(x) = \frac{3}{10}(x + 8)(x + 5)(x + 1)$

O. $f(x) = -\frac{5}{4}x(x + 3)(x - 2)$

A. $f(x) = 2x^3 + 2x^2 - 8x - 8$

N. $f(x) = 2(x - 1)(x + 1)(x - 2)$

H. $f(x) = -\frac{1}{4}(x - 2)(x - 1)(x - 4)$

C. $f(x) = -\frac{1}{36}(x + 10)(x + 5)(x - 4)$

1	2	3	4	5	6

Name ______________________________ Date __________

Chapter 3 Cumulative Review

Simplify the expression. Write your answer in standard form.

1. $4(x^2 + 7x + 3)$
2. $-5(2x^2 + 5x - 10)$
3. $(x + 12)(x + 4)$
4. $-8(x - 9)(x - 11)$
5. $3(x - 1)(x + 3)$
6. $-4(-4x + 7)(x + 1)$
7. $6(-2x - 11)(8x - 7)$
8. $2(x^2 + 10x - 11) + 12$
9. $(x + 4)(7x - 5) + 3$
10. $9(6x^2 + x + 4) - 5$
11. $(2x + 9)(-3x - 4) - 12x + 8$
12. $2(x + 7)(2x - 5) + 3x - 17$
13. You walk your dog each night. To walk the same route and distance, it takes your sister 3 additional minutes and your brother 2 fewer minutes. If it takes you 9 minutes to walk your dog, how long does it take your brother and sister to walk the dog?
14. You are building a fence around a garden. The length is 5 feet more than the width.
    - **a.** The perimeter is 66 feet. What are the length and width of the fence?
    - **b.** Fence material is \$16 per linear foot. How much will the fence material cost?
    - **c.** Determine the area of the garden.

Factor the expression.

15. $x^2 + 10x + 24$
16. $x^2 - 8x - 33$
17. $x^2 + 5x - 84$
18. $x^2 + 16x + 55$
19. $x^2 - 9x + 14$
20. $x^2 - 15x + 50$
21. $5x^2 + 47x + 56$
22. $9x^2 - 53x - 6$
23. $2x^2 + 3x - 20$
24. $3x^2 + 13x + 12$
25. $2x^2 - x - 3$
26. $2x^2 - x - 28$
27. Your family is purchasing new carpet for the living room. The length of the room is 14 feet and the width is 18 feet.
    - **a.** How much carpet is needed?
    - **b.** The carpet your family wants is \$3.29 per square foot. How much will the carpet for the living room cost?

Name______________________________ Date__________

Chapter 3 Cumulative Review (continued)

28. The length (in feet) of a fence is $(x + 4)$, and the width (in feet) is $(2x - 23)$.

a. Write an equation for the perimeter p of the fence.

b. Write an equation for the area a of the fence.

c. The perimeter of the fence is 76 feet. Find the dimensions of the length and the width.

d. What is the area of the fence?

Solve the quadratic equation by factoring.

29. $x^2 - x - 6 = 0$

30. $x^2 - 8x - 9 = 0$

31. $x^2 + 12x + 35 = 0$

32. $x^2 - 6x + 9 = 0$

33. $x^2 + 7x - 44 = 0$

34. $x^2 - 8x + 7 = 0$

35. $6x^2 + 35x + 36 = 0$

36. $2x^2 + x - 21 = 0$

37. $5x^2 - 19x - 4 = 0$

38. $8x^2 - 14x - 9 = 0$

Use the Quadratic Formula to solve the equation.

39. $x^2 - 7x + 8 = 0$

40. $x^2 + 13x - 1 = 0$

41. $x^2 - 14x = -2$

42. $x^2 + 9x = 4$

43. $3x^2 - 24x + 7 = 0$

44. $2x^2 + 9x - 3 = 0$

45. $4x^2 - 36x + 1 = 0$

46. $5x^2 - 42x - 4 = 0$

Find all the zeros of the function.

47. $f(x) = x^2 - 9x + 21$

48. $g(x) = x^2 - x - 20$

49. $h(x) = x^2 - 3x - 1$

50. $f(x) = x^2 - 5x - 19$

51. $g(x) = 4x^2 - 5x + 10$

52. $h(x) = 3x^2 - 7x - 21$

53. $f(x) = 5x^2 - 11x + 8$

54. $g(x) = 2x^2 - 9x - 7$

55. Your class's midterm grades are modeled by the equation $y = -0.0705x^2 + 10.460x - 384.21$, where x represents the midterm grade (percent) and y represents the number of students who earned the grade.

a. Approximately how many students earned an 80%?

b. Approximately how many students earned a 75%?

Name ______________________________ Date __________

Chapter 3 Cumulative Review (continued)

56. Your class's semester grades are modeled by the equation $y = -0.0680x^2 + 11.845x - 511.86,$ where x represents the semester grade (percent) and y represents the number of students who earned the grade.

a. Approximately how many students earned an 85%?

b. Approximately how many students earned a 93%?

Describe the transformation of $f(x) = x^2$ represented by g.

57. $g(x) = (x - 4)^2$

58. $g(x) = x^2 + 8$

59. $g(x) = (x - 3)^2 + 5$

60. $g(x) = (x + 1)^2 - 4$

61. $g(x) = -x^2 - 3$

62. $g(x) = -(x - 1)^2 + 5$

63. $g(x) = -5x^2 - 2$

64. $g(x) = \frac{1}{3}(x + 5)^2 - 3$

Write a function whose graph represents the indicated transformation of the graph of f.

65. $f(x) = x - 1$; reflection in the y-axis and a translation 4 units up

66. $f(x) = |x + 3|$; vertical shrink by a factor of $\frac{1}{4}$ and a translation 1 unit up

67. $f(x) = \sqrt{x} + 4$; vertical stretch by a factor of 4, followed by a translation 2 units right

68. $f(x) = x^2 - 3$; reflection in the x-axis, followed by a translation 2 units left and 5 units up

69. You are playing baseball. The ball follows a parabolic path modeled by the equation $y = -2x^2 + 19x + 2,$ where x represents the time (in seconds) after the ball is hit and y represents the height of the ball (in feet) from the ground.

a. How high is the ball after 3 seconds?

b. How high is the ball after 6 seconds?

c. What was the original starting height of the ball? What does this mean, according to the situation described?

d. Does this function have a maximum or a minimum?

e. What is the maximum or minimum, and when does this maximum or minimum occur?

Name ______________________________ Date __________

Chapter 3 Cumulative Review (continued)

Find the sum or difference.

70. $(19x^2 + 12x + 5) + (-23 - 12x^2 + 10)$

71. $(x - 9) - (-x^2 - 19x + 6)$

72. $(18x + 2x - 21) + (13x + 8 - 11)$

73. $(-16x^2 - 14) - (21 - 25x - 15x^2)$

74. $(24x^3 + 23x - 7) + (18x - 7x^2 + 3x^3 + 16)$

75. $(21x^5 + 5x - 17) - (12x + 16x^2 - 4x^4 - 11 + 22x^5)$

76. $(9 + x^3 + 22x + 25) + (-15 + 10x^2 + 7x^3)$

77. $(-10x^4 - 19x^2 + 8x + 15x^3 + 3) - (-6x^2 - 20x^3 + 23 + 6x^2)$

Find the product.

78. $4x^3(x^2 - 10x + 11)$

79. $-5x^5(2x^2 - x - 3)$

80. $(3x^2 - 11)(2x - 7)$

81. $(12x + 2)(x^2 - 12x + 7)$

82. $(-5x^2 - 1)(-6x^2 + 8x - 3)$

83. $(-7x - 9)(12x^2 + 8x + 1)$

84. $(x^2 - 8x + 11)(x^2 + 4x - 9)$

85. $(-2x^2 + 12x + 1)(x^3 + 3x - 11)$

86. $(x^2 + 9x - 6)(6x^2 + 11x - 10)$

87. $(x + 4)(x + 2)(x - 6)$

88. $(x - 1)(x - 2)(1 - 3x)$

89. $(5x + 10)(-4x + 9)(6x - 1)$

Solve the equation.

90. $x^3 - 4x^2 - 12x = 0$

91. $4y^4 + 12y^3 - 9y^2 - 8y = 0$

92. $w^2 + 2 = -w^4$

93. $2a^3 + a^2 - 60 = 43a$

94. $2p^3 + p^2 = 8p + 4$

95. $2r^5 + 5r^4 = 20r^2 + 8r^3$

Write a polynomial function *f* of least degree that has rational coefficients, a leading coefficient of 1, and the given zeros.

96. $-5, 0, 3$

97. $-4, -1, 7$

98. $2, 7i$

99. $-3i, 2 - i$

100. $5, -\sqrt{6}, 1 - i$

101. $-7, 3 + \sqrt{2}, -4 + i$

Chapter 4

Name____ Date__________

Chapter 4 Rational Exponents and Radical Functions

Dear Family,

In this chapter, you will solve equations to determine depreciation values of items. When an item depreciates, it loses value over time. You may have heard that when you buy a new car, the car loses value once you drive it off the lot. According to a certain website, the average depreciation rate for a new car is about 11%. That means, if you buy a brand new car for $30,000, the car is worth $26,700 once you drive away. The website also says that after 5 years, your car is only worth 37% of what you originally paid. Not all cars depreciate at the same rate, so these numbers are averages. Sources such as Kelley Blue Book give the market value prices for all new and used vehicles. When you use that website, you can give specific information about your vehicle, so you have an accurate value to use when pricing to sell or buying new.

Working together, choose three different cars to research. Complete the table by finding the price of each car for five years. For a fair comparison of each car, choose a constant annual mileage of 10,000 miles.

Car Values			
Year	Make/Model	Make/Model	Make/Model
	$	$	$
	$	$	$
	$	$	$
	$	$	$
	$	$	$

Use the completed table to answer the following questions.

- Which car has the lowest depreciation rate? the highest?
- Which car would you be most likely to buy? Explain.
- What factors contribute to a vehicle's depreciation rate?

When buying a new or a used car, you may also want to use the Internet to compare the reliability of each car. This may end up saving you time and money in the long run.

Remember to always buckle up!

Nombre ______________________________ Fecha __________

Capítulo 4 Exponentes racionales y funciones radicales

Estimada familia:

En este capítulo, resolverán ecuaciones para determinar los valores de depreciación de objetos. Cuando un objeto se deprecia, pierde su valor con el transcurso del tiempo. Tal vez hayan escuchado que cuando compran un carro nuevo, el carro pierde su valor cuando lo sacan del auto lote. De acuerdo con un sitio web, la tasa de depreciación promedio de un carro nuevo es de aproximadamente el 11%. Eso significa que si compran un carro nuevo por $30,000, el carro vale $26,700 cuando se lo llevan. El sitio web también dice que después de 5 años, su carro solo vale el 37% de lo que pagaron originalmente. No todos los carros se deprecian a la misma tasa, entonces estos números son un promedio. Fuentes tal como Kelley Blue Book dan los precios de valor de mercado de todos los vehículos nuevos y usados. Cuando usan ese sitio web, pueden dar información específica sobre su vehículo, así tienen un valor preciso cuando coticen para vender o para comprar un carro nuevo.

Trabajen juntos para elegir tres carros diferentes para investigar. Para completar la tabla, hallen el precio de cada carro durante cinco años. A fin de hacer una comparación equitativa de cada carro, elijan un millaje constante anual de 10,000 millas.

Valores de carros			
Año	**Marca/Modelo**	**Marca/Modelo**	**Marca/Modelo**
	$	$	$
	$	$	$
	$	$	$
	$	$	$
	$	$	$

Usen la tabla completada para responder las siguientes preguntas.

- ¿Cuál carro tiene la menor tasa de depreciación? ¿La mayor?
- ¿Cuál carro es más probable que compren? Expliquen.
- ¿Qué factores contribuyen a la tasa de depreciación de un vehículo?

Cuando compren un carro nuevo o usado, también pueden buscar en Internet para comparar la confiabilidad de cada carro. Esto puede ahorrarles tiempo y dinero a largo plazo.

¡ Recuerden siempre abrocharse el cinturón de seguridad!

4.1 Start Thinking

An exponent n, in the term x^n, indicates how many times x is a factor of the product. Any product in the form x^n may be written in expanded form. For example, $x^3 = x \bullet x \bullet x$.

Use expanded form to complete the table.

Example	Expanded Form	Simplest Form
$x^2 + x^2$		
$x^4 \bullet x^4$		
$\dfrac{x^8}{x^5}$		

Enter 2^{-2} into a scientific calculator. What is the answer in fraction form? How does this compare to 2^2?

4.1 Warm Up

Simplify.

1. $k(k^4)$

2. $(4u^5v)(6u^5v^2)$

3. $(5a^3b^{10}c)^2$

4. $(3x^3y)(3xy^2z)^4(3xyz)$

5. $(-g^2h)(-2gj^3)^3(-ghj)^4$

6. $(2xy^5)(-y)^4$

4.1 Cumulative Review Warm Up

Write an equation of the parabola in vertex form.

1. passes through: $(14, 7)$; vertex: $(4, 1)$

2. passes through: $(-6, -16)$; vertex: $(-4, 8)$

Name ______________________________ Date ________

4.1 Practice A

In Exercises 1–3, find the indicated real *n*th root(s) of *a*.

1. $n = 3, a = 125$

2. $n = 2, a = 49$

3. $n = 4, a = 81$

In Exercises 4–9, evaluate the expression without using a calculator.

4. $27^{1/3}$

5. $16^{1/4}$

6. $4^{3/2}$

7. $625^{3/4}$

8. $(-1000)^{2/3}$

9. $32^{1/5}$

In Exercises 10–15, evaluate the expression using a calculator. Round your answer to two decimal places when appropriate.

10. $\sqrt[5]{16,807}$

11. $\sqrt[6]{15,625}$

12. $12^{-1/3}$

13. $92^{1/5}$

14. $6561^{5/4}$

15. $113^{-3/4}$

In Exercises 16 and 17, find the radius of the figure with the given volume.

16. $V = 1726 \text{ in.}^3$

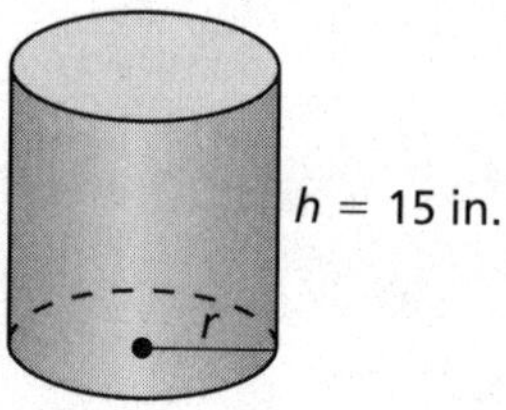

17. $V = 734 \text{ m}^3$

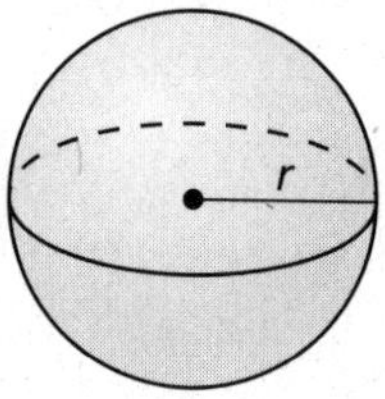

In Exercises 18–23, find the real solution(s) of the equation. Round your answer to two decimal places when appropriate.

18. $x^4 = 256$

19. $3x^3 = 375$

20. $(x - 6)^2 = 40$

21. $(x + 7)^3 = 1000$

22. $x^5 = -112$

23. $9x^4 = 54$

24. When the average price of an item increases from p_1 to p_2 over a period of n years, the price p_2 is given by $p_2 = p_1(r + 1)^n$, where r is the annual rate of inflation (in decimal form). Find the annual rate of inflation when the price of a loaf of bread was \$1.19 in 1970 and \$3.29 in 2010.

Name____ Date__________

4.1 Practice B

In Exercises 1–3, find the indicated real *n*th root(s) of *a*.

1. $n = 3, a = 343$ **2.** $n = 6, a = -64$ **3.** $n = 5, a = -243$

In Exercises 4–9, evaluate the expression without using a calculator.

4. $36^{3/2}$ **5.** $16^{3/4}$ **6.** $(-32)^{2/5}$

7. $(-125)^{5/3}$ **8.** $256^{-5/4}$ **9.** $27^{-4/3}$

In Exercises 10–15, evaluate the expression using a calculator. Round your answer to two decimal places when appropriate.

10. $28^{-1/5}$ **11.** $150^{2/5}$ **12.** $40{,}351^{6/7}$

13. $750^{-2/5}$ **14.** $\left(\sqrt[5]{223}\right)^3$ **15.** $\left(\sqrt[7]{-34}\right)^5$

In Exercises 16 and 17, find the radius of the figure with the given volume.

16. $V = 425 \text{ in.}^3$ **17.** $V = 1458 \text{ m}^3$

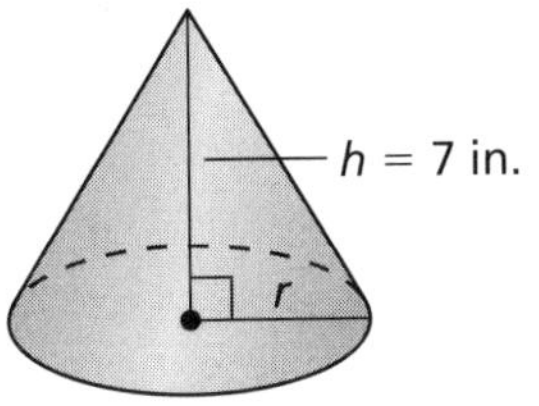

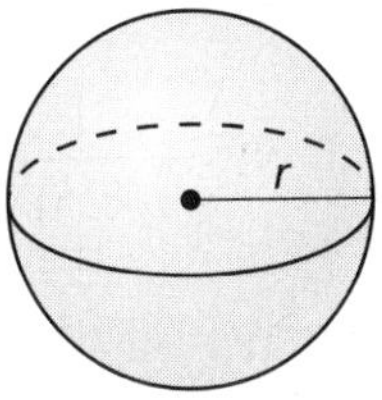

In Exercises 18–23, find the real solution(s) of the equation. Round your answer to two decimal places when appropriate.

18. $6x^4 = 60$ **19.** $x^5 = -233$ **20.** $x^4 + 19 = 100$

21. $x^3 + 17 = 57$ **22.** $\frac{1}{5}x^4 = 125$ **23.** $\frac{1}{7}x^3 = -49$

24. Kepler's third law states that the relationship between the mean distance d (in astronomical units) of a planet from the Sun and the time t (in years) it takes the planet to orbit the Sun can be given by $d^3 = t^2$.

a. It takes Venus 0.616 year to orbit the Sun. Find the mean distance of Venus from the Sun (in astronomical units).

b. The mean distance of Jupiter from the Sun is 5.24 astronomical units. How many years does it take Jupiter to orbit the Sun?

Name ______________________________ Date ________

4.1 Enrichment and Extension

Approximating Square Roots

When the quantity $\left(a + \frac{b}{2a}\right)^2$ is multiplied, it equals $a^2 + b + \frac{b^2}{4a^2}$.

Notice that the greater a is compared to b, the more the quantity $\frac{b^2}{4a^2}$ approaches zero.

So, if $a > b$, then $a^2 + b + \frac{b^2}{4a^2} \approx a^2 + b$ and $\left(a + \frac{b}{2a}\right)^2 \approx a^2 + b$.

If you take the positive square root of each side, then $a + \frac{b}{2a} \approx \sqrt{a^2 + b}$.

The same process can be used to find the approximate value of $\left(a - \frac{b}{2a}\right)^2$.

Example:

Use the formula $\sqrt{a^2 \pm b} \approx a \pm \frac{b}{2a}$ to approximate each square root.

a. $\sqrt{26} = \sqrt{25 + 1} = \sqrt{5^2 + 1} \approx 5 + \frac{1}{2(5)} \approx 5.1$

b. $\sqrt{34} = \sqrt{36 - 2} = \sqrt{6^2 - 2} \approx 6 - \frac{2}{2(6)} \approx 5.83$

In Exercises 1–12, use the formula $\sqrt{a^2 \pm b} \approx a \pm \frac{b}{2a}$ to approximate the square root to the nearest hundredth. Check your answer with a calculator. Some of your answers may not be the same as those obtained with a calculator due to rounding.

1. $\sqrt{10}$

2. $\sqrt{38}$

3. $\sqrt{104}$

4. $\sqrt{126}$

5. $\sqrt{83}$

6. $\sqrt{52}$

7. $\sqrt{141}$

8. $\sqrt{164}$

9. $\sqrt{13}$

10. $\sqrt{215}$

11. $\sqrt{249}$

12. $\sqrt{395}$

Name________________________________ Date__________

4.1 Puzzle Time

What Did The Waiter Say When The Customer Asked, "What Is This Insect In My Soup?"

A	B	C	D	E	F
G	H	I	J	K	L
M	N	O	P		

Complete each exercise. Find the answer in the answer column. Write the word under the answer in the box containing the exercise letter.

16 NOT
±5 WISH
±6 ME
−2 WOULD
3 I
$\frac{1}{16}$ ASK
−32 THAT
−16 YOU

Evaluate the expression. Round your answer to two decimal places when appropriate.

A $27^{1/3}$

B. $\sqrt[4]{625}$

C. $-64^{2/3}$

D. $\sqrt[5]{-32}$

E. $(-1024)^{2/5}$

F. 2^{-4}

G. $-\sqrt[4]{1296}$

H. $\left(\sqrt[3]{-8}\right)^5$

I. $343^{4/3}$

J. $225^{3/2}$

K. $\left(\sqrt[7]{16{,}384}\right)^{1/2}$

L. $-4096^{1/4}$

M. $54^{1/3}$

N. $14^{-1/2}$

O. $\sqrt[5]{100}$

P. $\left(-\sqrt[3]{-18}\right)^2$

3.78 ONE
0.27 BUG
2 NOT
−8 KNOW
2401 I
3375 DO
2.51 FROM
6.87 ANOTHER

4.2 Start Thinking

Simplify $x^2 \bullet x^2$. What operation did you perform on the exponents?

Consider $\sqrt{x} \bullet \sqrt{x}$. Without simplifying, do you expect the answer to be greater than or less than x? Simplify the expression first by using rational number rules and then by using exponent rules. Is the answer greater than or less than x? Explain how the answer relates to the hypothesis you made before simplifying.

4.2 Warm Up

Simplify.

1. x^{-1}

2. $3x^{-6}$

3. $\dfrac{4}{b^{-7}}$

4. $\left(ab^4\right)^1$

5. $\dfrac{y^6}{y^4}$

6. $\dfrac{p^4}{p}$

4.2 Cumulative Review Warm Up

Find the discriminant of the quadratic equation. Then describe the number and type of solutions of the equation.

1. $x^2 - 13x - 35 = 0$

2. $5n^2 + 3n + 25 = 0$

3. $3x^2 = 6x + 9$

4. $25x = -49 + 4x^2$

5. $x^2 - 5x - 16 = 0$

6. $x^2 + 13x = -48$

Name______________________________ Date__________

4.2 Practice A

In Exercises 1–6, use the properties of rational exponents to simplify the expression.

1. $\left(7^2\right)^{1/4}$

2. $\left(14^3\right)^{1/2}$

3. $\dfrac{5^{1/5}}{5}$

4. $\dfrac{10}{10^{1/4}}$

5. $\left(\dfrac{6^5}{9^5}\right)^{-1/5}$

6. $\left(7^{-3/4} \bullet 7^{1/4}\right)^{-1}$

In Exercises 7–12, use the properties of radicals to simplify the expression.

7. $\sqrt{3} \bullet \sqrt{75}$

8. $\sqrt[3]{81} \bullet \sqrt[3]{9}$

9. $\sqrt[4]{12} \bullet \sqrt[4]{8}$

10. $\sqrt[4]{9} \bullet \sqrt[4]{9}$

11. $\dfrac{\sqrt[5]{128}}{\sqrt[5]{4}}$

12. $\dfrac{\sqrt{5}}{\sqrt{80}}$

In Exercises 13–18, write the expression in simplest form.

13. $\sqrt[4]{208}$

14. $\dfrac{\sqrt[3]{9}}{\sqrt[3]{4}}$

15. $\sqrt{\dfrac{5}{27}}$

16. $\dfrac{1}{2 + \sqrt{3}}$

17. $\dfrac{6}{4 - \sqrt{5}}$

18. $\dfrac{8}{\sqrt{2} + \sqrt{5}}$

In Exercises 19–24, simplify the expression.

19. $8\sqrt[4]{2} + 5\sqrt[4]{2}$

20. $7\sqrt[5]{13} - 17\sqrt[5]{13}$

21. $4\left(9^{1/4}\right) + 7\left(9^{1/4}\right)$

22. $4\sqrt{18} - 15\sqrt{2}$

23. $8\sqrt{7} + 12\sqrt{63}$

24. $\sqrt[4]{405} + 2\sqrt[4]{5}$

25. The volume of a cube is 80 cubic centimeters.

a. Use exponents to solve the formula for the volume V of a cube with side length s, $V = s^3$, for s.

b. Substitute the expression for s from part (a) into the formula for the surface area of a cube, $S = 6s^2$.

c. Substitute the volume of the given cube into the formula found in part (b) to find the surface area, S. Simplify, if possible.

Name ______________________________ Date ________

4.2 Practice B

In Exercises 1–6, use the properties of rational exponents to simplify the expression.

1. $\dfrac{2^{2/5}}{2}$

2. $\left(\dfrac{3^6}{12^6}\right)^{-1/6}$

3. $\left(11^{3/2} \bullet 11^{-5/2}\right)^{-1/3}$

4. $\left(9^{-3/5} \bullet 9^{1/5}\right)^{-1}$

5. $\dfrac{3^{3/4} \bullet 27^{3/4}}{9^{3/4}}$

6. $\dfrac{25^{5/9} \bullet 25^{7/9}}{5^{4/3}}$

In Exercises 7–12, use the properties of radicals to simplify the expression.

7. $\sqrt[3]{25} \bullet \sqrt[3]{625}$

8. $\sqrt[5]{6} \bullet \sqrt[5]{81}$

9. $\dfrac{\sqrt[4]{176}}{\sqrt[4]{11}}$

10. $\dfrac{\sqrt{7}}{\sqrt{700}}$

11. $\dfrac{\sqrt[3]{5} \bullet \sqrt[3]{50}}{\sqrt[3]{2}}$

12. $\dfrac{\sqrt[4]{4} \bullet \sqrt[4]{12}}{\sqrt[8]{3} \bullet \sqrt[8]{3}}$

In Exercises 13–18, write the expression in simplest form.

13. $\dfrac{\sqrt[3]{4}}{\sqrt[3]{9}}$

14. $\sqrt[3]{\dfrac{4}{25}}$

15. $\sqrt[4]{\dfrac{2401}{4}}$

16. $\dfrac{7}{5 - \sqrt{3}}$

17. $\dfrac{6}{\sqrt{2} + \sqrt{7}}$

18. $\dfrac{\sqrt{2}}{\sqrt{15} - \sqrt{3}}$

In Exercises 19–24, simplify the expression.

19. $10\left(25^{2/3}\right) - 6\left(25^{2/3}\right)$

20. $2\sqrt{54} - 11\sqrt{6}$

21. $13\sqrt[3]{3} - \sqrt[3]{375}$

22. $\sqrt[5]{486} + 10\sqrt[5]{2}$

23. $4\left(48^{1/4}\right) - 3\left(3^{1/4}\right)$

24. $\left(7^{1/3}\right) + 4\left(189^{1/3}\right)$

25. The volume of a right circular cylinder is $V = 9\pi r^2$, where r is the radius.

    a. Use radicals to solve $V = 9\pi r^2$ for r. Simplify, if possible.

    b. Substitute the expression for r from part (a) into the formula for the surface area of a right cylinder, $S = 18\pi r + \pi r^2$.

    c. Use the answer to part (b) to find the surface area of a right cylinder when the volume is 108 cubic meters.

Name____ Date__________

4.2 Enrichment and Extension

Factoring Expressions with Rational Exponents

You can factor expressions involving rational exponents using the factoring techniques you have learned previously and the properties of rational exponents.

Example: Factor the expression $x^{1/2} - x^{5/2}$ completely.

Step 1 Find the greatest common factor of the terms.

$$x^{1/2}$$

$$x^{5/2} = x^{1/2} \bullet x^{4/2} = x^{1/2} \bullet x^2$$

The GCF is $x^{1/2}$. Notice that the GCF is the lesser power of x.

Step 2 Write the expression as a product of the GCF and its remaining factors.

$$x^{1/2} - x^{5/2} = x^{1/2}(1) - x^{1/2}(x^2) \qquad \text{Factor common monomial.}$$

$$= x^{1/2}(1 - x^2) \qquad \text{Distributive Property}$$

$$= x^{1/2}(1 - x)(1 + x) \qquad \text{Difference of Two Squares Pattern}$$

Example: Factor the expression $(x - 2)^{5/2} + (x + 2)(x - 2)^{3/2}$ completely.

Step 1 Find the greatest common factor of the terms.

The GCF is $(x - 2)^{3/2}$, the lesser power of $(x - 2)$.

Step 2 Write the expression as a product of the GCF and its remaining factors.

$$(x - 2)^{5/2} + (x + 2)(x - 2)^{3/2} = (x - 2)^{3/2}(x - 2)^{2/2} + (x - 2)^{3/2}[(x + 2)(1)] \qquad \text{Factor common binomial.}$$

$$= (x - 2)^{3/2}[(x - 2) + (x + 2)] \qquad \text{Distributive Property}$$

$$= (x - 2)^{3/2}(2x) \qquad \text{Combine like terms.}$$

Factor the expression completely.

1. $x^{3/2} - 8x^{1/2}$

2. $12x^{4/3} - 5x^{2/3} - 2$

3. $10x^{3/8} - 15x^{1/3} + 20x^{5/6}$

4. $8x^{7/3} - 4x^{11/3} + 12x^{5/3}$

5. $x^{3/5} + 5x^{2/5} - 3x^{1/5} - 15$

6. $x^{1/2} - 3x^{1/3} - 3x^{1/6} - 9$

7. $(x + 2)^{7/2} - (x + 2)^{3/2}$

8. $4(x + 1)^{1/2} - 5(x + 1)^{3/2} + (x + 1)^{5/2}$

9. $3x^{3/2} - 9x^{1/2} + 6x^{-1/2}$

10. $2x^{1/2} + 5x^{-1/2} + 2x^{-3/2}$

Name ______________________________ Date __________

4.2 Puzzle Time

Why Did The 25-Watt Bulb Flunk Out Of School?

Write the letter of each answer in the box containing the exercise number.

Simplify the expression.

1. $\left(8^2\right)^{1/3}$
2. $5^{3/4} \bullet 5^{1/2}$
3. $\left(\dfrac{64}{125}\right)^{1/3}$
4. $\left(2^4 \bullet 6^4\right)^{-1/4}$
5. $\left(3^{2/3} \bullet 5^{1/4}\right)^6$
6. $\dfrac{4}{4^{1/2}}$
7. $\sqrt[3]{27} \bullet \sqrt[3]{729}$
8. $\dfrac{\sqrt[4]{32}}{\sqrt[4]{256}}$
9. $\dfrac{\sqrt[3]{24} \bullet \sqrt[3]{12}}{\sqrt[3]{2}}$
10. $\dfrac{25^{1/6} \bullet 25^{1/3}}{5^{6/5}}$
11. $\dfrac{4}{8 - \sqrt{3}}$
12. $10\sqrt[4]{5} - 3\sqrt[4]{5}$
13. $12\left(4^{3/4}\right) + 5\left(4^{3/4}\right)$
14. $\sqrt[5]{3^5}$
15. $\left(3^{2/5}\right)^{1/2}$
16. $-3\left(2^{1/5}\right) - 2\left(2^{1/5}\right)$

Answers

E. $5^{5/4}$ **W.** $\dfrac{4}{5}$

H. 4 **A.** $\dfrac{1}{12}$

N. 2 **S.** $81 \bullet 5^{3/2}$

H. $3^{1/5}\,3^{1/5}$ **T.** $\left(\dfrac{1}{8}\right)^{1/4}$

S. $2\sqrt[3]{18}$ **O.** 27

O. $\dfrac{1}{5^{1/5}}$ **R.** $7\sqrt[4]{5}$

G. 3 **I.** $17\left(4^{3/4}\right)$

T. $-5\left(2^{1/5}\right)$ **B.** $\dfrac{32 + 4\sqrt{3}}{61}$

1	2		3	4	5		6	7	8		9	10		11	12	13	14	15	16

4.3 Start Thinking

Create a function with a domain that consists of all real numbers, and a range that only consists of non-negative real numbers. Explain the thought process you went through when creating the function.

Consider the function $f(x) = \sqrt{x}$. What is its domain? What is its range?

4.3 Warm Up

Describe the transformation of $f(x) = x^2$ represented by g.

1. $g(x) = x^2 - 9$
2. $g(x) = \frac{1}{4}x^2 + 5$
3. $g(x) = \frac{1}{2}x^2$
4. $g(x) = x^2 - \frac{1}{2}$
5. $g(x) = x^2 + 3$
6. $g(x) = -2x^2 + 1$

4.3 Cumulative Review Warm Up

Determine whether the binomial is a factor of the polynomial function.

1. $g(x) = x^4 - 11x^2 + 21; x^2 - 3$
2. $t(x) = x^4 + 4x^3 + 8x - 46; x^3 + 2$
3. $f(x) = x^5 + 7x^4 + 7x + 49; x + 7$
4. $s(x) = x^3 - 5x^2 - 51x + 255; x^2 - 51$

Name ______________________ Date ________

4.3 Practice A

In Exercises 1–6, graph the function. Identify the domain and range of the function.

1. $g(x) = \sqrt{x} + 4$

2. $h(x) = \sqrt{x} - 2$

3. $f(x) = -\sqrt[3]{4x}$

4. $h(x) = \sqrt[3]{-2x}$

5. $f(x) = \frac{1}{3}\sqrt{x - 2}$

6. $g(x) = \frac{1}{4}\sqrt{x + 5}$

In Exercises 7–12, describe the transformation of *f* represented by *g*. Then graph each function.

7. $f(x) = \sqrt{x}; g(x) = \sqrt{x - 1} + 4$

8. $f(x) = \sqrt{x}; g(x) = 3\sqrt{x + 2}$

9. $f(x) = \sqrt[3]{x}; g(x) = -2\sqrt[3]{x}$

10. $f(x) = \sqrt[3]{x}; g(x) = \sqrt[3]{x - 1} + 3$

11. $f(x) = x^{1/2}; g(x) = 3(-x)^{1/2}$

12. $f(x) = x^{1/3}; g(x) = -\frac{1}{3}x^{1/3}$

In Exercises 13–15, use a graphing calculator to graph the function. Then identify the domain and range of the function.

13. $f(x) = \sqrt{x^2 - x}$

14. $g(x) = \sqrt[3]{x^2 - x}$

15. $h(x) = \sqrt[3]{2x^2 + 3x}$

In Exercises 16 and 17, write a rule for *g* described by the transformations of the graph of *f*.

16. Let g be a vertical shrink by a factor of $\frac{1}{3}$, followed by a translation 3 units right of the graph of $f(x) = \sqrt{x + 5}$.

17. Let g be a reflection in the x-axis, followed by a translation 2 units down of the graph of $f(x) = 5\sqrt{x} + 3$.

In Exercises 18 and 19, use a graphing calculator to graph the equation of the parabola. Identify the vertex and the direction that the parabola opens.

18. $\frac{1}{2}y^2 = x$

19. $-3y^2 = x + 6$

In Exercises 20 and 21, use a graphing calculator to graph the equation of the circle. Identify the center, radius, and intercepts.

20. $x^2 + y^2 = 16$

21. $25 - (y - 2)^2 = x^2$

Name______________________________ Date__________

4.3 Practice B

In Exercises 1–6, graph the function. Identify the domain and range of the function.

1. $g(x) = -\sqrt{x} + 2$

2. $f(x) = \sqrt[3]{-4x}$

3. $f(x) = \frac{1}{4}\sqrt{x + 5}$

4. $h(x) = (5x)^{1/2} - 2$

5. $g(x) = -2(x - 3)^{1/3}$

6. $h(x) = -\sqrt[5]{x}$

In Exercises 7–12, describe the transformation of *f* represented by *g*. Then graph each function.

7. $f(x) = \sqrt{x};\ g(x) = 4\sqrt{x - 2}$

8. $f(x) = \sqrt[3]{x};\ g(x) = \sqrt[3]{x - 5} - 1$

9. $f(x) = x^{1/4};\ g(x) = \frac{1}{3}(-x)^{1/4}$

10. $f(x) = x^{1/3};\ g(x) = \frac{1}{2}x^{1/3} - 3$

11. $f(x) = \sqrt[4]{x};\ g(x) = -\sqrt[4]{x - 1} + 3$

12. $f(x) = \sqrt[5]{x};\ g(x) = \sqrt[5]{-243x} - 2$

In Exercises 13–15, use a graphing calculator to graph the function. Then identify the domain and range of the function.

13. $g(x) = \sqrt[3]{2x^2 - 3x}$

14. $f(x) = \sqrt{\frac{1}{3}x^2 - x + 2}$

15. $h(x) = \sqrt[3]{3x^2 - 6x + 2}$

In Exercises 16 and 17, write a rule for *g* described by the transformations of the graph of *f*.

16. Let g be a horizontal stretch by a factor of 2, followed by a translation 2 units up of the graph of $f(x) = \sqrt{3x}$.

17. Let g be a translation 1 unit up and 4 units left, followed by a reflection in the y-axis of the graph of $f(x) = \sqrt{-x} - \frac{1}{2}$.

In Exercises 18 and 19, use a graphing calculator to graph the equation of the parabola. Identify the vertex and the direction that the parabola opens.

18. $3y^2 + 5 = x$

19. $x - 3 = -\frac{1}{2}y^2$

In Exercises 20 and 21, use a graphing calculator to graph the equation of the circle. Identify the center, radius, and intercepts.

20. $y^2 = 81 - (x + 3)^2$

21. $x^2 + y^2 + 8y + 15 = 0$

Name ____ Date __________

4.3 Enrichment and Extension

Polar Coordinates

So far, you have represented graphs of equations as collections of points (x, y) in the Cartesian coordinate plane. You can use *polar coordinates* to represent the same points.

For example, the point $(3, 4)$ can be represented in the Cartesian plane by moving 3 units right on the x-axis and 4 units up on the y-axis. This point can also be represented by the distance r directly from the origin and the angle θ measured counterclockwise from the positive x-axis.

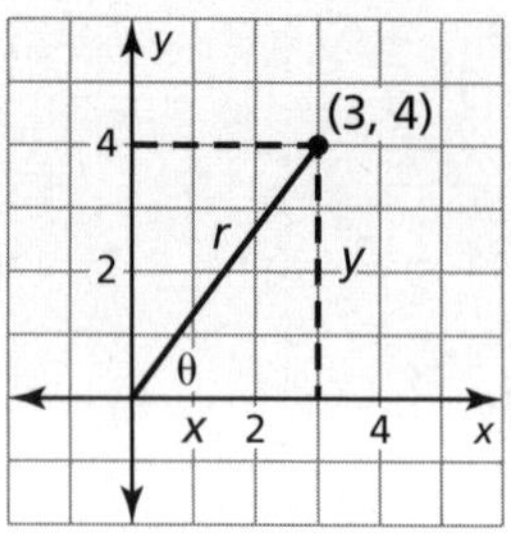

You can see that the sides x, y, and r form a right triangle. So, you can solve for r using the Pythagorean Theorem.

$$x^2 + y^2 = r^2$$
$$3^2 + 4^2 = r^2$$
$$25 = r^2$$
$$5 = r$$

To find the number of degrees θ from the x-axis, you can use trigonometric ratios.

$$\sin \theta = \frac{y}{r} \qquad \cos \theta = \frac{x}{r} \qquad \tan \theta = \frac{y}{x}$$

You could use all three ratios above, but the tangent relates x and y, which were given.

$$\tan \theta = \frac{y}{x} \qquad \tan \theta = \frac{4}{3} \qquad \tan^{-1}\left(\frac{4}{3}\right) = \theta \qquad \theta \approx 53.13°$$

So, the Cartesian coordinates $(x, y) = (3, 4)$ can also be represented by the polar coordinates $(r, \theta) \approx (5, 53.13°)$.

In Exercises 1–6, convert the Cartesian coordinates to approximate polar coordinates. Graph the point in a coordinate plane, if needed. Round your answer to two decimal places when necessary.

1. $(6, 8)$

2. $(1, 1)$

3. $(8, 4)$

4. $(-10, 2)$

5. $(-3, -4)$

6. $(-5, -7)$

Name______________________________ Date__________

4.3 Puzzle Time

What Has Lots Of Eyes But Cannot See?

Write the letter of each answer in the box containing the exercise number.

Match the graph with its function or equation.

1. 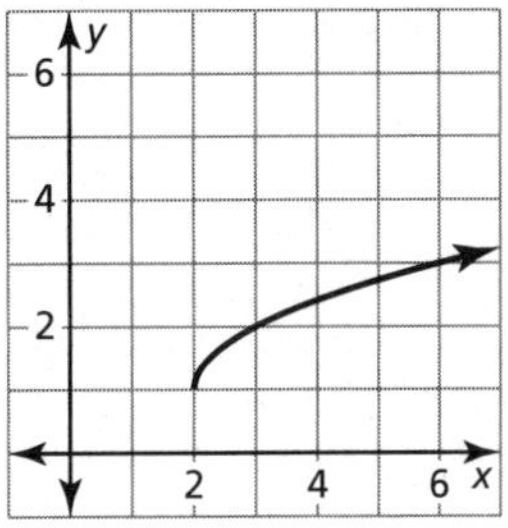

2. 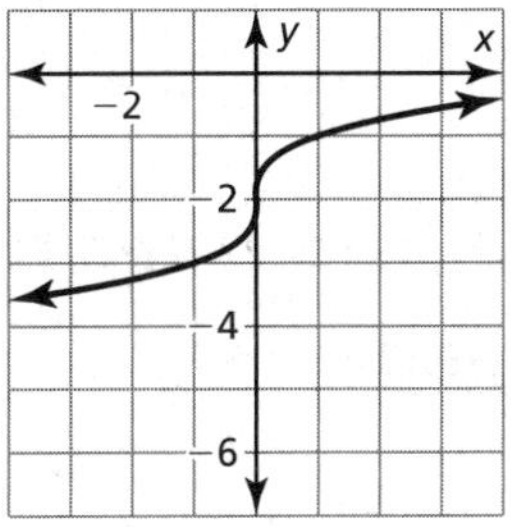

3. 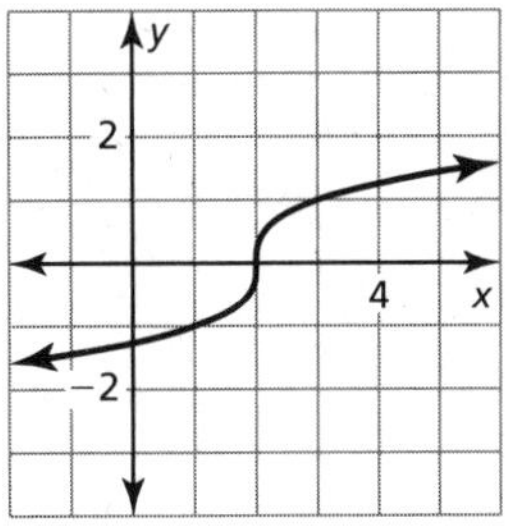

4. 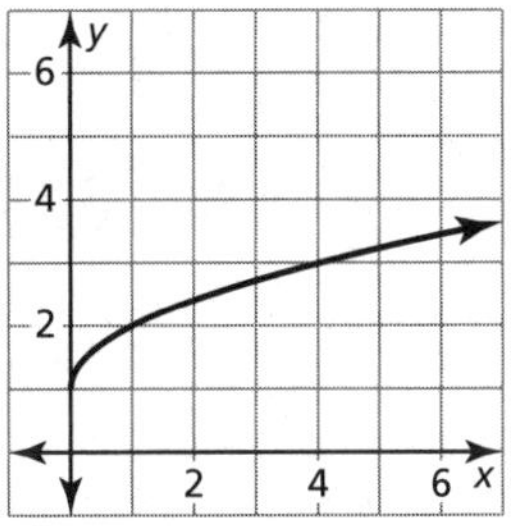

5. 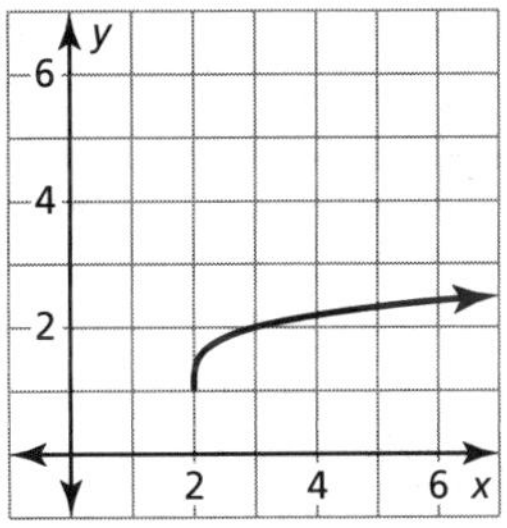

6. 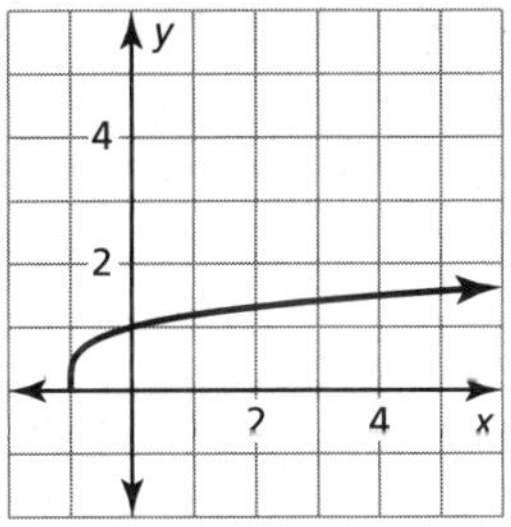

7. 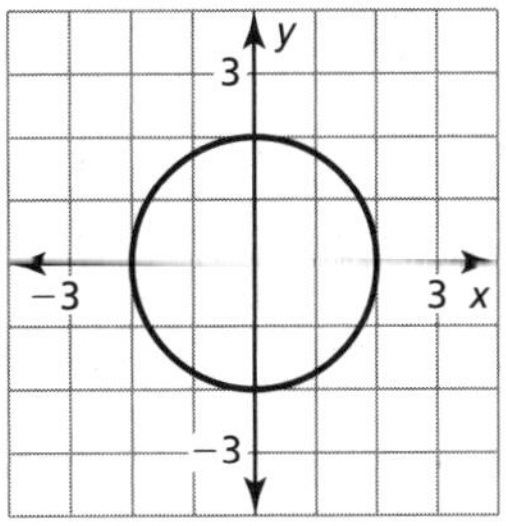

Answers

O. $f(x) = \sqrt[3]{x - 2}$

A. $f(x) = \sqrt{x - 2} + 1$

L. $x^2 - y^2 = 4$

T. $f(x) = x^{1/2} + 1$

A. $f(x) = \sqrt[4]{x - 2} + 1$

O. $x^2 + y^2 = 4$

T. $f(x) = (x + 1)^{1/4} - 2$

P. $f(x) = x^{1/3} - 2$

1		2	3	4	5	6	7

4.4 Start Thinking

Suppose the right triangle shown has hypotenuse length $c = 13$. Use the given information to write the Pythagorean Theorem in simplest form.

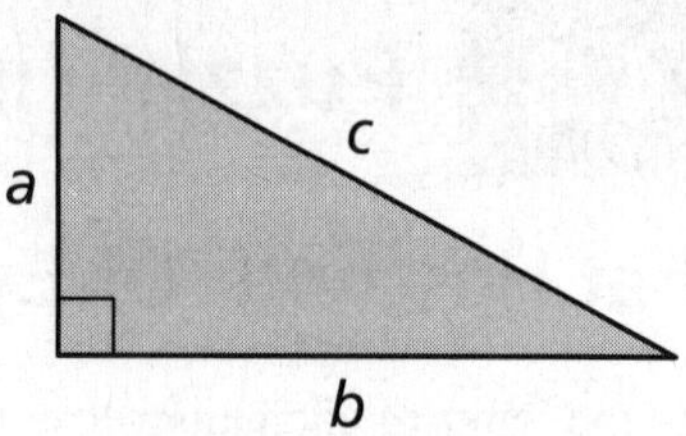

4.4 Warm Up

Determine whether the given numbers could be the lengths of the sides of a right triangle.

1. $a = 9, b = 40, c = 41$

2. $a = \sqrt{6}, b = 6, c = 15$

3. $a = 18, b = 24, c = 30$

4. $a = 10, b = 20, c = 24$

5. $a = 8, b = 15, c = 17$

6. $a = 10, b = 28, c = 29$

4.4 Cumulative Review Warm Up

Use the graph to write an equation of the line and interpret the slope.

1.

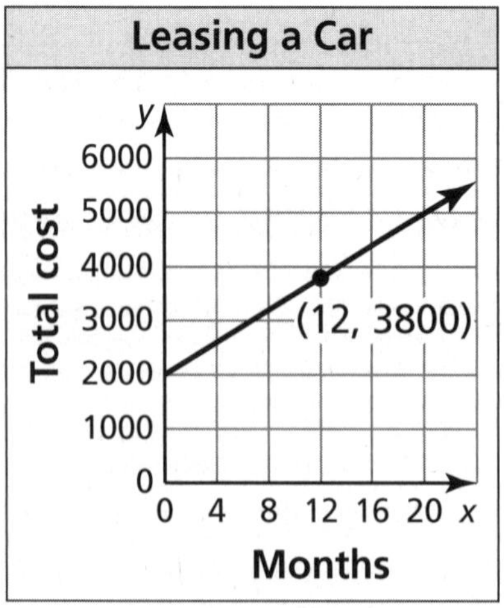

2.

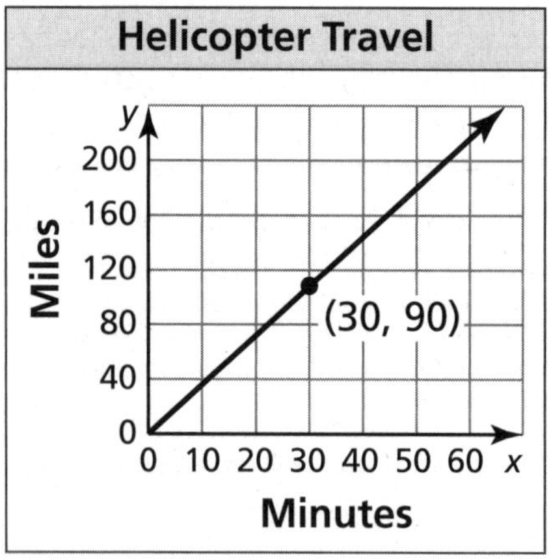

Name____ Date__________

4.4 Practice A

In Exercises 1–6, solve the equation. Check your solution.

1. $\sqrt{3x - 2} = 5$ **2.** $\sqrt{6x + 1} = 9$ **3.** $\sqrt[3]{x + 10} = 4$

4. $\sqrt[3]{x} - 8 = -2$ **5.** $-3\sqrt{16x} + 14 = -10$ **6.** $6\sqrt[3]{25x} - 16 = 14$

7. Biologists have discovered that the shoulder height h (in centimeters) of a male Asian elephant can be modeled by $h = 62.5\sqrt[3]{t} + 75.8$, where t is the age (in years) of the elephant. Determine the age of an elephant with a shoulder height of 300 centimeters.

In Exercises 8–13, solve the equation. Check your solution(s).

8. $x - 8 = \sqrt{4x}$ **9.** $\sqrt{2x - 14} = x - 7$

10. $\sqrt{x + 22} = x + 2$ **11.** $\sqrt[3]{8x^3 + 27} = 2x + 3$

12. $\sqrt[4]{2 - 9x^2} = 3x$ **13.** $\sqrt{3x - 5} = \sqrt{x + 9}$

In Exercises 14–16, solve the equation. Check your solution(s).

14. $2x^{2/3} = 18$ **15.** $x^{3/4} + 10 = 0$ **16.** $(x + 12)^{1/2} = x$

17. Describe and correct the error in solving the equation.

$$\sqrt[3]{2x + 1} = 8$$
$$2x + 1 = 2$$
$$2x = 1$$
$$x = \frac{1}{2}$$

In Exercises 18–20, solve the inequality.

18. $3\sqrt{x} - 4 \geq 5$ **19.** $\sqrt{x - 3} \leq 7$ **20.** $5\sqrt{x - 1} > 10$

21. The length ℓ (in inches) of a standard nail can be modeled by $\ell = 54d^{3/2}$, where d is the diameter (in inches) of the nail.

a. What is the diameter of a standard nail that is 2 inches long?

b. What is the diameter of a standard nail that is 4 inches long?

c. The nail in part (b) is twice as long as the nail in part (a). Is the diameter twice as long? Explain.

Name ______________________________ Date __________

4.4 Practice B

In Exercises 1–6, solve the equation. Check your solution.

1. $\sqrt[3]{x - 14} = -2$

2. $-5\sqrt{16x} + 17 = -8$

3. $\frac{1}{4}\sqrt[3]{2x} + 8 = 6$

4. $\sqrt{3x} - \frac{3}{4} = 0$

5. $3\sqrt[5]{x} + 9 = 15$

6. $\sqrt[4]{8x} - 16 = -12$

In Exercises 7–12, solve the equation. Check your solution(s).

7. $\sqrt{10x + 24} = x + 12$

8. $x + 3 = \sqrt{\frac{22}{3}x + 9}$

9. $\sqrt[4]{2 - 25x^2} = 5x$

10. $\sqrt{4x - 4} - \sqrt{x + 8} = 0$

11. $\sqrt[3]{4x - 1} = \sqrt[3]{6x + 5}$

12. $\sqrt{4x - 10} = \sqrt{2x - 13} + 1$

In Exercises 13–15, solve the equation. Check your solution(s).

13. $3x^{2/3} - 30 = 18$

14. $(6x + 8)^{1/2} - 3x = 0$

15. $(2x^2 + 8)^{1/4} = x$

In Exercises 16–18, solve the inequality.

16. $4\sqrt{x} + 3 \le 23$

17. $\sqrt{x + 10} \ge 6$

18. $-3\sqrt{x + 2} < 15$

19. "Hang time" is the time you are suspended in the air during a jump. Your hang time t in seconds is given by the function $t = 0.5\sqrt{h}$, where h is the height (in feet) of the jump. A kite sailor has a hang time of 2.5 seconds. Find the height of the kite sailor's jump.

In Exercises 20–23, solve the nonlinear system. Justify your answer with a graph.

20. $y^2 = x + 2$
$y = x + 2$

21. $y^2 = -x + 7$
$y = x - 1$

22. $x^2 + y^2 = 9$
$y = x - 3$

23. $x^2 + y^2 = 16$
$y = x + 4$

24. The speed s (in miles per hour) of a car can be given by $s = \sqrt{30fd}$, where f is the coefficient of friction and d is the stopping distance (in feet). The coefficient of friction for a snowy road is 0.30. You are driving 20 miles per hour and approaching an intersection. How far away from the intersection must you begin to brake?

Name____ Date__________

4.4 Enrichment and Extension

Solving Radical Equations and Inequalities

You can use a graphing calculator to solve radical equations and inequalities.

Example: Use a graphing calculator to solve $\sqrt{x+1} = 5 - \sqrt{x+6}$.

$0 = \sqrt{x+6} + \sqrt{x+1} - 5$ Rewrite the equation so that 0 is on one side.

$y = \sqrt{x+6} + \sqrt{x+1} - 5$ Write a related function.

Use a table to locate the solution(s). Y1 will either be 0, meaning that the corresponding x-value is a solution, or there will be a change in sign in Y1, meaning that a solution lies between the two corresponding x-values. Graph the related function and use the *zero* feature to find the solution of the equation. The table and graph below show that the solution is $x = 3$. Check this in the original equation.

```
TABLE SETUP
\TblStart=0
\deltaTbl=1
\Ind:AUTO
\Dep:auto
```

X	Y_1
0	-1.551
1	-.94
2	-.4395
3	0
4	.39835
5	.76611
6	1.1099
7	1.434
8	1.7417
9	2.0353
10	2.3166

X=0

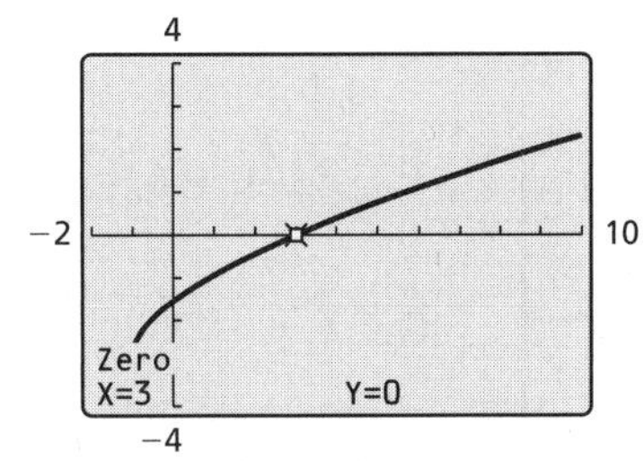

Example: Use a graphing calculator to solve $2\sqrt{x+3} \geq 6$.

Write two functions using each side of the original inequality to obtain $y_1 = 2\sqrt{x+3}$ and $y_2 = 6$.

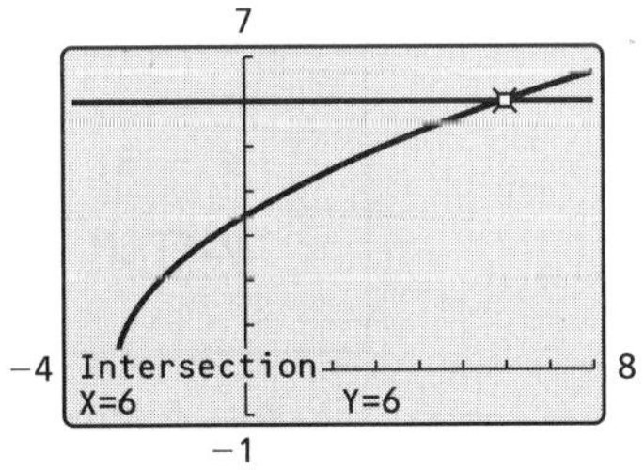

Graph each function. Use the *intersect* feature to approximate where the graphs intersect.

The figure shows that $y_1 \geq y_2$ when $x \geq 6$, so the solution of the inequality is $x \geq 6$. Check this in the original inequality.

In Exercises 1–6, use a graphing calculator to solve the radical equation or inequality. Round your answer to two decimal places when necessary.

1. $\sqrt{x+25} = 2$
2. $\sqrt{x+1} = 5 - \sqrt{x+6}$
3. $\sqrt{2x+5} = \sqrt{3x-3} - 1$
4. $\sqrt[3]{x+5} = 2\sqrt[3]{2x+6}$
5. $\sqrt{x-3} \geq 4$
6. $\sqrt{x+4} \geq 2\sqrt{x}$

Name ______________________ Date __________

4.4 Puzzle Time

What Did The Cucumber Say To The Vinegar?

A	B	C	D	E	F
G	H	I	J		

Complete each exercise. Find the answer in the answer column. Write the word under the answer in the box containing the exercise letter.

$x = 14$ WELL
$x = 16$ US
$x = 32$ FINE
$x = 128$ A
$x = 4$ YOU'VE
$x = 4$ and $x = -2$ SOUR

Solve the equation. Check your solution(s).

A. $\sqrt{2x - 3} = 5$

B. $\sqrt[3]{x - 4} = -2$

C. $3\sqrt{2x} + 10 = 22$

D. $\frac{1}{4}\sqrt[3]{4x} - 1 = 1$

E. $2\sqrt[5]{x} + 6 = 10$

F. $\left(\sqrt{x - 27}\right)^3 = 64$

G. $x = \sqrt{2x + 8}$

H. $\sqrt{5x - 3} = \sqrt{2x + 12}$

I. $\sqrt{x - 7} = 7 - \sqrt{x}$

J. $(x + 30)^{1/2} = x$

$x = 43$ PICKLE
$x = 6$ INTO
$x = 8$ IS
$x = -4$ THIS
$x = 5$ GOTTEN
$x = 6$ and $x = -5$ SALAD

4.5 Start Thinking

The president of a company is calculating the company's yearly profit. The equation $f(x) = 3x - 750$ represents the company's earnings on bracelets when x is the number of bracelets sold. The equation $g(x) = 15x - 1300$ represents the company's earnings on designed T-shirts when x is the number of designed T-shirts sold.

In 2014, the company sold the same number of bracelets and T-shirts. Write the equations as a system. The company's president combines the equations to form $h(x) = 18x - 2050$. Is this equation logical? What does $h(x)$ represent?

4.5 Warm Up

Simplify.

1. $3x(x^3 + 2x)$

2. $\dfrac{14x}{-2x^7}$

3. $\dfrac{x^3}{x^2 - x}$

4. $(ab)^4$

5. $(a + b)(a - 6b)$

6. $(x - 7x + 6)(x)$

4.5 Cumulative Review Warm Up

Write an equation of the parabola in intercept form.

1. x-intercepts: 11 and -7; passes through: $(15, 3)$

2. x-intercepts: 10 and 0; passes through: $(1, -19)$

3. x-intercepts: -15 and -3; passes through: $(-17, 71)$

4. x-intercepts: -6 and -4; passes through: $(-1, -3)$

Name ______________________ Date ________

4.5 Practice A

In Exercises 1 and 2, find $(f + g)(x)$ and $(f - g)(x)$ and state the domain of each. Then evaluate $f + g$ and $f - g$ for the given value of x.

1. $f(x) = -3\sqrt[4]{x}; g(x) = 15\sqrt[4]{x}; x = 81$

2. $f(x) = 9x + 2x^2; g(x) = x^2 - 3x + 7; x = 1$

In Exercises 3–5, find $(fg)(x)$ and $\left(\frac{f}{g}\right)(x)$ and state the domain of each. Then evaluate fg and $\frac{f}{g}$ for the given value of x.

3. $f(x) = x^2; g(x) = 2\sqrt{x}; x = 9$

4. $f(x) = 10x^3; g(x) = 4x^{5/3}; x = 8$

5. $f(x) = 4x^{2/3}; g(x) = 2x^{1/3}; x = -27$

In Exercises 6 and 7, use a graphing calculator to evaluate $(f + g)(x)$, $(f - g)(x)$, $(fg)(x)$, and $\left(\frac{f}{g}\right)(x)$ when $x = 5$. Round your answers to two decimal places.

6. $f(x) = 5x^3; g(x) = 20x^{1/4}$

7. $f(x) = 4x^{2/3}; g(x) = 16x^{4/3}$

8. Describe and correct the error in stating the domain.

> ✗ $f(x) = 4x^{1/2} + 2$ and $g(x) = -4x^{1/2}$
>
> The domain of $(f + g)(x)$ is all real numbers.

9. The growth of mold in Specimen A can be modeled by $A(t) = \frac{5}{6}t^{2/3}$. The growth of mold in Specimen B can be modeled by $B(t) = \frac{1}{3}t^{2/3}$.

   a. Find $(A - B)(t)$.

   b. Explain what the function $(A - B)(t)$ represents.

Name____ Date__________

4.5 Practice B

In Exercises 1 and 2, find $(f + g)(x)$ and $(f - g)(x)$ and state the domain of each. Then evaluate $f + g$ and $f - g$ for the given value of x.

1. $f(x) = \sqrt[3]{4x}; g(x) = -9\sqrt[3]{4x}; x = -2$

2. $f(x) = 3x - 5x^2 - x^3; g(x) = 6x^2 - 4x; x = -1$

In Exercises 3–5, find $(fg)(x)$ and $\left(\frac{f}{g}\right)(x)$ and state the domain of each. Then evaluate fg and $\frac{f}{g}$ for the given value of x.

3. $f(x) = 3x^3; g(x) = \sqrt[3]{x^2}; x = -8$

4. $f(x) = 3x^2; g(x) = 5x^{1/4}; x = 16$

5. $f(x) = 10x^{5/6}; g(x) = 2x^{1/3}; x = 64$

In Exercises 6 and 7, use a graphing calculator to evaluate $(f + g)(x)$, $(f - g)(x)$, $(fg)(x)$, and $\left(\frac{f}{g}\right)(x)$ when $x = 5$. Round your answers to two decimal places.

6. $f(x) = -3x^{1/3}; g(x) = 4x^{1/2}$

7. $f(x) = 6x^{3/4}; g(x) = 3x^{1/2}$

8. Describe and correct the error in stating the domain.

> X $f(x) = 4x^{7/3}$ and $g(x) = 2x^{2/3}$
>
> The domain of $\left(\frac{f}{g}\right)(x)$ is all real numbers.

9. The table shows the outputs of the two functions f and g. Use the table to evaluate $(f + g)(5)$, $(f - g)(0)$, $(fg)(3)$, and $\left(\frac{f}{g}\right)(2)$.

x	0	1	2	3	4	5
f(x)	18	13	8	3	−2	−7
g(x)	64	32	16	8	4	2

Name ______________________________ Date ________

4.5 Enrichment and Extension

Modular Arithmetic

Modular arithmetic is a system of arithmetic that is useful in situations where numbers repeat in cycles, such as remainders in a division problem.

If $a \equiv b \bmod n$, then a and b have the same remainder when divided by n, where a and b are whole numbers and n is a natural number. The statement $a \equiv b \bmod n$ is called a *congruence,* and the number n is the *modulus*.

In Exercises 1–4, determine whether the congruence is *true* or *false*. Explain.

1. $18 \equiv 60 \bmod 7$

2. $21 \equiv 44 \bmod 6$

3. $15 \equiv 25 \bmod 8$

4. $10 \equiv 14 \bmod 4$

Arithmetic modulo n uses the basic operations of addition, subtraction, multiplication, and division. For whole numbers, you perform the arithmetic operation and then divide by the modulus. The answer is the remainder.

Example: Evaluate $(33 - 16) \bmod 6$.

Solution: $(33 - 16) \bmod 6 = 17 \bmod 6 = 5$

In Exercises 5–8, perform the modular arithmetic.

5. $(25 - 10) \bmod 4$

6. $(4 \bullet 22) \bmod 3$

7. $(42 + 35) \bmod 7$

8. $(27 \div 3) \bmod 5$

The day of the week for any date on the Gregorian calendar can be determined by evaluating Zeller's congruence formula given by

$$x \equiv \left(\left\lfloor \frac{13m - 1}{5} \right\rfloor + \left\lfloor \frac{c}{4} \right\rfloor + \left\lfloor \frac{y}{4} \right\rfloor + d + y - 2c\right) \bmod 7$$

where x is the day of the week with $x = 0$ representing Sunday, m is the month with $m = 1$ representing March, d is the day of the month, c is the first two digits of the year, and y is the last two digits of the year (for dates in January and February, subtract 1 from y). The floor function $\lfloor z \rfloor$ gives the largest integer less than or equal to z.

In Exercises 9–11, use the information above to determine the day of the week for the given date in history.

9. July 20, 1969: Apollo 11 reaches the moon.

10. August 12, 1981: IBM launches the first personal computer.

11. The day of the week you were born.

Name____ Date__________

4.5 Puzzle Time

What Did One Plate Say To The Other?

Write the letter of each answer in the box containing the exercise number.

Find $(f + g)(x)$.

1. $f(x) = 3\sqrt{x}, g(x) = -8\sqrt{x}$
2. $f(x) = -\sqrt[3]{5x}, g(x) = -3\sqrt[3]{5x}$
3. $f(x) = 12\sqrt[4]{x-1}, g(x) = -3\sqrt[4]{x-1}$
4. $f(x) = 8x - 2x^2, g(x) = x^2 - 3x + 4$

Find $(f - g)(x)$.

5. $f(x) = 3\sqrt{x}, g(x) = -8\sqrt{x}$
6. $f(x) = -\sqrt[3]{5x}, g(x) = -3\sqrt[3]{5x}$
7. $f(x) = 12\sqrt[4]{x-1}, g(x) = -3\sqrt[4]{x-1}$
8. $f(x) = 8x - 2x^2, g(x) = x^2 - 3x + 4$

Find $(fg)(x)$.

9. $f(x) = 3x^2, g(x) = \sqrt{x}$
10. $f(x) = 2x^3, g(x) = 4x^{5/4}$

Find $\left(\frac{f}{g}\right)(x)$.

11. $f(x) = 9x^{5/4}, g(x) = 3x^{1/2}$
12. $f(x) = -2x^{2/3}, g(x) = 4x^2$

Answers

N. $8x^{17/4}$

N. $-x^2 + 5x + 4$

D. $-5\sqrt{x}$

N. $9\sqrt[4]{x-1}$

E. $11\sqrt{x}$

M. $3x^{3/4}, x \neq 0$

I. $15\sqrt[4]{x-1}$

I. $-4\sqrt[3]{5x}$

S. $-3x^2 + 11x - 4$

R. $2\sqrt[3]{5x}$

O. $3x^{5/2}$

E. $-\dfrac{1}{2x^{4/3}}$

1	2	3	4	5	6		7	8		9	10		11	12

4.6 Start Thinking

Consider the equation $y = -\frac{2}{3}x + 4$. Use a graphing calculator to graph the equation. Solve the equation for x. Then switch the x- and y-values in the new equation. This creates a system of equations. Graph the second equation in the same window of the graphing calculator and find the point of intersection. What does the point of intersection represent?

4.6 Warm Up

Solve the literal equation for *y*.

1. $3x - y = 4$

2. $3x - 2y = 10$

3. $5x + 6y = 9$

4. $4x + y - 6 = 1$

5. $x - 5y = 11$

6. $x - 6y = 9$

4.6 Cumulative Review Warm Up

Find the square root of the number.

1. $\sqrt{-25}$

2. $\sqrt{-20}$

3. $3\sqrt{-18}$

4. $-3\sqrt{-36}$

5. $\sqrt{-4}$

6. $\sqrt{-44}$

Name____________________ Date__________

4.6 Practice A

In Exercises 1–3, solve $y = f(x)$ for x. Then find the input(s) when the output is −3.

1. $f(x) = 2x + 3$
2. $f(x) = \frac{1}{3}x - 2$
3. $f(x) = 8x^3$

In Exercises 4–6, find the inverse of the function. Then graph the function and its inverse.

4. $f(x) = 4x$
5. $f(x) = 4x - 1$
6. $f(x) = \frac{1}{2}x - 5$

7. Find the inverse of the function $f(x) = \frac{1}{5}x - 2$ by switching the roles of x and y and solving for y. Then find the inverse of the function f by using inverse operations in the reverse order. Which method do you prefer? Explain.

8. Determine whether each pair of functions f and g are inverses. Explain your reasoning.

a.

x	−2	−1	0	1	2
f(x)	−3	3	9	15	21

x	−3	3	0	15	21
g(x)	−2	−1	0	1	2

b.

x	1	2	3	4	5
f(x)	9	7	5	3	1

x	9	7	5	3	1
g(x)	1	2	3	4	5

In Exercises 9–11, find the inverse of the function. Then graph the function and its inverse.

9. $f(x) = 9x^2, x \geq 0$
10. $f(x) = 16x^2, x \leq 0$
11. $f(x) = (x + 2)^3$

In Exercises 12 and 13, use the graph to determine whether the inverse of f is a function. Explain your reasoning.

12.

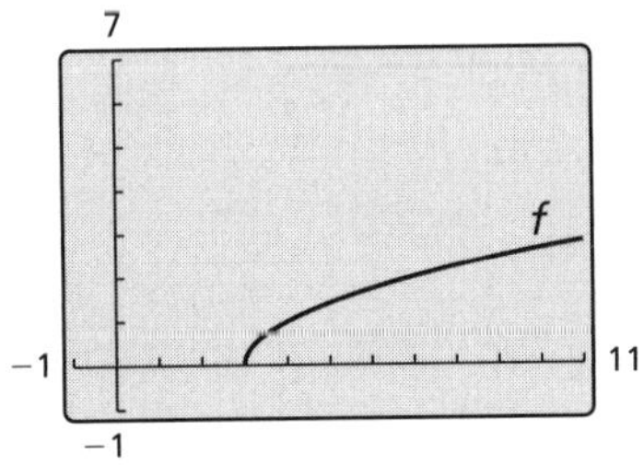

13.

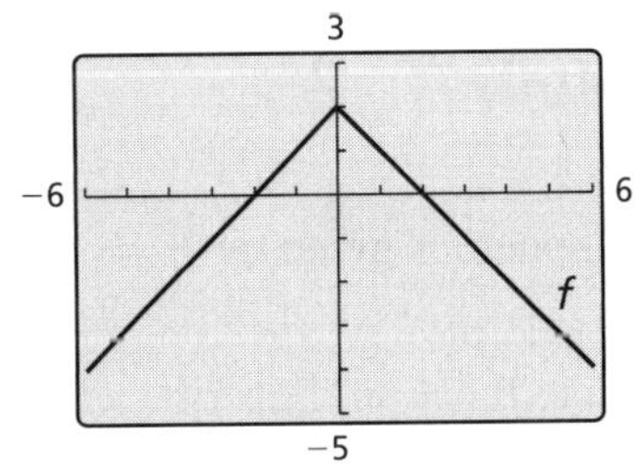

Name ______________________________ Date __________

4.6 Practice B

In Exercises 1–3, solve $y = f(x)$ for x. Then find the input(s) when the output is -3.

1. $f(x) = -\frac{4}{3}x + 2$ **2.** $f(x) = 25x^4$ **3.** $f(x) = (x - 3)^2 - 4$

In Exercises 4–6, find the inverse of the function. Then graph the function and its inverse.

4. $f(x) = -3x + 4$ **5.** $f(x) = -\frac{1}{3}x + 1$ **6.** $f(x) = \frac{2}{5}x - \frac{1}{5}$

7. Describe and correct the error in finding the inverse function.

✗
$f(x) = 3x - 8$
$y = 3x - 8$
$x = 3y - 8$
$g(x) = 3x - 8$

In Exercises 8–10, find the inverse function. Then graph the function and its inverse.

8. $f(x) = -9x^2, x \le 0$ **9.** $f(x) = (x - 1)^3$ **10.** $f(x) = x^6, x \le 0$

11. Find the inverse of the function $f(x) = 8x^3$ by switching the roles of x and y and solving for y. Then find the inverse of the function f by using inverse operations in the reverse order. Which method do you prefer? Explain.

In Exercises 12–15, determine whether the functions are inverse functions.

12. $f(x) = 6x + 1$; $g(x) = 6x - 1$ **13.** $f(x) = \frac{\sqrt[3]{x - 6}}{2}$; $g(x) = 8x^3 + 6$

14. $f(x) = \frac{5 - x}{2}$; $g(x) = 5 - 2x$ **15.** $f(x) = 4x^2 + 3$; $g(x) = -\frac{x - 3}{4}$

16. The volume of a sphere is given by $V = \frac{4}{3}\pi r^3$, where r is the radius.

a. Find the inverse function. Describe what it represents.

b. Find the radius of a sphere with a volume of 146 cubic meters.

Name____ Date__________

4.6 Enrichment and Extension

Inverse of a Function

Find the inverse of $f(x) = x^2 - 6x + 1, x \geq 3$.

$y = x^2 - 6x + 1$	Set y equal to $f(x)$.
$x = y^2 - 6y + 1$	Switch x and y.
$x - 1 = y^2 - 6y$	Subtract 1 from each side.

At this step, it is necessary to complete the square to solve for y.

$x - 1 + 9 = y^2 - 6y + 9$	Add $\left(\frac{b}{2}\right)^2 = \left(\frac{-6}{2}\right)^2 = 9$ to each side.
$x + 8 = (y - 3)^2$	Write the right side as a binomial squared.
$\pm\sqrt{x + 8} = y - 3$	Take the square root of each side.
$3 \pm \sqrt{x + 8} = y$	Add 3 to each side.

The domain of f is restricted to values of $x \geq 3$. So, the range of the inverse must also be restricted to these values.

The inverse of f is $g(x) = 3 + \sqrt{x + 8}$.

In Exercises 1–12, find the inverse of the function.

1. $f(x) = x^2 - 2x, x \geq 1$

2. $f(x) = x^2 + 8x, x \geq -4$

3. $f(x) = x^2 - 3x, x \geq \frac{3}{2}$

4. $f(x) = x^2 - 8x + 12, x \geq 4$

5. $f(x) = x^2 - 12x - 5, x \geq 6$

6. $f(x) = x^2 + 10x + 15, x \geq -5$

7. $f(x) = x^2 + 6x + 1, x \geq -\frac{3}{2}$

8. $f(x) = x^2 - 4x + 12, x \geq 2$

9. $f(x) = 4x^2 - 4x, x \geq \frac{1}{2}$

10. $f(x) = 6x^2 + 12x, x \geq -1$

11. $f(x) = 4x^2 - 16x - 15, x \geq 2$

12. $f(x) = 9x^2 - 18x + 5, x \geq 1$

Name ______________________________ Date __________

4.6 Puzzle Time

What Do You Get When You Cross A SWAT Team With An Octopus?

Write the letter of each answer in the box containing the exercise number.

Find the inverse of the function.

1. $f(x) = 3x + 1$
2. $f(x) = -4x - 8$
3. $f(x) = \frac{x}{7} - 2$
4. $f(x) = \frac{2x}{3} - 4$
5. $f(x) = \frac{x - 1}{5}$
6. $f(x) = -2x^2, x \geq 0$
7. $f(x) = 5x^4, x \geq 0$
8. $f(x) = (x + 1)^3$
9. $f(x) = \sqrt{8x - 3}$
10. $f(x) = (x - 21)^3$

Answers

A. $g(x) = \frac{x - 1}{3}$

B. $g(x) = 5x + 1$

B. $g(x) = \frac{x + 8}{-4}$

O. $g(x) = 7(x + 2)$

D. $g(x) = \sqrt[3]{x} + 21$

S. $g(x) = \sqrt{-\frac{x}{2}}$

Q. $g(x) = \sqrt[4]{\frac{x}{5}}$

M. $g(x) = \frac{3(x + 4)}{2}$

U. $g(x) = \sqrt[3]{x} - 1$

I. $g(x) = \frac{x^2 + 3}{8}, x \geq 0$

1		2	3	4	5		6	7	8	9	10

Name________________________________ Date__________

Chapter 4 Cumulative Review

In Exercises 1–18, simplify the expression.

1. $w^9 \bullet w^2$

2. $b^2 \bullet b^{-1}$

3. $g^{-5} \bullet g^3$

4. $\dfrac{c^7}{c^2}$

5. $\dfrac{m^9}{m^2}$

6. $\dfrac{v^{12}}{v^6}$

7. $\dfrac{j^4}{j^3 \bullet j^1}$

8. $\dfrac{p^7}{p^3 \bullet p^9}$

9. $\dfrac{x^8}{x^2 \bullet x^4}$

10. $\dfrac{a^3}{a} \bullet 5a^6$

11. $\dfrac{y^8}{y^5} \bullet -4y^2$

12. $3n^3 \bullet \dfrac{n^7}{n^5}$

13. $\left(\dfrac{6m^3}{3n}\right)^5$

14. $\left(\dfrac{7y}{q^4}\right)^3$

15. $\left(\dfrac{8x^5}{2j^3}\right)^4$

16. $\left(\dfrac{x^3 \bullet x}{b^2 \bullet x^2}\right)^5$

17. $\left(\dfrac{3r^5 \bullet c^2}{r^3 \bullet r}\right)^4$

18. $\left(\dfrac{d^4 \bullet d^7}{d^6 \bullet w^2}\right)^3$

In Exercises 19–30, solve the literal equation for *y*.

19. $3x + y = 7$

20. $y - 8 = 1$

21. $-14x + 7y = 28$

22. $y - 3x = 4$

23. $2x - 10y = 40$

24. $x - \frac{1}{2}y = 6$

25. $4x - 2y = -12$

26. $-6x + 3y = -24$

27. $\frac{2}{3}x + \frac{1}{6}y = 1$

28. $4x - \frac{1}{4}y = -2$

29. $3xy + 9y = 12$

30. $4xy - 2y = 16$

31. You painted a picture that measures 18 inches by 24 inches.

a. Find the perimeter of the picture.

b. Find the area of the picture.

32. You are purchasing movies on Blu-ray that cost \$18 each. You have \$100 to spend.

a. Write an expression that shows how much money you have left after purchasing *m* movies.

b. Evaluate the expression when $m = 4$. What does this mean?

c. Evaluate the expression when $m = 5$. What does this mean?

Name ______________________________ Date __________

Chapter 4 Cumulative Review (continued)

In Exercises 33–48, simplify the expression. Write your answers using only positive exponents.

33. $4^2 \bullet 4$ **34.** $5^3 \bullet 5^2$ **35.** $6^5 \bullet 6^{-2}$ **36.** $3^{-4} \bullet 3^6$

37. $\frac{3^3}{3^4}$ **38.** $\frac{7^4}{7^6}$ **39.** $\frac{4^2}{4^{-2}}$ **40.** $\frac{9^5}{9^3}$

41. $(m^4)^{-3}$ **42.** $(p^{-2})^5$ **43.** $(k^{-4})^{-7}$ **44.** $(j^3)^8$

45. $\left(\frac{4y}{5}\right)^2$ **46.** $\left(\frac{2}{3z}\right)^3$ **47.** $\left(\frac{5w}{6}\right)^{-2}$ **48.** $\left(\frac{3}{4x}\right)^{-3}$

In Exercises 49–60, solve the equation. Check your solutions.

49. $|5x - 12| = 3$ **50.** $|3x + 6| = 24$ **51.** $|-7x + 12| = 40$

52. $|5x + 13| = 38$ **53.** $|6x - 17| = -21$ **54.** $|3x + 12| = -14$

55. $|4x - 12| = |x|$ **56.** $|x| = |3x + 16|$ **57.** $|4x - 24| = |2x|$

58. $|2x - 4| = |x + 9|$ **59.** $|4x - 13| = |5x + 16|$ **60.** $|6x - 12| = |4x - 6|$

In Exercises 61–68, determine whether the relation is a function.

61. $(3, -2), (-1, 2), (1, 7), (2, -7)$ **62.** $(4, -6), (-7, 1), (-7, 9), (-6, 7)$

63. $(6, -3), (8, -4), (-8, -4), (8, 1)$ **64.** $(2, 3), (-2, 5), (2, -3), (1, -4)$

65. $(3, 4), (9, -1), (6, -4), (-3, 6)$ **66.** $(5, -3), (5, -7), (-9, 3), (-6, -7)$

67. $(2, -1), (6, 5), (-2, 4), (4, 4)$ **68.** $(7, -8), (-5, 6), (5, 8), (6, -8)$

69. You want to buy either a Blu-ray or a DVD for 15 of your friends. Blu-rays cost \$18 each and DVDs cost \$15 each.

a. Write an expression for the total amount you must spend.

b. Evaluate the expression when six people get Blu-rays.

70. You decide to purchase a membership to a local art supply store in order to purchase acrylic paint at a discount. The monthly membership costs \$5.85, and each tube of acrylic paint is \$28.84.

a. Write an expression for the total amount you spend on m months and t tubes of paint.

b. Evaluate the expression for 5 months and 3 tubes of paint.

Name_______________________________________ Date__________

Chapter 4 Cumulative Review (continued)

In Exercises 71–77, write a function *g* whose graph represents the indicated transformation of the graph of *f*.

71. $f(x) = x + 7$; translation 5 units down

72. $f(x) = x - 1$; translation 3 units up

73. $f(x) = |2x - 1| + 3$; translation 4 units left

74. $f(x) = |3x + 4| - 1$; translation 6 units right

75. $f(x) = 3x + 4$; translation 1 unit left

76. $f(x) = -2x + 5$; reflection in the x-axis

77. $f(x) = -\frac{1}{4}x - 1$; translation 3 units left, followed by a translation 4 units up

In Exercises 78–86, solve the system. Check your solution, if possible.

78. $x - y + z = 9$
$2x - y + z = 12$
$-x + y + 3z = 11$

79. $x + y + z = 3$
$2x + y + 2z = 2$
$-2x - y + z = -11$

80. $x + y + z = 0$
$-x - y + z = 2$
$2x + 2y + z = -1$

81. $x - y - z = -10$
$2x - y + z = -5$
$-x + 2y - z = 5$

82. $2x - y + z = -11$
$-2x + y + z = 5$
$x + 2y + 3z = -3$

83. $2x - y + 2z = -15$
$x - y + 3z = -23$
$-x + y - z = 11$

84. $2x + 2y - 2z = 4$
$4x + 4y - 4z = 8$
$-3x - 3y + 3z = -6$

85. $2x + y + 2z = -5$
$-x + 2y - z = -4$
$-x - y + z = 9$

86. $x + y - z = -4$
$2x - 2y + 2z = 12$
$-x - y - z = -2$

87. A car salesperson's base salary is $28,000. She earns a 7% commission on total car sales. How much did she sell in car sales if she earns a total of $37,975 this year?

88. Your grocery bill came to a total of $61.29. The groceries cost $56.70. What percent sales tax did you pay? Round your answer to the nearest percent.

89. Your total bill for a new outfit is $131.25. There was 5% sales tax added. How much is the outfit without the sales tax?

90. Your total bill for a new pair of boots is $72.08. There was 6% sales tax added. How much are the boots without the sales tax?

91. You have to pay sales tax when you purchase a new vehicle. The cost of your new car is $32,560, and the cost after taxes are included is $34,513.60. What percent sales tax do you pay?

Name ______________________________ Date __________

Chapter 4 Cumulative Review (continued)

In Exercises 92–106, solve the equation. Check your solutions.

92. $x^2 = 36$

93. $4x^2 = 256$

94. $5x^2 = 245$

95. $x^2 - 5 = 11$

96. $x^2 - 8 = 28$

97. $2x^2 - 4 = 124$

98. $x^2 - 3x - 10 = 0$

99. $x^2 - 2x - 3 = 0$

100. $x^2 - 7x + 6 = 0$

101. $x^2 - 11x + 28 = 0$

102. $x^2 - 9x + 8 = 0$

103. $x^2 + 10x + 21 = 0$

104. $2x^2 + 7x - 4 = 0$

105. $3x^2 + 10x - 8 = 0$

106. $5x^2 + 8x + 3 = 0$

In Exercises 107–110, add or subtract. Write the answer in standard form.

107. $(15 + 4i) + (10 - 13i)$

108. $(-7 + 2i) + (-14 + i)$

109. $(11 + 7i) - (2 - 12i)$

110. $(-2 + 5i) - (13 + 7i)$

In Exercises 111–116, multiply. Write the answer in standard form.

111. $2i(5 + 3i)$

112. $-4i(4 - 2i)$

113. $(-5 - 3i)(-8 - 7i)$

114. $(4 + 2i)(7 - 8i)$

115. $(6 + 2i)^2$

116. $(3 - 4i)^2$

117. You travel in a car at an average rate of 60 miles per hour on the highway and 40 miles per hour in the city.

a. How long does it take you to travel 45 highway miles and 20 city miles?

b. How long does it take you to travel 36 highway miles and 16 city miles?

118. You travel in a car at an average rate of 55 miles per hour on the highway and 35 miles per hour in the city.

a. How long does it take you to travel 16.5 highway miles and 7 city miles?

b. How long does it take you to travel 44 highway miles and 28 city miles?

In Exercises 119–122, determine whether the inverse of *f* is a function. Then find the inverse.

119. $f(x) = x^3 - 2$

120. $f(x) = \sqrt{x + 3}$

121. $f(x) = 3\sqrt[3]{x - 2}$

122. $f(x) = \sqrt[3]{\dfrac{4x + 2}{3}}$

Chapter 5

Name ______________________ Date __________

Chapter 5 Exponential and Logarithmic Functions

Dear Family,

What are some things you will need to save money for in the future? In the short term, you may be thinking about buying a car and what to do after high school. In the long term, things to think about are buying a house, starting a family, saving for retirement, or even a nice vacation.

The most important thing you can do with your money is set up a budget. You can see where your money goes each month. This also gives you a chance to see how you can start to save money each month. When you get into the habit of living within a budget instead of borrowing money through financing and credit cards, you end up saving a lot of money in the long run and are able to start investing money for the future. For example, if you do not have to finance a car, you may be able to negotiate a better deal and will not be spending money on interest. As a family, talk about your short- and long-term financial goals. How is your family's budget set up? Does your family use a financial advisor?

Choose one goal that you would like to be saving money toward.

- How much money will you need to meet your goal?
- When will you need the money?
- What type of investment options will meet your needs?
- Use the Internet to find the interest rate for your investment.
- Use an online financial calculator to find how much money you need to save each month to meet your goal.
- If you wait one year to start saving, how much additional money will you need to save each month to meet your goal?

Studies have shown that you are more likely to save money for retirement if you calculate different hypothetical situations than if you don't. When you get a job, see if your company has a retirement savings plan. Some companies even match what you invest up to a certain percentage. When the money is taken directly from your paycheck, you are less likely to miss it.

In this chapter, you will use exponential growth equations to determine the amount of money you have in your savings account. If you make it a habit to start saving money early in life, you are more likely to be prepared for the future.

The sooner you start saving, the better!

Nombre ____ Fecha _________

Capítulo 5 Funciones exponenciales y logarítmicas

Estimada familia:

¿Cuáles son algunas cosas para las que tendrán que ahorrar dinero en el futuro? A corto plazo, quizás piensen en comprar un carro y qué hacer después de la escuela secundaria. A largo plazo, cosas a considerar son comprar una casa, formar una familia, ahorrar para cuando se retiren o incluso unas lindas vacaciones.

Lo más importante que pueden hacer con su dinero es establecer un presupuesto. Pueden ver dónde va el dinero todos los meses. Además, les da una oportunidad para ver cómo pueden comenzar a ahorrar dinero todos los meses. Cuando adquieren el hábito de vivir de acuerdo a un presupuesto en lugar de pedir préstamos mediante financiación o tarjetas de crédito, terminan ahorrando mucho dinero a largo plazo y son capaces de comenzar a invertir dinero para el futuro. Por ejemplo, si no tienen que financiar un carro, quizás puedan negociar un precio mejor y no gastarán dinero en intereses. En familia, hablen sobre sus metas financieras a corto y largo plazo. ¿Cómo está establecido su presupuesto familiar? ¿Su familia tiene un asesor financiero?

Elijan una meta para la que les gustaría ahorrar dinero.

- ¿Cuánto dinero necesitarán para lograr su meta?
- ¿Cuándo necesitarán el dinero?
- ¿Qué clase de opciones de inversión cumplirán sus necesidades?
- Busquen en Internet la tasa de interés de su inversión.
- Usen una calculadora financiera en línea para hallar cuánto dinero necesitan ahorrar por mes para lograr su meta.
- Si esperan un año más para comenzar a ahorrar, ¿cuánto dinero adicional necesitarán ahorrar por mes para lograr su meta?

Los estudios han demostrado que tienen más probabilidades de ahorrar dinero para su jubilación si calculan diferentes situaciones hipotéticas que si no lo hacen. Cuando consigan un trabajo, fíjense si la compañía tiene un plan de ahorros para su retiro. Algunas compañías incluso igualan la cantidad que ustedes han invertido hasta cierto porcentaje. Cuando el dinero se saca directamente de su cheque de pago, tienen menos probabilidades de olvidarse.

En este capítulo, usarán ecuaciones de crecimiento exponencial para determinar la cantidad de dinero que tienen en su cuenta de ahorros. Si adquieren el hábito de comenzar a ahorrar dinero desde una temprana edad, tienen más probabilidades de estar preparados para el futuro.

¡Cuanto antes comiencen a ahorrar, mejor!

5.1 Start Thinking

You deposit \$7000 in an account that pays 2.25% annual interest. Compare what the balance will be after 5 years if the interest is compounded semi-annually, quarterly, monthly, and daily. What do you think will happen to the balance if the number of compoundings continues to increase?

5.1 Warm Up

Determine if the model represents *exponential growth* or *exponential decay*. Then identify the initial amount and the percent increase or decrease.

1. $y = 1.2^x$
2. $y = 0.78^x$
3. $y = \left(\frac{5}{8}\right)^x$
4. $y = 28(1.03)^x$
5. $y = 25{,}000(0.95)^x$
6. $y = 2^x$

5.1 Cumulative Review Warm Up

Find the inverse of the function. Then graph the function and its inverse in the same coordinate plane.

1. $y = \frac{1}{2}x - 5$
2. $y = -3x + 7$
3. $y = \dfrac{x - 9}{5}$
4. $y = (x + 2)^3$
5. $y = \dfrac{x^3 - 6}{2}$
6. $y = -2x^2,\ x \geq 0$

Name____ Date __________

5.1 Practice A

In Exercises 1–6, simplify the expression.

1. $e^2 \bullet e^5$

2. $e^{-3} \bullet e^8$

3. $\dfrac{12e^5}{36e^2}$

4. $\dfrac{15e^4}{3e^9}$

5. $\left(3e^{3x}\right)^2$

6. $\sqrt{16e^{10x}}$

7. Describe and correct the error in simplifying the expression.

$$\times \quad \left(2e^{3x}\right)^2 = (2)^2\left(e^{3x}\right)^2$$
$$= 4e^{9x^2}$$

In Exercises 8–10, tell whether the function represents *exponential growth* or *exponential decay*. Then graph the function.

8. $y = e^{4x}$

9. $y = e^{-x}$

10. $y = 4e^{-2x}$

In Exercises 11–13, use the properties of exponents to rewrite the function in the form $y = a(1 + r)^t$ or $y = a(1 - r)^t$. Then find the percent rate of change.

11. $y = e^{-0.5x}$

12. $y = 2e^{-0.2x}$

13. $y = 5e^{0.6x}$

In Exercises 14–16, use a table of values or a graphing calculator to graph the function. Then identify the domain and range.

14. $y = e^{x-1}$

15. $y = e^{x+2}$

16. $y = 3e^x + 2$

17. You invest $4000 in an account to save for college.

   a. Option 1 pays 5% annual interest compounded semi-annually. What would be the balance in the account after 2 years?

   b. Option 2 pays 4.5% annual interest compounded continuously. What would be the balance in the account after 2 years?

   c. At what time t (in years) would Option 1 give you $100 more than Option 2?

Name ______________________________ Date __________

5.1 Practice B

In Exercises 1–6, simplify the expression.

1. $e^{-9} \cdot e^{7}$
2. $\dfrac{27e^{4}}{18e^{7}}$
3. $\left(5e^{-4x}\right)^{3}$
4. $\sqrt{20e^{8x}}$
5. $\sqrt[3]{64e^{9x}}$
6. $e^{2x} \bullet e^{5} \bullet e^{x-2}$

7. Describe and correct the error in simplifying the expression.

$$\times \quad \left(2e^{-3x}\right)^{4} = \frac{1}{16e^{12x}}$$

In Exercises 8–10, tell whether the function represents *exponential growth* or *exponential decay*. Then graph the function.

8. $y = 2e^{3x}$
9. $y = 0.5e^{-2x}$
10. $y = 0.4e^{0.5x}$

In Exercises 11–13, use the properties of exponents to rewrite the function in the form $y = a(1 + r)^{t}$ or $y = a(1 - r)^{t}$. Then find the percent rate of change.

11. $y = e^{0.25x}$
12. $y = 3e^{-0.65x}$
13. $y = 0.25e^{0.9x}$

In Exercises 14–16, use a table of values or a graphing calculator to graph the function. Then identify the domain and range.

14. $y = e^{x-4}$
15. $y = 4e^{x} - 1$
16. $y = 2e^{x} + 5$

17. You invest \$5000 in an account to save for college.

    a. Option 1 pays 4% annual interest compounded monthly. What would be the balance in the account after 2 years?

    b. Option 2 pays 4% annual interest compounded continuously. What would be the balance in the account after 2 years?

    c. What is the difference between the two options after 10 years?

    d. How would your answer to part (c) change if you invested \$50,000?

Name____ Date__________

5.1 Enrichment and Extension

The Natural Base e

The natural base e is an irrational number like pi (π), so you cannot find its exact value. However, you have discovered one way to approximate e by increasing the value of x in the expression $\left(1 + \frac{1}{x}\right)^x$. As x approaches infinity, $\left(1 + \frac{1}{x}\right)^x$ approaches $e \approx 2.71828182846$.

Another way to find e is to approximate the sum $1 + \frac{1}{1} + \frac{1}{1 \bullet 2} + \frac{1}{1 \bullet 2 \bullet 3} + \frac{1}{1 \bullet 2 \bullet 3 \bullet 4} + \ldots.$

The value of π can be approximated many ways as well. One way is to approximate the sum $\sqrt{12}\left(1 - \frac{1}{3 \bullet 3} + \frac{1}{5 \bullet 3^2} - \frac{1}{7 \bullet 3^3} + \ldots\right)$. Use the table to approximate π.

Sum	Approximation
$\sqrt{12}\left(1 - \frac{1}{3 \bullet 3} + \frac{1}{5 \bullet 3^2} - \frac{1}{7 \bullet 3^3} + \ldots\right)$	
$\sqrt{12}\left(1 - \frac{1}{3 \bullet 3} + \frac{1}{5 \bullet 3^2} - \frac{1}{7 \bullet 3^3} + \frac{1}{9 \bullet 3^4} - \ldots\right)$	
$\sqrt{12}\left(1 - \frac{1}{3 \bullet 3} + \frac{1}{5 \bullet 3^2} - \frac{1}{7 \bullet 3^3} + \frac{1}{9 \bullet 3^4} - \frac{1}{11 \bullet 3^5} + \ldots\right)$	
$\sqrt{12}\left(1 - \frac{1}{3 \bullet 3} + \frac{1}{5 \bullet 3^2} - \frac{1}{7 \bullet 3^3} + \frac{1}{9 \bullet 3^4} - \frac{1}{11 \bullet 3^5} + \frac{1}{13 \bullet 3^6} \ldots\right)$	

In Exercises 1–6, use your textbook or another source to answer the question(s) about π and e.

1. What is the definition of π?

2. Approximate the value of π to 11 decimal places.

3. Because π is an irrational number, it cannot be written as a ratio of two integers. Why is π represented sometimes as $\frac{22}{7}$?

4. π is a transcendental number. What does this mean? Is e transcendental?

5. Are all transcendental numbers irrational? Are all irrational number transcendental? Explain.

6. An equation that relates π, e, and the imaginary number $i = \sqrt{-1}$ is $e^{i\pi} + 1 = 0$. What is this identity called?

Name ______________________________ Date __________

5.1 Puzzle Time

What Does An Astronaut Use To Dust Those Hard-To-Reach Black Holes?

Write the letter of each answer in the box containing the exercise number.

Simplify the expression.

1. $e^{-2} \bullet e^{8}$

2. $e^{7} \bullet e \bullet e^{3}$

3. $2e^{4} \bullet 3e^{2}$

4. $\dfrac{12e^{7}}{8e^{5}}$

5. $\dfrac{5e^{11}}{10e^{14}}$

6. $\sqrt{64e^{10x}}$

7. $\sqrt[5]{32e^{10x}}$

8. $\left(3e^{2x}\right)^{2}$

9. $e^{-2x} \bullet e^{x} \bullet e^{6x+2}$

10. $\left(2e^{-5x}\right)^{4}$

Match the graph with its function.

11.

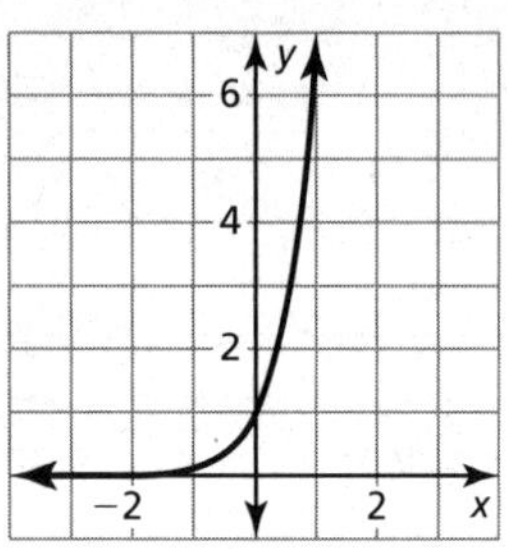

12.

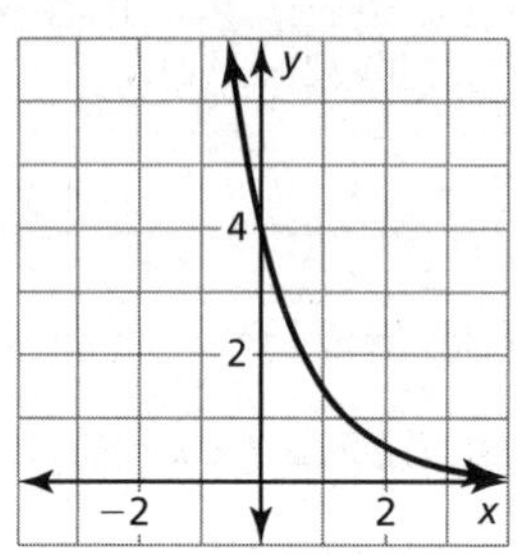

13.

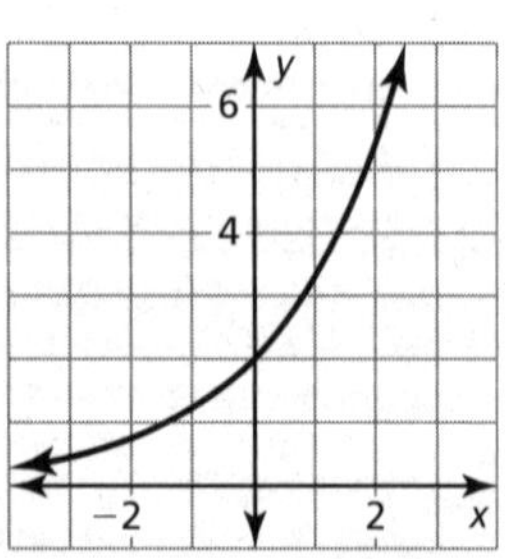

Answers

M. $8e^{5x}$	**A.** e^{11}
V. e^{6}	**N.** $y = e^{2x}$
C. $6e^{6}$	**U.** $\dfrac{3e^{2}}{2}$
C. $2e^{2x}$	**L.** $9e^{4x}$
R. $y = 2e^{0.5x}$	**U.** $\dfrac{1}{2e^{3}}$
E. e^{5x+2}	**A.** $\dfrac{16}{e^{20x}}$
E. $y = 4e^{-x}$	

1	2	3	4	5	6		7	8	9	10	11	12	13

5.2 Start Thinking

A function and its inverse are sketched in the same coordinate plane. How are the two graphs related? How do the domain and range of $f(x)$ relate to the domain and range of $f^{-1}(x)$?

5.2 Warm Up

Determine the value of *x* in the expression.

1. $2^x = 4$ **2.** $8^x = 1$ **3.** $\left(\frac{2}{3}\right)^x = \frac{8}{27}$

4. $4^x = \frac{1}{4}$ **5.** $4^x = 2$ **6.** $4^x = \frac{1}{2}$

5.2 Cumulative Review Warm Up

Use the parent function to sketch the graph of each function.

1. Parent function: $y = x^2$

a. $y = x^2 - 2$ **b.** $y = (x + 3)^2$ **c.** $y = -(x - 1)^2$

2. Parent function: $y = \sqrt{x}$

a. $y = 2\sqrt{x}$ **b.** $y = \sqrt{-x}$ **c.** $y = -\sqrt{x - \frac{1}{2}} + 3$

Name ____ Date __________

5.2 Practice A

In Exercises 1–3, rewrite the equation in exponential form.

1. $\log_2 8 = 3$
2. $\log_7 7 = 1$
3. $\log_5 25 = 2$

In Exercises 4–6, rewrite the equation in logarithmic form.

4. $4^2 = 16$
5. $5^0 = 1$
6. $6^{-1} = \frac{1}{6}$

In Exercises 7–12, evaluate the logarithm.

7. $\log_2 16$
8. $\log_5 125$
9. $\log_6 6$
10. $\log_5 \frac{1}{5}$
11. $\log_9 1$
12. $\log_2 \frac{1}{8}$

In Exercises 13–15, evaluate the logarithm using a calculator. Round your answer to three decimal places.

13. $\log 5$
14. $\ln 14$
15. $\log \frac{1}{4}$

16. The decibel level D of sound is given by the equation $D = 10 \log\left(\frac{I}{10^{-12}}\right)$, where I is the intensity of the sound. What is the decibel level when the intensity of the sound is 10^{-9}?

In Exercises 17–19, simply the expression.

17. $5^{\log_5 x}$
18. $8^{\log_8 2x}$
19. $\log_4 4^{3x}$

In Exercises 20–25, find the inverse of the function.

20. $y = 1.1^x$
21. $y = 3^x$
22. $y = \log_3 x$
23. $y = \log\left(\frac{1}{3}x\right)$
24. $y = \ln(3x)$
25. $y = e^{5x}$

26. The wind speed s (in miles per hour) near the center of a tornado can be modeled by $s = 93 \log d + 65$, where d is the distance (in miles) that the tornado travels.

    a. A tornado traveled 35 miles. Estimate the wind speed near the center of the tornado.

    b. The wind speed near the center of a tornado was 150 miles per hour. Find the distance that the tornado traveled.

Name____ Date __________

5.2 Practice B

In Exercises 1–3, rewrite the equation in exponential form.

1. $\log_9 1 = 0$

2. $\log_6 216 = 3$

3. $\log_2 \frac{1}{4} = -2$

In Exercises 4–6, rewrite the equation in logarithmic form.

4. $13^{-2} = \frac{1}{169}$

5. $4^{3/2} = 8$

6. $81^{1/2} = 9$

In Exercises 7–12, evaluate the logarithm.

7. $\log_8 64$

8. $\log_2 32$

9. $\log_{10} 1$

10. $\log_3 \frac{1}{81}$

11. $\log_2 0.125$

12. $\log_{10} 0.01$

In Exercises 13–15, evaluate the logarithm using a calculator. Round your answer to three decimal places.

13. $\log\left(\frac{1}{5}\right)$

14. $2\ln(1.4)$

15. $\ln(0.4) - 2$

16. The decibel level D of sound is given by the equation $D = 10\log\left(\frac{I}{10^{-12}}\right)$, where I is the intensity of the sound. The pain threshold for sound is 125 decibels. Does a sound with an intensity of 10 exceed the pain threshold? Explain.

In Exercises 17–19, simply the expression.

17. $e^{\ln 7x}$

18. $10^{\log 18}$

19. $\log\left(10^{3x}\right)$

In Exercises 20–25, find the inverse of the function.

20. $y = 0.75^x$

21. $y = \log_{3/4} x$

22. $y = \log\left(\frac{x}{2}\right)$

23. $y = \ln(x + 2)$

24. $y = e^{x-3}$

25. $y = 6^x + 2$

26. The length ℓ (in inches) of an alligator and its weight w (in pounds) are related by the function $\ell = 27.1\ln w - 32.8$.

a. Estimate the length (in inches) of an alligator that weighs 250 pounds. What is its length in feet?

b. Find the inverse of the given function. Use the inverse function to find the weight of a 14-foot alligator. (*Hint*: Convert to inches first.)

Name ______________________________ Date __________

5.2 Enrichment and Extension

Benford's Law

Benford's Law states that given a wide range of data values, the probability P that a randomly selected number has first digit d can be modeled by

$$P = \log\left(1 + \frac{1}{d}\right).$$

A data analyst can use Benford's Law to help identify fraudulent recordkeeping. When valid numbers are replaced with random numbers, the probability model above may no longer fit the data.

In Exercises 1–7, use Benford's Law to answer the question.

1. Complete the table below. Round your answers to four decimal places.

First digit, *d*	1	2	3	4	5	6	7	8	9
Probability, *P*									

2. Construct a histogram for the probability distribution in Exercise 1.

3. Find the probability that a randomly-selected debit card purchase begins with the number 9.

4. Find the probability that a randomly-selected debit card purchase begins with a number other than 1.

5. Find the probability that a randomly-selected debit card purchase begins with either the number 2 or the number 8.

6. While analyzing a large number of debit card purchases, you discover there are about 7 times as many purchases that begin with a 1 than purchases that begin with a 9. Do your findings warrant an audit? Explain.

7. Do you think Benford's Law can be used to analyze the debit card purchases at an automated car wash? Explain your reasoning.

8. Research the Internet or some other source to find other data sets for which Benford's Law can be applied.

Name______________________________ Date__________

5.2 Puzzle Time

Where Do Spies Do Their Shopping?

Write the letter of each answer in the box containing the exercise number.

Rewrite the equation in exponential form.

1. $\log_3 9 = 2$
2. $\log_{25} 25 = 1$
3. $\log_2 32 = 5$
4. $\log_{1/8} 8 = -1$

Rewrite the equation in logarithmic form.

5. $7^2 = 49$
6. $3^{-1} = \frac{1}{3}$
7. $16^{3/4} = 8$
8. $7^{-3} = \frac{1}{343}$

Evaluate the logarithm.

9. $\log_2 64$
10. $\log_9 81$
11. $\log_8 2$
12. $\log_3 \frac{1}{27}$
13. $\log_{1/2} 16$
14. $\log_2 0.5$
15. $\log_{36} 6$
16. $\log_4 64$

Answers

A. $3^2 = 9$

A. $2^5 = 32$

N. $\log_7 49 = 2$

O. $\log_{16} 8 = \frac{3}{4}$

S. $\left(\frac{1}{8}\right)^{-1} = 8$

R. -4

O. $\log_3 \frac{1}{3} = -1$

P. $\log_7 \frac{1}{343} = -3$

R. 2

T. $25^1 = 25$

T. 3 **M.** $\frac{1}{3}$

A. -3 **K.** -1

E. $\frac{1}{2}$ **E.** 6

1	2		3		4	5	6	7	8	9	10		11	12	13	14	15	16

5.3 Start Thinking

A window manufacturing company trains new employees to assemble double-hung windows. The company has determined that the learning curve for the average employee is given by $W = 18 - 18e^{-0.08t}$, where W is the number of windows assembled in t days of training.

1. Use a graphing calculator to graph the function. What is an appropriate window setting to view the graph?

2. Approximate the number of windows a new employee should be able to assemble after 5 days of training, and then after 10 days of training.

3. What appears to be the maximum number of windows an employee would be able to assemble in a day? How many days of training does it take for the average employee to reach this maximum?

5.3 Warm Up

Describe the transformation of *f* represented by *g*.

1. $f(x) = x^2, g(x) = -x^2 + 3$
2. $f(x) = x^3, g(x) = \frac{1}{2}(x + 2)^3 - 5$
3. $f(x) = 2^x, g(x) = 2^{-3x}$
4. $f(x) = \left(\frac{2}{3}\right)^x, g(x) = 4\left(\frac{2}{3}\right)^{x-2}$

5.3 Cumulative Review Warm Up

Use the information to find an equation of the line.

1. slope: 3

   y-intercept: $(0, 5)$

2. slope: -5

   x-intercept: $\left(-\frac{2}{5}, 0\right)$

3. passes through $(1, 7)$ and $(0, 9)$
4. passes through $(-2, 8)$ and $(-7, -3)$

Name______________________________ Date__________

5.3 Practice A

In Exercises 1–8, describe the transformation of *f* represented by *g*. Then graph each function.

1. $f(x) = 2^x, g(x) = 2^x + 3$
2. $f(x) = e^x, g(x) = e^x - 2$
3. $f(x) = 3^x, g(x) = 3^{x-1}$
4. $f(x) = e^{-x}, g(x) = e^{-x} + 4$
5. $f(x) = e^x, g(x) = e^{3x}$
6. $f(x) = e^x, g(x) = \frac{3}{2}e^x$
7. $f(x) = 3^x, g(x) = -3^{x+2}$
8. $f(x) = e^{-x}, g(x) = 2e^{-5x}$
9. Describe and correct the error in graphing the function $f(x) = 2e^x$.

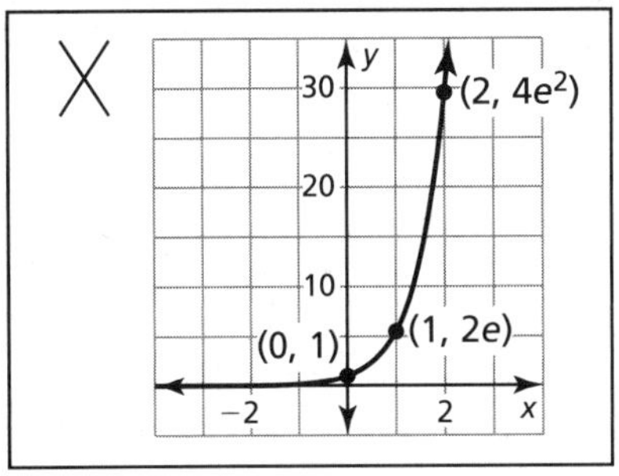

In Exercises 10 and 11, describe the transformation of *f* represented by *g*. Then graph each function.

10. $f(x) = \log_2 x, g(x) = 4 \log_2 x - 1$
11. $f(x) = \log_{1/2} x, g(x) = -\log_{1/2} x + 3$

In Exercises 12–15, write a rule for *g* that represents the indicated transformation of the graph of *f*.

12. $f(x) = 3^x$; reflection in the x-axis, followed by a translation 3 units left and 1 unit down
13. $f(x) = e^x$; vertical shrink by a factor of $\frac{1}{4}$, followed by a translation 5 units up
14. $f(x) = \log_8 x$; reflection in the y-axis, followed by a translation 4 units left
15. $f(x) = \log_{1/6} x$; vertical stretch by a factor of 9, followed by translations 2 units right and 3 units down

Name ______________________________ Date ________

5.3 Practice B

In Exercises 1–8, describe the transformation of *f* represented by *g*. Then graph each function.

1. $f(x) = e^x, g(x) = e^x - 4$

2. $f(x) = 4^x, g(x) = 4^{x+2}$

3. $f(x) = e^{-x}, g(x) = e^{-x} - 5$

4. $f(x) = \left(\frac{1}{3}\right)^x, g(x) = \left(\frac{1}{3}\right)^x + 2$

5. $f(x) = 3^x, g(x) = 3^{2x} - 1$

6. $f(x) = e^x, g(x) = -e^{x+2}$

7. $f(x) = e^{-x}, g(x) = e^{-4x+1}$

8. $f(x) = \left(\frac{1}{3}\right)^x, g(x) = \left(\frac{1}{3}\right)^{x-2} + 3$

9. Describe and correct the error in graphing the function $f(x) = 2^{x+3}$.

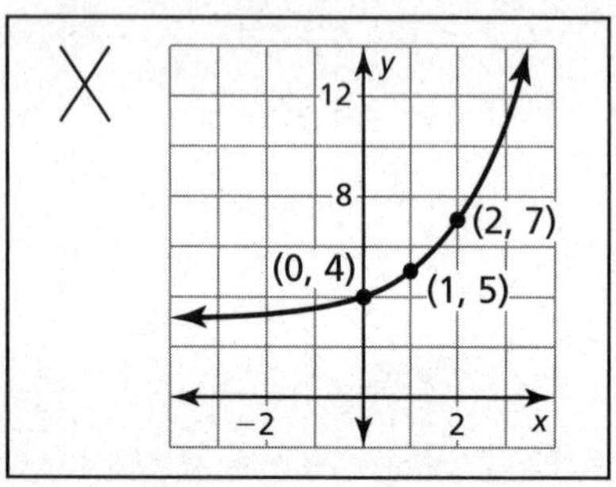

In Exercises 10 and 11, describe the transformation of *f* represented by *g*. Then graph each function.

10. $f(x) = \log_4 x, g(x) = \log_4(x - 2) + 4$

11. $f(x) = \log_{1/3} x, g(x) = -\log_{1/3}(-x)$

In Exercises 12–14, write a rule for *g* that represents the indicated transformation of the graph of *f*.

12. $f(x) = \left(\frac{2}{5}\right)^x$; reflection in the y-axis, followed by a horizontal shrink by a factor of 2 and a translation 4 units down

13. $f(x) = e^{-x}$; translation 2 units left and 3 units up, followed by a vertical stretch by a factor of 2

14. $f(x) = \log_{12} x$; translation 5 units right and 2 units down, followed by a reflection in the x-axis

Name______________________________ Date__________

5.3 Enrichment and Extension

Transformations of Exponential and Logarithmic Functions

In Exercises 1–14, match the function with its graph.

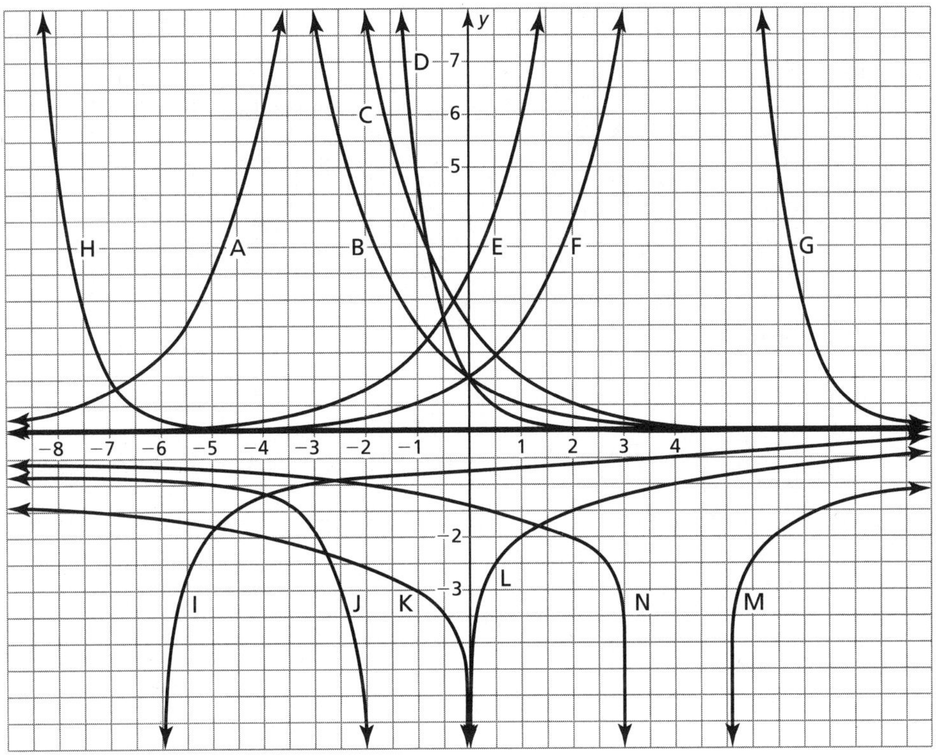

1. $f(x) = \log_4(-x) - 3$

2. $f(x) = \left(\frac{1}{5}\right)^x$

3. $f(x) = 3(2)^x$

4. $f(x) = \log_5(x - 5) - 2$

5. $f(x) = \log_4 x - 2$

6. $f(x) = \left(\frac{1}{5}\right)^{x-7}$

7. $f(x) = \log_5(x + 6) - 2$

8. $f(x) = \log_5(-x - 2) - 2$

9. $f(x) = 2^x$

10. $f(x) = 3(2)^{x+5}$

11. $f(x) = \log_5(-x + 3) - 2$

12. $f(x) = \left(\frac{1}{2}\right)^{x-1}$

13. $f(x) = \left(\frac{1}{2}\right)^x$

14. $f(x) = \left(\frac{1}{5}\right)^{x+7}$

Name ______________________________ Date ________

5.3 Puzzle Time

What Do You Call A Mouse Who Hangs Out With A Bunch Of Pythons?

Write the letter of each answer in the box containing the exercise number.

Match the graph with its function.

1.

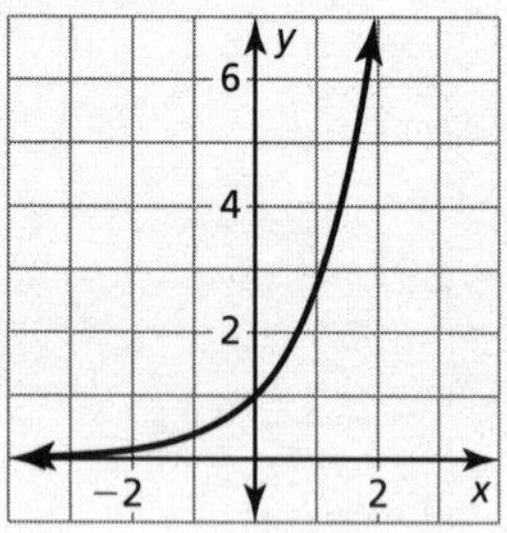

2.

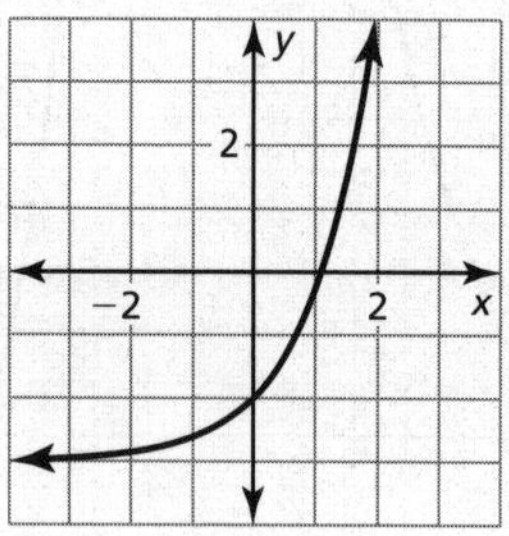

3.

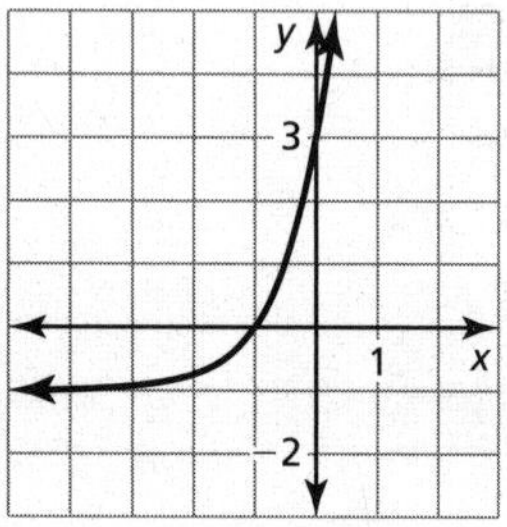

4.

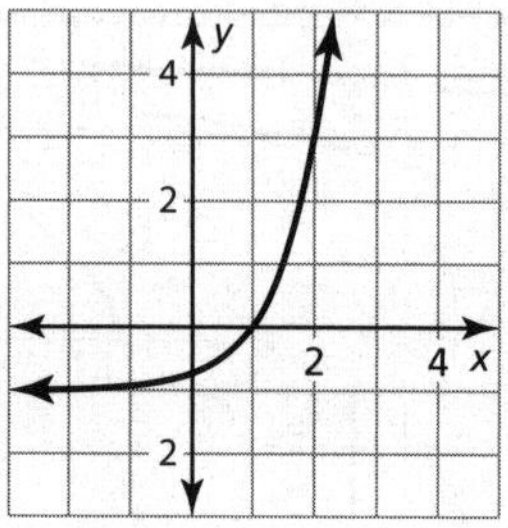

5.

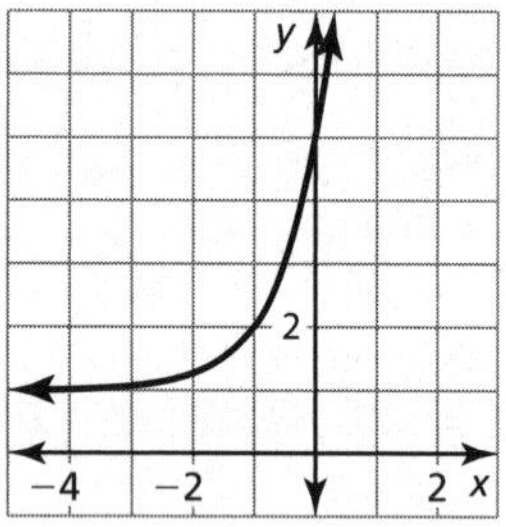

Answers

U. $f(x) = e^x - 3$

L. $f(x) = e^x$

H. $f(x) = 4^{x+1} + 1$

C. $f(x) = 4^{x-1} - 1$

N. $f(x) = 4^{x+1} - 1$

1	2	3	4	5

5.4 Start Thinking

Examine each group of logarithmic equations below. Try to identify any patterns that relate the three equations in each group. Then use your conclusion to make a prediction for the value of $\log_5 25x$ if you know $\log_5 x \approx 1.4$.

$\log_2 3 \approx 1.6$	$\log_3 15 \approx 2.5$	$\log_4 9 \approx 1.6$
$\log_2 4 = 2$	$\log_3 \frac{1}{3} = -1$	$\log_4 2 = 0.5$
$\log_2 12 \approx 3.6$	$\log_3 5 \approx 1.5$	$\log_4 18 \approx 2.1$

5.4 Warm Up

Determine the value of *n* in the expression.

1. $\dfrac{3^n}{3^2} = 3^5$

2. $\dfrac{x^4}{x^n} = x^{-2}$

3. $x^2x^n = x^{5/2}$

4. $\left(2^8\right)^n = 4$

5. $\dfrac{\left(y^n\right)^2}{y^3} = y$

6. $y^{2n}y = y^{-5}$

5.4 Cumulative Review Warm Up

Solve the equation. Write your answer in simplest form.

1. $2(x - 3)^2 = 8$

2. $6x^2 - 7x - 20 = 0$

3. $x^2 - 4x - 8 = 0$

4. $6x - 3x^2 = 15$

5. $x^2 = 6x - 9$

6. $-\frac{1}{2}(3x + 1)^2 = -5$

Name ____________________ Date ________

5.4 Practice A

In Exercises 1–3, use $\log_5 3 \approx 0.683$ and $\log_5 6 \approx 1.113$ to evaluate the logarithm.

1. $\log_5 2$
2. $\log_5 18$
3. $\log_5 9$

In Exercises 4–6, expand the logarithmic expression.

4. $\log_2 5x$
5. $\log 7x^4$
6. $\log_6 \dfrac{2x}{y}$

7. Describe and correct the error in expanding the logarithmic expression.

✗ $\log_4 3x = 3 \log_4 x$

In Exercises 8–11, condense the logarithmic expression.

8. $\log_7 3 - \log_7 5$
9. $\log 10 - \log 5$
10. $3 \ln x + 9 \ln y$
11. $\log_2 9 + \frac{1}{2} \log_2 y$

In Exercises 12–14, use the change-of-base formula to evaluate the logarithm.

12. $\log_5 3$
13. $\log_2 11$
14. $\log_6 10$

15. Your friend claims that you can use the change-of-base formula to write the expression $\ln x$ as a common logarithm. Is your friend correct? Explain your reasoning.

16. For a sound with intensity I (in watts per square meter), the loudness $L(I)$ of the sound (in decibels) is given by the function $L(I) = 10 \log \dfrac{I}{I_0}$, where I_0 is the intensity of a barely audible sound (about 10^{-12} watts per square meter). The sound of a coach's whistle is five times greater than the intensity of the referee's whistle. Find the difference in the decibel levels of the sounds made by the coach and the referee.

Name________________________________ Date__________

5.4 Practice B

In Exercises 1–3, use $\log_5 3 \approx 0.683$ and $\log_5 6 \approx 1.113$ to evaluate the logarithm.

1. $\log_5 81$ **2.** $\log_5 \frac{1}{6}$ **3.** $\log_5 \frac{1}{2}$

In Exercises 4–6, expand the logarithmic expression.

4. $\log_3 12x^7$ **5.** $\log_6 \frac{5x^2}{y^3}$ **6.** $\log_8 6\sqrt{xy}$

7. Describe and correct the error in expanding the logarithmic expression.

> ✗ $\ln \sqrt[3]{xy} = \frac{1}{3} \ln x + \ln y$

In Exercises 8–11, condense the logarithmic expression.

8. $5 \log_9 x - \log_9 4$ **9.** $\log_8 5 + \frac{1}{4} \log_8 x$

10. $2 \ln 4 + 5 \ln x + 3 \ln y$ **11.** $\log_6 9 + 2 \log_6 \frac{1}{3} - 3 \log_6 x$

In Exercises 12–14, use the change-of-base formula to evaluate the logarithm.

12. $\log_8 15$ **13.** $\log_3 30$ **14.** $\log_4 \frac{8}{17}$

15. Your friend claims you can use the change-of-base formula to write the expression $\frac{\ln y}{\ln 3}$ as a logarithm with base 3. Is your friend correct? Explain your reasoning.

16. For a sound with intensity I (in watts per square meter), the loudness $L(I)$ of the sound (in decibels) is given by the function $L(I) = 10 \log \frac{I}{I_0}$, where I_0 is the intensity of a barely audible sound $\left(\text{about } 10^{-12} \text{ watts per square meter}\right)$. The bass guitar player in a band turns up the volume of the speaker so that the intensity of the sound triples. By how many decibels does the loudness increase?

Name ______________________________ Date ________

5.4 Enrichment and Extension

Properties of Logarithms

Recall the Exponential-Logarithmic Inverse Properties:

For $b > 0$ and $b \neq 1$:

$\log_b b^x = x$ and $b^{\log_b x} = x$ for $x > 0$.

Example: Use the properties of logarithms to evaluate $\log_2 32 - 5^{\log_5 3}$.

Solution: $\log_2 32 - 5^{\log_5 3} = 5 - 5^{\log_5 3} = 5 - 3 = 2$

In Exercises 1–16, use the properties of logarithms and the table approximations to evaluate the expression.

$\log_3 10 \approx 2.0959$	$\log_8 3 \approx 0.5283$	$\log_5 10 \approx 1.4307$
$\log_3 2 \approx 0.6309$	$\log_5 2 \approx 0.4307$	$\log_5 3 \approx 0.6826$

1. $4^{\log_4 8} + \log_2 2$
2. $5^{\log_5 5} + \log_5 45 - \log_5 9$
3. $\log_2 6 + \log_2 4 - \log_2 3$
4. $\log_3 30 - \log_4 4^5$
5. $2^{\log_2 3} - \log_8 64^7 + \log_8 24$
6. $\log_6 \frac{1}{36} + \log_6 36 + 5^{\log_6 1}$
7. $\log_5 2 + \log_5 20 - \log_5 4$
8. $9^{\log_3 20}$
9. $16^{\log_4 10} - 2^{\log_2 50}$
10. $\log_2 \frac{1}{8} + \log_2 4 + \log_6 \frac{1}{216}$
11. $\log_5 20 + \log_5 40 - \log_5 30$
12. $\log_5 50 + 2^{\log_5 1/5}$
13. $\log_3 162 + 27^{\log_3 4}$
14. $\log_4 6 + \log_4 3 - \log_4 9$
15. $25^{\log_5 10} - \log_4 64 + 1000^{\log_{10} 4}$
16. $81^{\log_3 10} - \log_{125} 5 - 81^{\log_9 4}$

Name______________________________ Date__________

5.4 Puzzle Time

What Type Of Lizard Loves To Tell Jokes?

Write the letter of each answer in the box containing the exercise number.

Match the expression with the logarithm that has the same value.

1. $\log_2 6 + \log_2 8$
2. $\log_3 10 - \log_3 5$
3. $4 \log_{1/2} 2$
4. $\log_2 2 - \log_2 3$
5. $\log_3 4 + \log_3 2$
6. $6 \log 8$

Condense the logarithmic expression.

7. $4 \log 2 - \log 5$
8. $\log 6 + \log 2 - \log 7$
9. $\frac{1}{2} \log 4 + \log 3$
10. $3 \log 2 - 2 \log 2$
11. $1 - \log 4$
12. $\ln 3 + 4 \ln x \quad \ln y$

Answers

I. $\log_{1/2} 16$

A. $\log_2 48$

M. $\log \frac{16}{5}$

Y. $\log\left(8^6\right)$

E. $\log \frac{5}{2}$

S. $\log_3 2$

N. $\log 6$

L. $\log_2 \frac{2}{3}$

L. $\log_3 8$

A. $\log \frac{12}{7}$

D. $\log 2$

R. $\ln \frac{3x^4}{y}$

1		2	3	4	5	6	7	8	9	10	11	12

5.5 Start Thinking

An investment account is created to provide funds for a student's college tuition. The balance B (in dollars) of the education fund after t years can be modeled by $B = 5000e^{0.0525t}$. What equation would you have to solve to determine how long it will take for the initial amount to double? Use a graphing calculator to solve the equation.

5.5 Warm Up

Rewrite the exponential equations in logarithmic form and the logarithmic equations in exponential form. Then solve for *x* and check your answer.

1. $2^{x-1} = 5$

2. $e^{-2x} = 11$

3. $3^{3-2x} = 4$

4. $\log_2(x + 5) = -1$

5. $\log(3 - x) = 2$

6. $\ln(2x) = \frac{1}{3}$

5.5 Cumulative Review Warm Up

Describe the end behavior of the graph of the polynomial function.

1. $f(x) = -3x^4 + 2x - 8$

2. $g(x) = 5x - x^2 - 9x^3$

3. $h(x) = x^5 - 5x^4 + 3x^2 - 15$

4. $p(x) = 1.8 - 2.5x^7 + 5.7x^{10}$

Name______________________________ Date__________

5.5 Practice A

In Exercises 1–6, solve the equation.

1. $6^{x-7} = 6^{2x+3}$

2. $e^{5x} = e^{3x+4}$

3. $3^{x+1} = 9^{x-3}$

4. $2^x = 5$

5. $8^x = 35$

6. $16^{3x-2} = \left(\frac{1}{4}\right)^{5-x}$

7. The length ℓ (in centimeters) of a scalloped hammerhead shark can be modeled by the function $\ell = 266 - 219e^{-0.05t}$, where t is the age (in years) of the shark.

a. How old is a shark that is 200 centimeters long?

b. How long is a shark that is twice as old as the shark in part (a)?

In Exercises 8–13, solve the equation.

8. $\ln(3x - 8) = \ln(x + 6)$

9. $\log_3(9x - 2) = \log_3(4x + 3)$

10. $\log(4x + 1) = \log 25$

11. $\log_6(5x + 4) = 2$

12. $\log(10x - 7) = 3$

13. $\log_3(4x + 2) = \log_3 6x$

In Exercises 14–17, solve the equation. Check for extraneous solutions.

14. $\log_2 x + \log_2(x - 3) = 2$

15. $\log_3 3x + \log_3(2x + 1) = 2$

16. $\ln x + \ln(x + 4) = 3$

17. $\log_6 2x^2 + \log_6 3 = 2$

18. You deposit \$400 in an account that pays 5% annual interest. How long will it take for the balance to double for each frequency of compounding?

a. annually

b. quarterly

c. daily

d. continuously

In Exercises 19–21, solve the inequality.

19. $7^x < 42$

20. $3^x \geq 24$

21. $\log_3 x > 2$

In Exercises 22 and 23, use a graphing calculator to solve the equation.

22. $\ln 3x = 4^{-x+5}$

23. $\log x = 9^{-2x}$

Name ______________________ Date ________

5.5 Practice B

In Exercises 1–6, solve the equation.

1. $9^{3x-5} = 81^{3x+2}$

2. $7^x = 32$

3. $9^{3x+6} = \left(\frac{1}{3}\right)^{8-x}$

4. $6^{4x} = 13$

5. $2e^{3x} + 6 = 10$

6. $4e^{2x} - 7 = 1$

7. Fifty grams of radium are stored in a container. The amount R (in grams) of radium present after t years can be modeled by $R = 50e^{-0.00043t}$.

a. After how many years will only 20 grams of radium be present?

b. Seventy-five grams of radium are stored in a different container. The amount R (in grams) of radium present after t years can be modeled by $R = 75e^{-0.00043t}$. Will it take *more years* or *fewer years* for only 20 grams of the radium in this container to be present, compared to the answer in part (a)? Explain.

In Exercises 8–13, solve the equation.

8. $\ln(5x - 2) = \ln(x + 6)$

9. $\log(3x + 5) = \log 6$

10. $\log_2(3x + 12) = 4$

11. $\log_3(3x + 7) = \log_3(10x)$

12. $\log_2(x^2 - 2x + 1) = 4$

13. $\log_3(x^2 + x + 7) = 3$

In Exercises 14–17, solve the equation. Check for extraneous solutions.

14. $\ln x + \ln(x - 2) = 5$

15. $\log_5 2x^2 + \log_5 8 = 2$

16. $\log_3(-x) + \log_3(x + 8) = 2$

17. $\log_2(x + 2) + \log_2(x + 5) = 4$

In Exercises 18–20, solve the inequality.

18. $e^{x-2} < 8$

19. $\ln x > 5$

20. $-2 \log_3 x + 2 \le 10$

21. You deposit \$2000 in Account A, which pays 2.25% annual interest compounded monthly. You deposit another \$2000 in Account B, which pays 3% annual interest compounded monthly. When is the sum of the balance in both accounts at least \$5000?

Name____ Date __________

5.5 Enrichment and Extension

Solving Exponential and Logarithmic Equations

In this section, you learned how to solve exponential equations by taking the logarithm of each side and using the inverse properties of exponential functions and logarithmic functions. In a similar manner, you learned to solve logarithmic equations by exponentiating each side and using inverse properties.

Another way to solve exponential equations is to rewrite the equation in logarithmic form. Similarly, to solve logarithmic equations, you can rewrite the equation in exponential form. These techniques are illustrated in the following example.

Example: Solve each equation.

a. $9^x = 3$

b. $\log_2(5x - 17) = 3$

Solution:

a.

$9^x = 3$	Write original equation.
$\log_9 3 = x$	Rewrite in logarithmic form.
$\frac{1}{2} = x$	Evaluate logarithm.

b.

$\log_2(5x - 17) = 3$	Write original equation.
$2^3 = 5x - 17$	Rewrite in exponential form.
$8 = 5x - 17$	Evaluate power.
$25 = 5x$	Add 17 to each side.
$5 = x$	Divide each side by 5.

In Exercises 1–8, solve the equation. Round decimal answers to the nearest thousandth.

1. $\log_3 x = \frac{1}{3}$

2. $e^{3x} = 12$

3. $\ln x = 2$

4. $6^x = \frac{1}{216}$

5. $\log_5 4x = 2$

6. $\left(\frac{1}{2}\right)^x - 2 = 30$

7. $\log 5x + \log(x - 1) = 2$

8. $3^x = 20$

Name ______________________________ Date ________

5.5 Puzzle Time

Why Do Elephants Have Big Trunks?

A	B	C	D	E	F
G	H	I	J	K	L

Complete each exercise. Find the answer in the answer column. Write the word under the answer in the box containing the exercise letter.

−2 THE
−6 SHOPPING
$-\frac{2}{3}$ HAVE
−1 GO
2 TO
−100 GROCERIES

Solve the equation.

A. $5^{3-2x} = 5^{-x}$

B. $3^{2x} = 3^{-x}$

C. $3^{-2x+1} \bullet 3^{-2x-3} = 3^{-x}$

D. $10^{1-x} = 10^4$

E. $3^{2x-1} = 27$

F. $4^{x+1} = \frac{1}{64}$

G. $\log_3(x + 3) = 0$

H. $\log(-x) + 2 = 4$

I. $\log_3(x - 3) = 4$

J. $2 \log_7(-2x) = 0$

K. $\log(-2x + 9) = \log(7 - 4x)$

L. $\log_{12}(x^2 + 35) = \log_{12}(-12x - 1)$

0 THEY
$-\frac{1}{2}$ THEY
−4 PUT
−3 SOMEWHERE
84 WHEN
3 SO

5.6 Start Thinking

Determine if the scatter plot is best modeled by a *linear*, a *quadratic*, or an *exponential* model. Then discuss a real-life situation that could be modeled by each plot.

1. 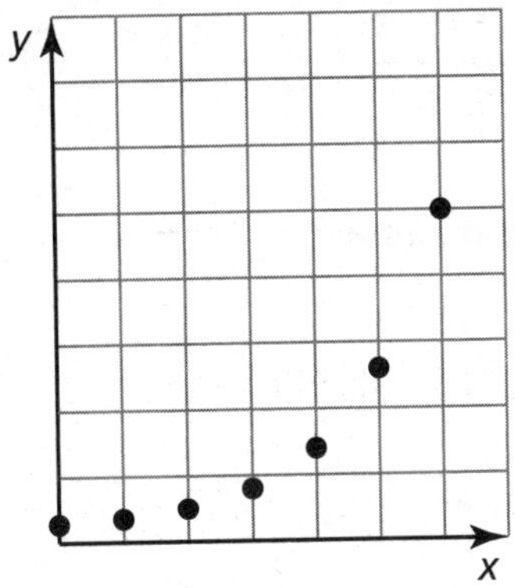

2. 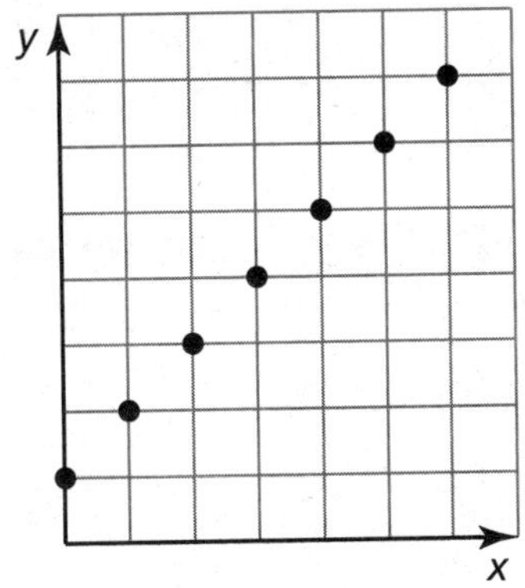

3. 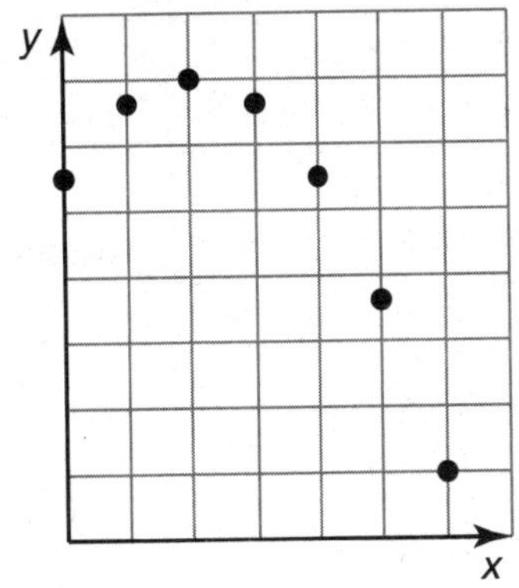

5.6 Warm Up

Use the *linear regression* feature on a graphing calculator to determine the line of best fit for the data set.

1.

x	0	2	4	6	8	10
y	−4	1	5	6	11	15

2.

x	2009	2010	2011	2012	2013	2014
y	122.8	124.2	123.7	120.5	118.3	115.6

5.6 Cumulative Review Warm Up

Solve the radical equation.

1. $\sqrt{2x - 1} = 4$

2. $4 + \sqrt{2x} = x$

3. $\sqrt{4 - 3x} - 2\sqrt{x} = 0$

4. $\sqrt{x + 3} - \sqrt{x - 1} = 1$

Name ____ Date __________

5.6 Practice A

In Exercises 1 and 2, determine the type of function represented by the table. Explain your reasoning.

1.

x	1	3	5	7	9
y	81	27	9	3	1

2.

x	1	2	3	4	5
y	3	1	1	3	7

In Exercises 3–8, write an exponential function $y = ab^x$ whose graph passes through the given points.

3. (1, 6), (2, 12)

4. (1, 20), (2, 80)

5. (2, 18), (3, 54)

6. (3, 250), (4, 1250)

7. (3, 56), (5, 224)

8. (2, 45), (4, 405)

9. Describe and correct the error in determining the type of function represented by the data.

✗

x	0	1	3	5	7
y	2	4	8	16	32

×2 ×2 ×2 ×2

The outputs have a common ratio of 2, so the data represent an exponential function.

In Exercises 10 and 11, determine whether the data show an exponential relationship. Then write a function that models the data.

10.

x	−4	−2	0	2	4
y	6	24	96	384	1536

11.

x	0	1	2	3	4
y	0	5	10	15	20

12. Use a graphing calculator to find an exponential model for the data in the table.

x	1	3	7	10	12
y	6	8.64	17.916	30.959	44.581

Name____ Date__________

5.6 Practice B

In Exercises 1 and 2, determine the type of function represented by the table. Explain your reasoning.

1.

x	0	2	4	6	8
y	$\frac{1}{8}$	$\frac{1}{2}$	2	8	32

2.

x	0	1	2	3
y	8	12	18	27

In Exercises 3–8, write an exponential function $y = ab^x$ whose graph passes through the given points.

3. (1, 10), (2, 20)

4. (1, 18), (3, 162)

5. (2, 36), (3, 72)

6. (3, 375), (4, 1875)

7. (2, 3.6), (5, 777.6)

8. (2, 8), (5, 512)

9. Describe and correct the error in determining the type of function represented by the data.

x	0	2	4	6	8
y	−2	−4	−8	−16	−32

×2 ×2 ×2 ×2

The outputs have a common ratio of 2, but the outputs are negative, so the data does not represent a recognizable function.

In Exercises 10 and 11, determine whether the data show an exponential relationship. Then write a function that models the data.

10.

x	1	3	5	7	9
y	64	32	16	8	4

11.

x	0	10	20	30	40
y	0	15	30	45	60

12. Use a graphing calculator to find an exponential model for the data in the table.

x	2	5	6	8	9
y	7.65	25.819	38.728	87.138	130.71

Name ______________________________ Date __________

5.6 Enrichment and Extension

Using Semi-Logarithmic Graph Paper

You can use semi-logarithmic graph paper to graph exponential data in a straight line without transforming the data pairs. Semi-logarithmic graph paper has one arithmetic axis and one logarithmic axis. Notice that the tick marks on the y-axis are not evenly spaced. The horizontal lines start out spread apart and get closer and closer until the pattern repeats itself.

The semi-logarithmic graph paper shown below has a vertical logarithmic axis with three cycles of base 10. Notice that the vertical axis begins at 1 because the logarithm of zero is undefined.

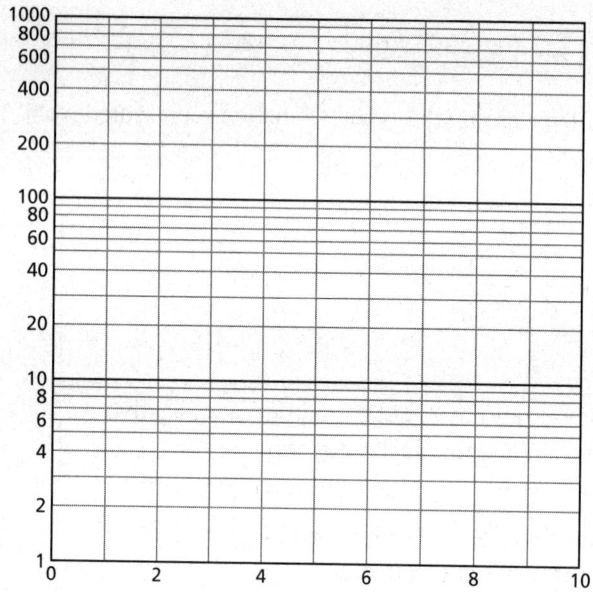

In Exercises 1 and 2, graph the exponential data using semi-logarithmic graph paper.

1. The table shows the cost y (in dollars) of disk storage space (in megabytes) during a 20-year period.

Year, *x*	2	7	12	17	22
Cost, *y*	1000	50.33	2.53	0.13	0.01

2. The table shows the saturation of water in air y (in millimeters of water per cubic meter of air) at different air temperatures x (in degrees Celsius).

Temperature, *x*	0	5	10	20	25	30
Saturation, *y*	4.8	6.8	9.4	17.3	23.1	30.4

Name____ Date__________

5.6 Puzzle Time

What Do You Call A Lamb Who Does Aerobics?

Write the letter of each answer in the box containing the exercise number.

Write an exponential function $y = ab^x$ whose graph passes through the given points.

1. $(0, 4), (2, 16)$
2. $(-1, 2), (2, 54)$
3. $(-2, 8), (-3, 16)$
4. $(1, 6), (3, 24)$
5. $(-2, 16), (-3, 32)$
6. $(0, 1), (1, 0.75)$
7. $(-2, 9), (-3, 27)$
8. $(0, 2), (1, 5)$

9.

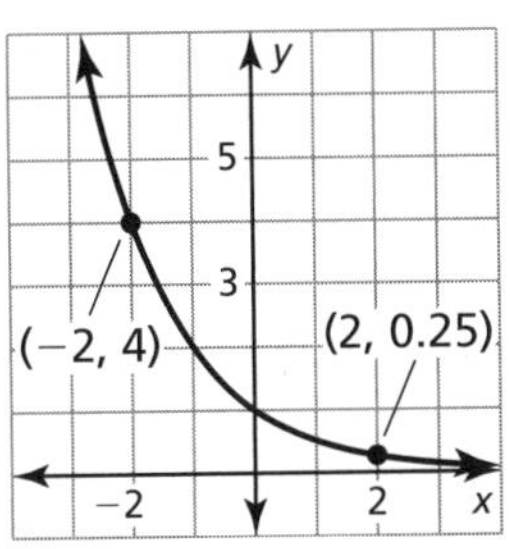

10.

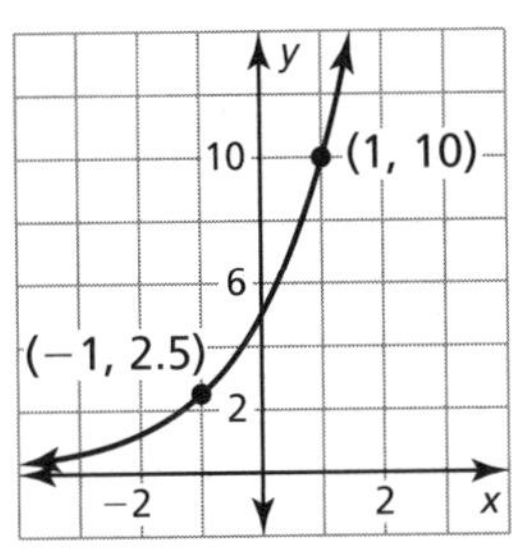

Answers

E. $y = 3(2^x)$

S. $y = 4(2^x)$

P. $y = \left(\frac{1}{2}\right)^x$

E. $y = 2\left(\frac{1}{2}\right)^x$

A. $y = 2\left(\frac{5}{2}\right)^x$

P. $y = 4\left(\frac{1}{2}\right)^x$

H. $y = 6(3^x)$

S. $y = \left(\frac{3}{4}\right)^x$

H. $y = \left(\frac{1}{3}\right)^x$

E. $y = 5(2^x)$

1	2	3	4	5		6	7	8	9	10

Name ______________________ Date ________

Chapter 5 Cumulative Review

Evaluate the expression.

1. $5 \bullet 7^2$

2. $3 \bullet 4^2$

3. $6 \bullet 3^3$

4. $-4 \bullet 2^3$

5. $-7 \bullet 2^5$

6. $5 \bullet -4^3$

7. $(-2)^3$

8. $(-3)^2$

9. $(-2)^4$

10. $\left(\frac{2}{3}\right)^3$

11. $\left(\frac{4}{5}\right)^2$

12. $\left(\frac{1}{3}\right)^4$

13. $\left(-\frac{2}{5}\right)^3$

14. $\left(-\frac{1}{4}\right)^4$

15. $\left(-\frac{3}{4}\right)^3$

16. $-\left(\frac{1}{2}\right)^5$

17. $-\left(\frac{2}{3}\right)^4$

18. $-\left(\frac{3}{5}\right)^3$

Write the number in scientific notation.

19. 0.0005

20. 382,000,000,000

21. −0.0132

22. 0.0000057

23. 2,300,000

24. −0.00000000092

25. −0.000359

26. 51,300

27. 0.000041

28. 4500

29. 12,000,000

30. 731,000

31. −0.000000817

32. 417,000,000,000

33. −62,000

34. 8,300,000,000,000

35. 0.000000089

36. 0.002

37. Bowling Alley A charges $5 admission and $3.50 per game. Bowling Alley B charges $25 admission, but offers unlimited games with no additional charge.

a. How much does it cost to play four games at each bowling alley?

b. How much does it cost to play five games at each bowling alley?

c. When is it financially more beneficial to bowl at Bowling Alley B?

38. Admission to the local fair is $15. Ride tickets are $1.25 each.

a. How much does it cost for admission and six ride tickets?

b. How much does it cost for admission and eight ride tickets?

c. You and three friends want to go to the fair and need eight ride tickets each. How much does that cost?

Name________________________ Date__________

Chapter 5 Cumulative Review (continued)

Write the number in standard form.

39. 1.85×10^{-3} **40.** -8.9×10^{4} **41.** 3.84×10^{-8}

42. 2.57×10^{-7} **43.** 4.6×10^{5} **44.** -3.34×10^{-3}

45. 3.05×10^{-8} **46.** 4.49×10^{4} **47.** 5.9×10^{5}

48. -2.25×10^{9} **49.** 3.08×10^{-1} **50.** -1.06×10^{-8}

51. 5.51×10^{7} **52.** 6.18×10^{8} **53.** 1.84×10^{-7}

54. 8.9×10^{5} **55.** 9.2×10^{-8} **56.** -7.32×10^{-5}

Write an equation of the line that passes through the given point and has the given slope.

57. $(-5, -10);\ m = -\frac{2}{5}$ **58.** $(11, 1);\ m = \frac{5}{11}$

59. $(5, -6);\ m = \frac{3}{5}$ **60.** $(12, -1);\ m = \frac{1}{4}$

61. $(-8, 10);\ m = -\frac{7}{8}$ **62.** $(0, -12);\ m = \frac{3}{5}$

63. $(7, -7);\ m = -2$ **64.** $(8, 4);\ m = 3$

65. $(-2, -9);\ m = -\frac{1}{2}$ **66.** $(-3, 3);\ m = \frac{5}{6}$

67. $(7, -8);\ m = -3$ **68.** $(-6, 0);\ m = 4$

Find the zero(s) of the function.

69. $f(x) = x^2 - 7x$ **70.** $h(x) = x^2 + 3x$

71. $g(x) = x^2 + 4x$ **72.** $h(x) = x^2 - 8x$

73. $g(x) = x^2 - 3x + 2$ **74.** $h(x) = x^2 + 6x - 27$

75. $f(x) = x^2 + 11x + 30$ **76.** $h(x) = x^2 - 6x - 16$

77. $g(x) = 2x^2 - 3x - 35$ **78.** $f(x) = 3x^2 + 14x - 5$

79. $h(x) = 2x^2 + x - 28$ **80.** $g(x) = 3x^2 - 20x - 7$

81. $f(x) = 5x^2 - 14x - 3$ **82.** $g(x) = 3x^2 - 34x + 63$

Name _______________________________ Date _________

Chapter 5 Cumulative Review (continued)

Add or subtract. Write the answer in standard form.

83. $(8 + i) + (-4 + 3i)$

84. $(-1 - i) + (5 - 2i)$

85. $(5 - 8i) + (6 - 4i)$

86. $(-3 + 2i) + (-9 - 3i)$

87. $(4 - i) + (6 + 7i)$

88. $(11 - 8i) + (-3 - 12i)$

89. $(4 + 7i) + (-3 - 10i)$

90. $(2 + 12i) - (4 - 2i)$

91. $(-9 + 2i) - (5 + 5i)$

92. $(12 - 2i) - (-9 + 3i)$

93. $(8 + 10i) - (3 + 2i)$

94. $(-11 - 2i) - (-1 - 8i)$

95. $(10 + 11i) - (-8 - 10i)$

96. $(-7 - 9i) - (10 - 4i)$

Multiply. Write the answer in standard form.

97. $3i(4 + 2i)$

98. $5i(-1 + 2i)$

99. $-7i(3 - 4i)$

100. $-i(7 - 6i)$

101. $(12 + 3i)(-5 + 7i)$

102. $(6 - 3i)(-10 + 6i)$

103. $(-4 + 10i)(4 + 2i)$

104. $(-9 - 2i)(6 + 2i)$

105. $(-10 - 9i)(-4 - 9i)$

106. $(-1 + 8i)(7 + 3i)$

107. $(11 + 6i)(-5 + 4i)$

108. $(3 - 3i)(4 + 5i)$

Find the zero(s) of the function.

109. $f(x) = 2x^2 + 6x + 5$

110. $f(x) = 3x^2 - x + 3$

111. $f(x) = 4x^2 + 3x + 6$

112. $f(x) = 6x^2 - 5x + 8$

113. $f(x) = -x^2 + 4x - 7$

114. $f(x) = 5x^2 - 6x + 3$

115. $f(x) = 4x^2 - 9x + 6$

116. $f(x) = -2x^2 + 5x - 4$

117. A rectangular swimming pool has a length of $4x - 10$ feet and a width of $x + 10$ feet.

a. Write an equation for the perimeter of the swimming pool.

b. Write an equation for the area of the swimming pool.

c. You know that the total perimeter is 120 feet. What is the value of x?

d. What is the area of the swimming pool?

Name____ Date__________

Chapter 5 Cumulative Review (continued)

Solve the equation.

118. $x^2 + 5x - 4 = 0$

119. $2x^2 + 4x + 1 = 0$

120. $-x^2 + 8x + 5 = 0$

121. $x^2 - 3x + 2 = 0$

122. $2x^2 - 5x + 3 = 0$

123. $x^2 - 5x + 2 = 0$

124. $-3x^2 + 6x + 1 = 0$

125. $-5x^2 - 3x + 4 = 0$

126. $6x^2 - 5x + 1 = 0$

127. $-2x^2 + 7x + 3 = 0$

128. $x^2 + 7x + 3 = 0$

129. $-4x^2 + x + 8 = 0$

130. $3x^2 - 5x + 1 = 0$

131. $-2x^2 + 9x + 6 = 0$

132. A rectangular parking pad has a length of $2x$ feet and a width of $x + 4$ feet.

a. Write an equation for the perimeter of the parking pad.

b. Write an equation for the area of the parking pad.

c. You know that the total perimeter is 56 feet. What is the value of x?

d. What is the area of the parking pad?

133. You want to tile a bathroom. You know that the width of the room is 13.5 feet and the length is $3x$ feet.

a. Write an equation for the area of the bathroom.

b. You know the perimeter of the room is 63 feet. What is the value of x?

c. What is the area of the bathroom?

d. The tile you want to purchase is \$2.49 per square foot. How much will it cost for the tile in the bathroom?

134. You want to put hardwood floor down in your living room. You know that the width of the room is 16 feet and the length is $2x$ feet.

a. Write an equation for the area of the living room.

b. You know the perimeter of the room is 76 feet. What is the value of x?

c. What is the area of the room?

d. The hardwood you want to purchase is \$4.59 per square foot. How much will it cost for the hardwood in the living room?

Chapter 6

Name____ Date__________

Chapter 6 Rational Functions

Dear Family,

Do you have a cell phone? If so, are you getting the most from your plan? Are you paying too much for your plan?

When choosing a service contract, you should compare different providers and their coverage areas and the different prices for talk, text, and data plans. You want to make sure your plan has what you need, so you don't end up paying fees for going over what your plan allows. The bigger cell phone companies offer a subsidized plan with a two-year contract in which you pay a smaller amount for a phone and then a monthly charge for service. Another option is an unlocked phone in which you buy the phone outright with a monthly fee for services and no fixed contract.

Use the Internet to compare two different plans for a two-year subsidized contract for the first 24 months of service. Then compare two cell phone plans for an unlocked phone for the first 24 months of service. Remember to include the initial cost of the phone in the total cost for the first month. Then find the average monthly cost for each plan. For a fair comparison of plans, choose the same cell phone for each plan. It may be helpful to use a spreadsheet to organize and calculate your data.

	Company A		Company B	
Month	Total cost (in dollars)	Cost per month (in dollars)	Total cost (in dollars)	Cost per month (in dollars)
1				
2				
⋮	⋮	⋮	⋮	⋮
24				

In this chapter, you will solve rational equations for situations in which you want to find the average monthly cost of service. Before you sign a contract with a cell phone provider, researching different plans may end up saving you time and money in the long run.

Have fun talking!

Nombre ______________________________ Fecha ________

Capítulo 6 Funciones racionales

Estimada familia:

¿Tienen teléfono celular? Si lo tienen, ¿aprovechan su plan al máximo? ¿Pagan demasiado por su plan?

Cuando elijan un contrato de servicio, deberían comparar diferentes proveedores y sus áreas de cobertura y los diferentes precios de los planes de datos para hablar y mandar mensajes de texto. Les conviene asegurarse de que su plan tenga que lo necesitan, así no terminan pagando tarifas por pasarse del límite permitido por su plan. Las compañías de telefonía celular más grandes ofrecen un plan subvencionado con un contrato de dos años donde pagan una cantidad menor por un teléfono y luego, un cargo mensual por el servicio. Otra opción es un teléfono desbloqueado donde compran el teléfono en el acto con una tarifa mensual por los servicios sin contrato fijo.

Usen Internet para comparar dos planes diferentes con un contrato de dos años subvencionado para los primeros 24 meses del servicio. Luego, comparen dos planes de telefonía celular para un teléfono desbloqueado para los primeros 24 meses de servicio. Recuerden incluir el costo inicial del teléfono en el costo total del primer mes. Luego, hallen el costo mensual promedio de cada plan. Para hacer una comparación justa de los planes, elijan el mismo teléfono celular para cada plan. Quizás sea útil usar una hoja de cálculo para organizar y calcular sus datos.

	Compañía A		Compañía B	
Mes	Costo total (en dólares)	Costo por mes (en dólares)	Costo total (en dólares)	Costo por mes (en dólares)
1				
2				
⋮	⋮	⋮	⋮	⋮
24				

En este capítulo, resolverán ecuaciones racionales en situaciones donde quieran hallar el costo mensual promedio de un servicio. Antes de firmar un contrato con un proveedor de telefonía celular, investigar diferentes planes puede terminar ahorrándoles tiempo y dinero a largo plazo.

¡Diviértanse hablando!

6.1 Start Thinking

Determine a relationship between x and y in the table. Then create a scatter plot for each set of points. What conclusions can you make about the graphs and the relationship you discovered between x and y?

1.

x	y
1	2
3	6
4	8
6	12

2.

x	y
1	2
2	1
4	$\frac{1}{2}$
6	$\frac{1}{3}$

3.

x	y
1	3
3	1
5	$\frac{3}{5}$
7	$\frac{3}{7}$

6.1 Warm Up

Use the given values of x and y to find the value of k in the equation.

1. $y = kx$
$x = 4, y = -3$

2. $y = kx$
$x = \frac{1}{2}, y = 5$

3. $y = kx$
$x = -3.5, y = -2.8$

4. $y = \frac{k}{x}$
$x = -2, y = 13$

5. $y = \frac{k}{x}$
$x = 5.2, y = -5$

6. $y = \frac{k}{x}$
$x = -\frac{3}{5}, y = \frac{5}{2}$

6.1 Cumulative Review Warm Up

Solve the system.

1. $-4x - 3y = -7$
$6x + 2y = 3$

2. $y = -\frac{2}{3}x + 4$
$-2y = 6$

3. $y = x^2$
$y = 6 - x^2$

Name ______________________________ Date ________

6.1 Practice A

In Exercises 1–6, tell whether *x* and *y* show *direct variation, inverse variation,* or *neither.*

1. $y = \frac{5}{x}$

2. $xy = 7$

3. $6x = y$

4. $\frac{y}{x} = 10$

5. $x + y = 8$

6. $2y = x$

In Exercises 7–10, tell whether *x* and *y* show *direct variation, inverse variation,* or *neither.*

7.

x	2	4	8	10
y	38	19	9.5	7.6

8.

x	3	5	8	10
y	15	9	6	5.5

9.

x	1.5	4	6.5	10
y	9	24	39	60

10.

x	1.5	4	6	12
y	84	31.5	21	10.5

In Exercises 11–13, the variables *x* and *y* vary inversely. Use the given values to write an equation relating *x* and *y*. Then find *y* when $x = 3$.

11. $x = 6, y = -5$

12. $x = 1, y = 7$

13. $x = 3, y = \frac{2}{3}$

14. The variables *x* and *y* vary inversely. Describe and correct the error in writing an equation relating *x* and *y*.

> ✗ $x = 6, y = 5$
>
> $\frac{y}{x} = a$
>
> $\frac{5}{6} = a$
>
> So, $y = \frac{5}{6x}$.

15. The number *y* of songs that can be stored on an MP3 player varies inversely with the average size *x* of a song. A certain MP3 player can store 3000 songs when the average size of a song is 5 megabytes. Find the number of songs that will fit on the MP3 player when the average size of a song is 4 megabytes.

Name____ Date__________

6.1 Practice B

In Exercises 1–6, tell whether *x* and *y* show *direct variation*, *inverse variation*, or *neither*.

1. $y = \dfrac{12}{x}$

2. $xy = 15$

3. $9x = y$

4. $y = x - 3$

5. $\dfrac{y}{x} = 9$

6. $xy = \dfrac{1}{3}$

In Exercises 7–10, tell whether *x* and *y* show *direct variation*, *inverse variation*, or *neither*.

7.

x	2.5	4	7.5	9
y	30	48	90	108

8.

x	12	5	2.5	1.5
y	35	84	168	280

9.

x	2.5	3	6	10
y	8	9.6	1.6	6

10.

x	2.5	10	16	21
y	672	168	105	80

In Exercises 11–13, the variables *x* and *y* vary inversely. Use the given values to write an equation relating *x* and *y*. Then find *y* when $x = 3$.

11. $x = 4, y = -3$

12. $x = \frac{2}{3}, y = -5$

13. $x = -10, y = -\frac{1}{5}$

14. The variables *x* and *y* vary inversely. Describe and correct the error in writing an equation relating *x* and *y*.

X $x = \frac{1}{3}, y = 2$

$xy - a$

$\dfrac{1}{3} \bullet 2 = a$

$a = \dfrac{2}{3}$

So, $y = \dfrac{3x}{2}$.

15. The current *y* in a certain circuit varies inversely with the resistance *x* in the circuit. If the current is 8 amperes when the resistance is 20 ohms, what will the current be when the resistance increases to 25 ohms?

Name ______________________________ Date __________

6.1 Enrichment and Extension

Combined Variation

Combined variation can be used to describe the relationship between three or more variables that vary directly and inversely with one another. You can solve combined variation problems by writing a combination of variation equations and using the given values of the variables to solve for the constant of variation.

Example: The variable y varies directly with x and w, inversely with the square of z, and $y = 100$ when $x = 2$, $w = 4$, and $z = 20$. Write an equation relating w, x, y, and z.

Solution:

$y = \dfrac{axw}{z^2}$ Given the direct variation with y, constant a, and x and w are in the numerator. Given the inverse variation with y, z^2 is in the denominator.

$100 = \dfrac{a(2)(4)}{(20)^2}$ Substitute.

$a = 5000$ Solve for a.

$y = \dfrac{5000xw}{z^2}$ Now you can solve for one of the variables when given the values of the other three variables.

In Exercises 1–4, solve the combined variation problem.

1. The variable x varies directly with y and the square root of z, and $z = 9$ when $x = -18$ and $y = 2$. Find y when $x = 10$ and $z = 4$.

2. The variable w varies inversely with z and the cube root of v, directly with y, and $z = 5$ when $w = 4$, $v = 27$, and $y = 2$. Find w when $y = 3$, $v = 64$, and $z = 6$.

3. The volume V of wood in a tree varies directly with the height h and inversely with the square of the girth g. The volume of a tree is 144 cubic meters when the height is 20 meters and the girth is 1.5 meters. What is the height of a tree with a volume of 100 cubic meters and girth of 2 meters?

4. The pressure P of a gas varies directly with the number of moles n and temperature T of the gas and inversely with volume V. Given the equation of the situation described, as the pressure of a gas increases, what happens to the volume of the gas? Write the number of moles as a function of pressure, volume, and temperature.

Name________________________ Date__________

6.1 Puzzle Time

What Part Of A Computer Keyboard Do Astronauts Like The Best?

Write the letter of each answer in the box containing the exercise number.

Tell whether *x* and *y* show *direct variation*, *inverse variation*, or *neither*.

1. $xy = 3$
2. $x + y = 12$
3. $\frac{y}{x} = 6$

The variables *x* and *y* vary inversely. Use the given values to write an equation relating *x* and *y*.

4. $x = 5, y = -2$
5. $x = 8, y = 2$
6. $x = \frac{1}{2}, y = 6$
7. $x = -12, y = \frac{2}{3}$
8. $x = -\frac{1}{3}, y = -\frac{9}{2}$
9. $x = 2, y = -\frac{3}{4}$
10. $x = -2.4, y = 8.6$
11. $x = -15, y = -\frac{2}{5}$

Answers

P. $y = \frac{16}{x}$

T. inverse variation

S. $y = -\frac{10}{x}$

R. $y = -\frac{6}{x}$

E. direct variation

A. $y = \frac{3}{x}$

A. $y = -\frac{20.64}{x}$

C. $y = -\frac{8}{x}$

E. $y = \frac{3}{2x}$

H. neither

B. $y = -\frac{3}{2x}$

1	2	3		4	5	6	7	8		9	10	11

6.2 Start Thinking

Complete the table for the function $g(x) = \frac{1}{x}$. Review the values in the table to make conclusions about the nature of the graph of g as x-values approach zero and as x-values approach both positive and negative infinity.

1.

x	-1	$-\frac{1}{2}$	$-\frac{1}{10}$	$-\frac{1}{100}$	$\frac{1}{100}$	$\frac{1}{10}$	$\frac{1}{2}$	1
g(x)								

2.

x	-1000	-100	-10	10	100	1000
g(x)						

6.2 Warm Up

Use the graph of *f* to sketch the transformation.

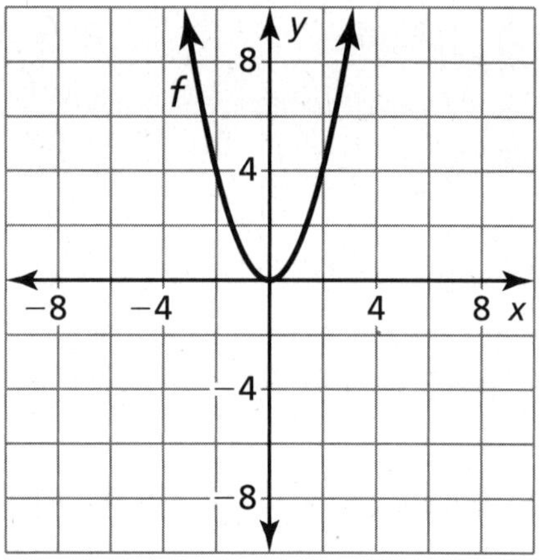

1. $f(x) + 2$

2. $f(x - 2)$

3. $f(-x)$

4. $-f(x)$

5. $2f(x)$

6. $-f(x + 3) - 1$

6.2 Cumulative Review Warm Up

Determine the maximum or minimum value of the function.

1. $y = -2x^2 + 5x$

2. $y = 3x^2 - 7x + 8$

3. $y = 4 - \frac{3}{2}x - \frac{1}{2}x^2$

Name____ Date__________

6.2 Practice A

In Exercises 1–3, graph the function. Compare the graph with the graph of $f(x) = \frac{1}{x}$.

1. $h(x) = \frac{2}{x}$

2. $g(x) = \frac{9}{x}$

3. $h(x) = \frac{-4}{x}$

In Exercises 4–15, graph the function. State the domain and range.

4. $f(x) = \frac{3}{x} + 2$

5. $y = \frac{5}{x} - 1$

6. $g(x) = \frac{4}{x - 3}$

7. $y = \frac{1}{x + 4}$

8. $h(x) = \frac{-1}{x + 3}$

9. $y = \frac{-4}{x - 5}$

10. $f(x) = \frac{x + 3}{x - 2}$

11. $y = \frac{x - 5}{x + 3}$

12. $g(x) = \frac{x + 4}{2x - 6}$

13. $y = \frac{5x + 2}{3x - 9}$

14. $h(x) = \frac{-2x + 3}{3x + 4}$

15. $y = \frac{8x - 1}{5x - 1}$

In Exercises 16–21, rewrite the function in the form $g(x) = \frac{a}{x - h} + k$. Graph the function. Describe the graph of g as a transformation of the graph of $f(x) = \frac{a}{x}$.

16. $g(x) = \frac{4x + 5}{x + 1}$

17. $g(x) = \frac{6x + 5}{x - 2}$

18. $g(x) = \frac{3x - 6}{x - 4}$

19. $g(x) = \frac{5x - 12}{x + 2}$

20. $g(x) = \frac{x + 15}{x - 5}$

21. $g(x) = \frac{x + 3}{x - 9}$

22. Your choir is taking a trip. The trip has an initial cost of \$500, plus \$150 for each student.

    a. Estimate how many students must go on the trip for the average cost per student to fall to \$175.

    b. What happens to the average cost as more students go on the trip?

In Exercises 23–25, use a graphing calculator to graph the function. Then determine whether the function is *even*, *odd*, or *neither*.

23. $f(x) = \frac{5}{x^2 - 1}$

24. $g(x) = \frac{3x^2}{x^2 + 4}$

25. $h(x) = \frac{x^3}{2x^2 + x^4}$

Name ______________________________ Date __________

6.2 Practice B

In Exercises 1–3, graph the function. Compare the graph with the graph of $f(x) = \frac{1}{x}$.

1. $h(x) = \frac{12}{x}$
2. $g(x) = \frac{-8}{x}$
3. $h(x) = \frac{0.2}{x}$

In Exercises 4–15, graph the function. State the domain and range.

4. $f(x) = \frac{5}{x} - 2$
5. $g(x) = \frac{3}{x + 4}$
6. $y = \frac{-8}{x - 3}$
7. $h(x) = \frac{-1}{x + 5}$
8. $y = \frac{-2}{x + 1} + 3$
9. $y = \frac{9}{x - 4} - 2$
10. $f(x) = \frac{x + 5}{x - 4}$
11. $g(x) = \frac{x - 3}{2x + 8}$
12. $h(x) = \frac{-8x + 3}{5x + 2}$
13. $y = \frac{3x - 1}{5x - 1}$
14. $y = \frac{-3x}{-4x - 1}$
15. $y = \frac{-2x + 5}{-x + 8}$

In Exercises 16–21, rewrite the function in the form $g(x) = \frac{a}{x - h} + k$. Graph the function. Describe the graph of g as a transformation of the graph of $f(x) = \frac{a}{x}$.

16. $g(x) = \frac{3x + 7}{x + 2}$
17. $g(x) = \frac{4x - 2}{x - 3}$
18. $g(x) = \frac{4x - 10}{x + 5}$
19. $g(x) = \frac{x + 12}{x - 3}$
20. $g(x) = \frac{5x - 30}{x + 4}$
21. $g(x) = \frac{7x - 2}{x + 6}$

22. You are creating statues made of cement. The mold costs \$300. The material for each statue costs \$22.

    a. Estimate how many statues must be made for the average cost per statue to fall below \$30.

    b. What happens to the average cost as more statues are created?

23. The concentration c of a certain drug in a patient's bloodstream t hours after an injection is given by $c(t) = \frac{t}{4t^2 + 1}$.

    a. Use a graphing calculator to graph the function. Describe a reasonable domain and range.

    b. Determine the time at which the concentration is the highest.

Name____ Date__________

6.2 Enrichment and Extension

Arrow Notation

Arrow notation is a convenient way to describe the end behavior of a function or its possible asymptotic behavior.

Symbol	*Meaning*
$x \to \infty$	x approaches infinity or x increases without bound.
$x \to -\infty$	x approaches negative infinity or x decreases without bound.
$x \to a^+$	x approaches a from the right.
$x \to a^-$	x approaches a from the left.

Use the graph of the rational function to complete each statement.

1. 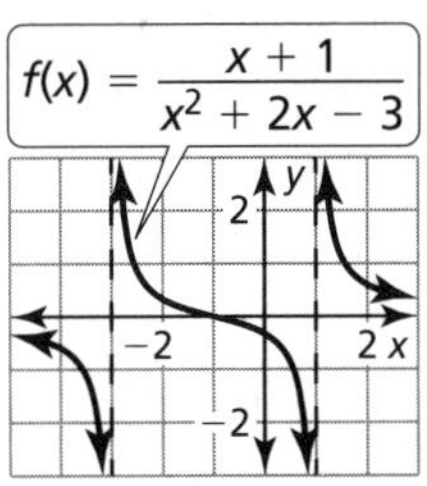

a. As $x \to -3^-, f(x) \to$ ______.

b. As $x \to -3^+, f(x) \to$ ______.

c. As $x \to 1^-, f(x) \to$ ______.

d. As $x \to 1^+, f(x) \to$ ______.

e. As $x \to -\infty, f(x) \to$ ______.

f. As $x \to \infty, f(x) \to$ ______.

2. 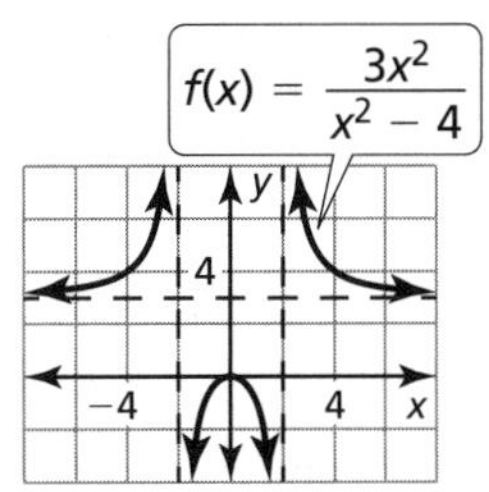

a. As $x \to -2^-, f(x) \to$ ______.

b. As $x \to -2^+, f(x) \to$ ______.

c. As $x \to 2^-, f(x) \to$ ______.

d. As $x \to 2^+, f(x) \to$ ______.

e. As $x \to -\infty, f(x) \to$ ______.

f. As $x \to \infty, f(x) \to$ ______.

3. 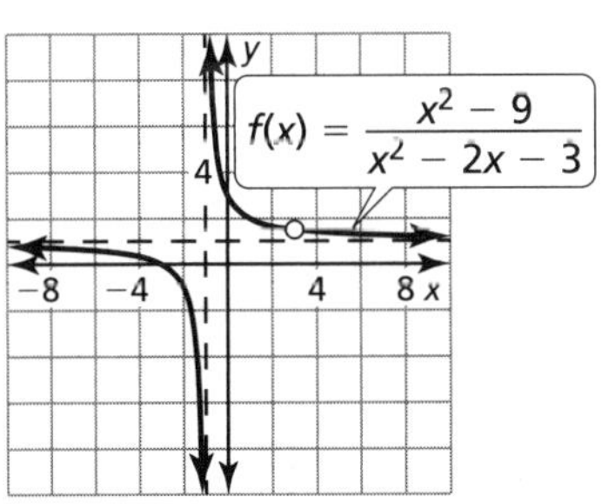

a. As $x \to -1^-, f(x) \to$ ______.

b. As $x \to -1^+, f(x) \to$ ______.

c. As $x \to 3^-, f(x) \to$ ______.

d. As $x \to 3^+, f(x) \to$ ______.

e. As $x \to -\infty, f(x) \to$ ______.

f. As $x \to \infty, f(x) \to$ ______.

4. 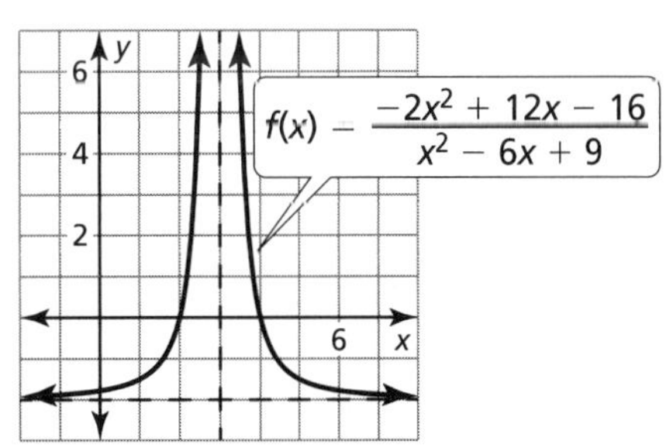

a. As $x \to 3^-, f(x) \to$ ______.

b. As $x \to 3^+, f(x) \to$ ______.

c. As $x \to -\infty, f(x) \to$ ______.

d. As $x \to \infty, f(x) \to$ ______.

Name ______________________ Date ________

6.2 Puzzle Time

What Has A Head And A Tail But No Body?

Write the letter of each answer in the box containing the exercise number.

Match the graph with its function.

1.

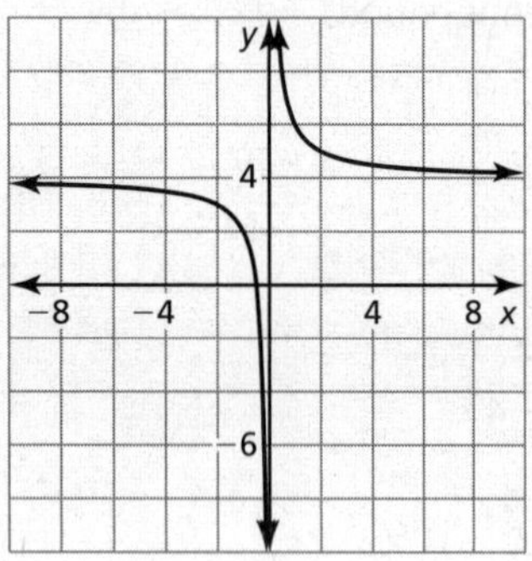

2.

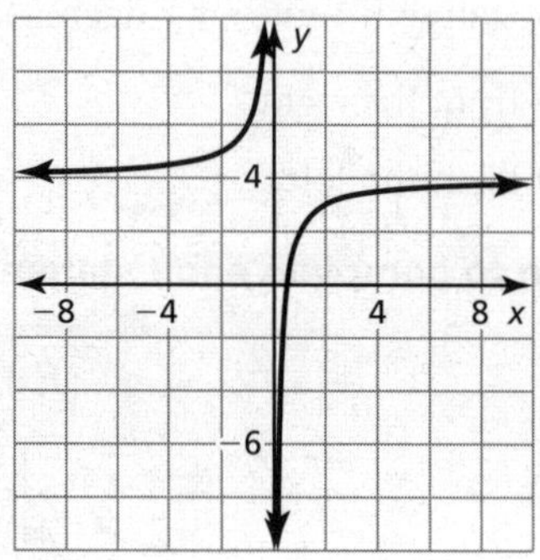

3.

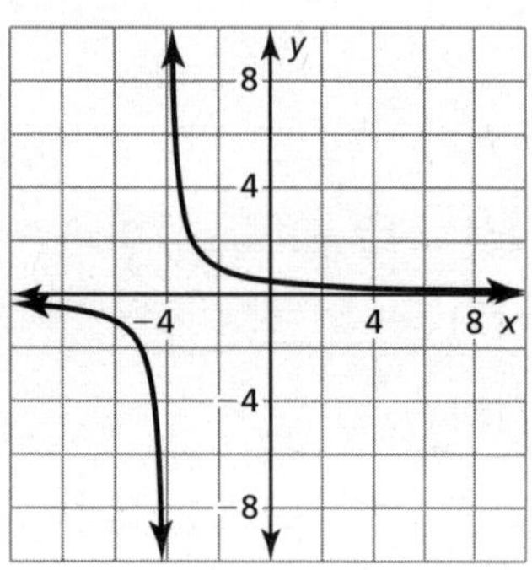

4.

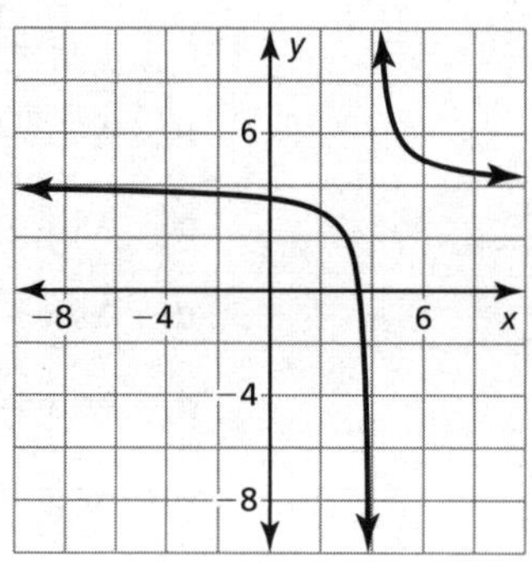

5.

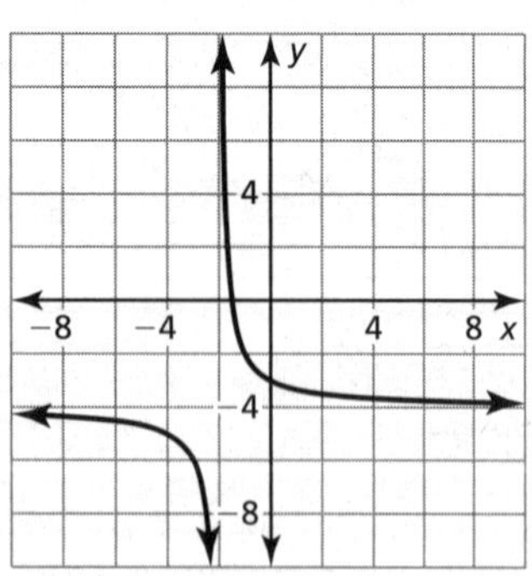

Answers

I. $y = \dfrac{2}{x - 4} + 4$

A. $y = \dfrac{2}{x} + 4$

O. $y = \dfrac{2}{x + 4}$

N. $y = \dfrac{4}{x + 2} - 4$

C. $y = -\dfrac{2}{x} + 4$

1		2	3	4	5

6.3 Start Thinking

In algebraic expressions, values that make a denominator zero are called *excluded values*. Determine the excluded values in the expression.

1. $\dfrac{1}{x + 8}$

2. $\dfrac{1}{x^2 - 5x}$

3. $\dfrac{1}{6x^2 - x - 15}$

6.3 Warm Up

Perform the indicated operation. Simplify your answer and indicate any excluded values of *x*.

1. $\dfrac{2}{7} \bullet \dfrac{5}{3}$

2. $\dfrac{9}{10} \bullet \dfrac{8}{5}$

3. $\dfrac{4}{11} \div \dfrac{2}{3}$

4. $\dfrac{x - 1}{5} \bullet \dfrac{2}{5}$

5. $\dfrac{x}{x - 1} \bullet \dfrac{x}{x + 1}$

6. $\dfrac{-3}{2x + 1} \div \dfrac{x}{x + 1}$

6.3 Cumulative Review Warm Up

You deposit \$7000 in an account that pays annual interest. Find the balance in the account after 3 years if the interest is compounded as described below at the given rate.

1. quarterly; 1.26%

2. monthly; 3.25%

3. daily; 2.8%

4. continuously; 2.3%

5. annually; 4.125%

6. continuously; 1.86%

Name ______________________________ Date ________

6.3 Practice A

In Exercises 1–6, simplify the expression, if possible.

1. $\dfrac{3x^2}{5x^2 + 2x}$

2. $\dfrac{6x^4 - x^3}{2x^4}$

3. $\dfrac{x^2 - 4x - 5}{x^2 - 7x + 10}$

4. $\dfrac{x^2 - 3x}{x^2 + 5x + 6}$

5. $\dfrac{x^2 - x - 2}{x^3 - 8}$

6. $\dfrac{x^2 - 3x - 4}{x^3 + 1}$

In Exercises 7–12, find the product.

7. $\dfrac{54x^4y^2}{y^4} \bullet \dfrac{x^3y^2}{9x^5y^3}$

8. $\dfrac{x^3(x + 2)}{x - 1} \bullet \dfrac{(x - 1)(x - 3)}{x^4}$

9. $\dfrac{x^2(x - 5)}{x + 7} \bullet \dfrac{(x + 7)(x - 1)}{4x^2}$

10. $\dfrac{x^2 - 5x}{x + 3} \bullet \dfrac{x^2 + 4x + 3}{x}$

11. $\dfrac{x^2 + 3x}{x - 2} \bullet \dfrac{x^2 - 5x + 6}{4x}$

12. $\dfrac{x^2 - 4x - 5}{x^2 + 6x + 9} \bullet \dfrac{2x^2 + 6x}{x^2 + 3x + 2}$

13. Compare the function $f(x) = \dfrac{(4x + 1)(x - 5)}{(4x + 1)}$ to the function $g(x) = x - 5$.

In Exercises 14–17, find the quotient.

14. $\dfrac{28x^4y}{y^7} \div \dfrac{y^9}{2x^5}$

15. $\dfrac{x^2 - x - 6}{3x^4 + 6x^3} \div \dfrac{x - 3}{6x^3}$

16. $\dfrac{4x^2 + 12x}{x^2 + 2x - 3} \div \dfrac{4x}{5x - 5}$

17. $\dfrac{x^2 + 5x - 14}{x + 3} \div (x^2 - 4x + 4)$

18. Manufacturers often package products in a way that uses the least amount of material. One measure of the efficiency of a package is the ratio of its surface area S to its volume V. The smaller the ratio, the more efficient the packaging. A company makes a cylindrical can to hold popcorn. The company is designing a new can with the same height h and twice the radius r of the old can.

a. Write an expression for the efficiency ratio $\dfrac{S}{V}$.

b. Find the efficiency ratio for each can.

c. Did the company make a good decision by creating the new can? Explain.

Name______________________________ Date__________

6.3 Practice B

In Exercises 1–6, simplify the expression, if possible.

1. $\dfrac{4x^3}{3x^3 + 7x}$

2. $\dfrac{x^2 + 5x + 6}{x^2 + 2x - 3}$

3. $\dfrac{2x^2 - 5x}{x^2 + 7x + 12}$

4. $\dfrac{x^2 - x - 20}{x^3 + 64}$

5. $\dfrac{x^4 - 16}{5x^3 - 3x^2 + 20x - 12}$

6. $\dfrac{6x^3 - 6x^2 + 5x - 5}{72x^4 - 50}$

In Exercises 7–12, find the product.

7. $\dfrac{x^4(x - 4)}{x + 3} \bullet \dfrac{(x + 3)(x - 2)}{x^5}$

8. $\dfrac{x^2 + 6x}{x - 4} \bullet \dfrac{x^2 - 2x - 8}{x}$

9. $\dfrac{x^2 - 2x}{x + 5} \bullet \dfrac{x^2 + 6x + 5}{3x}$

10. $\dfrac{x^2 - x - 6}{x^2 + 8x + 16} \bullet \dfrac{3x^2 + 12x}{x^2 - 2x - 3}$

11. $\dfrac{x^2 + 3x - 28}{x^2 - 25} \bullet (x^2 - 8x + 15)$

12. $\dfrac{x^2 + 2x - 15}{x^2 - 9} \bullet (x^2 - x - 12)$

In Exercises 13–16, find the quotient.

13. $\dfrac{2x^3 + 10x^2}{x^2 + x - 20} \div \dfrac{2x^2}{x - 4}$

14. $\dfrac{x^2 - 10x + 21}{x + 2} \div (x^2 - 14x + 49)$

15. $\dfrac{x^2 - 2x - 3}{x^2 + 2x - 8} \div \dfrac{x^2 + 4x + 3}{x^2 + 6x + 8}$

16. $\dfrac{x^2 + x - 6}{x^2 + 7x + 12} \div \dfrac{x^2 - 5x + 6}{x^2 + x - 12}$

17. Find the ratio of the perimeter to the area of the square shown.

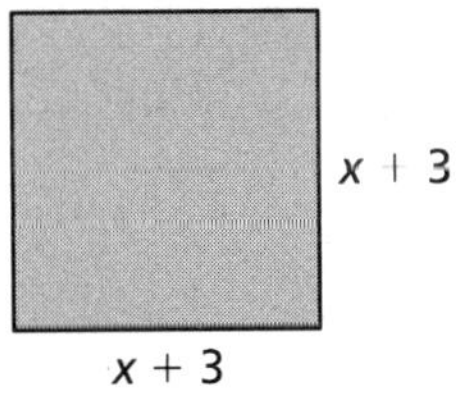

18. Find the expression that makes the following statement true. Assume $x \neq -5$ and $x \neq -3$.

$$\frac{x + 3}{x^2 - 8x + 12} \div \frac{\boxed{}}{x^2 + 3x - 10} = \frac{x + 5}{x - 6}$$

Name ______________________________ Date __________

6.3 Enrichment and Extension

Multiplying and Dividing Rational Expressions

Let f be the rational function

$$f(x) = \frac{a_n x^n + a_{n-1}x^{n-1} + \ldots + a_0}{b_m x^m + b_{m-1}x^{m-1} + \ldots + b_0}$$

where the numerator and denominator have no common factors.

You can determine the horizontal asymptote of the graph of f by comparing the degrees of the numerator and denominator.

- If $n < m$, then the x-axis is the horizontal asymptote of the graph of f.
- If $n = m$, then the line $y = \frac{a}{b}$ is the horizontal asymptote of the graph of f.
- If $n > m$, then the graph of f has no horizontal asymptote.

In Exercises 1–5, simplify the rational function. Then find the horizontal asymptote, if any, of the graph of the rational function.

1. $f(x) = \frac{x^2 - x - 2}{x^2 - 3x - 4} \bullet \frac{x^2 - x - 12}{x^2 - 2x - 35}$

2. $h(x) = \frac{x^2 + x - 2}{x^2 - 9} \div \frac{x^2 + 2x - 3}{x^3 - 3x^2 - 16x + 48}$

3. $k(x) = \frac{5x^2 - 4x^3 + 2}{5x + 2x^2 + 7x^3 - 9}$

4. $m(x) = \frac{2x^2 - 7x + 6}{2x^2 - 3x - 2} \bullet \frac{2x + 1}{3x^2 + 5x - 2}$

5. $t(x) = \frac{3x^2 + 5x - 2}{5x - 4} \div \frac{2x^2 + 7x + 6}{x - 7}$

Name ______________________________ Date __________

6.3 Puzzle Time

What Is Black And White And Red All Over?

Write the letter of each answer in the box containing the exercise number.

Simplify the expression, if possible.

1. $\dfrac{45}{10x - 15}$

2. $\dfrac{x - 7}{x^2 - 3x - 28}$

3. $\dfrac{2x}{3x^2 + 8}$

Find the product or quotient.

4. $\dfrac{y}{x^2 - 1} \bullet \dfrac{x - 1}{2y}$

5. $\dfrac{3x}{x + 1} \bullet \left(x^2 + 2x + 1\right)$

6. $\dfrac{(x + 3)}{(x + 2)} \div \dfrac{(x - 1)(x + 3)}{(x - 1)^2}$

7. $\dfrac{1}{x + 9} \div \dfrac{6 - x}{3x - 18}$

8. $\dfrac{10x^2yz^4}{5xy^3} \div 2x^5y^2z$

9. $\dfrac{x - 8}{x^2 - 2x - 48} \bullet \dfrac{4x^2 + 40x}{x + 10}$

10. $\dfrac{1}{5x^2} \div \dfrac{9x - 36}{5x^3 - 35x^2}$

Answers

E. $\dfrac{4x}{x + 6}, x \neq -10, x \neq 8$

S. $3x(x + 1), x \neq -1$

A. $\dfrac{9}{2x - 3}$

P. $\dfrac{z^3}{x^4y^4}, z \neq 0$

N. $\dfrac{1}{x + 4}, x \neq 7$

W. $\dfrac{1}{2(x + 1)}, x \neq 1, y \neq 0$

P. $\dfrac{x - 1}{x + 2}, x \neq -3, x \neq 1$

R. $\dfrac{x - 7}{9(x - 4)}, x \neq 0, x \neq 7$

E. $\dfrac{2x}{3x^2 + 8}$

A. $-\dfrac{3}{x + 9}, x \neq 6$

1		2	3	4	5	6	7	8	9	10

6.4 Start Thinking

To add or subtract rational numbers, you must first find a common denominator. The same is true for rational expressions in algebra. Determine the least common denominator (LCD) for the three rational expressions shown below. Then determine the values of x that make the LCD zero.

$$\frac{1}{3x^2 - 6x}, \quad \frac{4}{x + 3}, \quad \frac{5}{x^2 + x - 6}$$

6.4 Warm Up

Simplify the expression.

1. $\frac{3}{7} + \frac{2}{3}$
2. $\frac{7}{10} + \frac{8}{5}$
3. $\frac{3}{28} - \frac{1}{4}$
4. $\dfrac{\frac{2}{3} - \frac{1}{2}}{\frac{3}{5}}$
5. $\dfrac{-\frac{5}{8} - \frac{3}{4}}{2}$
6. $\dfrac{\frac{5}{2} - \frac{7}{5}}{-\frac{2}{3} + 4}$

6.4 Cumulative Review Warm Up

Given $f(x) = 2x - x^2$ and $g(x) = 4 - 3x$, determine the value of the expression.

1. $(f + g)(x)$
2. $f(x) \bullet g(x)$
3. $\left(\frac{g}{f}\right)(x)$
4. $f(g(x))$
5. $(f - g)(2)$
6. $g(f(3))$

Name________________________________ Date__________

6.4 Practice A

In Exercises 1–3, find the sum or difference.

1. $\frac{12}{5x} + \frac{3}{5x}$
2. $\frac{x}{9x^2} - \frac{3}{9x^2}$
3. $\frac{7}{x-2} - \frac{3x}{x-2}$

In Exercises 4–7, find the least common multiple of the expressions.

4. $3x^2, 6x - 18$
5. $5x, 5x(x+2)$
6. $x^2 - 9, x + 3$
7. $x^2 - 3x - 10, x + 2$
8. Describe and correct the error in finding the sum.

   ✗ $\frac{x}{x+3} - \frac{2}{x-1} = \frac{x-2}{(x+3)(x-1)}$

In Exercises 9–12, find the sum or difference.

9. $\frac{7}{2x^2} - \frac{4}{3x}$
10. $\frac{2}{x-1} + \frac{4}{x+2}$
11. $\frac{6}{x+4} - \frac{5x}{x-3}$
12. $\frac{14}{x^2 + 7x - 18} + \frac{6}{x+9}$

In Exercises 13 and 14, tell whether the statement is *always*, *sometimes*, or *never* true. Explain.

13. The LCD of two rational functions is the sum of the denominators.
14. The LCD of two rational functions is equal to one of the denominators.

In Exercises 15–18, rewrite the function in the form $g(x) = \frac{a}{x-h} + k$. Graph the function. Describe the graph of g as a transformation of the graph of $f(x) = \frac{a}{x}$.

15. $g(x) = \frac{4x-5}{x-2}$
16. $g(x) = \frac{5x+3}{x+4}$
17. $g(x) = \frac{10x}{x-3}$
18. $g(x) = \frac{3x+4}{x}$

Name ______________________________ Date __________

6.4 Practice B

In Exercises 1–3, find the sum or difference.

1. $\dfrac{x}{25x^2} - \dfrac{5}{25x^2}$

2. $\dfrac{2x^2}{x+6} + \dfrac{8x}{x+6}$

3. $\dfrac{3x}{x-4} - \dfrac{12}{x-4}$

In Exercises 4–7, find the least common multiple of the expressions.

4. $36x^2, 9x^2 - 18x$

5. $x^2 - 100, x - 10$

6. $25x^2 - 4, 3x^2 - 10x - 8$

7. $x^2 + 7x - 18, x + 9$

8. Describe and correct the error in finding and simplifying the sum.

$$\times \quad \frac{4}{7x} + \frac{5}{x^3} = \frac{4(x^3)}{7x(x^3)} + \frac{5(7x)}{x^3(7x)} = \frac{4x^3 + 35x}{7x^4}$$

In Exercises 9–12, find the sum or difference.

9. $\dfrac{7}{x-5} + \dfrac{4x}{x+1}$

10. $\dfrac{7}{x^2 - 5x - 24} + \dfrac{3}{x-8}$

11. $\dfrac{x^2 - 3}{x^2 - 6x - 16} - \dfrac{x+5}{x+2}$

12. $\dfrac{x-2}{x-3} + \dfrac{3}{x} + \dfrac{6x}{2x+1}$

In Exercises 13 and 14, tell whether the statement is *always*, *sometimes*, or *never* true. Explain.

13. The LCD of two rational functions is one of the denominators when the other denominator is a factor.

14. The LCD of two rational functions will have a degree equal to that of the denominator with the higher degree.

In Exercises 15–18, rewrite the function in the form $g(x) = \dfrac{a}{x-h} + k$. Graph the function. Describe the graph of g as a transformation of the graph of $f(x) = \dfrac{a}{x}$.

15. $g(x) = \dfrac{5x+3}{x+4}$

16. $g(x) = \dfrac{9x}{x+12}$

17. $g(x) = \dfrac{5x-4}{x}$

18. $g(x) = \dfrac{8x+13}{x-6}$

Name________________________ Date __________

6.4 Enrichment and Extension

Partial Fraction Decomposition

You have learned how to combine rational expressions as follows.

$$\frac{3}{x+1} + \frac{2}{x-2} = \frac{3(x-2) + 2(x+1)}{(x+1)(x-2)} = \frac{3x - 6 + 2x + 2}{(x+1)(x-2)} = \frac{5x-4}{x^2 - x - 2}$$

The decomposition of a rational expression into partial fractions is the reverse of this process. The method shown in the example below can be used for rational expressions when the degree of the numerator is less than the degree of the denominator, and the denominator can be factored into distinct linear factors.

Example: Decompose $\frac{5x-4}{x^2 - x - 2}$ into partial fractions.

Solution: Factor the denominator. $\frac{5x-4}{x^2 - x - 2} = \frac{5x-4}{(x+1)(x-2)}$

Express the fraction as the sum of two fractions, with the individual factors as denominators and the unknown numerators A and B.

$$\frac{5x-4}{(x+1)(x-2)} = \frac{A}{x+1} + \frac{B}{x-2}$$

To clear fractions, multiply each side of the rational equation by the LCD, $(x+1)(x-2)$.

$$5x - 4 = A(x-2) + B(x+1)$$

To determine A and B, first eliminate B by substituting -1 for x.

$$\begin{aligned} 5(-1) - 4 &= A(-1-2) + B(-1+1) \\ -5 - 4 &= -3A \\ A &= 3 \end{aligned}$$

Then eliminate A by substituting 2 for x.

$$\begin{aligned} 5(2) - 4 &= A(2-2) + B(2+1) \\ 10 - 4 &= 3B \\ B &= 2 \end{aligned}$$

Use the values of A and B to write the original rational expression as the sum of two partial fractions.

$$\frac{5x-4}{(x+1)(x-2)} = \frac{3}{x+1} + \frac{2}{x-2}$$

In Exercises 1–6, decompose the rational expression into partial fractions.

1. $\frac{3x+18}{x^2 + 5x + 4}$

2. $\frac{6x^2 + 2x - 4}{x^3 - 4x}$

3. $\frac{6x^2 + 10x - 4}{x^3 + 2x^2 - 9x - 18}$

4. $\frac{2x+12}{x^2 - 3x - 28}$

5. $\frac{2x^2 + 7x - 4}{x^3 - 9x}$

6. $\frac{7x^2 + 18x - 19}{x^3 + 2x^2 - 5x - 6}$

Name ____________________ Date ________

6.4 Puzzle Time

What Do You Get When You Cross Cinderella With A Barber?

Write the letter of each answer in the box containing the exercise number.

Find the sum or difference.

1. $\dfrac{10}{2x} - \dfrac{3}{2x}$

2. $\dfrac{5 - x}{x + 1} + \dfrac{2 + 3x}{x + 1}$

3. $\dfrac{x}{2x - 4} - \dfrac{5}{x + 1}$

4. $\dfrac{x - 2}{3x} + \dfrac{2x}{x}$

5. $\dfrac{7}{x + 8} - \dfrac{2}{x - 1}$

6. $\dfrac{3x}{x - 4} + \dfrac{8}{x - 3}$

7. $\dfrac{1}{x^2 + 3x - 10} - \dfrac{4}{x - 2}$

Simplify the complex fraction.

8. $\dfrac{\frac{x}{25}}{\frac{4}{x^2}}$

9. $\dfrac{\frac{x^2}{4}}{\frac{x}{x + 4}}$

10. $\dfrac{\frac{4}{3x} + \frac{2}{x^2}}{\frac{x}{x + 1} - \frac{4}{x + 1}}$

11. $\dfrac{x^2}{\frac{4}{5} - \frac{4}{x}}$

12. $\dfrac{\frac{x^2}{9} + \frac{1}{4}}{6x}$

13. $\dfrac{\frac{1}{2} - \frac{x + 5}{4}}{\frac{x^2}{2} - \frac{5}{2}}$

Answers

R. $\dfrac{4x^2 + 9}{216x}$

L. $\dfrac{-4x - 19}{x^2 + 3x - 10}$

S. $\dfrac{5x - 23}{x^2 + 7x - 8}$

L. $\dfrac{2x + 7}{x + 1}$

P. $\dfrac{4x^2 + 10x + 6}{3x^3 - 12x^2}, x \neq -1$

S. $\dfrac{7x - 2}{3x}$

I. $\dfrac{x^3}{100}, x \neq 0$

G. $\dfrac{7}{2x}$

P. $\dfrac{x^2 + 4x}{4}, x \neq -4, x \neq 0$

C. $\dfrac{3x^2 - x - 32}{x^2 - 7x + 12}$

A. $\dfrac{x^2 - 9x + 20}{2x^2 - 2x - 4}$

E. $\dfrac{5x^3}{4x - 20}, x \neq 0$

S. $\dfrac{-x - 3}{2x^2 - 10}$

1	2	3	4	5		6	7	8	9	10	11	12	13

6.5 Start Thinking

A racquetball club offers a $200 annual membership that includes unlimited free court rental. They also offer a $120 membership with a $10 fee for each court rental. Write a rational equation that represents the cost per court rental for each membership. Then use a graphing calculator to graph both rational functions and determine the point of intersection. What does this point of intersection represent in the context of the problem? Which membership is a better deal?

6.5 Warm Up

Solve the equation.

1. $-3(x - 1) = 4(x + 1)$
2. $2(x - 1) = x^2 - x$
3. $4\left(x + \frac{3}{2}\right) = (x - 2)3$
4. $-2(x + 1) = (x - 8)(x - 1)$
5. $3x(2x) = 8(x + 1)$
6. $x(x - 3) + 4x(x - 4) = 3x(x - 4)$

6.5 Cumulative Review Warm Up

Perform the indicated operation. Write your answer in standard form.

1. $(3 + 2i) - (5 + i)$
2. $-2i(11 + 7i)$
3. $(-5 + 3i)^2$
4. $(-4 + 9i)(5 + 6i)$
5. $3i^5(4i^2 - 5i + 8)$
6. $(5 - 3i^7) - (11 + 2i^6)$

Name ______________________________ Date __________

6.5 Practice A

In Exercises 1–3, solve the equation by cross multiplying. Check your solution(s).

1. $\dfrac{3}{4x} = \dfrac{1}{x-2}$

2. $\dfrac{4}{x+2} = \dfrac{6}{x-2}$

3. $\dfrac{-3}{x+1} = \dfrac{x-5}{x-5}$

4. So far in baseball practice, you have pitched 47 strikes out of 61 pitches. Solve the equation $\dfrac{80}{100} = \dfrac{47+x}{61+x}$ to find the number x of consecutive strikes you need to pitch to raise your strike percentage to 80%.

In Exercises 5 and 6, identify the LCD of the rational expressions in the equation.

5. $\dfrac{x}{x-2} + \dfrac{2}{x} = \dfrac{5}{x}$

6. $\dfrac{3x}{x+5} - \dfrac{8}{x} = \dfrac{2}{x}$

In Exercises 7–12, solve the equation by using the LCD. Check your solution(s).

7. $\dfrac{4}{3} + \dfrac{2}{x} = 4$

8. $\dfrac{5}{2x} + \dfrac{1}{4} = \dfrac{9}{2x}$

9. $\dfrac{x-2}{x-3} + 3 = \dfrac{2x}{x}$

10. $\dfrac{4}{x-5} + \dfrac{1}{x} = \dfrac{x-1}{x-5}$

11. $\dfrac{8}{x} + 3 = \dfrac{x+8}{x-4}$

12. $\dfrac{12}{x^2-2x} - \dfrac{3}{x-2} = \dfrac{3}{x}$

13. Describe and correct the error in the first step of solving the equation.

> ✗
> $$\frac{4}{x} + \frac{1}{2} = 1$$
> $$2x \bullet \frac{4}{x} + 2x \bullet \frac{1}{2} = 1$$

14. You can clean the gutters of your house in 5 hours. Working together, you and your friend can clean the gutters in 3 hours. Let t be the time (in hours) your friend would take to clean the gutters when working alone. Write and solve an equation to find how long your friend would take to clean the gutters when working alone. $\left(\textit{Hint: }(\text{Work done}) = (\text{Work rate}) \times (\text{Time})\right)$

In Exercises 15 and 16, determine whether the inverse of *f* is a function. Then find the inverse.

15. $f(x) = \dfrac{5}{x-2}$

16. $f(x) = \dfrac{4}{x} - 1$

Name______________________________ Date__________

6.5 Practice B

In Exercises 1–3, solve the equation by cross multiplying. Check your solution(s).

1. $\dfrac{3}{x+2} = \dfrac{5}{x-2}$

2. $\dfrac{2}{x-4} = \dfrac{x-3}{x-1}$

3. $\dfrac{x-5}{4} = \dfrac{x^2-5}{x+4}$

4. So far in soccer practice, you have made 10 out of 32 goal attempts. Solve the equation $0.45 = \dfrac{10+x}{32+x}$ to find the number x of consecutive goals you need to make to raise your goal average to 0.45.

In Exercises 5 and 6, identify the LCD of the rational expressions in the equation.

5. $\dfrac{6}{x+3} + \dfrac{x}{x+2} = \dfrac{4}{5}$

6. $\dfrac{6}{x-8} - \dfrac{2x}{3x-2} = \dfrac{9}{4}$

In Exercises 7–12, solve the equation by using the LCD. Check your solution(s).

7. $\dfrac{3}{4x} + \dfrac{1}{8} = \dfrac{7}{4x}$

8. $\dfrac{5}{x-6} + \dfrac{1}{x} = \dfrac{x-1}{x-6}$

9. $\dfrac{x-4}{x-5} + 5 = \dfrac{4x}{x}$

10. $\dfrac{16}{x^2-4x} - \dfrac{8}{x-4} = \dfrac{4}{x}$

11. $\dfrac{x+1}{x+2} + \dfrac{1}{x} = \dfrac{2x+1}{x+2}$

12. $\dfrac{4}{x} - 1 = \dfrac{4}{x+2}$

13. Describe and correct the error in the first step of solving the equation.

X

$$\frac{3}{x+2} + 5 = \frac{1}{x}$$

$$(x+2) \bullet \frac{3}{x+2} + (x+2) \bullet 5 = (x+2) \bullet \frac{1}{x}$$

14. You can kayak around a certain island in 3 hours. Kayaking together, you and your friend can kayak around the island in 1.4 hours. Let t be the time (in hours) your friend would take to kayak around the island when kayaking alone. Write and solve an equation to find how long your friend would take to kayak around the island when kayaking alone.

(*Hint:* (Work done) = (Work rate) × (Time))

In Exercises 15 and 16, determine whether the inverse of *f* is a function. Then find the inverse.

15. $f(x) = \dfrac{3}{2x-7}$

16. $f(x) = \dfrac{1}{x^3} + 9$

Name ______________________________ Date __________

6.5 Enrichment and Extension

Solving Rational Equations

In Exercises 1–12, use the function $f(x) = \dfrac{2x + 9}{x + 2}$ to answer the question.

1. Name three points that lie on the graph of the inverse function f^{-1}. Explain your reasoning.

2. Does the point $\left(\dfrac{11}{3}, 2\right)$ lie on the graph of f^{-1}? Explain your reasoning.

3. Algebraically find $f^{-1}(x)$.

4. Prove that the functions are inverses by algebraically showing that $f\left(f^{-1}(x)\right) = f^{-1}\left(f(x)\right) = x$.

5. Graph both f and f^{-1} in the same coordinate plane.

6. Write the domain and range of f.

7. Write the domain and range of f^{-1}. How do the domain and range of f^{-1} compare with the domain and range of f?

8. What do you notice about the vertical asymptotes and horizontal asymptotes of the graphs of f and f^{-1}?

9. What is significant about where the graphs of f and f^{-1} intersect?

10. Is there a relationship between the asymptotes of the graph of a function and the zeros of its inverse? Explain.

11. Can you find the inverse of $g(x) = \dfrac{x - 7}{2x + 1}$? If so, find $g^{-1}(x)$.

12. Do all rational functions have an inverse? Explain.

Name________________________________ Date__________

6.5 Puzzle Time

What Is Green And Sings?

Write the letter of each answer in the box containing the exercise number.

Solve the equation by cross multiplying. Check your solution(s).

1. $\frac{x-5}{15} = \frac{4}{5}$

2. $\frac{x-6}{3} = \frac{-2x-2}{15}$

3. $\frac{x+3}{x+1} = \frac{15}{x+7}$

4. $\frac{5x}{x-1} = 4$

5. $\frac{x+2}{x-2} = \frac{2x+4}{x+1}$

6. $\frac{3}{x} = \frac{2}{5-x}$

Solve the equation by using the LCD. Check your solution(s).

7. $\frac{5}{x} + \frac{1}{3} = 1$

8. $\frac{2}{5x} + \frac{1}{3} = \frac{4}{15x}$

9. $\frac{1}{x} - 3 = \frac{4x}{x+2}$

10. $\frac{3}{x-1} + \frac{1}{x+1} = \frac{10}{x^2-1}$

11. $\frac{12}{x^2+4x} + \frac{3}{x} = \frac{1}{x+4}$

12. $\frac{8}{x+2} - \frac{2}{x} = \frac{x-2}{x^2+2x}$

Answers

L. $x = 2$

S. $x = -2, x = 5$

A. $x = \frac{15}{2}$

E. $x = 17$

P. $x = 3$

S. $x = \frac{2}{7}, x = -1$

E. $x = -12$

I. $x = -4$

R. $x = -\frac{2}{5}$

Y. $x = \frac{2}{5}$

L. $x = 4$

V. $x = 3, x = 2$

1	2	3	4	5		6	7	8	9	10	11	12

Name ______________________ Date __________

Chapter 6 Cumulative Review

In Exercises 1–18, factor the polynomial.

1. $x^2 - 4x - 21$
2. $x^2 + 8x - 48$
3. $x^2 + 9x + 8$

4. $x^2 - 9x + 20$
5. $x^2 - 6x - 27$
6. $x^2 - 12x + 32$

7. $2x^2 + 5x - 3$
8. $2x^2 - 2x - 24$
9. $3x^2 + 23x - 36$

10. $5x^2 + 23x + 12$
11. $3x^2 - 25x + 42$
12. $4x^2 - 7x + 3$

13. $4x^2 + 30x + 56$
14. $6x^2 - 7x - 5$
15. $8x^2 - 2x - 3$

16. $5x^2 - 2x - 7$
17. $8x^2 + 22x - 6$
18. $12x^2 - 18x - 12$

In Exercises 19–34, write the prime factorization of the number. If the number is prime, then write *prime*.

19. 12
20. 201
21. 88
22. 56

23. 75
24. 300
25. 2
26. 99

27. 188
28. 199
29. 85
30. 41

31. 17
32. 169
33. 131
34. 65

In Exercises 35–46, solve the equation. Check your solution.

35. $\frac{1}{2}x + 6 = \frac{5}{2}x$
36. $\frac{3}{4}x - 7 = \frac{11}{4}x$
37. $\frac{5}{7}x - 4 = \frac{2}{5}$

38. $\frac{3}{5}x + 9 = \frac{2}{5}x - 7$
39. $\frac{1}{4}x + 5 = \frac{5}{4}x - 1$
40. $\frac{1}{2}x - 3 = 8x$

41. $\frac{1}{3}x + 5 = \frac{3}{5}x$
42. $\frac{2}{7}x - 3 = \frac{5}{8}x$
43. $\frac{7}{2}x - 3 = \frac{2}{5}x - \frac{3}{4}$

44. $\frac{4}{5}x - \frac{2}{7} = \frac{1}{4}x + \frac{2}{3}$
45. $\frac{5}{9}x - \frac{1}{4} = \frac{7}{8}x - \frac{2}{5}$
46. $\frac{3}{5}x + \frac{2}{7} = \frac{3}{4}x - \frac{7}{8}$

47. The expression for the area of a rectangle is $x^2 - 5x - 24$. What are the expressions for the length and the width?

48. The expression for the area of a rectangle is $2x^2 + 3x - 20$. What are the expressions for the length and the width?

49. The expression for the area of a rectangle is $6x^2 + x - 15$. What are the expressions for the length and the width?

Name_______________________________ Date__________

Chapter 6

Cumulative Review (continued)

In Exercises 50–63, find the zero(s) of the function.

50. $f(x) = x^2 + 2x - 35$

51. $h(x) = x^2 + 3x - 18$

52. $f(x) = x^2 + 2x - 99$

53. $h(x) = x^2 - 2x - 48$

54. $h(x) = x^2 - 8x + 15$

55. $g(x) = x^2 - 4x - 45$

56. $g(x) = x^2 + 5x + 4$

57. $f(x) = x^2 - 25$

58. $h(x) = x^2 - 4x - 32$

59. $h(x) = x^2 + 3x$

60. $f(x) = 2x^2 + 17x + 8$

61. $g(x) = 3x^2 - 26x + 35$

62. $f(x) = 2x^2 + 7x - 15$

63. $f(x) = 4x^2 - 14x$

In Exercises 64–75, find the discriminant of the quadratic equation and describe the number and type of solutions of the equation.

64. $x^2 + 4x - 12 = 0$

65. $x^2 + 5x + 10 = 0$

66. $x^2 + 8x + 16 = 0$

67. $x^2 + 9x - 24 = 0$

68. $x^2 - 6x - 32 = 0$

69. $x^2 + 11x + 150 = 0$

70. $x^2 + 42 = 3x$

71. $x^2 + 14x = -49$

72. $2x^2 - 4x = 3$

73. $3x^2 = -6x + 8$

74. $4x^2 + 8 = -6x$

75. $x^2 + 25 = 10x$

In Exercises 76–87, solve the equation using the Quadratic Formula.

76. $2x^2 - 4x + 2 = 0$

77. $3x^2 - 5x + 2 = 0$

78. $5x^2 + 4x - 16 = 0$

79. $-3x^2 + 2x = 18$

80. $x^2 + 9x = 15$

81. $4x^2 + 5x - 2$

82. $6x^2 + 11 = 8x$

83. $-3x^2 - 24 = 7x$

84. $5x^2 + 36 = -2x$

85. $-4x^2 + 12x = -64$

86. $8x^2 - 7x = -5$

87. $x^2 = x - 4$

88. The expression for the area of a square is $x^2 + 16x + 64$. What is the expression for the length of one side?

89. The expression for the area of a square is $4x^2 + 20x + 25$. What is the expression for the length of one side?

90. The expression for the area of a square is $16x^2 - 8x + 1$. What is the expression for the length of one side?

Name ______________________________ Date __________

Chapter 6 Cumulative Review (continued)

In Exercises 91–102, solve the system by substitution.

91. $y = x - 9$
$y = 12$

92. $y = x + 14$
$x = -3$

93. $y = x + 2x - 9$
$x = -4$

94. $y = 2x + 8$
$y = -6$

95. $y = 2x - 17$
$y = 13$

96. $y = 2x + 7$
$y = -11$

97. $y = 4x + 12$
$y = 6x$

98. $y = 4x + 5$
$y = 9x$

99. $y = 3x - 8$
$y = -5x$

100. $y = x^2 + 5x - 6$
$y = x - 1$

101. $y = x^2 - 2x - 15$
$y = x + 3$

102. $y = x^2 + 10x + 16$
$y = x + 2$

In Exercises 103–112, solve the system by elimination.

103. $2x - y = 8$
$-8x + y = 4$

104. $3x + 8y = 20$
$-3x - 4y = 8$

105. $3x - 4y = 15$
$-3x + 3y = -13$

106. $-2x + 6y - 14 = 0$
$5x - 6y = 10$

107. $2x - 4y = -13$
$-4x + 12y = 22$

108. $-3x + 4y = -7$
$2x - 8y = -10$

109. $6x^2 - 8x + 12 + y = 14$
$-5x^2 + 7x - 14 - y = -12$

110. $-3x^2 + 2y + 7x - 12 = -8$
$4x^2 - 5x + 8 - 2y = 14$

111. $3x^2 + 7x - 8 = 3y$
$x^2 + 3y + 5x + 9 = -2$

112. $9x^2 - 14 - 3x + 8x - 4y = 18$
$-6x^2 + 4y - 7x + 21 + 5x = 23$

113. Your friend has a swimming pool that is in the shape of a rectangular prism. It measures 18 feet by 34 feet, and is 4 feet deep. What is the volume of the swimming pool?

114. A cube has an edge length of 3 inches.

a. Find the surface area of the cube.

b. Find the volume of the cube.

115. A sphere has a radius of 7 centimeters.

a. Find the surface area of the sphere.

b. Find the volume of the sphere.

Chapter 6 Cumulative Review (continued)

In Exercises 116–118, evaluate the function for the given value of *x*.

116. $f(x) = -4x^3 + 2x^2 - 6x + 14; x = -3$

117. $g(x) = -3x^4 + 2x^3 - 11x + 36; x = -5$

118. $g(x) = -2x^3 + 8x^2 - 12x + 52; x = -\frac{1}{4}$

In Exercises 119–121, find the sum.

119. $(3x^4 - 7x^2 + 12x - 9) + (7x^4 - 5x^2 - 14x + 8)$

120. $(9x^4 - 7x^3 + 3x - 11) + (5x^3 - x^2 - 8x + 15)$

121. $(8x^5 - 7x^4 + 5x^3 - 10x^2 - 3x + 1) + (3x^4 - 3x^2 - 5x + 9)$

In Exercises 122–124, find the difference.

122. $(-11x^3 + 10x - 1) - (-5x^3 - 8x + 9)$

123. $(-12x^4 + 4x^3 + 6x - 9) - (12x^4 - 10x + 2)$

124. $(-8x^5 + 4x^2 + 2x - 1) - (9x^5 - 2x^4 + 7x^3 + 4x^2 + 12x - 6)$

In Exercises 125–128, tell whether *x* and *y* show *direct variation*, *inverse variation*, or *neither*.

125. $4y = 5x$

126. $12 = \frac{x}{y}$

127. $x + 3 = 5y$

128. $xy = 8$

In Exercises 129 and 130, find the product or quotient.

129. $\frac{3x + 9}{2(x + 5)} \bullet \frac{2x^2 - 4x - 70}{x^2 - x - 12}$

130. $\frac{x^2 + 13x + 42}{4x^2 + 40x + 84} \div \frac{x - 11}{-5x - 15}$

In Exercises 131 and 132, find the sum or difference.

131. $\frac{x + 1}{x^2 + 7x + 10} + \frac{2x}{3x + 6}$

132. $\frac{x - 1}{x^2 + x - 42} - \frac{10x}{x^2 - 36}$

In Exercises 133 and 134, solve the equation. Check your solution.

133. $\frac{9}{x + 5} = \frac{4}{x}$

134. $\frac{x}{x^2 - 16} = \frac{2}{x - 4}$

Chapter 7

Name______________________________ Date__________

Chapter 7

Sequences and Series

Dear Family,

In this chapter, you will be learning about ordered lists of numbers called sequences. You can usually determine the pattern of a sequence after the first few numbers are given. You will be using sequences to find the next numbers in the pattern, write a rule, and graph the numbers.

Here is a fun activity involving a sequence that you can do as a family. Using eight coins of the same size, determine the number of ways to arrange the coins in rows. Each coin in a row must be touching another coin in the same row (if there are any in the same row) and must be touching two coins below (if there are any below). So, each row has a smaller number of coins than the row below it and there are no spaces between the coins.

Examples:

1 coin = 1 arrangement

2 coins = 1 arrangement

3 coins = 2 arrangements

Coins	Number of arrangements
1	1
2	1
3	2
4	
5	
6	
7	
8	

Complete the table for all eight coins. What pattern do you notice about the sequence? Without using coins, what are the next two numbers in the pattern?

The pattern above is called the Fibonacci sequence, named after Leonardo Fibonacci, written in his book published in 1202. Use the Internet to check the terms of your sequence and learn more about the applications of the Fibonacci sequence in math, history, art, and music.

Enjoy finding patterns in your daily life!

Nombre ______________________________ Fecha ________

Capítulo 7 Secuencias y series

Estimada familia:

En este capítulo, aprenderán sobre listas ordenadas de números llamadas secuencias. A menudo, se puede determinar el patrón de una secuencia después de que se den los primeros números. Usarán secuencias para hallar los próximos números en el patrón, escribir una regla y hacer la gráfica de los números.

He aquí una actividad divertida que incluye una secuencia que pueden hacer en familia. Con ocho monedas del mismo tamaño, determinen el número de maneras de colocar las monedas en filas. Cada moneda de una fila debe tocar otra moneda en la misma fila (si hay alguna en la misma fila) y debe tocar dos monedas en la fila inferior (si hay alguna fila inferior). Entonces, cada fila tiene un número más pequeño de monedas que la fila que está debajo y no hay espacios entre las monedas.

Ejemplos:

1 moneda = 1 distribución

2 monedas = 1 distribución

3 monedas = 2 distribuciones

Monedas	**Número de distribuciones**
1	1
2	
3	
4	
5	
6	
7	
8	

Completen la tabla para las ocho monedas. ¿Qué patrón observan sobre la secuencia? Sin usar monedas, ¿cuáles son los próximos dos números en el patrón?

El patrón que está arriba se llama secuencia de Fibonacci, que lleva el nombre de Leonardo Fibonacci, escrita en su libro publicado en 1202. Consulten en Internet para verificar los términos de su secuencia y aprender más sobre las aplicaciones de la secuencia de Fibonacci en matemáticas, historia, arte y música.

¡Disfruten al hallar patrones en su vida cotidiana!

7.1 Start Thinking

Review the list of numbers and then describe the pattern that relates the numbers in the list.

1. 5, 7, 9, 11, 13, ...

2. $\frac{2}{3}, \frac{4}{7}, \frac{8}{11}, \frac{16}{15}, \frac{32}{19}, \ldots$

3. $8, \frac{13}{2}, 5, \frac{7}{2}, 2, \ldots$

4. 3, −6, 12, −24, 48, ...

7.1 Warm Up

For the function, determine the value of $f(1)$, $f(2)$, $f(3)$, $f(4)$, and $f(5)$.

1. $f(x) = 2x - 2$

2. $f(x) = 2^x$

3. $f(x) = 2x - 1$

4. $f(x) = (-1)^x(5)$

5. $f(x) = \frac{2^{x-1}}{3}$

6. $f(x) = (-1)^{x-1}x^2$

7.1 Cumulative Review Warm Up

Solve the equation. Round to the nearest thousandth, if necessary.

1. $2^{x-1} + 5 = 13$

2. $3^{x+2} = 9^{x-1}$

3. $3e^{2x} - 4 = 8$

4. $3 + \log_2(4x) = 5$

5. $3\ln(2 - x) = -6$

6. $\log_3(x^2 - 3) = \log_3(2x)$

Name ______________________________ Date __________

7.1 Practice A

In Exercises 1–6, write the first six terms of the sequence.

1. $a_n = n - 3$ **2.** $a_n = 4 - n$ **3.** $a_n = n^3$

4. $a_n = n^2 - 5$ **5.** $a_n = 2^{2n}$ **6.** $a_n = -n^2 + 1$

In Exercises 7–14, describe the pattern, write the next term, and write a rule for the *n*th term of the sequence.

7. 1, 4, 7, 10, ... **8.** 1, 3, 9, 27, ...

9. 1.5, 3, 4.5, 6, ... **10.** 4.2, 5.8, 7.4, 9, ...

11. 4.7, 3.4, 2.1, 0.8, ... **12.** –5, 10, –15, 20, ...

13. $\frac{1}{6}, \frac{2}{6}, \frac{3}{6}, \frac{4}{6}, \ldots$ **14.** $\frac{3}{2}, \frac{3}{4}, \frac{3}{6}, \frac{3}{8}, \ldots$

15. You agree to work for your uncle. You earn \$10 the first day, \$20 the second day, \$40 the third day, and \$80 the fourth day. Write a rule for the number of dollars that you will earn on the *n*th day. Then graph the sequence.

In Exercises 16–21, write the series using summation.

16. $4 + 8 + 12 + 16 + 20$ **17.** $3 + 9 + 15 + 21 + 27$

18. $1 + 7 + 13 + 19 + \ldots$ **19.** $-1 + 1 + 3 + 5 + \ldots$

20. $\frac{1}{5} + \frac{1}{25} + \frac{1}{125} + \frac{1}{625} + \ldots$ **21.** $\frac{1}{7} + \frac{2}{8} + \frac{3}{9} + \frac{4}{10} + \ldots$

In Exercises 22–27, find the sum.

22. $\sum_{i=1}^{4} 3i$ **23.** $\sum_{i=1}^{5} 6i$ **24.** $\sum_{n=0}^{5} n^2$

25. $\sum_{n=2}^{7} (3n + 2)$ **26.** $\sum_{k=1}^{6} (k^2 - 3)$ **27.** $\sum_{i=3}^{7} \frac{5}{i}$

28. You are building a brick garden wall six rows high. The bottom row has 25 bricks. Each of the other rows has three fewer bricks than the one below it. How many bricks will you need to build the garden wall? Justify your answer.

Name______________________________ Date__________

7.1 Practice B

In Exercises 1–6, write the first six terms of the sequence.

1. $a_n = n^2 + 4$

2. $a_n = 3^{n-2}$

3. $a_n = (n - 2)^2$

4. $a_n = -n^2 + 3$

5. $f(n) = \dfrac{-n}{n + 5}$

6. $a_n = \dfrac{n}{3n - 1}$

In Exercises 7–14, describe the pattern, write the next term, and write a rule for the *n*th term of the sequence.

7. 2, 7, 12, 17, ...

8. 2.7, 6.2, 9.7, 13.2, ...

9. 6.4, 3.5, 0.6, −2.3, ...

10. $-\frac{1}{7}, -\frac{2}{7}, -\frac{3}{7}, -\frac{4}{7}, \ldots$

11. $\frac{5}{2}, \frac{5}{4}, \frac{5}{8}, \frac{5}{16}, \ldots$

12. $\frac{3}{1}, \frac{9}{2}, \frac{27}{3}, \frac{81}{4}, \ldots$

13. $\frac{2}{3}, \frac{4}{9}, \frac{8}{27}, \frac{16}{81}, \ldots$

14. 2, 20, 200, 2000, ...

15. You are renting tables for a school event. There is a set-up and delivery fee of $100. The rental fee per table is $7. Write a rule for the cost of renting *n* tables. Then graph the sequence.

In Exercises 16–21, write the series using summation.

16. 6 + 13 + 20 + 27 + 34

17. 10 + 13 + 16 + 19 + ...

18. −3 + 0 + 5 + 12 + ...

19. $\frac{1}{4} + \frac{2}{16} + \frac{3}{64} + \frac{4}{256} + \ldots$

20. $\frac{2}{5} + \frac{2}{6} + \frac{2}{7} + \frac{2}{8} + \ldots$

21. −1 + 3 − 5 + 7 − 9

In Exercises 22–27, find the sum.

22. $\sum_{i=1}^{6} 9i$

23. $\sum_{i=0}^{5} 2i^3$

24. $\sum_{n=1}^{6} (n^2 + 2)$

25. $\sum_{k=3}^{5} \frac{1}{k+2}$

26. $\sum_{k=1}^{15} 3$

27. $\sum_{i=13}^{27} i$

28. You are making party favors. In the first hour you made 36 party favors. Each hour you make two more party favors than in the previous hour. How many party favors will you have made after 7 hours? Justify your answer.

Name ______________________________ Date __________

7.1 Enrichment and Extension

Defining and Using Sequences and Series

The sum of the first n terms of an arithmetic series can be found using the formula

$$S_n = \frac{n}{2}(a_1 + a_n).$$

To use this formula, you need to substitute values for n, a_1, and a_n. You can write a general rule for the sum of an arithmetic series so that you only need to substitute the value of n.

For example, consider the sum of the first n even positive integers, shown below. Because there is a common difference of 2, the series is arithmetic. Apply the sum formula above, using $a_1 = 2$ and $a_n = 2n$.

$$2 + 4 + 6 + \cdots + 2n$$

$$S_n = \frac{n}{2}(a_1 + a_n) = \frac{n}{2}(2 + 2n) = n(n + 1)$$

So, a general rule for the sum of the first n even positive integers is $S_n = n(n + 1)$.

In Exercises 1–3, (a) verify that the sequence of numbers is arithmetic by stating the common difference, (b) write a rule for the *n*th term of the sequence, and (c) write a general rule for the sum of the first *n* terms.

1. odd positive integers

2. positive multiples of 3

3. positive multiples of k, with $k > 0$

4. Explain how to use the result of Exercise 3 to write a general rule for the sum of the first n terms of the series $\log 10 + \log 100 + \log 1000 + \log 10{,}000 + \ldots$.

In Exercises 5 and 6, use the results of Exercises 3 and 4 to write a general rule for the sum of the first *n* terms of the series.

5. $\ln e + \ln e^2 + \ln e^3 + \ln e^4 + \ldots$

6. $\log_b a + \log_b a^2 + \log_b a^3 + \log_b a^4 + \ldots$

Name____ Date__________

7.1 Puzzle Time

What Do You Get When You Cross A Judge With Poison Ivy?

Write the letter of each answer in the box containing the exercise number.

Write the first four terms of the sequence.

1. $a_n = n^2 + 2$

2. $a_n = 2n - 4$

3. $a_n = 4 - n$

4. $a_n = \dfrac{n}{n+1}$

5. $f(n) = 2^{2n-2}$

6. $a_n = (n + 2)^2$

Write a rule for the *n*th term of the sequence.

7. $-2, 4, -8, 16, \ldots$

8. $-1, 2, 5, 8, \ldots$

9. $1, \frac{1}{2}, \frac{1}{4}, \frac{1}{8}, \ldots$

10. $-0.2, 0.8, 1.8, 2.8, \ldots$

11. $1.5, 3, 4.5, 6, \ldots$

12. $5, -7, 17, -31, \ldots$

13. $-5, -6, -7, -8, \ldots$

Answers

S. 3, 2, 1, 0

N. $f(n) = (-2)^{n+1} + 1$

R. 3, 6, 11, 18

E. 9, 16, 25, 36

O. $a_n = \dfrac{3n}{2}$

H. $\frac{1}{2}, \frac{2}{3}, \frac{3}{4}, \frac{4}{5}$

D. 1, 4, 16, 64

C. $f(n) = (-2)^n$

I. $a_n = 3n - 4$

S. $a_n = -n - 4$

S. $f(n) = \left(\dfrac{1}{2}\right)^{n-1}$

A. $-2, 0, 2, 4$

I. $a_n = n - 1.2$

1	2	3	4		5	6	7	8	9	10	11	12	13

7.2 Start Thinking

List the five terms of the sequence shown in the graph. Then write a rule for the nth term of the sequence. Describe some of the common features of the graphs and the rules.

1. 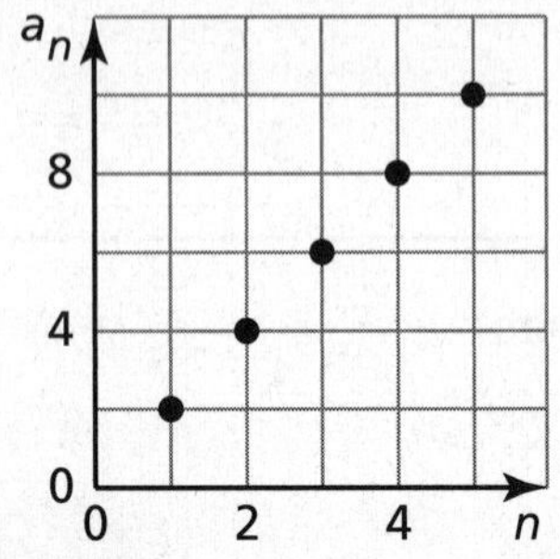

2. 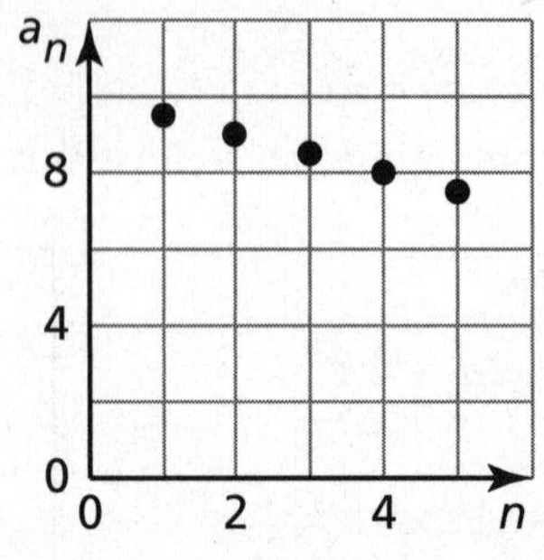

3. 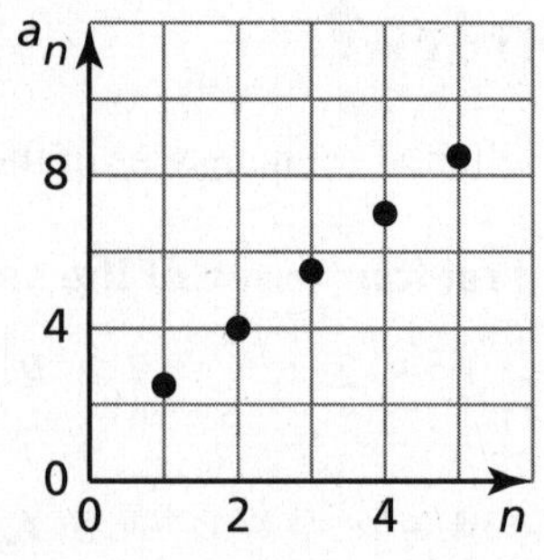

7.2 Warm Up

Solve the system.

1. $x - 2y = 8$
$3x + y = -11$

2. $9 = 5x - 3y$
$16 = 2x + 5y$

3. $-12 = 5n + p$
$32 = -2n + 3p$

4. $8 = a + 5d$
$2 = a + 2d$

5. $-36 = a + 15d$
$-6 = a + 3d$

6. $6 = a + 36d$
$4 = a + 24d$

7.2 Cumulative Review Warm Up

Write a polynomial function *f* of the least degree that has rational coefficients and the given zeros.

1. $-3, \sqrt{2}$

2. $1, -2 + \sqrt{3}$

3. $0, \frac{1}{2}, -\sqrt{6}$

4. $-\frac{3}{2}, -4, 4, 0$

5. $1 + \sqrt{2}, -3 - \sqrt{2}$

6. $0, \frac{5}{2}, -\frac{3}{5}, 5$

Name______________________________ Date__________

7.2 Practice A

In Exercises 1–4, tell whether the sequence is arithmetic. Explain your reasoning.

1. $5, 2, -1, -4, -7, \ldots$

2. $9, 7, 4, 0, -5, \ldots$

3. $\frac{1}{3}, \frac{2}{3}, \frac{3}{3}, \frac{4}{3}, \frac{5}{3}, \ldots$

4. $1, 3, 9, 27, 81, \ldots$

5. Write a rule for the arithmetic sequence with the given description.

a. The first term is -5 and each term is 4 more than the previous term.

b. The first term is 9 and each term is 3 less than the previous term.

In Exercises 6–9, write a rule for the *n*th term of the sequence. Then find a_{20}.

6. $15, 22, 29, 36, \ldots$

7. $62, 53, 44, 35, \ldots$

8. $-25, -10, 5, 20, \ldots$

9. $-3, -\frac{3}{2}, 0, \frac{3}{2}, \ldots$

10. Describe and correct the error in writing a rule for the *n*th term of the arithmetic sequence $-27, -12, 3, 18, 33, \ldots$.

> ✗ Use $a_1 = 27$ and $d = 15$.
>
> $a_n = 27 + (n - 1)15$
>
> $a_n = 12 + 15n$

In Exercises 11 and 12, write a rule for the *n*th term of the sequence. Then graph the first six terms of the sequence.

11. $a_9 = 35, d = 4$

12. $a_{15} = -32, d = -4$

In Exercises 13–16, write a rule for the *n*th term of the sequence.

13. $a_6 = 37, a_{10} = 53$

14. $a_8 = 66, a_{13} = 96$

15. $a_5 = 22, a_{12} = -48$

16. $a_{13} = -76, a_{16} = -97$

17. Find the sum of the positive even integers less than 250. Explain your reasoning.

Name ______________________________ Date __________

7.2 Practice B

In Exercises 1–4, tell whether the sequence is arithmetic. Explain your reasoning.

1. 100, 50, 25, 12.5, 6.25, …

2. 0, −4, −8, −12, −16, …

3. $\frac{1}{6}, \frac{1}{3}, \frac{1}{2}, \frac{2}{3}, \frac{5}{6}, \ldots$

4. $\frac{3}{10}, \frac{3}{5}, \frac{9}{10}, \frac{6}{5}, \frac{3}{2}, \ldots$

5. Write a rule for the arithmetic sequence with the given description.

a. The first term is 12 and each term is 7 less than the previous term.

b. The first term is –8 and each term is 10 more than the previous term.

In Exercises 6–9, write a rule for the *n*th term of the sequence. Then find a_{20}.

6. 37, 29, 21, 13, …

7. $-4, -\frac{8}{3}, -\frac{4}{3}, 0, \ldots$

8. 0.2, 2.3, 4.4, 6.5, …

9. 2.2, 1.5, 0.8, 0.1, …

10. Describe and correct the error in writing a rule for the *n*th term of the arithmetic sequence −27, −12, 3, 18, 33, ….

> ✗ Use $a_1 = -27$ and $d = 15$.
>
> $a_n = a_1 - (n-1)d$
>
> $a_n = -27 - (n-1)15$
>
> $a_n = -12 - 15n$

In Exercises 11 and 12, write a rule for the *n*th term of the sequence. Then graph the first six terms of the sequence.

11. $a_{23} = 107, d = 4$

12. $a_{13} = 12, d = \frac{1}{2}$

In Exercises 13–16, write a rule for the *n*th term of the sequence.

13. $a_4 = 44, a_9 = 69$

14. $a_9 = -73, a_{14} = -158$

15. $a_{15} = 63, a_{21} = 99$

16. $a_{15} = 28, a_{24} = 34$

17. Find the sum of the positive odd integers less than 500. Explain your reasoning.

Name____ Date__________

7.2 Enrichment and Extension

Analyzing Arithmetic Sequences

In Exercises 1–11, use your knowledge of arithmetic sequences.

1. How many terms are in the arithmetic sequence 18, 24, …, 336?

2. How many terms are in the arithmetic sequence 178, 170, …, 2?

3. How many terms are in the arithmetic sequence 506, 500, …, −4?

4. Find the number of multiples of 7 between 30 and 300.

5. Find the number of multiples of 6 between 28 and 280.

6. Find the number of multiples of 11 between 999 and 1999.

7. How many 3-digit numbers are divisible by 4 and 6?

8. How many 4-digit numbers are not divisible by 11?

9. How many 4-digit numbers are not divisible by 2 and 8?

10. Write a rule for t_n, the nth term of the arithmetic sequence $2a - 2b, 3a - b, 4a, 5a + b, \ldots$.

11. Find the values of x and y for the arithmetic sequence $y, 2x + y, 7y, 20, \ldots$.

Name ______________________________ Date __________

7.2 Puzzle Time

What Do You Call A Haunted Chicken?

Write the letter of each answer in the box containing the exercise number.

Write a rule for the *n*th term of the arithmetic sequence.

1. 3, 7, 11, 15, ...
2. 25, 16, 7, −2, ...
3. −10, −7, −4, −1, ...
4. −5, −9, −13, −17, ...
5. $a_8 = 12, d = 7$
6. $a_{20} = -6, d = -3$
7. $a_{14} = 54, d = 8$
8. $a_{30} = -72, d = -15$
9. $a_2 = 11, a_{12} = 51$
10. $a_{15} = 73, a_{21} = 55$
11. $a_6 = -8, a_{20} = -50$
12. $a_{22} = 17, a_{30} = -23$

Answers

T. $a_n = -5n + 127$

P. $a_n = 4n - 1$

U. $a_n = 3n - 13$

T. $a_n = 7n - 44$

O. $a_n = -9n + 34$

I. $a_n = -3n + 118$

R. $a_n = -3n + 54$

L. $a_n = -4n - 1$

Y. $a_n = 8n - 58$

G. $a_n = -15n + 378$

E. $a_n = 4n + 3$

S. $a_n = -3n + 10$

1	2	3	4	5	6	7	8	9	10	11	12

7.3 Start Thinking

List the four terms of the sequence shown in the graph. Then write a rule for the nth term of the sequence. Describe some of the common features of the graphs and the rules.

1. 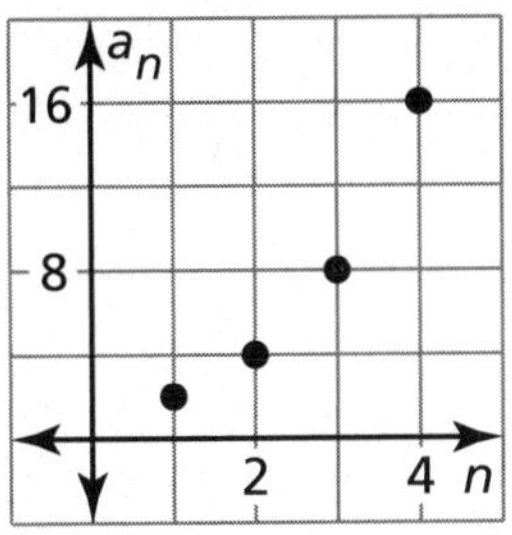

2. 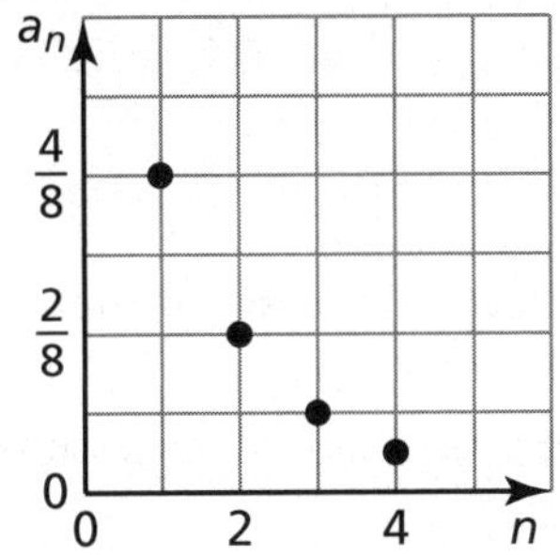

3. 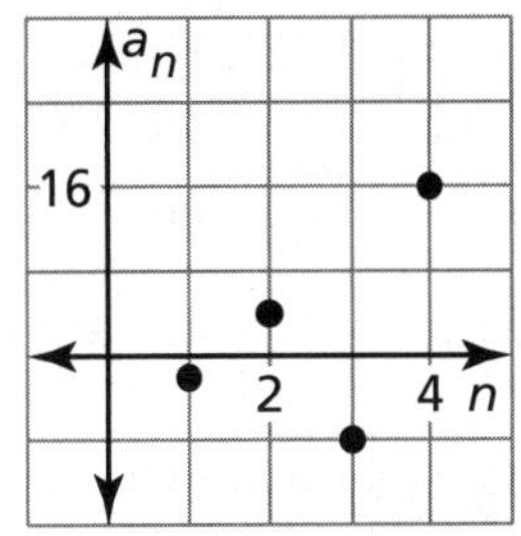

7.3 Warm Up

Find the sum.

1. $\sum_{i=1}^{4} 3i$
2. $\sum_{i=1}^{9} (2 - i)$
3. $\sum_{i=1}^{15} (i^2 + 2)$
4. $\sum_{k=5}^{7} (10 - 2k)$
5. $\sum_{i=1}^{4} 2^{i-1}$
6. $\sum_{i=1}^{5} 2\left(-\frac{1}{2}\right)^{i}$

7.3 Cumulative Review Warm Up

Sketch the graph of the function.

1. $f(x) = 2^{x-1}$
2. $f(x) = 2^x + 3$
3. $f(x) = 2^{-x}$
4. $f(x) = \log_2 x$
5. $f(x) = 3 + \log_2 x$
6. $f(x) = \log_2(x + 1)$

Name ______________________________ Date __________

7.3 Practice A

In Exercises 1–4, tell whether the sequence is geometric. Explain your reasoning.

1. 64, 32, 16, 8, 4, ...
2. 88, 66, 44, 22, 0, ...
3. 0.3, 1.2, 2.1, 3, 3.9, ...
4. 0.8, 4.8, 28.8, 172.8, ...

5. Write a rule for the geometric sequence with the given description.

   a. The first term is −5, and each term is 3 times the previous term.

   b. The first term is 54, and each term is $\frac{1}{6}$ times the previous term.

In Exercises 6–9, write a rule for the *n*th term of the sequence. Then find a_7.

6. 3, 6, 12, 24, ...
7. 7, 21, 63, 189, ...
8. 192, 96, 48, 24, ...
9. 36, 24, 16, $\frac{32}{3}$, ...

In Exercises 10–13, write a rule for the *n*th term of the sequence. Then graph the first six terms of the sequence.

10. $a_3 = 9,\ r = 3$
11. $a_2 = 12,\ r = 4$
12. $a_4 = 5,\ r = \frac{1}{2}$
13. $a_5 = -208,\ r = 2$

14. Describe and correct the error in writing a rule for the *n*th term of the geometric sequence for which $a_3 = 147,\ r = 7$.

$$\begin{aligned} a_n &= r a_1^{\,n-1} \\ 147 &= 7a_1^2 \\ 21 &= a_1^2 \\ \sqrt{21} &= a_1 \\ a_n &= 7\sqrt{21}^{\,n-1} \end{aligned}$$

15. You are buying a new car. You take out a 3-year loan for $10,000. The annual interest rate of the loan is 6%. You can calculate the monthly payment M (in dollars) for a loan using the formula $M = \dfrac{L}{\sum_{k=1}^{t}\left(\frac{1}{1+i}\right)^k}$, where L is the loan amount (in dollars), i is the monthly interest rate (in decimal form), and t is the term (in months). Calculate the monthly payment.

Name________________________ Date__________

7.3 Practice B

In Exercises 1–4, tell whether the sequence is geometric. Explain your reasoning.

1. 3, 6, 18, 72, 360, ...

2. 162, 54, 18, 6, 2, ...

3. 0.7, 3.5, 17.5, 87.5, 437.5, ...

4. $\frac{5}{3}, \frac{10}{3}, \frac{20}{3}, \frac{40}{3}, \frac{80}{3}, \ldots$

5. Write a rule for the geometric sequence with the given description.

a. The first term is −12, and each term is 7 times the previous term.

b. The first term is 62, and each term is $\frac{1}{2}$ times the previous term.

In Exercises 6–9, write a rule for the *n*th term of the sequence. Then find a_7.

6. 9, 18, 36, 72, ...

7. 80, 20, 5, $\frac{5}{4}$, ...

8. $3, \frac{6}{5}, \frac{12}{25}, \frac{24}{125}, \ldots$

9. 1.2, −2.4, 4.8, −9.6, ...

In Exercises 10–13, write a rule for the *n*th term of the sequence. Then graph the first six terms of the sequence.

10. $a_3 = 50, r = 5$

11. $a_2 = 18, r = \frac{1}{3}$

12. $a_4 = -378, r = 3$

13. $a_5 = 1, r = -\frac{1}{4}$

14. Describe and correct the error in writing a rule for the *n*th term of the geometric sequence for which $a_3 = 147, r = 7$.

✗
$$a_n = a_1 r^{n-1}$$
$$a_n = 147(7)^{n-1}$$

15. You are buying a new boat. You take out a 5-year loan for $20,000. The annual interest rate of the loan is 4%. You can calculate the monthly payment M (in dollars) for a loan using the formula $M = \frac{L}{\sum_{k=1}^{t}\left(\frac{1}{1+i}\right)^k}$, where L is the loan amount (in dollars), i is the monthly interest rate (in decimal form), and t is the term (in months). Calculate the monthly payment.

Name ______________________________ Date __________

7.3 Enrichment and Extension

Analyzing Geometric Sequences and Series

The ideas of arithmetic and geometric sequences can be combined, as shown below.

Example: Three numbers form a geometric sequence whose common ratio is 3. When the first number is decreased by 1, the second is decreased by 3, and the third is reduced to 2 less than one-half of its value, the resulting three numbers form an arithmetic sequence. Determine the original three numbers.

$x, 3x, 9x$	Represent the three numbers that form a geometric sequence.
$x - 1, 3x - 3, \frac{9}{2}x - 2$	Represent the three numbers that form an arithmetic sequence.
$(3x - 3) - (x - 1) = \left(\frac{9}{2}x - 2\right) - (3x - 3)$	Write an equation using the constant difference of consecutive terms in an arithmetic sequence.
$x = 6$	Solve for x.

Evaluate x, $3x$, and $9x$ when $x = 6$ to obtain the original numbers. The original numbers are 6, 18, and 54.

1. Three numbers form a geometric sequence whose common ratio is $\frac{1}{2}$. When the first number is reduced to 10 more than one-fourth of its value, the second is decreased by 10, and the third is increased by 10 more than twice its value, the resulting three numbers form an arithmetic sequence. Determine the original three numbers.

2. Three numbers form an arithmetic sequence whose common difference is 3. When the first number is increased by 1, the second is increased by 6, and the third is increased by 19, the resulting three numbers form a geometric sequence. Determine the original three numbers.

3. Insert two real numbers between 2 and 9 so that the first three terms form an arithmetic sequence and the last three terms form a geometric sequence.

4. Determine the second term of an arithmetic sequence when the first term is 2 and the first, third, and seventh terms form a geometric sequence.

5. Let a, b, and c represent real numbers that are not consecutive terms of an arithmetic sequence or a geometric sequence. If $a < b < c$, determine a number that when added to a, b, and c yields consecutive terms of a geometric sequence.

6. Show that if a and b are real numbers, then the arithmetic mean of a^2 and b^2 is greater than or equal to the absolute value of their geometric means. $\left(\textit{Hint: } (a - b)^2 \geq 0 \text{ and } (a + b)^2 \geq 0\right)$

Name____ Date__________

7.3 Puzzle Time

What Does The Cow Like To Do On Her Day Off?

Write the letter of each answer in the box containing the exercise number.

Write a rule for the *n*th term of the geometric sequence.

1. 3, 9, 27, 81, ...
2. 7, 14, 28, 56, ...
3. 100, −50, 25, −12.5, ...
4. 125, 25, 5, 1, ...
5. $a_3 = 12, r = 2$
6. $a_4 = 40, r = \frac{1}{2}$
7. $a_2 = -27, r = 3$
8. $a_5 = 240, r = -2$
9. $a_2 = 8, a_4 = 32$
10. $a_4 = 16, a_7 = 1024$
11. $a_3 = -24, a_5 = -6$

Answers

I. $a_n = \frac{1}{4}(4)^{n-1}$

V. $a_n = 4(2)^{n-1}$

S. $a_n = 3(3)^{n-1}$

E. $a_n = 7(2)^{n-1}$

A. $a_n = 125\left(\frac{1}{5}\right)^{n-1}$

O. $a_n = 320\left(\frac{1}{2}\right)^{n-1}$

E. $a_n = 100\left(-\frac{1}{2}\right)^{n-1}$

O. $a_n = -9(3)^{n-1}$

O. $a_n = 15(-2)^{n-1}$

M. $a_n = 3(2)^{n-1}$

E. $a_n = -96\left(-\frac{1}{2}\right)^{n-1}$

1	2	3		4		5	6	7	8	9	10	11

7.4 Start Thinking

List the first 10 terms of the geometric sequence $a_n = 0.1(0.1)^{n-1}$.

Then find the value of $\sum_{n=1}^{10} 0.1(0.1)^{n-1}$ and make a conjecture about the value of $\sum_{n=1}^{\infty} 0.1(0.1)^{n-1}$.

7.4 Warm Up

Find the sum.

1. $\sum_{n=1}^{5} 2^{n-1}$

2. $\sum_{n=1}^{5} 2n - 1$

3. $\sum_{n=1}^{8} 4\left(\frac{2}{3}\right)^{n-1}$

4. $\sum_{n=1}^{12} -2(3)^{n-1}$

5. $\sum_{n=1}^{200} 5 - \frac{1}{2}n$

6. $\sum_{n=1}^{15} \frac{1}{2}(3)^{n-1}$

7.4 Cumulative Review Warm Up

Graph the function. State the domain and range.

1. $f(x) = \frac{1}{x - 1} + 3$

2. $f(x) = \frac{-2}{x + 3}$

3. $f(x) = \frac{3}{x - 2} + 4$

4. $f(x) = \frac{x}{x + 5}$

5. $f(x) = \frac{3x - 1}{2x - 1}$

6. $f(x) = \frac{-2x + 5}{x - 3}$

Name_______________________________ Date__________

7.4 Practice A

In Exercises 1 and 2, consider the infinite geometric series. Find the partial sums S_n for $n = 1, 2, 3, 4,$ and 5. Then describe what happens to S_n as n increases.

1. $\frac{1}{3} + \frac{1}{6} + \frac{1}{12} + \frac{1}{24} + \frac{1}{48} + \dots$

2. $5 + \frac{10}{3} + \frac{20}{9} + \frac{40}{27} + \frac{80}{81} + \dots$

In Exercises 3–6, find the sum of the infinite geometric series, if it exists.

3. $\sum_{n=1}^{\infty} 7\left(\frac{1}{4}\right)^{n-1}$

4. $\sum_{n=1}^{\infty} 3\left(\frac{5}{4}\right)^{n-1}$

5. $3 + \frac{9}{5} + \frac{27}{25} + \frac{81}{125} + \dots$

6. $-6 - 4 - \frac{8}{3} - \frac{16}{9} - \dots$

7. Describe and correct the error in finding the sum of the infinite geometric series.

> X $\sum_{n=1}^{\infty} \frac{5}{2}\left(\frac{1}{3}\right)^{n-1}$
>
> For this series, $a_1 = \frac{5}{2}$ and $r = \frac{1}{3}$.
>
> Because $|a_1| \geq 1$, this series does not have a sum.

8. You push your younger sister on a swing one time and then allow your sister to swing freely. On the first swing, your sister travels a distance of 8 feet. On each successive swing, your sister travels 80% of the distance of the previous swing. What is the total distance your sister swings?

In Exercises 9–11, write the repeating decimal as a fraction in simplest form.

9. 0.18181818...

10. 0.5555...

11. 1.6666...

12. A company had a profit of \$500,000 in its first year. Since then, the company's profit has decreased by 6% each year. Assuming this trend continues, what is the total profit the company can make over the course of its lifetime?

Name ______________________________ Date __________

7.4 Practice B

In Exercises 1 and 2, consider the infinite geometric series. Find the partial sums S_n for $n = 1, 2, 3, 4$, and 5. Then describe what happens to S_n as n increases.

1. $\frac{3}{4} + \frac{1}{4} + \frac{1}{12} + \frac{1}{36} + \frac{1}{108} + \dots$

2. $6 + 4 + \frac{8}{3} + \frac{16}{9} + \frac{32}{27} + \dots$

In Exercises 3–6, find the sum of the infinite geometric series, if it exists.

3. $\sum_{n=1}^{\infty} \frac{5}{3}\left(\frac{3}{4}\right)^{n-1}$

4. $\sum_{n=1}^{\infty} \frac{3}{7}\left(\frac{7}{2}\right)^{n-1}$

5. $8 - 10 + \frac{25}{2} - \frac{125}{4} + \dots$

6. $\frac{1}{5} + \frac{2}{15} + \frac{4}{45} + \frac{8}{135} + \dots$

7. Describe and correct the error in finding the sum of the infinite geometric series.

> ✗ $\sum_{n=1}^{\infty} \frac{5}{2}\left(\frac{1}{3}\right)^{n-1}$
>
> For this series, $a_1 = \frac{5}{2}$ and $r = \frac{1}{3}$.
>
> $$S = \frac{a_1}{1 - r} = \frac{\frac{5}{2}}{\frac{1}{3}} = \frac{5}{2} \bullet \frac{3}{1} = \frac{15}{2}$$

8. You are going for a 4-mile run. You know that you can run half the distance, and you successfully run 2 miles. There are 2 miles to go, and you know that you can run half that distance. You successfully run that next mile. Now there is 1 mile to go, and you know that you can run half that distance. You successfully run that next half mile. This process continues. Will you cover the 4 miles over the course of your run? Explain your answer.

In Exercises 9–11, write the repeating decimal as a fraction in simplest form.

9. 0.45454545…

10. 0.05050505…

11. 1.4444…

12. A radio station has a daily contest in which a random listener is asked a trivia question. On the first day, the station gives $300 to the first listener who answers correctly. On each successive day, the winner receives 95% of the winnings from the previous day. What is the total amount of prize money the radio station gives away during the contest?

Name ______________________________ Date __________

7.4 Enrichment and Extension

Finding Sums of Infinite Geometric Series

Example: For what value(s) of x does the following infinite series have a sum, or converge?

$$1 + (x - 2) + (x - 2)^2 + (x - 2)^3 + \dots$$

Solution: This is an infinite geometric series with common ratio $r = x - 2$. You have learned that such a series converges when $|r| < 1$; that is, when $|x - 2| < 1$, or $1 < x < 3$. The interval for which a series converges is called the *interval of convergence* for the series.

In Exercises 1–6, find (a) the interval of convergence and (b) an expression for the sum of the series.

1. $1 + x^2 + x^4 + x^6 + \dots$

2. $1 + 3x + 9x^2 + \dots$

3. $1 + (x - 3) + (x - 3)^2 + (x - 3)^3 + \dots$

4. $1 - (x - 1) + (x - 1)^2 - (x - 1)^3 + \dots$

5. $1 - \frac{2}{x} + \frac{4}{x^2} - \frac{8}{x^3} + \dots$

6. $\frac{x^2}{3} - \frac{x^4}{6} + \frac{x^6}{12} - \dots$

7. Show that the series $\sin^2 x + \sin^4 x + \sin^6 x + \dots$ converges to $\tan^2 x$ when $x \neq \frac{\pi}{2} + n\pi$, where n is an integer.

8. Show that the series $\tan^2 x - \tan^4 x + \tan^6 x - \dots$ converges to $\sin^2 x$ when $-\frac{\pi}{4} < x < \frac{\pi}{4}$. Are there other values of x for which the series converges?

9. Explain why there is no infinite geometric series with first term 10 and sum 4.

Name ______________________________ Date __________

7.4 Puzzle Time

What Kind Of Dog Does Dracula Have?

Write the letter of each answer in the box containing the exercise number.

Find the sum of the infinite geometric series, if it exists.

1. $1 + 0.5 + 0.25 + 0.125 + \dots$

2. $3 - \frac{9}{4} + \frac{27}{16} - \frac{81}{64} + \dots$

3. $1 - 0.6 + 0.36 - 0.216 + \dots$

4. $\sum_{i=1}^{\infty} 5\left(-\frac{1}{5}\right)^{i-1}$

5. $\sum_{i=1}^{\infty} \left(\frac{1}{3}\right)^{i-1}$

6. $\sum_{i=1}^{\infty} -6\left(-\frac{1}{2}\right)^{i-1}$

7. $\sum_{i=1}^{\infty} 4^{i-1}$

8. $\sum_{i=1}^{\infty} 42\left(\frac{1}{2}\right)^{i-1}$

9. $\sum_{i=1}^{\infty} 8\left(-\frac{7}{9}\right)^{i-1}$

10. $\sum_{i=1}^{\infty} \frac{7}{6^i}$

Answers

H. -4

N. $\frac{9}{2}$

B. 2

O. $\frac{5}{8}$

D. $\frac{3}{2}$

O. $\frac{25}{6}$

O. The sum does not exist.

U. 84

L. $\frac{12}{7}$

D. $\frac{7}{5}$

1	2	3	4	5	6	7	8	9	10

7.5 Start Thinking

Examine the sequence. Identify the pattern used to determine each consecutive term. Then find the next term in the sequence.

1. 1, 2, 3, 5, 8, 13, …

2. 1, 2, 2, 4, 8, 32, …

3. 10, 5, 5, 0, 5, −5, …

4. −2, −1, 2, −2, −4, 8, …

7.5 Warm Up

Write the first six terms of the sequence.

1. $a_n = 4 - \frac{1}{2}n$

2. $f(n) = \frac{3n + 1}{n}$

3. $a_n = n^2 + n$

4. $f(n) = 2\left(\frac{3}{2}\right)^{n-1}$

5. $a_n = n^3 - 10$

6. $a_1 = 1, a_n = a_{n-1} + 3$

7.5 Cumulative Review Warm Up

Solve the quadratic equation.

1. $3x^2 - 2x - 2 = 0$

2. $-x^2 + 3x = -5x + 2$

3. $5 - 2x = x^2 - 5x + 7$

4. $7x^2 + 3 = x^2 + 15$

5. $\frac{1}{2}x^2 - 5x + 3 = 0$

6. $\frac{2}{3} - \frac{1}{2}x^2 = -\frac{1}{3}x - 2$

Name ______________________________ Date __________

7.5 Practice A

In Exercises 1–6, write the first six terms of the sequence.

1. $a_1 = 1$
$a_n = a_{n-1} + 5$

2. $a_1 = 1$
$a_n = a_{n-1} - 4$

3. $f(0) = 3$
$f(n) = 4f(n-1)$

4. $f(0) = 12$
$f(n) = \frac{1}{3}f(n-1)$

5. $a_1 = 1$
$a_n = (a_{n-1})^2 + 2$

6. $a_1 = 2$
$a_n = \frac{1}{2}(a_{n-1})^2$

In Exercises 7–14, write a recursive rule for the sequence.

7. 32, 24, 16, 8, 0, ...

8. −47, −35, −23, −11, 1, ...

9. 2, 6, 18, 54, 162, ...

10. 5, −10, 20, −40, 80, ...

11. 21, 7, $\frac{7}{3}$, $\frac{7}{9}$, $\frac{7}{27}$, ...

12. 1, 7, 13, 19, 25, ...

13. 2, 3, 5, 8, 13, ...

14. 2, 3, 6, 18, 108, ...

In Exercises 15–20, write a recursive rule for the sequence.

15. $a_n = 5 + 2n$

16. $a_n = -4 - 3n$

17. $a_n = 15 - 13n$

18. $a_n = 8(10)^{n-1}$

19. $a_n = -2(7)^{n-1}$

20. $a_n = 1.8 - 0.8n$

21. The basic fee for a sailboat rental is $75. There is an additional $20 fee for each additional hour over 2 hours. The explicit rule $a_n = 75 + 20n$ gives the amount of the rental for n hours over 2 hours. Write a recursive rule for the amount of the rental for n hours over 2 hours.

In Exercises 22–25, write an explicit rule for the sequence.

22. $a_1 = 5,\ a_n = a_{n-1} - 3$

23. $a_1 = 14,\ a_n = a_{n-1} + 5$

24. $a_1 = -3,\ a_n = 2a_{n-1}$

25. $a_1 = 20,\ a_n = \frac{1}{2}a_{n-1}$

Name______________________________ Date __________

7.5 Practice B

In Exercises 1–6, write the first six terms of the sequence.

1. $a_1 = 1$
$a_n = a_{n-1} + 9$

2. $f(0) = 32$
$f(n) = \frac{1}{4}f(n-1)$

3. $f(0) = 24$
$f(n) = \frac{3}{2}f(n-1)$

4. $a_1 = 1$
$a_n = (a_{n-1})^2 - 1$

5. $f(0) = 1,\ f(1) = 4$
$f(n) = f(n-2) - f(n-1)$

6. $f(1) = 256,\ f(2) = 2$
$f(n) = \dfrac{f(n-2)}{f(n-1)}$

In Exercises 7–14, write a recursive rule for the sequence.

7. $30, 21, 12, 3, -6, \ldots$

8. $3, -15, 75, -375, \ldots$

9. $28, 4, \frac{4}{7}, \frac{4}{49}, \frac{4}{343}, \ldots$

10. $1, 12, 23, 34, 45, \ldots$

11. $2, 6, 12, 72, 864, \ldots$

12. $1, 7, 8, 15, 23, \ldots$

13. $61, 39, 22, 17, 5, \ldots$

14. $-5, -3, 0, 4, 9, \ldots$

In Exercises 15–20, write a recursive rule for the sequence.

15. $a_n = -7 + 3n$

16. $a_n = 6(15)^{n-1}$

17. $a_n = -16(9)^{n-1}$

18. $a_n = -2.4 + 0.3n$

19. $a_n = -\frac{1}{3}\left(\frac{1}{5}\right)^{n-1}$

20. $a_n = \frac{1}{2}(7)^{n-1}$

21. The rate of growth of an organism is given by the explicit rule $a_n = 26(1.002)^{n-1}$, where n is the number of hours in an incubator. Write a recursive rule for the rate of growth of the organism.

In Exercises 22–25, write an explicit rule for the sequence.

22. $a_1 = -19,\ a_n = a_{n-1} + 7.2$

23. $a_1 = -7,\ a_n = 0.45a_{n-1}$

24. $a_1 = 4,\ a_n = a_{n-1} + \frac{1}{6}$

25. $a_1 = -9,\ a_n = \frac{1}{3}a_{n-1}$

Name ______________________________ Date __________

7.5 Enrichment and Extension

Using Recursive Rules with Sequences

In Exercises 1 and 2, use the sequences defined recursively to solve the problem.

1. Write the first eight terms of the sequence defined recursively by $t_1 = 3$, $t_2 = 5$, and $t_n = t_{n-1} - t_{n-2}$. Then find the 100th term of the sequence.

2. Write the first eight terms of the sequence defined recursively by $t_1 = 4$, $t_2 = 8$, and $t_n = \frac{t_{n-1}}{t_{n-2}}$. Then find the 100th term of the sequence.

3. Let d_n represent the number of diagonals that can be drawn in an n-sided polygon. The leftmost diagram below shows that a hexagon has nine diagonals, and so $d_6 = 9$.

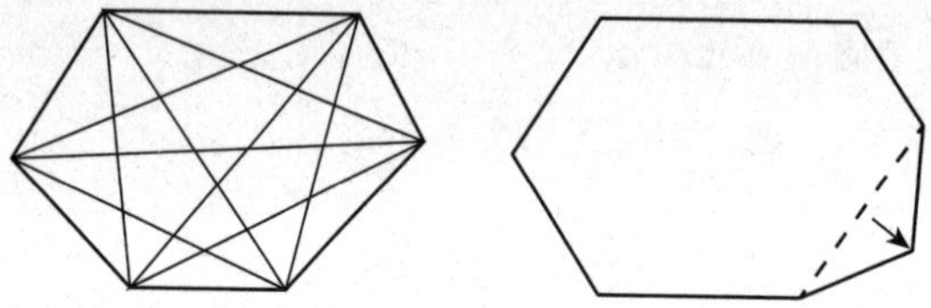

   a. Imagine pushing out one side of the hexagon so that a polygon of seven sides is formed, as shown above at the right. How many additional diagonals can be drawn?

   b. Imagine pushing out one side of a polygon with $n - 1$ sides so that an n-sided polygon is formed. Write an expression for how many additional diagonals can be drawn. Then write a recursive rule for the sequence of additional diagonals drawn.

4. Let S_n represent the number of dots in an n-by-n square array. Write a recursive equation that shows how S_{n+1} is related to S_n by reasoning how many additional dots are needed to form the $(n + 1)$ square from the previous nth square array. Illustrate your answer with a diagram.

Name________________________________ Date__________

7.5 Puzzle Time

What Do You Call A Large Gorilla Who Likes To Dance?

Write the letter of each answer in the box containing the exercise number.

Write the first six terms of the sequence.

1. $a_1 = 2$
   $a_n = a_{n-1} + 4$

2. $a_1 = 2$
   $a_n = a_{n-1} - 3$

3. $a_1 = 2$
   $a_n = (a_{n-1})^2 - 1$

Write a recursive rule for the sequence.

4. $a_n = 5 - 3n$

5. $a_n = -185 + 200n$

6. $a_n = 2\left(\frac{3}{2}\right)^{n-1}$

7. $a_n = -6(-2)^{n-1}$

8. $a_n = 5(3)^{n-1}$

9. $a_n = 297\left(\frac{1}{3}\right)^{n-1}$

Answers

C. $a_1 = 15$
$a_n = a_{n-1} + 200$

K. 2, 6, 10, 14, 18, 22

N. 2; 3; 8; 63; 3968; 15,745,023

O. $a_1 = 2$
$a_n = \frac{3}{2}a_{n-1}$

G. $a_1 = 2$
$a_n = a_{n-1} - 3$

N. $a_1 = -6$
$a_n = -2 \bullet a_{n-1}$

I. 2, −1, −4, −7, −10, −13

G. $a_1 = 5$
$a_n = 3 \bullet a_{n-1}$

A. $a_1 = 297$
$a_n = \frac{1}{3}a_{n-1}$

1	2	3	4		5	6	7	8	9

Name ______________________ Date __________

Chapter 7 Cumulative Review

In Exercises 1–9, copy and complete the table to evaluate the function.

1. $y = -11x + 26$

x	y
4	
8	
12	
16	

2. $y = 18x - 34$

x	y
2	
4	
6	
8	

3. $y = 6x + 13$

x	y
6	
12	
18	
24	

4. $y = 5x^2 - 8$

x	y
2	
3	
4	
5	

5. $y = -3x^2 + 4$

x	y
2	
3	
4	
5	

6. $y = 4x^2 - 19$

x	y
2	
3	
4	
5	

7. $y = 2^x - 3$

x	y
0	
1	
2	
3	

8. $y = 5 - 2^x$

x	y
0	
1	
2	
3	

9. $y = 3^x - 12$

x	y
0	
1	
2	
3	

In Exercises 10–21, solve the equation. Check your solution.

10. $8x + 17 = 73$

11. $-17x + 8 = -43$

12. $-7x + 2 = 79$

13. $\frac{1}{3}x - 6 = 14$

14. $-\frac{1}{2}x + 8 = -16$

15. $\frac{2}{3}x - 11 = 15$

16. $2x - 10 = 4x + 8$

17. $13x - 8 = 19x + 16$

18. $11x - 2 = -20x + 60$

19. $2^x - 3 = 61$

20. $3^x - 31 = 50$

21. $4 + 4^x = 68$

22. At 9:00 A.M. the temperature is 51°F. The temperature increases 3°F each hour for the next 6 hours. Write a function to model the temperature T at h hours after 9:00 A.M.

23. At 6:00 P.M. the temperature is 44°F. The temperature decreases 2°F each hour for the next 8 hours. Write a function to model the temperature T at h hours after 6:00 P.M.

Name____________________ Date__________

Chapter 7 Cumulative Review (continued)

In Exercises 24–29, evaluate the function for the given value of *x*.

24. $g(x) = 8x^4 - 9x^3 + x^2 + 4x - 14; x = -7$

25. $h(x) = -9x^4 + 12x^3 + 2x^2 - 3x - 1; x = 5$

26. $f(x) = 6x^4 + 10x^3 - 3x^2 - 8x + 7; x = 9$

27. $h(x) = 7x^5 - 15x^4 - 14x^3 + 14x^2 + 12x - 12; x = -5$

28. $p(x) = -x^5 - 8x^4 + 4x^3 + 14x^2 - 2x + 10; x = 8$

29. $g(x) = 2x^5 - 12x^4 - 8x^3 + 2x^2 - 9x + 5; x = 3$

In Exercises 30–33, find the sum.

30. $(6x^2 + 5x + 15) + (-8x^2 - x + 9)$

31. $(2x^3 - 6x^2 + 14x - 4) + (x^3 + 13x^2 - 11x + 9)$

32. $(9x^4 - 9x^5 + x^2 - x + 2) + (4x^3 - 2x^5 + x^4 + 7x - 6)$

33. $(8x^3 - 12x + 6x^4 + 13x^5 + 1) + (-6x^5 + x^4 - 6x^3 - 2x + 7x^2 - 7)$

In Exercises 34–37, find the difference.

34. $(10x^3 - 11x^2 + 9x - 8) - (8x^3 - 6x^2 - 6x + 1)$

35. $(3x^4 + 7x^3 - 3x^2 - 4x + 3) - (-3x^4 - 8x^3 + 5x^2 - 15x + 13)$

36. $(4x^4 - 13x^3 + 9x^2 - 11x - 6) - (10x^4 + 3x^2 - 14x + 2)$

37. $(5x^5 + 10x^4 - 10x^3 - 3x^2 + 12x - 3) - (10x^5 - 12x^4 + 10x^3 - 15x - 12)$

In Exercises 38–43, find the product.

38. $7x^2(3x^2 - 5x + 3)$

39. $8x(7x^5 - 15x^4 + 11x^3 + 3x^2 - 7x - 9)$

40. $(15x + 7)(-2x^2 + 3x + 2)$

41. $(11x^2 + x)(4x^2 + 13x - 7)$

42. $(7x - 1)(8x^3 - 2x^2 + 13x - 3)$

43. $(x^2 + x - 12)(2x^2 - 8x + 1)$

Name ______________________________ Date ________

Chapter 7 Cumulative Review (continued)

In Exercises 44–49, find the product of the binomials.

44. $(x + 5)(x + 2)(x - 4)$

45. $(x + 5)(x - 1)(x - 2)$

46. $(x + 4)(x - 4)(x + 4)$

47. $(x - 2)(x - 4)(x - 5)$

48. $(4x + 3)(x + 1)(5x - 2)$

49. $(3x - 1)(4x + 1)(2x + 5)$

In Exercises 50–55, find the product of the binomials.

50. $(x + 6)^2$

51. $(x - 7)(x + 7)$

52. $(x - 8)^2$

53. $(2x + 11)^2$

54. $(6x - 11)^2$

55. $(9x + 10)^2$

In Exercises 56–61, divide the polynomials.

56. $(2x^2 - 3x + 7) \div (x + 4)$

57. $(5x^2 + 6x - 7) \div (x + 7)$

58. $(9x^2 - x - 2) \div (x - 2)$

59. $(4x^3 + 5x^2 - 6x + 6) \div (x - 7)$

60. $(5x^4 - 3x^3 - 5x^2 + 2x - 8) \div (x + 4)$

61. $(3x^4 - 5x^3 + 4x - 5) \div (x + 5)$

62. A rectangle has an area expressed by $2x^2 + 10x + 12$ and a length expressed by $x + 3$.

a. Write the expression for the width of the rectangle.

b. The value of x is 2. Determine the area of the rectangle using the value of x.

63. A rectangle has an area expressed by $4x^2 + 29x + 30$ and a width expressed by $x + 6$.

a. Write the expression for the length of the rectangle.

b. The value of x is 3. Determine the area of the rectangle using the value of x.

c. Determine the width of the rectangle using the value of x given in part (b).

d. Determine the length of the rectangle using the expression from part (a) and the value of x given in part (b).

Name____ Date__________

Chapter 7 Cumulative Review (continued)

In Exercises 64–79, factor the polynomial completely.

64. $x^3 + 5x^2 - 24x$

65. $c^3 + 8c^2 - 9c$

66. $m^3 - 13m^2 + 40m$

67. $a^3 + 2a^2 - 3a$

68. $d^3 + 512$

69. $g^3 - 27$

70. $y^3 - 125$

71. $n^3 + 343$

72. $8b^3 + 125$

73. $27w^3 - 343$

74. $f^3 - 7f^2 + 3f - 21$

75. $3r^3 + 5r^2 - 6r - 10$

76. $5h^3 - 8h^2 - 5h + 8$

77. $3s^3 - 7s^2 + 12s - 28$

78. $4v^4 - v^3 + 28v - 7$

79. $7p^4 + 2p^3 - 63p - 18$

In Exercises 80–85, solve the equation.

80. $z^3 + 4z^2 - 12z = 0$

81. $x^4 = 8x^3 + 33x^2$

82. $m^3 = 7m^2 + 44m$

83. $h^3 + 13h^2 = -30h$

84. $-3v^3 = 42v^2 + 135v$

85. $5f^3 + 160f = 60f^2$

In Exercises 86–91, find the zeros of the function.

86. $f(x) = x^4 - 3x^3 - 28x^2$

87. $g(x) = x^4 + 5x^3 - 24x^2$

88. $h(x) = x^5 - 22x^4 + 120x^3$

89. $g(x) = x^3 - 7x^2 + 10x$

90. $h(x) = x^5 - 12x^4 + 20x^3$

91. $f(x) = x^4 - 8x^3 - 33x^2$

92. The area of a square picture is represented by the expression $9x^2 + 30x + 25$.

a. Write an expression for the side length of the square.

b. The value of x is 1. Determine the area of the square, using the value of x.

93. The area of a square picture is represented by the expression $36x^2 + 132x + 121$.

a. Write an expression for the side length of the square.

b. The value of x is 1.5. Determine the area of the square using the value of x.

c. Determine the side length of the square using the value of x given in part (b) and the expression found in part (a).

Chapter 8

Name____ Date__________

Chapter 8 Trigonometric Ratios and Functions

Dear Family,

Do you have a local coffee shop, diner, or café that you like to visit? What times of the day do you usually go? What hours are they open? Is the menu the same all day? Many restaurants collect data throughout the day to meet the needs of their customers, which makes their business more efficient and profitable.

The graph below shows the number of people in a café during a normal business day.

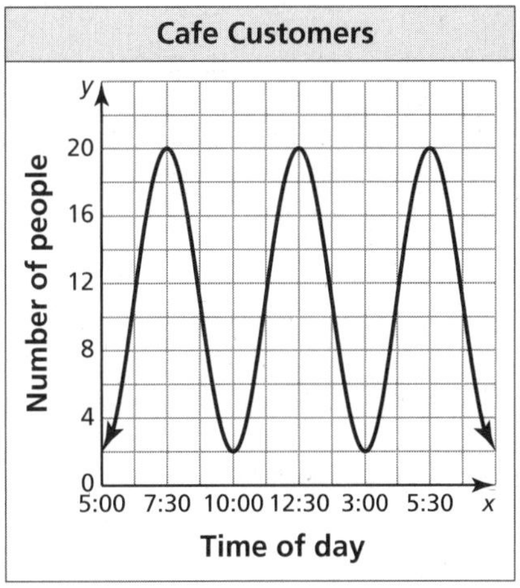

As a family, pretend you are the owners of the café. Use the graph to answer the following questions.

- When would you open and close the café?
- What hours would your employees work?
- At what times of the day would you need the most employees working for you?
- When would your employees change shifts?
- What type of menu would you have?
- What types of food would you serve at the busiest times of the day?
- What different types of music would you play throughout the day?
- How else could you use the graph above to help your business meet the needs of your customers?

The graph above is an example of a trigonometric function. What do you notice about the graph? Use the Internet to research other applications of trigonometric functions.

Enjoy your next experience at your favorite local coffee shop, diner, or café!

Nombre ______________________________ Fecha __________

Capítulo 8 Razones y funciones trigonométricas

Estimada familia:

¿Hay alguna cafetería, restaurante o café local donde les gusta ir? ¿En qué horario suelen ir? ¿En qué horario están abiertos? ¿El menú es el mismo todo el día? Muchos restaurantes recopilan datos durante el día para satisfacer las necesidades de sus clientes, que hace que su negocio sea más eficiente y rentable.

En la siguiente gráfica, se muestra el número de personas en un café durante un día hábil normal.

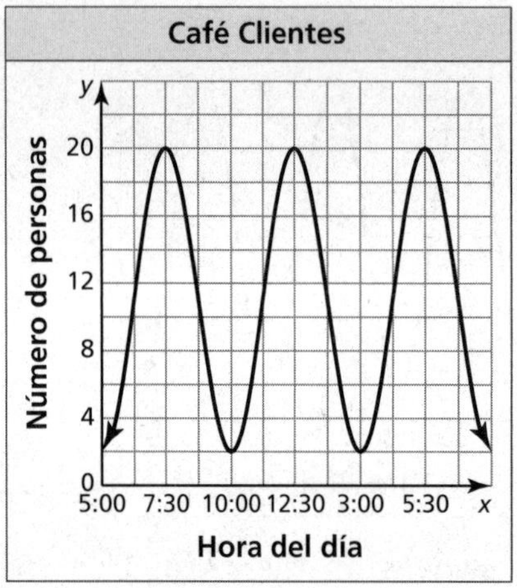

En familia, imaginen ser los dueños del café. Usen la gráfica para responder las siguientes preguntas:

- ¿Cuándo abrirían y cerrarían el café?
- ¿En qué horario trabajarían sus empleados?
- ¿A qué horas del día necesitarían que trabajen más empleados?
- ¿Cuándo sería el cambio de turno de sus empleados?
- ¿Qué tipo de menú tendrían?
- ¿Qué clase de comida servirían en los horarios más atareados?
- ¿Qué clases de música diferente podrían durante el día?
- ¿De qué otras maneras podrían usar la gráfica de arriba para ayudar a que su negocio satisfaga las necesidades de sus clientes?

La gráfica de arriba es un ejemplo de una función trigonométrica. ¿Qué observan sobre la gráfica? Investiguen en Internet otros usos de las funciones trigonométricas.

¡Disfruten su próxima visita a su cafetería, restaurante o café local favorito!

8.1 Start Thinking

Use a ruler and a protractor to draw two right triangles, as shown below. Make one hypotenuse exactly 3 inches long and the other exactly 4 inches long. Use a ruler to find the measure of the remaining sides of each triangle. Determine the value of each of the ratios below. Round to the nearest hundredth, when necessary. Compare the ratios and make a conjecture about them in general, for any 30°-60°-90° triangle.

1. $\dfrac{y_1}{3}$ **2.** $\dfrac{x_1}{3}$

3. $\dfrac{y_1}{x_1}$ **4.** $\dfrac{y_2}{4}$

5. $\dfrac{x_2}{4}$ **6.** $\dfrac{y_2}{x_2}$

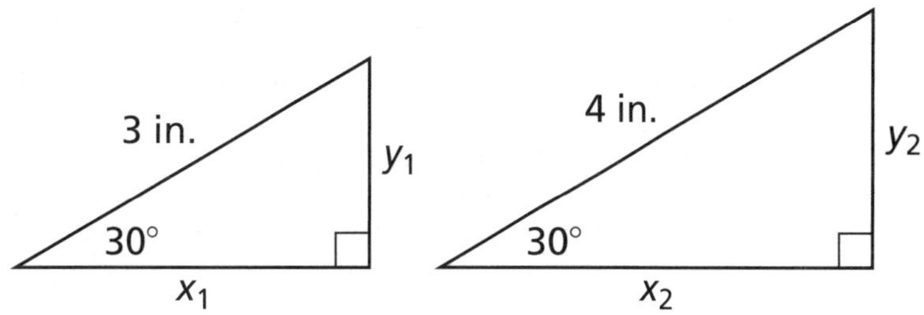

8.1 Warm Up

Use the Pythagorean Theorem to find the missing sides.

1.

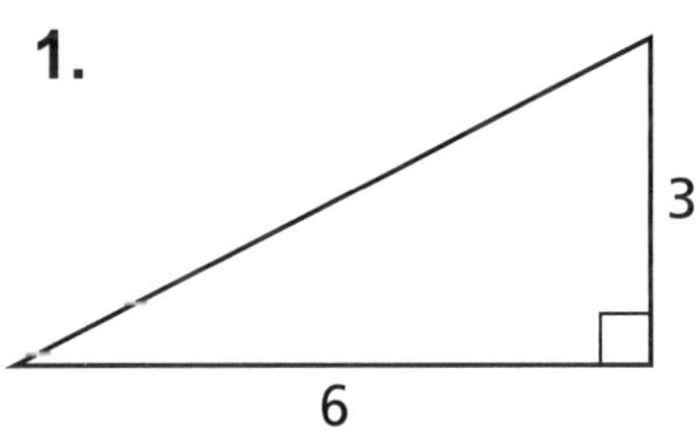

2.

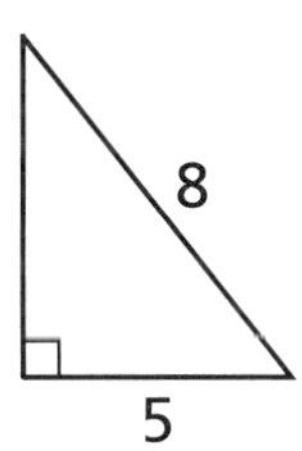

3.

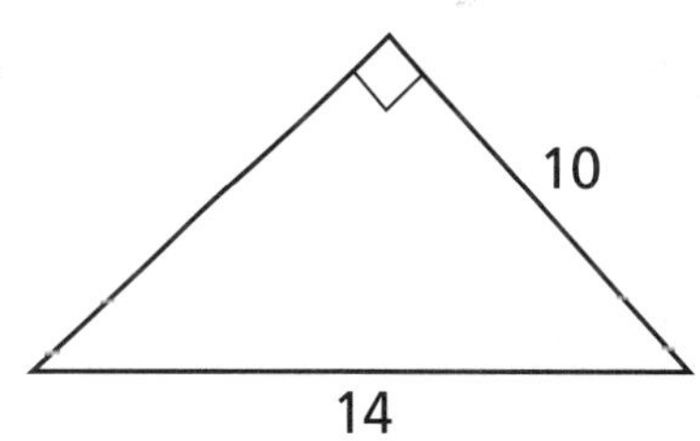

8.1 Cumulative Review Warm Up

Write a rule for the *n*th term of the sequence.

1. 11, 5, −1, −7, −13, ...

2. 4, −8, 16, −32, 64, ...

3. 0, 8, 16, 24, 32, ...

4. −1, 3, −5, 7, −9, ...

5. $\dfrac{2}{1}, \dfrac{3}{4}, \dfrac{4}{9}, \dfrac{5}{16}, \dfrac{6}{25}, \ldots$

6. $\dfrac{1}{2}, \dfrac{3}{5}, \dfrac{9}{8}, \dfrac{27}{11}, \dfrac{81}{14}, \ldots$

Name ______________________________ Date ________

8.1 Practice A

In Exercises 1 and 2, evaluate the six trigonometric functions of the angle θ.

1.

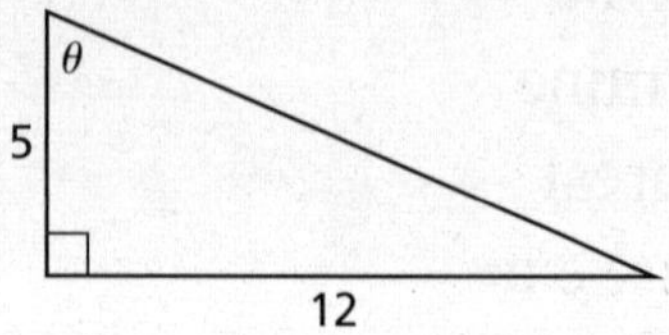

2.

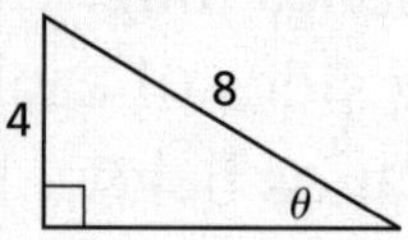

3. Let θ be an acute angle of a right triangle. Use the two trigonometric functions $\sin\theta = \frac{3}{7}$ and $\cot\theta = \frac{2\sqrt{10}}{3}$ to sketch and label the triangle. Then evaluate the other four trigonometric functions of θ.

In Exercises 4–6, let θ be an acute angle of a right triangle. Evaluate the other five trigonometric functions of θ.

4. $\sin\theta = \frac{4}{11}$ **5.** $\cos\theta = \frac{5}{6}$ **6.** $\tan\theta = \frac{3}{4}$

7. Describe and correct the error in finding $\tan\theta$ of the triangle below.

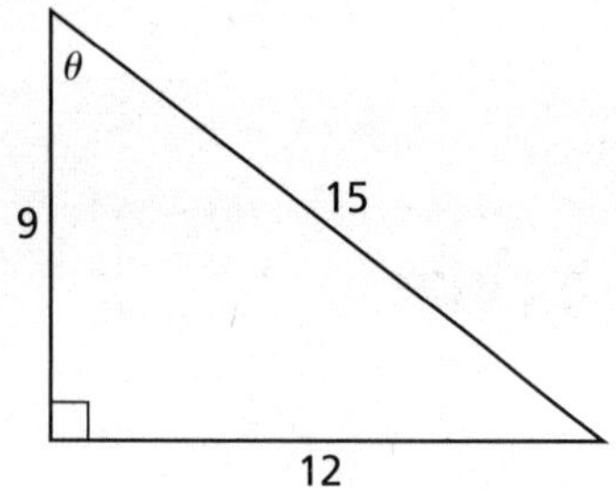

✗ $\tan\theta = \frac{\text{adj.}}{\text{opp.}} = \frac{9}{12} = \frac{3}{4}$

In Exercises 8 and 9, find the value of *x* for the right triangle.

8.

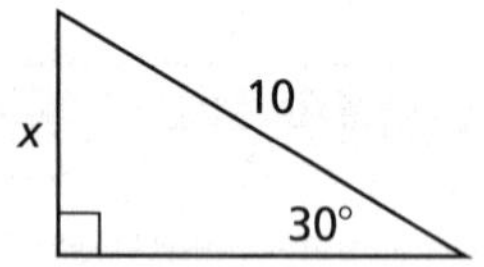

9.

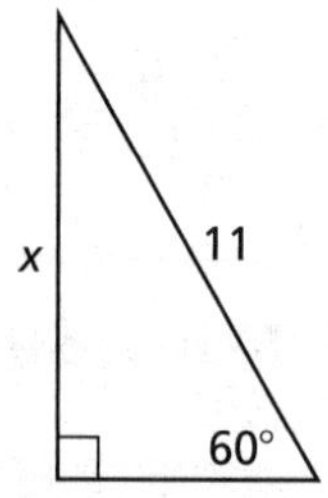

10. A parasail rider is attached to a boat with a rope that is 80 feet long. The angle of elevation from the boat to the parasail rider is 36°. Estimate the parasail rider's height above the boat. Round your answer to the nearest tenth.

Name______________________________ Date__________

8.1 Practice B

In Exercises 1 and 2, evaluate the six trigonometric functions of the angle θ.

1.

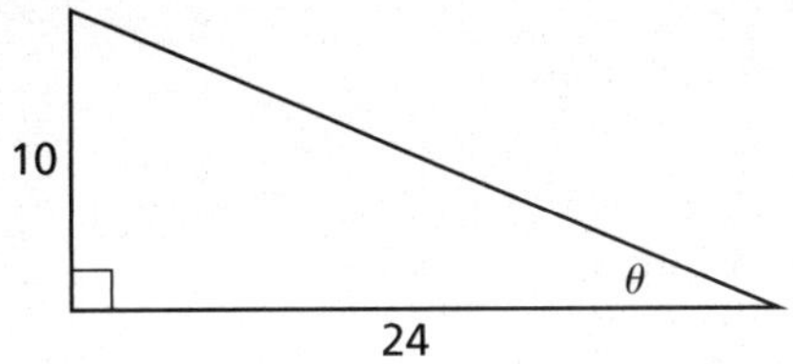

2.

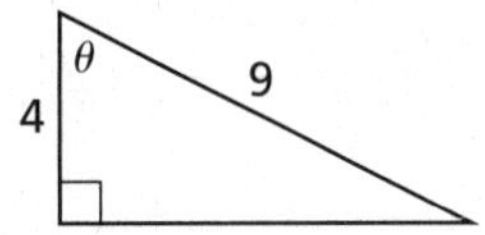

3. Evaluate the six trigonometric functions of the angle $90° - \theta$ in Exercise 1. Describe the relationships you notice.

In Exercises 4–6, let θ be an acute angle of a right triangle. Evaluate the other five trigonometric functions of θ.

4. $\cos\theta = \frac{5}{11}$

5. $\cot\theta = \frac{7}{8}$

6. $\sec\theta = \frac{11}{9}$

7. Describe and correct the error in finding $\csc\theta$ of the triangle below.

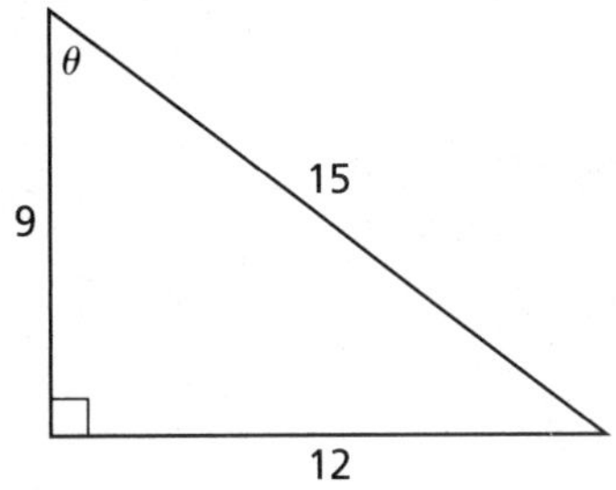

X $\sec\theta = \frac{\text{adj.}}{\text{hyp.}} = \frac{9}{15} = \frac{3}{5}$

In Exercises 8 and 9, find the value of *x* for the right triangle.

8.

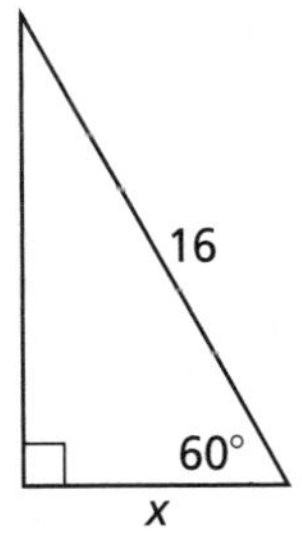

9.

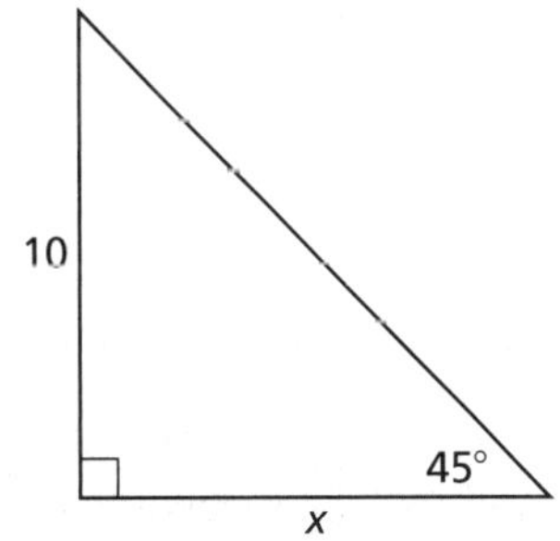

10. A cable is attached to the top of a pole and mounted to the ground 3 feet from the base of the pole. The angle of elevation from the mounting to the top of the pole is 78°. Estimate the height of the pole. Round your answer to the nearest tenth.

Name ______________________________ Date ________

8.1 Enrichment and Extension

Right Triangle Trigonometry

You can use right triangle trigonometry to solve problems involving one or more right triangles.

Example: As shown in the diagram, a pole is on the roof of a shed. From a point P 27 feet from the foot of the shed, the angle of elevations to the top of the pole and the top of the roof are 38° and 32°, respectively. Determine the height of the pole, to the nearest tenth of a foot.

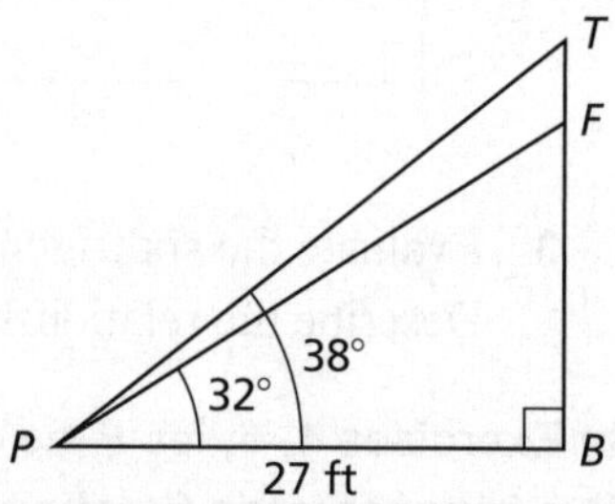

Solution: Because $\overline{TF}$ is not the side of a right triangle, first determine TB in ΔTBP and FB in ΔFBP, and then subtract to determine TF.

In ΔTBP:

$$\tan 38° = \frac{TB}{27}$$

$$TB = 27 \tan 38° \approx 21.09$$

In ΔFBP:

$$\tan 32° = \frac{FB}{27}$$

$$FB = 27 \tan 32° \approx 16.87$$

So, $TF = TB - FB \approx 21.09 - 16.87 \approx 4.2$ feet.

1. A ship located at point A is headed directly east toward a coastline formed by a vertical cliff that is 80 meters high. The angle of elevation to the top of the cliff is 10°. A few minutes later, at point D, the angle of elevation to the top of the cliff has increased to 20°. Find each measure to the nearest meter.

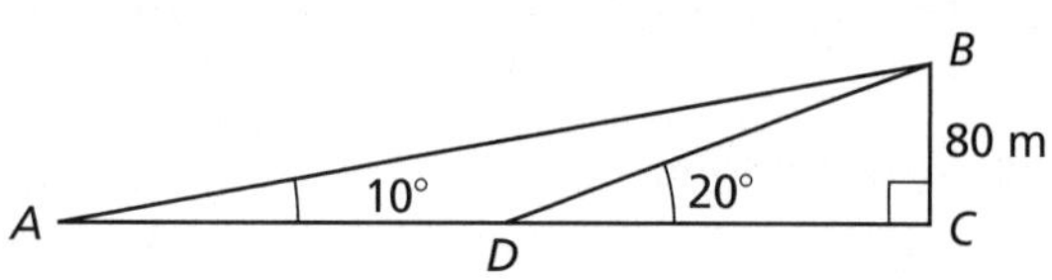

a. DC

b. AC

c. the distance between the two sightings

2. The heights of two buildings on opposite sides of a 90-foot-wide avenue are 100 feet and 50 feet, respectively. From a point E on the avenue, the angle of elevation to the top of the taller building is 55°. Find each measure to the nearest foot.

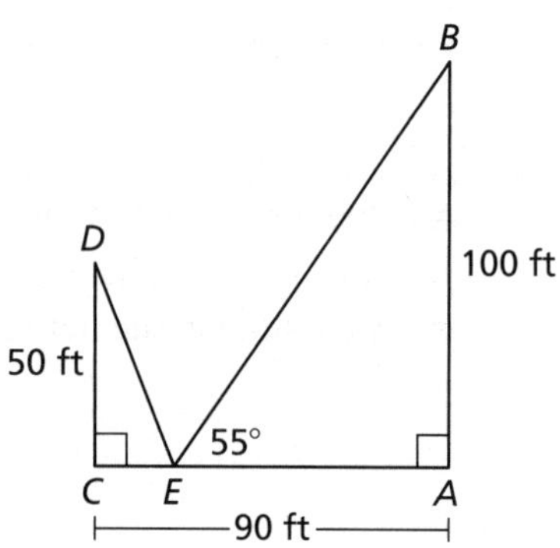

a. EA

b. ED

Name______________________________ Date__________

8.1 Puzzle Time

What Is A Chrysler's Favorite Game?

Write the letter of each answer in the box containing the exercise number.

Let θ be an acute angle of a right triangle. Evaluate $\csc\theta$.

1. $\sin\theta = \frac{2}{3}$

2. $\cos\theta = \frac{4}{7}$

3. $\tan\theta = \frac{10}{3}$

Let θ be an acute angle of a right triangle. Evaluate $\sin\theta$.

4. $\sec\theta = \frac{5}{2}$

5. $\cot\theta = \frac{2}{9}$

6. $\csc\theta = \frac{6}{5}$

Find the value of *x* for the right triangle.

7.

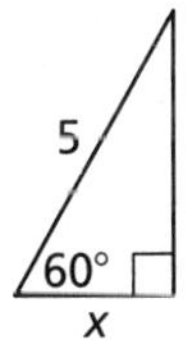

8.

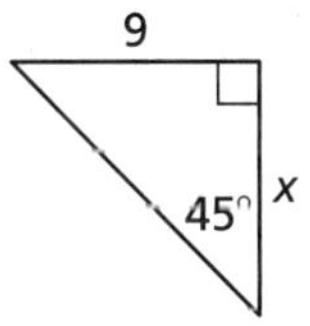

9.

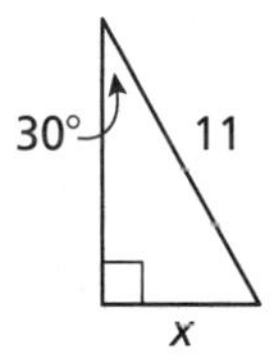

Answers

A. $\frac{5}{2}$

O. $\frac{7\sqrt{33}}{33}$

L. $\frac{11}{2}$

D. $\frac{\sqrt{109}}{10}$

D. $\frac{3}{2}$

E. $\frac{9\sqrt{85}}{85}$

B. $\frac{5}{6}$

G. $\frac{\sqrt{21}}{5}$

L. 9

1	2	3	4	5	6	7	8	9

8.2 Start Thinking

Use a compass and a ruler to draw three circles, one with a radius of 1 inch, one with a radius of 2 inches, and one with a radius of 4 inches. Use a piece of string to measure a 1-inch long arc length on the 1-inch circle. Then use a protractor to measure the central angle that would create this arc length. Repeat the process with the other two circles using a 2-inch string and a 4-inch string. What conclusion can you make about the central angle that is needed to create an arc length that is equal to the radius of a circle?

8.2 Warm Up

Use a protractor and a straightedge to draw the angle in the *x-y* coordinate plane according to the description given. One ray should align with the positive *x*-axis.

1. $45°$

2. $210°$

3. $300°$

8.2 Cumulative Review Warm Up

Solve the system.

1. $y = 2x + 5$

$y = -\frac{1}{2}x - 5$

2. $x - y = 4$

$2x + 3y = 3$

3. $-3x + 4y = 6$

$x - \frac{4}{3}y = -2$

4. $x^2 + y^2 = 9$

$2x - y = -6$

5. $x^2 - x - y = 0$

$x - y = -3$

6. $2x^2 - x - y = -1$

$3x + y = 5$

Name______________________________ Date__________

8.2 Practice A

In Exercises 1–6, draw an angle with the given measure in standard position.

1. 170°

2. 540°

3. -80°

4. 5π

5. $\frac{2\pi}{3}$

6. $-\frac{3\pi}{4}$

In Exercises 7–9, match the angle measure with the angle.

7. 500°

8. $\frac{7\pi}{3}$

9. -130°

A.

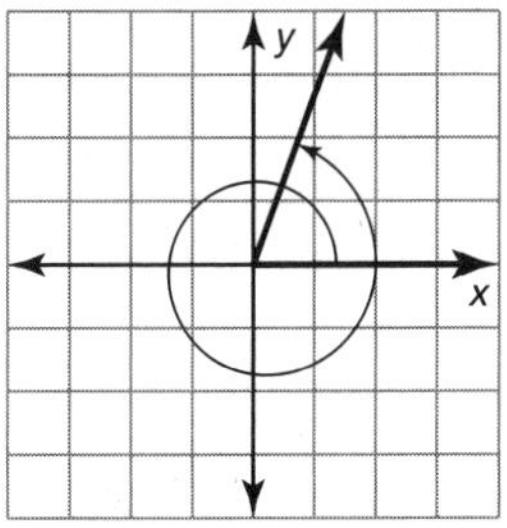

B.

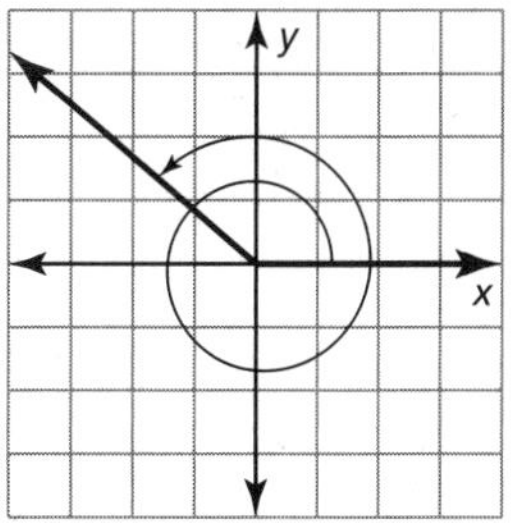

C.

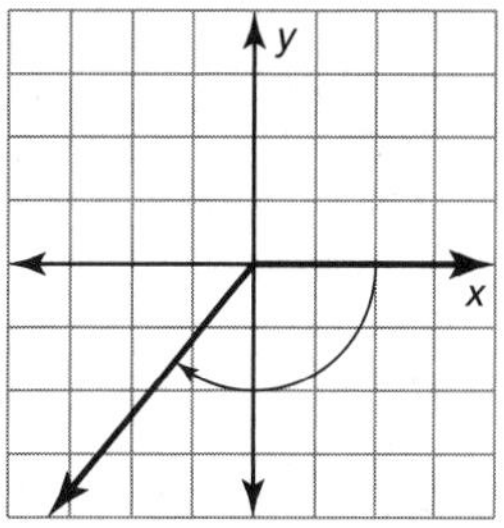

In Exercises 10–12, find one positive angle and one negative angle that are coterminal with the given angle.

10. 75°

11. -40°

12. $\frac{7\pi}{3}$

In Exercises 13–18, convert the degree measure to radians or the radian measure to degrees.

13. 225°

14. 50°

15. -160°

16. $\frac{\pi}{12}$

17. $\frac{5\pi}{9}$

18. $-\frac{\pi}{8}$

19. You work every Saturday in the yard from 8:00 A.M. to 11:30 A.M. Draw a diagram that shows the rotation completed by the hour hand of a clock during this time. Find the measure of the angle generated by the hour hand in both degrees and radians. Compare this angle with the angle generated by the minute hand from 8:00 A.M. to 11:30 A.M.

Name ______________________ Date ________

8.2 Practice B

In Exercises 1–6, draw an angle with the given measure in standard position.

1. 260°

2. 400°

3. -200°

4. $\frac{5\pi}{2}$

5. $\frac{7\pi}{6}$

6. -4π

In Exercises 7–9, match the angle measure with the angle.

7. -300°

8. $\frac{5\pi}{3}$

9. $-\frac{11\pi}{6}$

A.

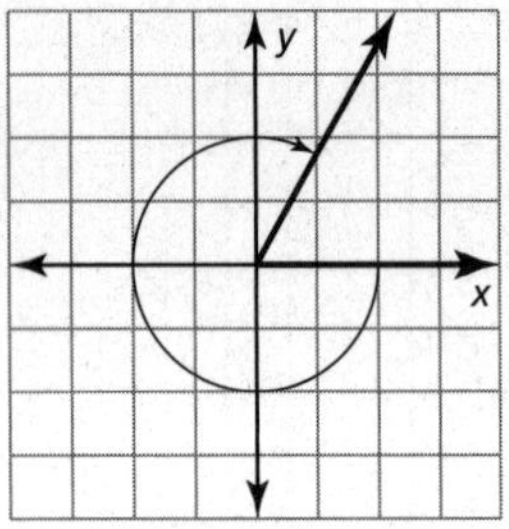

B.

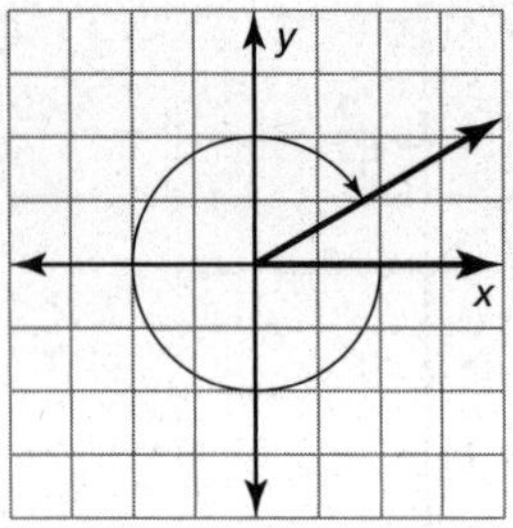

C.

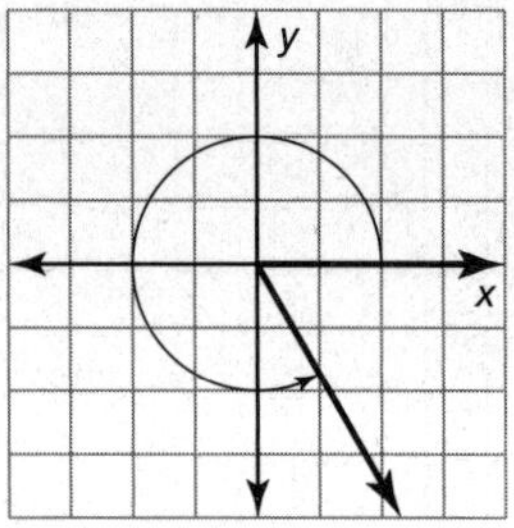

In Exercises 10–12, find one positive angle and one negative angle that are coterminal with the given angle.

10. $\frac{5\pi}{4}$

11. -420°

12. $-\frac{11\pi}{2}$

In Exercises 13–18, convert the degree measure to radians or the radian measure to degrees.

13. 200°

14. 1°

15. -475°

16. $\frac{3\pi}{10}$

17. $-\frac{5\pi}{12}$

18. 6

19. There are 60 minutes in 1 degree of arc, and 60 seconds in 1 minute of arc. The notation 50°30'10" represents an angle with a measure of 50°, 30 minutes, and 10 seconds.

a. Write the angle measure 160.44° using the notation above.

b. Write the angle measure $98^\circ 15' 45''$ to the nearest hundredth of a degree.

Name ____________________ Date __________

8.2 Enrichment and Extension

Angles and Radian Measures

In Exercises 1–12, without using a calculator or table, complete the statement with <, >, or =.

1. $\sin 40°$ ____ $\sin 30°$
2. $\cos 40°$ ____ $\cos 30°$
3. $\sin 172°$ ____ $\sin 8°$
4. $\sin 310°$ ____ $\sin 230°$
5. $\sin 130°$ ____ $\sin 50°$
6. $\cos 50°$ ____ $\cos(-50°)$
7. $\sin 214°$ ____ $\cos 213°$
8. $\sin 169°$ ____ $\sin 168°$
9. $\sin \frac{5\pi}{4}$ ____ $\cos\left(\frac{-3\pi}{4}\right)$
10. $\sin \frac{\pi}{4}$ ____ $\cos \frac{5\pi}{6}$
11. $\sin \frac{5\pi}{3}$ ____ $\cos \frac{3\pi}{4}$
12. $\sin 5$ ____ $\cos 6$

In Exercises 13 and 14, list the values in order from least to greatest.

13. $\sin 1,\ \sin 2,\ \sin 3,\ \sin 4$
14. $\cos 1,\ \cos 2,\ \cos 3,\ \cos 4$
15. A sector of a circle has a perimeter of 7 centimeters and an area of 3 square centimeters. Find all possible radii.
16. A sector of a circle has a perimeter of 12 centimeters and an area of 8 square centimeters. Find all possible radii.
17. The sector shown has a perimeter of 20 centimeters.

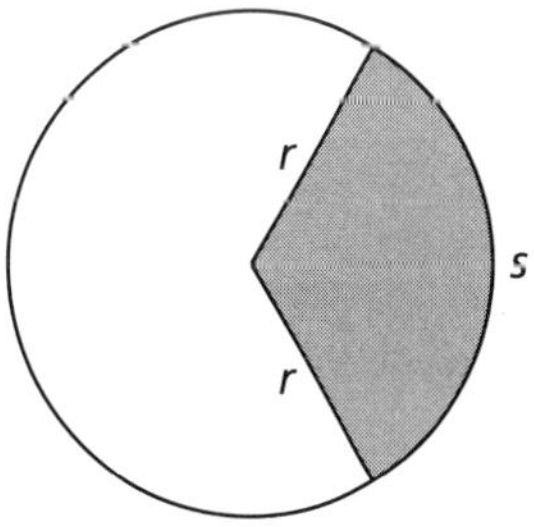

   a. Show that $\theta = \frac{20}{r} - 2$, and that the area of the sector is $A = 10r - r^2$.

   b. What value of r gives the maximum possible area of a sector of perimeter 20 centimeters?

   c. What is the measure of the central angle of the sector of maximum area?

Name ______________________________ Date __________

8.2 Puzzle Time

Why Did The Kangaroo Lose The Basketball Game?

Write the letter of each answer in the box containing the exercise number.

Find one positive angle and one negative angle that are coterminal with the given angle.

1. 80°

2. −124°

3. 112°

4. −31°

5. $\frac{3\pi}{2}$

6. $-\frac{5\pi}{3}$

7. $-\frac{\pi}{4}$

8. $\frac{12\pi}{7}$

Convert the degree measure to radians or the radian measure to degrees.

9. −425°

10. $\frac{7\pi}{2}$

11. 135°

12. $\frac{34\pi}{5}$

13. 74°

14. $-\frac{\pi}{4}$

15. −18°

16. $-\frac{2\pi}{3}$

Answers

E. 236°; −484° **A.** 329°; −391°

H. 440°; −280° **F.** 630°

O. 1224° **O.** $\frac{\pi}{3}; -\frac{11\pi}{3}$

R. 472°; −248° **D.** $-\frac{\pi}{10}$

U. $\frac{7\pi}{4}; -\frac{9\pi}{4}$ **T.** $\frac{26\pi}{7}; -\frac{2\pi}{7}$

O. $-\frac{85\pi}{36}$ **S.** −120°

B. $\frac{3\pi}{4}$ **U.** $\frac{37\pi}{90}$

N. $\frac{7\pi}{2}; -\frac{\pi}{2}$ **N.** −45°

1	2		3	4	5		6	7	8		9	10		11	12	13	14	15	16

8.3 Start Thinking

Diagram A shows a unit circle with five labeled points. Use the diagram to find the coordinates of the point (x, y). How could you define $\sin 45°$, $\cos 45°$, and $\tan 45°$ according to x and y? Use this definition to find the values of $\sin \theta$, $\cos \theta$, and $\tan \theta$ shown in Diagram B. Round to the nearest hundredth, when necessary. If this definition was applied to angles in the third and fourth quadrant, what conclusions could you make?

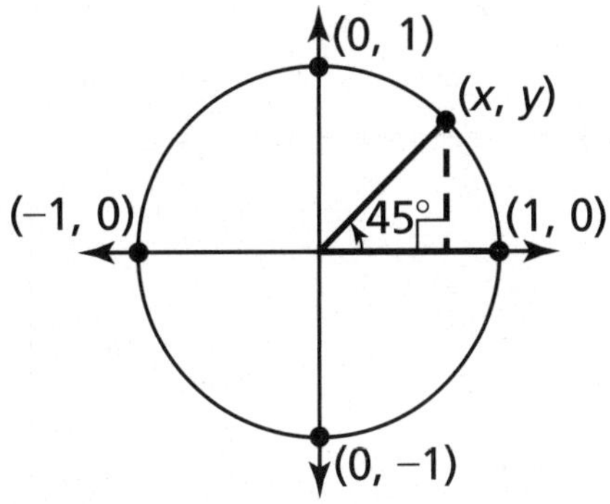

Diagram A

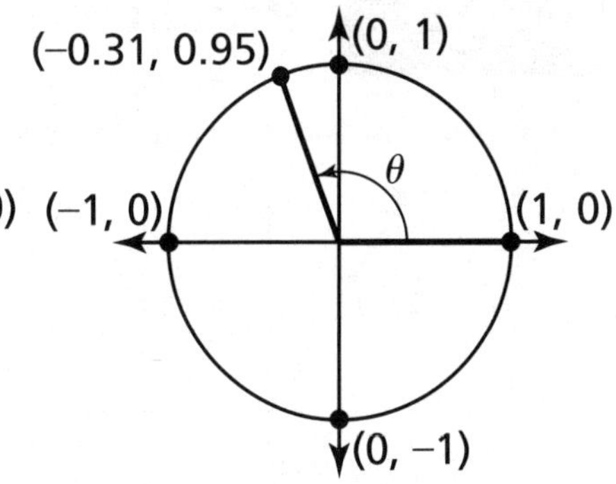

Diagram B

8.3 Warm Up

Find the measure of the hypotenuse of the triangle.

1.

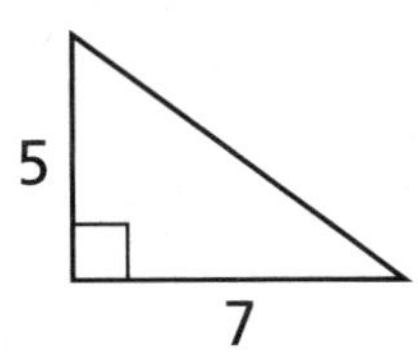

2.

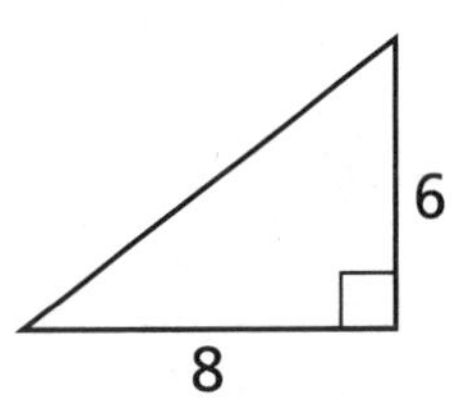

3.

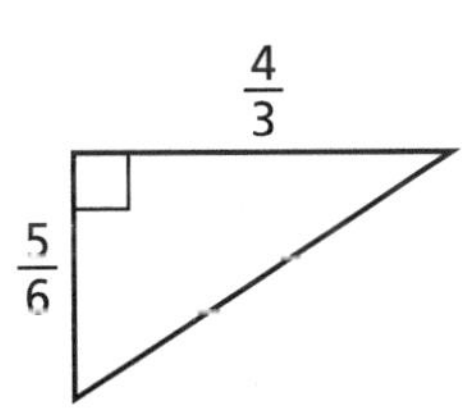

4.

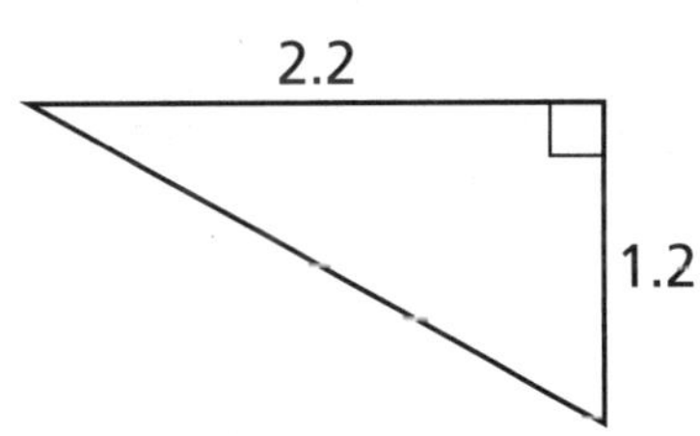

8.3 Cumulative Review Warm Up

Perform the indicated operation.

1. $(x^2 - 3x + 8) - (3x^2 - 5x)$

2. $(2x^2 + x - 5)(x - 7)$

3. $(2x - 1)^2$

4. $(3 - 5x)(3 + 5x)$

5. $(3 - x^2)^2$

6. $(3x - 2)^3$

Name ______________________________ Date __________

8.3 Practice A

In Exercises 1–4, evaluate the six trigonometric functions of θ.

1.

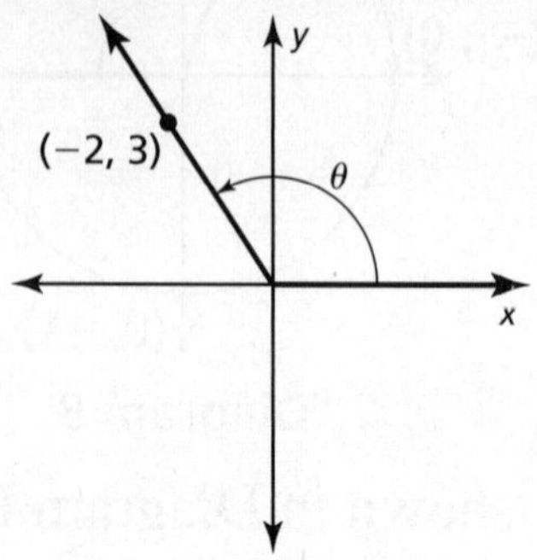

2.

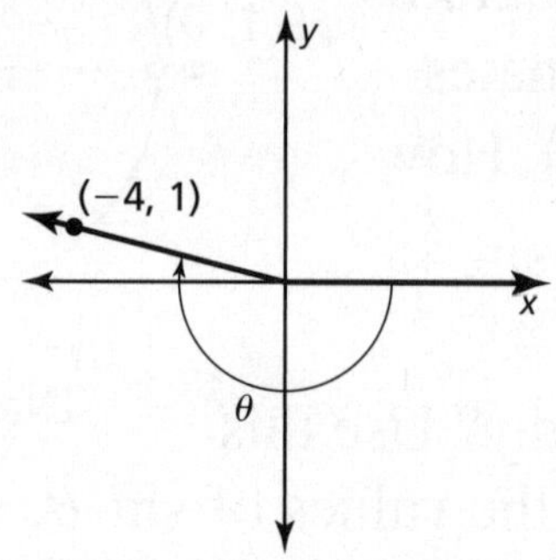

3.

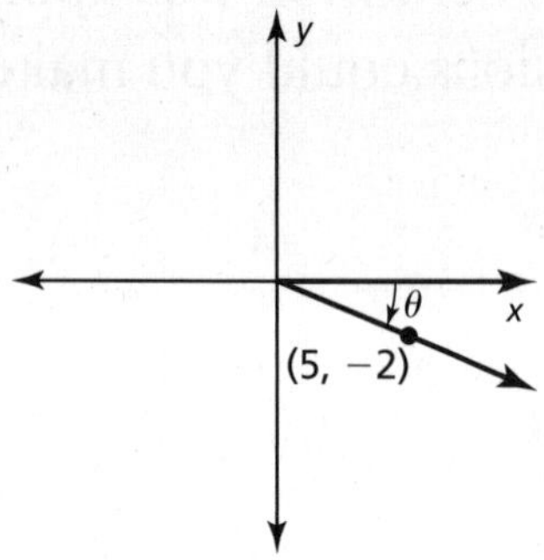

4.

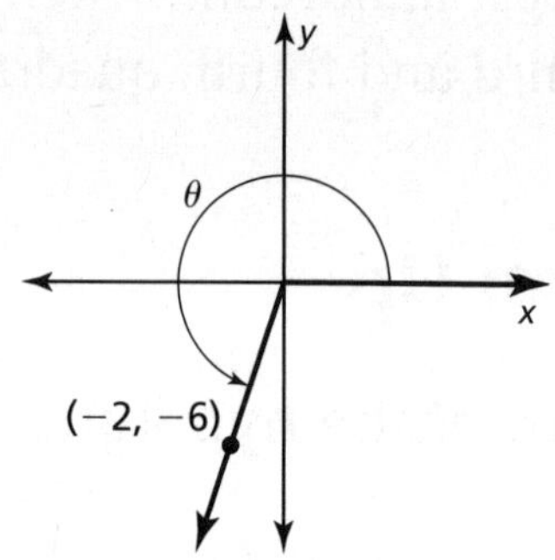

In Exercises 5–7, use the unit circle to evaluate the six trigonometric functions of θ.

5. 180°

6. 450°

7. $\dfrac{3\pi}{2}$

In Exercises 8–13, find the angle's reference angle.

8. -170°

9. 130°

10. 220°

11. $\dfrac{17\pi}{6}$

12. $\dfrac{15\pi}{4}$

13. $-\dfrac{7\pi}{3}$

In Exercises 14–16, evaluate the function without using a calculator.

14. $\csc 150^\circ$

15. $\tan \dfrac{5\pi}{4}$

16. $\cos(-210^\circ)$

17. The horizontal distance d (in feet) traveled by a projectile launched at an angle θ and with an initial speed v (in feet per second) is given by $d = \dfrac{v^2}{32}\sin 2\theta$. Estimate the horizontal distance (in feet) traveled by a football that is kicked at an angle of 60° with initial speed of $v = 80$ feet per second. What is the horizontal distance in yards? Round your answers to the nearest tenth.

Name______________________________ Date__________

8.3 Practice B

In Exercises 1–4, evaluate the six trigonometric functions of θ.

1.

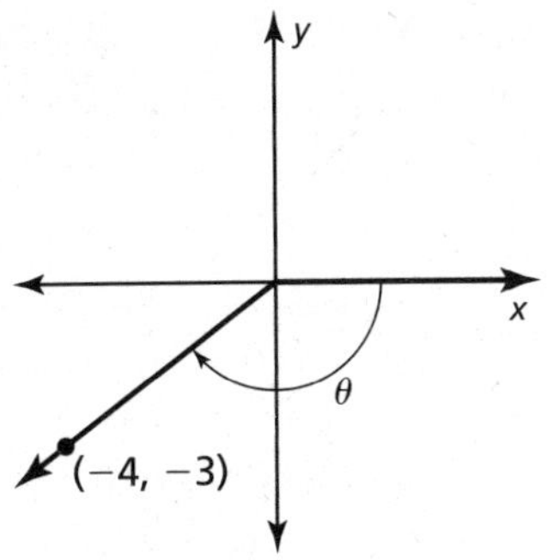

2.

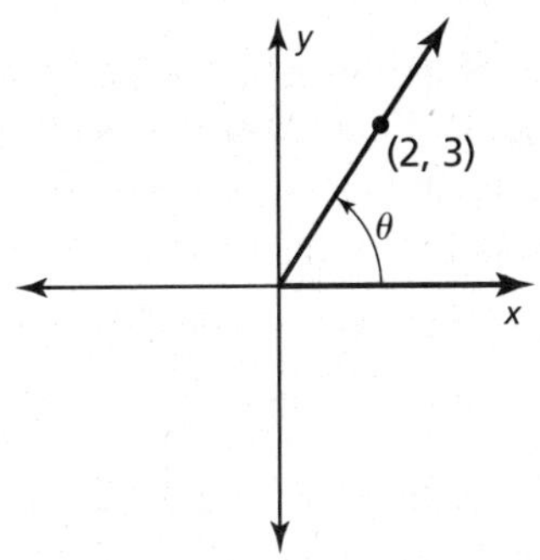

3.

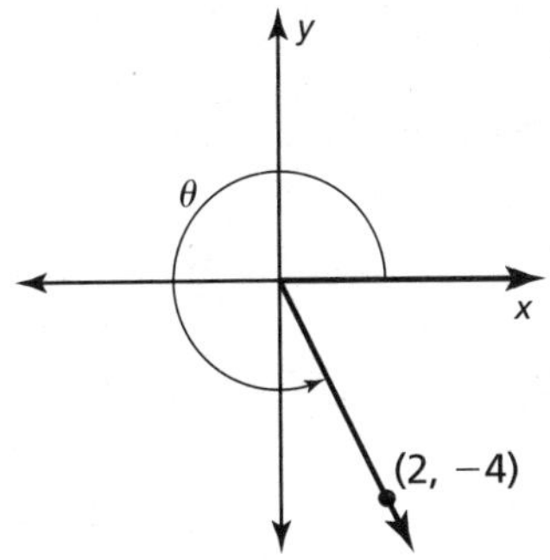

4.

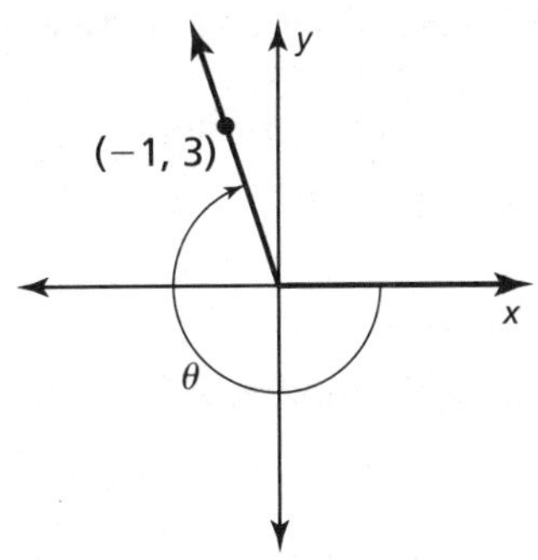

In Exercises 5–7, use the unit circle to evaluate the six trigonometric functions of θ.

5. 5π **6.** -720° **7.** $-\dfrac{5\pi}{2}$

In Exercises 8–13, find the angle's reference angle.

8. -250° **9.** 110° **10.** -310°

11. $\dfrac{13\pi}{4}$ **12.** $\dfrac{11\pi}{6}$ **13.** $-\dfrac{13\pi}{3}$

In Exercises 14–16, evaluate the function without using a calculator.

14. $\cot 240^\circ$ **15.** $\sin 315^\circ$ **16.** $\sec\left(-\dfrac{5\pi}{6}\right)$

17. The horizontal distance d (in feet) traveled by a projectile launched at an angle θ and with an initial speed v (in feet per second) is given by $d = \dfrac{v^2}{32}\sin 2\theta$. To win a shot-put competition, your last throw must travel a horizontal distance of at least 15 feet. You release the shot put at a 45° angle with an initial speed of 22 feet per second. Do you win the competition? Justify your answer.

Name ______________________________ Date __________

8.3 Enrichment and Extension

Trigonometric Functions of Any Angle

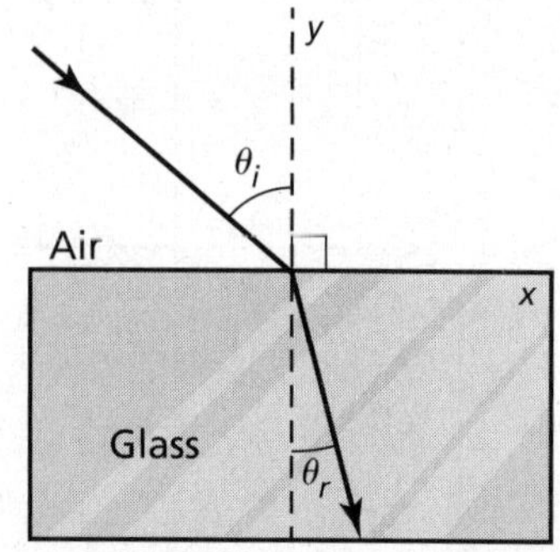

1. When a light ray passes from one medium, such as air, into another medium, such as glass, the light ray is bent, or refracted, at the boundary between the two media. The *index of refraction n* of a medium can be modeled by the equation below.

$$n = \frac{\sin \theta_i}{\sin \theta_r} \quad \begin{cases} \theta_i \text{ is the incoming } \textit{angle of incidence} \text{ of the light ray.} \\ \theta_r \text{ is the outgoing } \textit{angle of refraction.} \end{cases}$$

Measurements taken using crown glass gave an angle of incidence of 43.7° and an angle of refraction of 27.1°. Determine the index of refraction of crown glass. Round your answer to the nearest hundredth.

2. The Dutch physicist Willebrond Snell (1580–1626) discovered that the sine of the angle of incidence is proportional to the sine of the angle of refraction for any given interface between two media. As shown below, Snell's law may be stated in terms of the indices of refraction of the media.

$$n_1 \sin \theta_i = n_2 \sin \theta_r \quad \begin{cases} n_1 \text{ is the index of refraction of the first medium.} \\ n_2 \text{ is the index of refraction of the second medium.} \end{cases}$$

The measure of θ_i is 30° and the measure of θ_r for crown glass is 19.2°. Use the index of refraction of crown glass that you determined in Exercise 1 to find the index of refraction of air. Round your answer to the nearest tenth.

3. Snell's law may also be stated in terms of the velocity of light in the two media.

$$\frac{\sin \theta_i}{\sin \theta_r} = \frac{v_1}{v_2} \quad \begin{cases} v_1 \text{ is the speed of light in the first medium.} \\ v_2 \text{ is the speed of light in the second medium.} \end{cases}$$

The measure of the angle of incidence with which a ray of light strikes crown glass is 30° and the measure of the angle of refraction is 19.2°. The speed of light in air is 3.0×10^8 meters per second. Find the approximate speed of light in crown glass. Write your answer in scientific notation.

4. When light crosses a boundary and its speed increases in the second medium, the measure of the angle of refraction is greater than the measure of the angle of incidence. There is an angle of incidence, called the *critical angle* (θ_c), for which the corresponding angle of refraction is 90°, and the ray of light emerges parallel to the boundary. You can determine the measure of a critical angle when light passes from some medium into air by using the formula below.

$$\sin \theta_c = \frac{1}{n}, \text{ where } n \text{ is the index of refraction of the first medium.}$$

When light passes from diamond to air, the measure of the critical angle is 24.4°. Find the index of refraction of diamond. Round your answer to the nearest hundredth.

Name____ Date__________

8.3 Puzzle Time

What Kind Of Match Does Not Light On Fire?

Write the letter of each answer in the box containing the exercise number.

Use the unit circle to evaluate $\sin\theta$, $\cos\theta$, and $\tan\theta$.

1. $\theta = -90°$

2. $\theta = 180°$

3. $\theta = -360°$

4. $\theta = \dfrac{\pi}{2}$

Find the angle's reference angle.

5. $-230°$

6. $120°$

7. $\dfrac{3\pi}{5}$

8. $-\dfrac{11\pi}{4}$

Evaluate the function without using a calculator.

9. $\sin 120°$

10. $\tan(-105°)$

11. $\cos\dfrac{4\pi}{3}$

12. $\cot\dfrac{2\pi}{3}$

Answers

C. $-\dfrac{1}{2}$

T. $\sin\theta = 0$; $\cos\theta = -1$; $\tan\theta = 0$

A. $\dfrac{\sqrt{3}}{2}$

E. $\sin\theta = 0$; $\cos\theta = 1$; $\tan\theta = 0$

N. $50°$

A. $\sin\theta = -1$; $\cos\theta = 0$; $\tan\theta = \text{undefined}$

I. $60°$

H. $-\dfrac{\sqrt{3}}{3}$

S. $\dfrac{2\pi}{5}$

N. $\sin\theta = 1$; $\cos\theta = 0$; $\tan\theta = \text{undefined}$

M. $\dfrac{\pi}{4}$

T. $2 + \sqrt{3}$

1		2	3	4	5	6	7		8	9	10	11	12

8.4 Start Thinking

In Section 8.3, you learned to define the six trigonometric functions using the unit circle. Consider the function $f(\theta) = \sin \theta$. Use the unit circle to fill in the table of values. What is the highest value of $f(\theta)$? What is the lowest value of $f(\theta)$? How often do the high and low values repeat? What conclusions can you make about the graph of $f(\theta) = \sin \theta$?

θ	0	$\frac{\pi}{2}$	π	$\frac{3\pi}{2}$	2π	$\frac{5\pi}{2}$	3π	$\frac{7\pi}{2}$	4π
$f(\theta) = \sin \theta$									

8.4 Warm Up

Describe the transformation of *f* represented by *g*.

1. $f(x) = x^2, g(x) = 4x^2$
2. $f(x) = x^3, g(x) = -\frac{1}{2}x^3$
3. $f(x) = 2^x, g(x) = 2^{3x}$
4. $f(x) = \left(\frac{4}{5}\right)^x, g(x) = 7\left(\frac{4}{5}\right)^{x-2}$
5. $f(x) = \sin x, g(x) = 3 \sin x$
6. $f(x) = \cos x, g(x) = \cos\left(x + \frac{\pi}{2}\right)$

8.4 Cumulative Review Warm Up

Write a rule for the *n*th term of the sequence.

1. $0, \frac{1}{2}, 1, \frac{3}{2}, 2, \frac{5}{2}, \ldots$
2. $7, -2, -11, -20, -29, \ldots$
3. $\frac{5}{2}, \frac{7}{4}, \frac{9}{8}, \frac{11}{16}, \frac{13}{32}, \ldots$
4. $-3, 2, -\frac{4}{3}, \frac{8}{9}, -\frac{16}{27}, \ldots$

Name____ Date__________

8.4 Practice A

In Exercises 1 and 2, identify the amplitude and period of the graph of the function.

1. 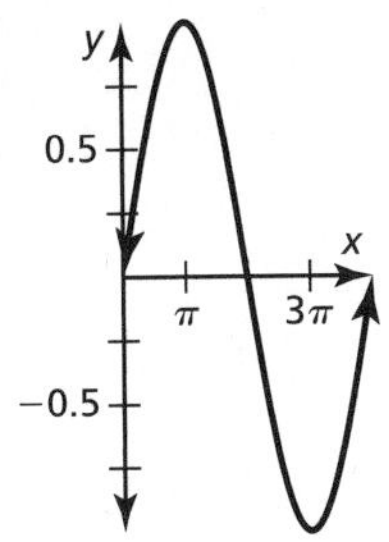

2. 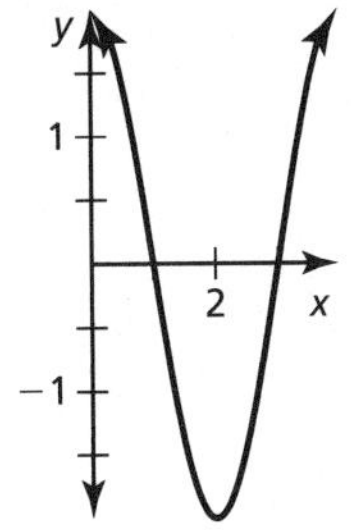

In Exercises 3–6, identify the amplitude and period of the function. Then graph the function and describe the graph of *g* as a transformation of the graph of its parent function.

3. $g(x) = 2 \sin x$

4. $g(x) = 4 \cos x$

5. $g(x) = \sin 2x$

6. $g(x) = 3 \cos \pi x$

7. Which functions have an amplitude of 2 and a period of π?

   A. $y = 2 \sin 2x$

   B. $y = -2 \cos \pi x$

   C. $y = 4 \sin 2x$

   D. $y = -2 \cos 2x$

8. The motion of a pendulum can be modeled by the function $d = 3 \cos 6\pi t$, where d is the horizontal displacement (in inches) of the pendulum relative to its position at rest and t is the time (in seconds). Find and interpret the period and amplitude in the context of this situation. Then graph the function.

In Exercises 9–12, graph the function.

9. $g(x) = \sin x - 3$

10. $g(x) = \cos\left(x + \frac{\pi}{2}\right)$

11. $g(x) = 2 \sin x + 1$

12. $g(x) = \cos 2(x + \pi)$

In Exercises 13 and 14, write a rule for *g* that represents the indicated transformations of the graph of *f*.

13. $f(x) = 2 \cos x$; translation 4 units down and π units left

14. $f(x) = \sin 4x$; translation 1 unit up and 2 units right

Name ______________________________ Date __________

8.4 Practice B

In Exercises 1 and 2, identify the amplitude and period of the graph of the function.

1.

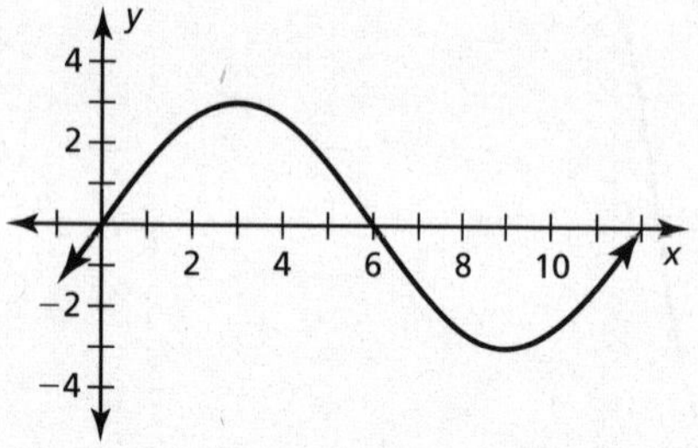

2.

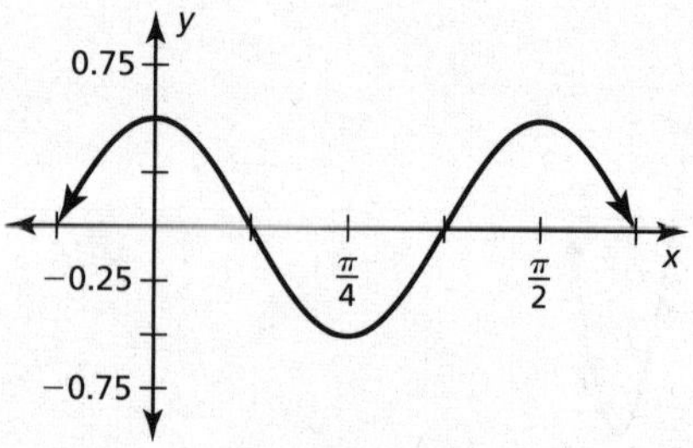

In Exercises 3–6, identify the amplitude and period of the function. Then graph the function and describe the graph of *g* as a transformation of the graph of its parent function.

3. $g(x) = 4 \sin x$

4. $g(x) = \cos \pi x$

5. $g(x) = 5 \sin 4x$

6. $g(x) = \frac{1}{4} \cos 2x$

7. Write an equation of the form $y = a \cos bx$, where $a > 0$ and $b > 0$, so that the graph has the given amplitude and period.

a. amplitude: 1
period: 3

b. amplitude: 3
period: 4

c. amplitude: 12
period: 2π

d. amplitude: $\frac{1}{3}$
period: π

In Exercises 8–11, graph the function.

8. $g(x) = \cos x + 3$

9. $g(x) = 2 \sin x - 1$

10. $g(x) = \sin \frac{1}{2}(x - \pi) - 2$

11. $g(x) = \cos \frac{1}{2}(x + \pi) - 4$

In Exercises 12 and 13, write a rule for *g* that represents the indicated transformations of the graph of *f*.

12. $f(x) = \frac{1}{2} \cos 3x$; translation 2 units up, followed by a reflection in the line $y = 2$

13. $f(x) = \frac{1}{3} \sin \pi x$; translation 3 units down, followed by a reflection in the line $y = -3$

Name______________________________ Date__________

8.4 Enrichment and Extension

Polar Graphs

You can analyze the behavior of the graph of $y = 3 \sin 2x$ in the rectangular coordinate system to graph the polar equation $r = 3 \sin 2\theta$ in the polar coordinate system, where r is the directed distance from the pole and θ is the directed angle from the polar axis.

Rectangular

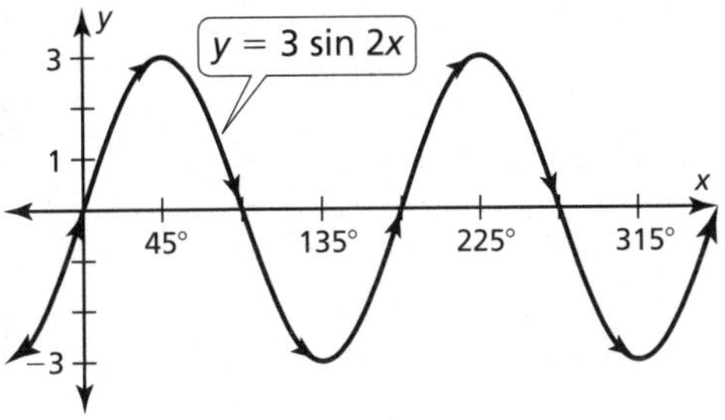

Polar

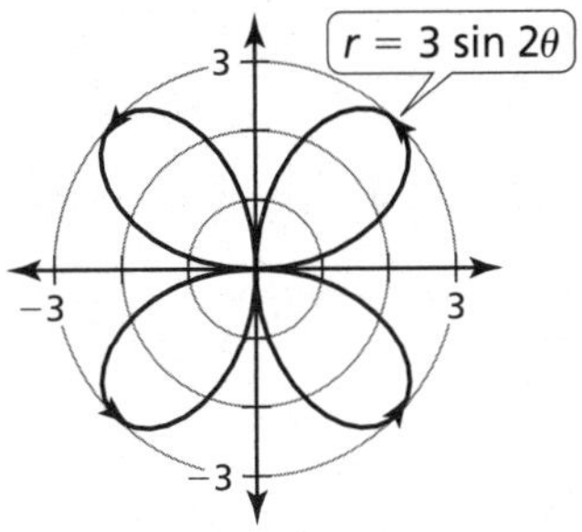

In the graph of the function $y = 3 \sin 2x$, notice that as x goes from $0°$ to $45°$, y increases from 0 to 3. Similarly, in the graph of the equation $r = 3 \sin 2\theta$, as θ rotates counterclockwise from $0°$ to $45°$, r increases from 0 to 3. As x continues from $45°$ to $90°$, y decreases from 3 to 0, and as θ rotates from $45°$ to $90°$, r decreases from 3 to 0.

When you continue this analysis for one complete revolution, or $360°$, you obtain a good sketch of the polar graph of $r = 3 \sin 2\theta$.

In Exercises 1–4, use the given graph to sketch a graph of the polar equation in the polar coordinate system.

1. $r = 3 \sin \theta$

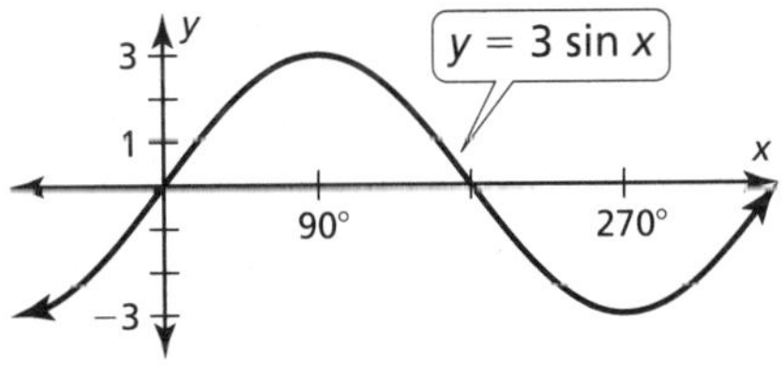

2. $r = 5 \cos 2\theta$

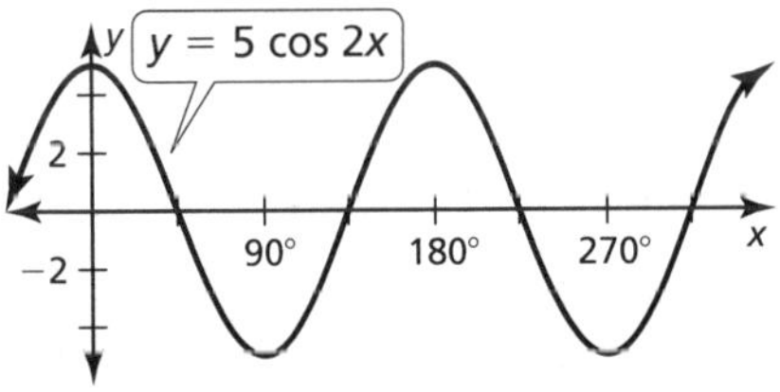

3. $r = 4 \cos 3\theta$

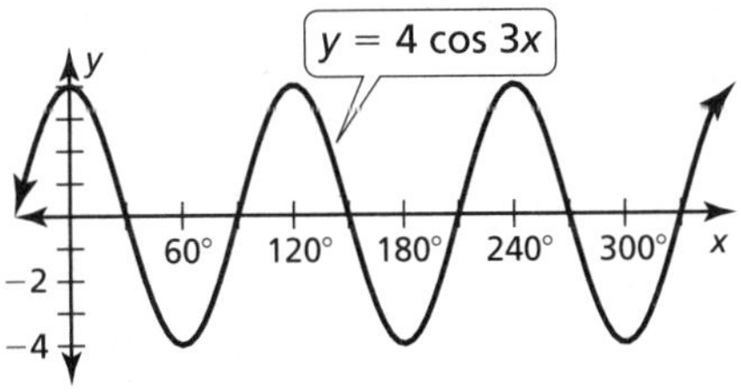

4. $r = 2 \sin 6\theta$

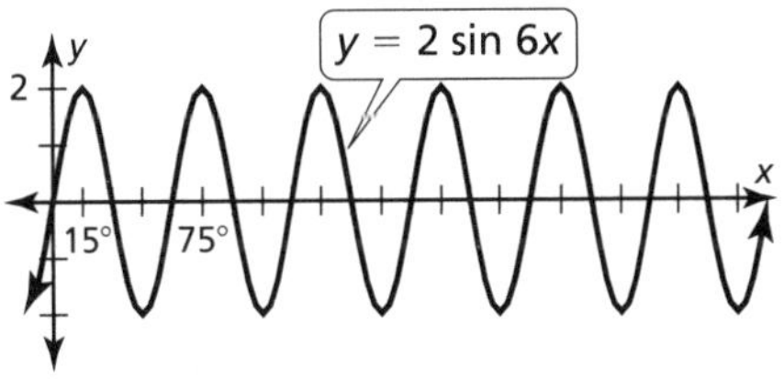

5. Rose curves are of the form $y = a \sin n\theta$ or $y = a \cos n\theta$, where a is a real number and n is a natural number. Make a conjecture about the number of petals and the value of n.

Name ______________________ Date ________

8.4 Puzzle Time

What Goes Best With Toast When You Are In A Car?

Write the letter of each answer in the box containing the exercise number.

Identify the amplitude and period of the function.

1. $g(x) = 2 \sin 3x$
2. $g(x) = \frac{1}{2} \sin 4x$
3. $g(x) = \cos \pi x$
4. $g(x) = 3 \cos 4x$
5. $g(x) = \sin x$
6. $g(x) = -\frac{1}{2} \cos 6x$
7. $g(x) = 2 \sin \pi x$
8. $g(x) = -2 \cos 2\pi x$

Match the graph with the correct function.

9.

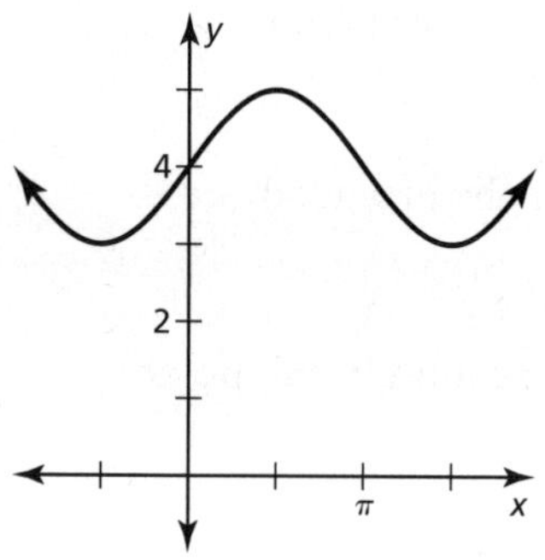

10.

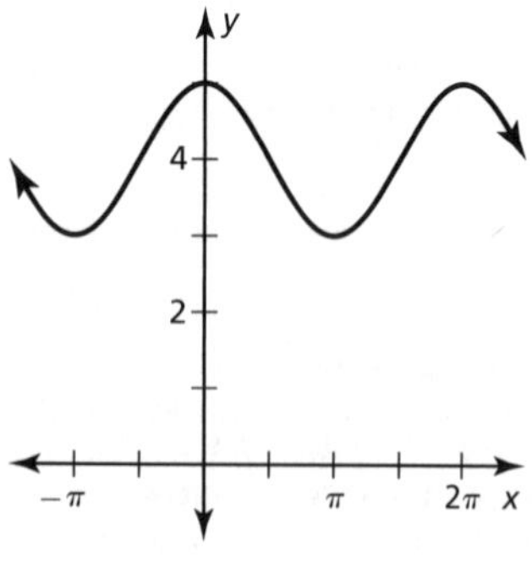

Answers

A. $a = 1$; period $= 2$

I. $a = \frac{1}{2}$; period $= \frac{\pi}{3}$

T. $a = 2$; period $= \frac{2\pi}{3}$

M. $g(x) = \cos x + 4$

J. $a = 2$; period $= 1$

F. $a = 3$; period $= \frac{\pi}{2}$

C. $a = 2$; period $= 2$

R. $a = \frac{1}{2}$; period $= \frac{\pi}{2}$

F. $a = 1$; period $= 2\pi$

A. $g(x) = \sin x + 4$

1	2	3	4	5	6	7		8	9	10

8.5 Start Thinking

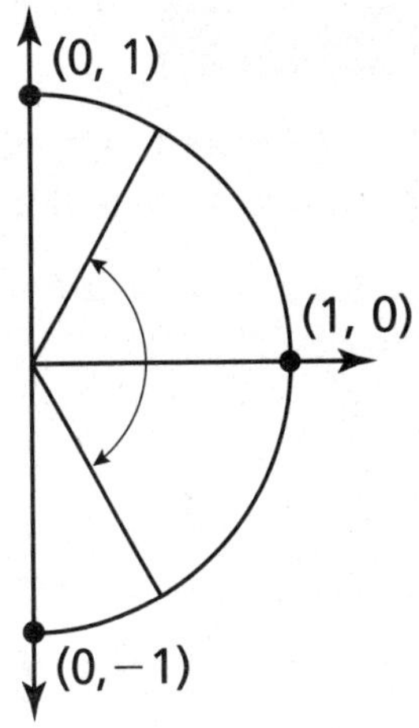

Consider the portion of the unit circle shown in the diagram. What angles are represented as you move from $(1, 0)$ to $(0, 1)$ in a counterclockwise direction? What angles are represented as you move from $(1, 0)$ to $(0, -1)$ in a clockwise direction? Describe what is happening to the values of x and y in both cases. Now consider the function $f(\theta) = \tan\theta$. How is this function defined using the unit circle? Describe the values of $f(\theta)$ as θ moves according to the descriptions above. What conclusions can you make about the graph of $f(\theta) = \tan\theta$?

8.5 Warm Up

Graph the function.

1. $f(x) = 3 - \sin x$

2. $f(x) = 2\cos 4x$

3. $f(x) = \frac{2}{3}\cos\left(x + \frac{\pi}{2}\right) - 3$

4. $f(x) = 2 - \frac{5}{2}\sin\left(x - \frac{\pi}{4}\right)$

8.5 Cumulative Review Warm Up

Expand the expression.

1. $\log_2 3x^2$

2. $\log\frac{2 - x}{x^3}$

3. $\ln\frac{\sqrt{2x^3}}{5}$

Condense the expression.

4. $\ln 8 - 2\ln x$

5. $\frac{1}{2}\log(x - 1) + \log 7$

6. $-2\log_3 x + 3\log_3(x - 2)$

Name ______________________ Date __________

8.5 Practice A

In Exercises 1–4, graph one period of the function. Describe the graph of g as a transformation of the graph of its parent function.

1. $g(x) = 4 \tan x$

2. $g(x) = 3 \cot x$

3. $g(x) = \tan 2x$

4. $g(x) = \cot 4x$

5. Describe and correct the error in finding the period of the function $f(x) = \tan 2\pi x$.

> ✗ Period: $\dfrac{\pi}{|b|} = \dfrac{\pi}{|2|} = \dfrac{\pi}{2}$

6. Use the given graph to graph each function.

a. $f(x) = 2 \sec 3x$

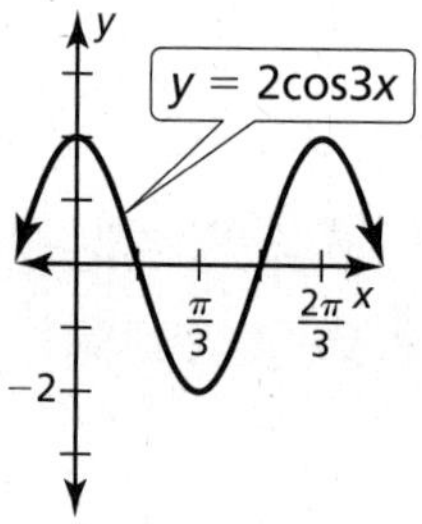

b. $f(x) = 3 \csc 2x$

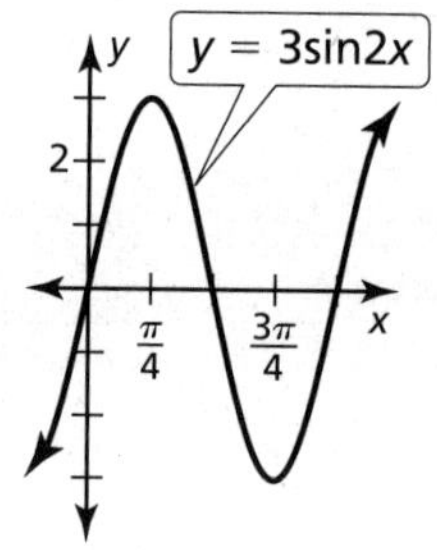

In Exercises 7–10, graph one period of the function. Describe the graph of g as a transformation of the graph of its parent function.

7. $g(x) = 4 \csc x$

8. $g(x) = \sec 2x$

9. $g(x) = \frac{1}{3} \sec 2\pi x$

10. $g(x) = 4 \csc \pi x$

Name______________________________ Date__________

8.5 Practice B

In Exercises 1–4, graph one period of the function. Describe the graph of *g* as a transformation of the graph of its parent function.

1. $g(x) = 2 \tan 4x$

2. $g(x) = 3 \cot \frac{1}{2}x$

3. $g(x) = \frac{1}{4} \tan 2\pi x$

4. $g(x) = \frac{1}{3} \cot \pi x$

5. Describe and correct the error in describing the transformation of $f(x) = \tan x$ represented by $g(x) = 4 \tan \frac{1}{2}x$.

> ✗ A vertical stretch by a factor of 4 and a horizontal shrink by a factor of $\frac{1}{2}$

6. Use the given graph to graph each function.

a. $f(x) = 4 \sec \frac{1}{2}x$

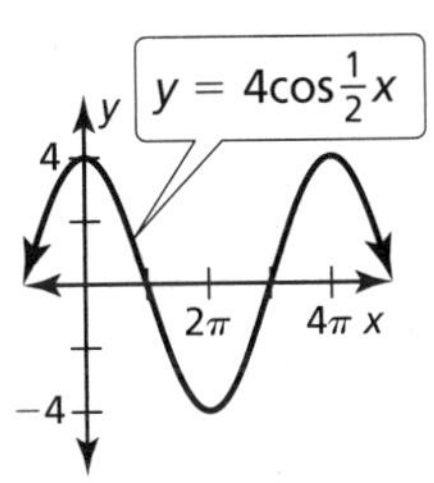

b. $f(x) = \frac{1}{2} \csc \pi x$

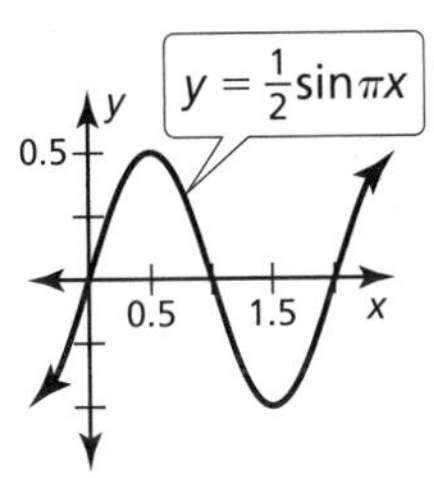

In Exercises 7–10, graph one period of the function. Describe the graph of *g* as a transformation of the graph of its parent function.

7. $g(x) = \frac{1}{3} \csc \pi x$

8. $g(x) = \frac{1}{2} \sec 6x$

9. $g(x) = \sec \frac{\pi}{2}x$

10. $g(x) = \csc \frac{\pi}{3}x$

Name ______________________ Date __________

8.5 Enrichment and Extension

Graphing Other Trigonometric Functions

Consider the graph at the right. It looks like a transformation of the graph of the parent tangent function. The graph of $y = \tan x$ has a midpoint of one cycle passing through the origin. The midpoint of a cycle of this graph passes through the point $(1, -2)$. This indicates that both a vertical translation and a phase shift are included in the equation of the function. The period is 2π which indicates a horizontal stretch in the period of the graph of the parent function, $y = \tan x$ by a factor of 2. Using these observations, an equation for the function is $y = \tan 0.5(x - 1) - 2$. Due to the periodic nature of the tangent function, other equations are possible.

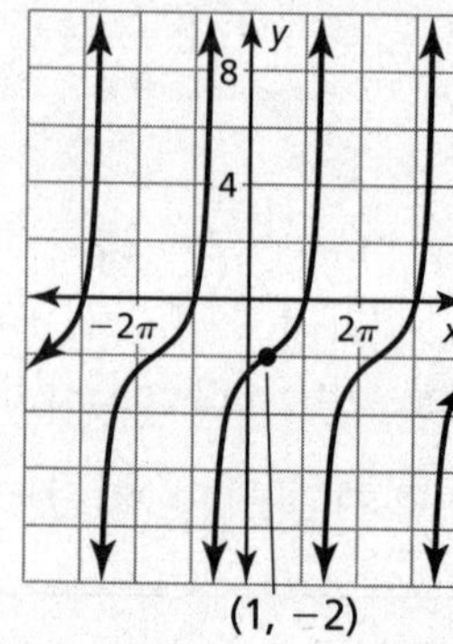

In Exercises 1–6, write an equation for the graph.

1.

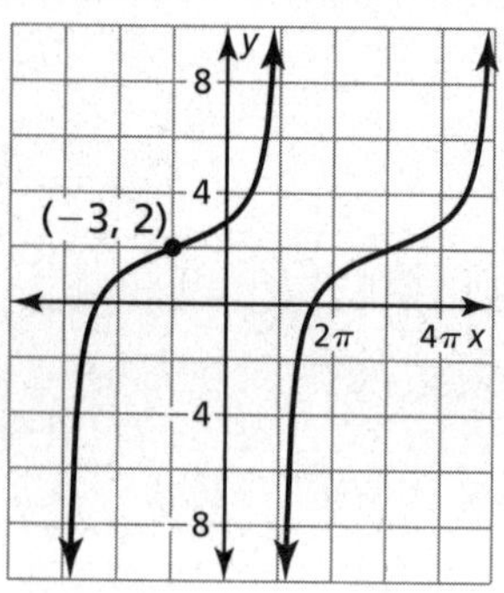

2.

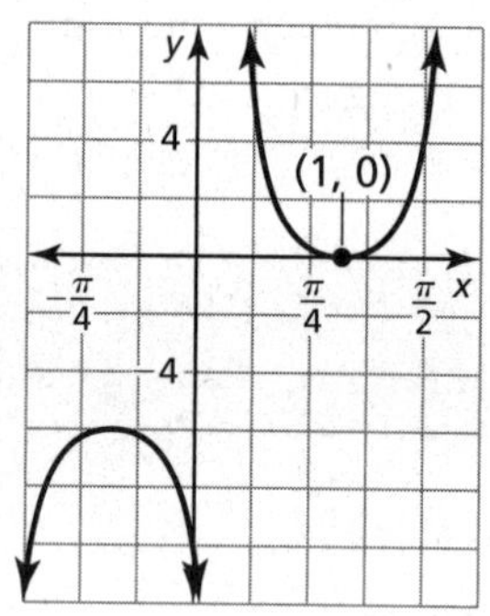

3.

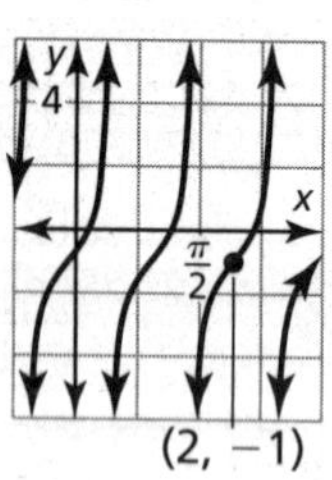

4.

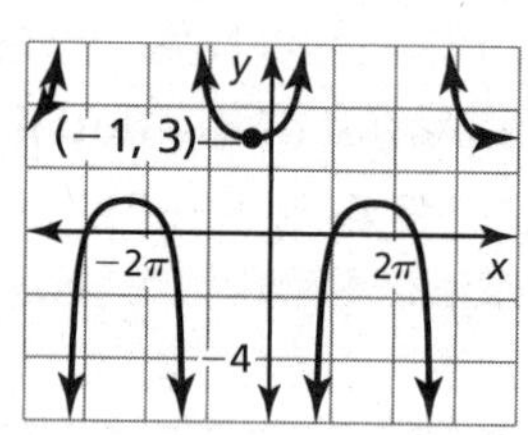

5.

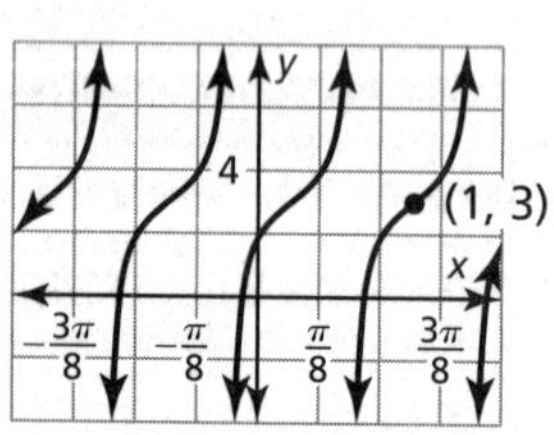

6.

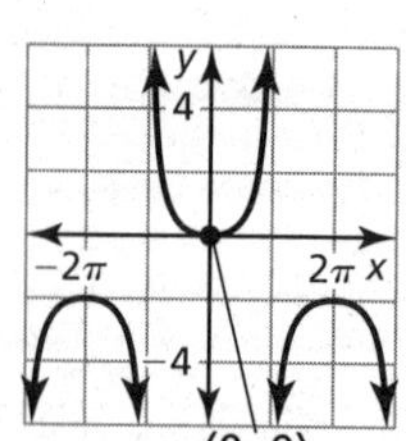

Name______________________________ Date__________

8.5 Puzzle Time

What Is An Elephant's Favorite Card Game?

Write the letter of each answer in the box containing the exercise number.

Match the graph with the correct equation.

1.

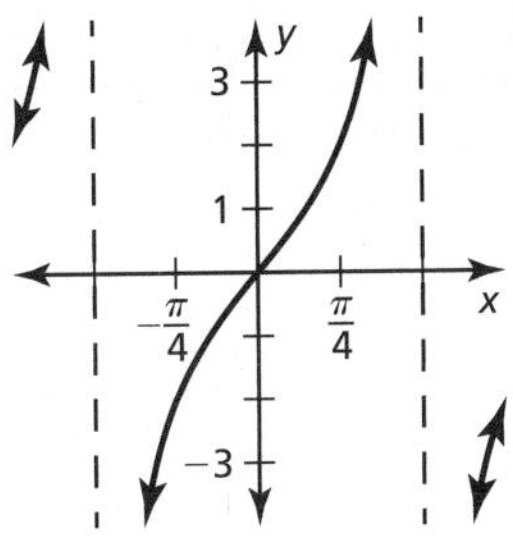

2.

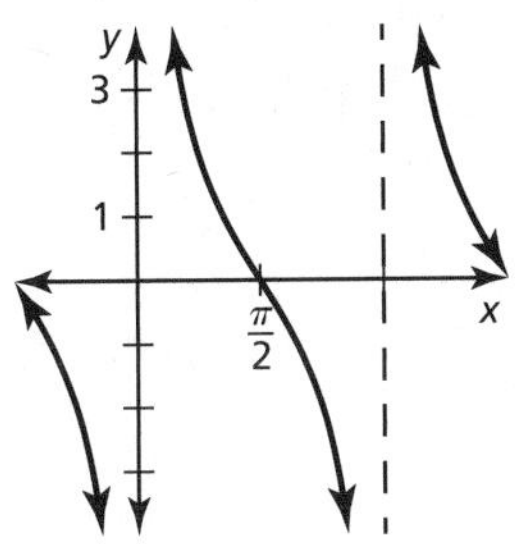

3.

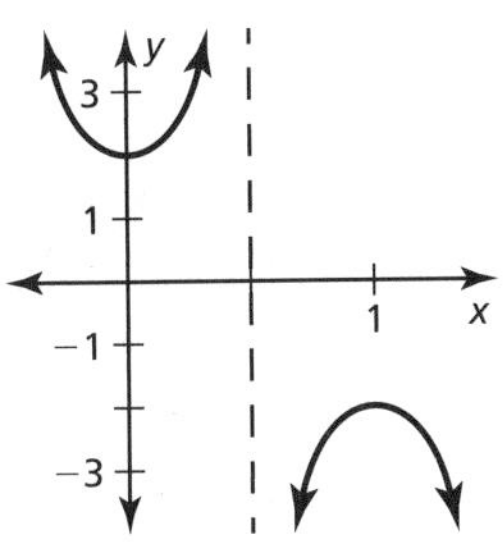

4.

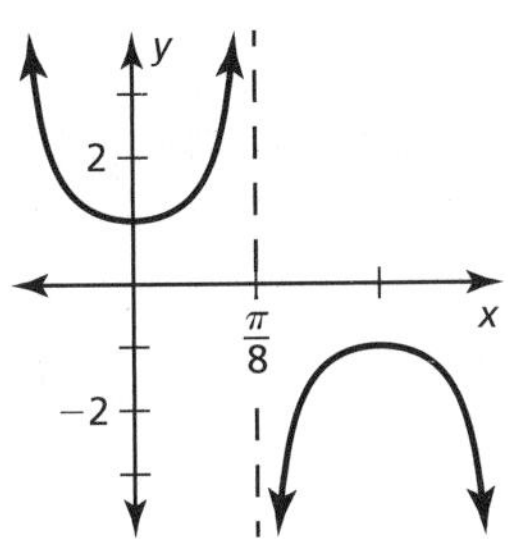

5.

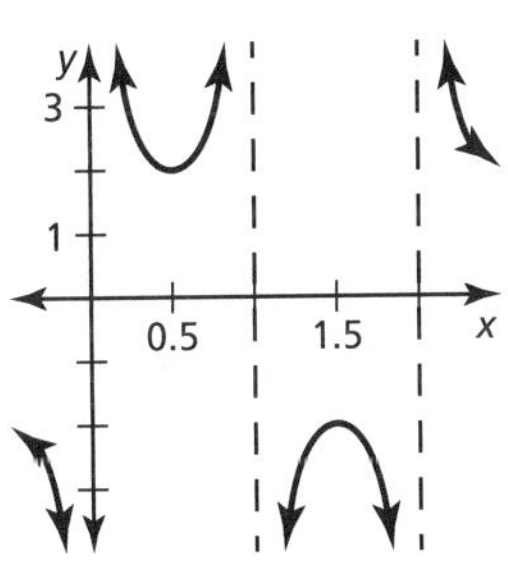

6.

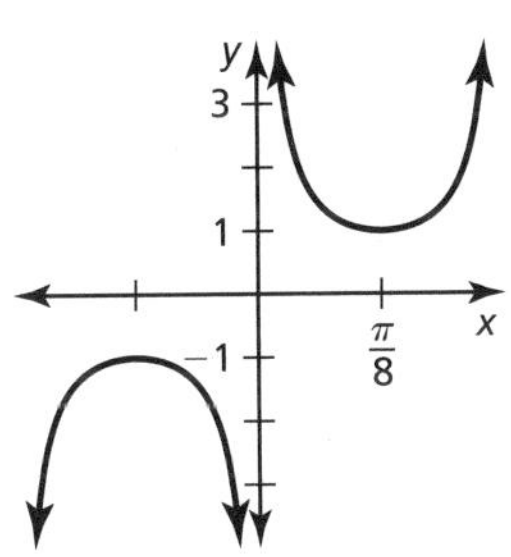

Answers

O. $g(x) = \sec 4x$

Y. $g(x) = \csc 4x$

M. $g(x) = 2 \tan x$

M. $g(x) = 2 \sec \pi x$

E. $g(x) = 2 \cot x$

R. $g(x) = 2 \csc \pi x$

1	2	3	4	5	6

8.6 Start Thinking

You are riding a Ferris wheel. The graph to the right represents your height h (in feet) above the ground atany time t (in seconds). What are the x-intercepts of the graph and what do they represent in relationship to your ride? Approximately how tall is the Ferris wheel? If one ride includes four complete revolutions, how long is your ride? Describe the function in terms of translations of the graph of $h = \cos t$.

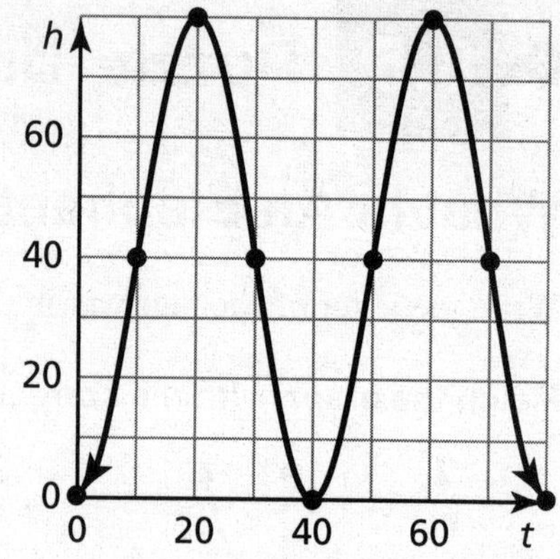

8.6 Warm Up

Identify the amplitude and the period of the function.

1. $y = -2\cos(x - \pi)$

2. $y = \frac{2}{3}\sin(2\pi x)$

3. $y = 4 - \sin\left(3x + \frac{\pi}{2}\right)$

4. $y = \frac{\sqrt{3}}{2}\cos(2x)$

5. $y = 1.4\sin(x - 0.3)$

6. $y = -2.4\cos(1.8x + 3)$

8.6 Cumulative Review Warm Up

Describe the domain and range of the function.

1. $f(x) = x^2 + 2x - 1$

2. $g(x) = 3 - \log_2(x - 2)$

3. $h(x) = \sqrt{x + 8} - 4$

4. $f(x) = 2x^3 - 5x^2 + x - 7$

5. $g(x) = -4 + 2\sin x$

6. $h(x) = 3\cos 2\pi x$

Name______________________________ Date__________

8.6 Practice A

In Exercises 1–4, find the frequency of the function.

1. $y = \sin 2x$

2. $y = \cos 3x - 1$

3. $y = -\sin 4x$

4. $y = \cos 5\pi x$

5. A middle-C tuning fork vibrates with a frequency f of 262 hertz (cycles per second). You strike a middle-C tuning fork with a force that produces a maximum pressure of 5 Pascals. Write and graph a sine model that gives the pressure P as a function of the time t (in seconds).

In Exercises 6 and 7, write a function for the sinusoid.

6.

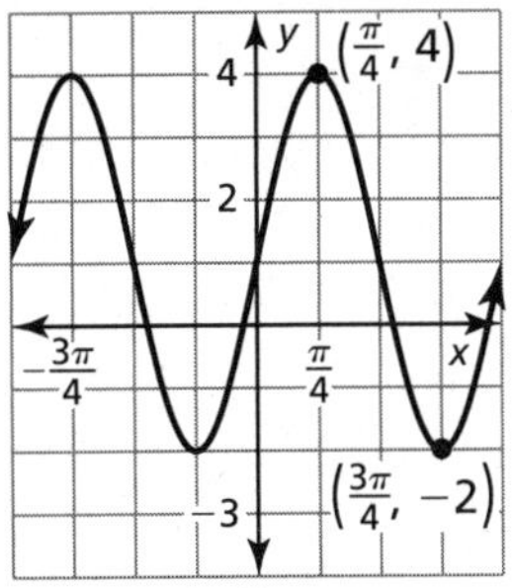

7.

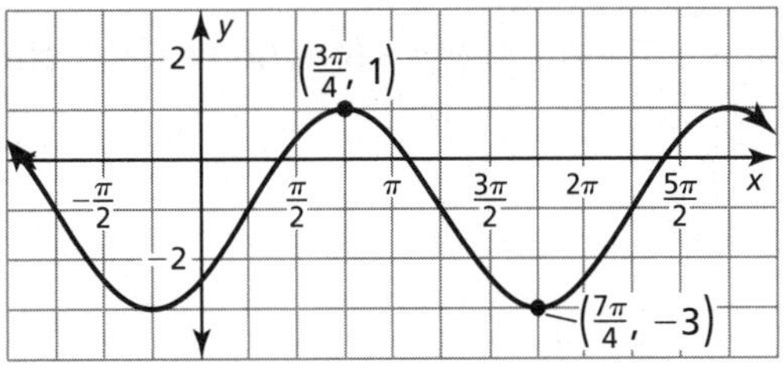

8. The table shows the depth d (in feet) of the water at the end of an inland dock that is located in a saltwater river that is affected by ocean tides. The time t is measured in hours, with $t = 0$ representing midnight.

t	Midnight	2 A.M.	4 A.M.	6 A.M.	8 A.M.	10 A.M.	Noon
d	2.55	3.80	4.40	3.80	2.55	1.80	2.27

a. Use sinusoidal regression to find a model that gives d as a function of t.

b. Predict the depth of the water at the end of the dock at 5 P.M.

Name ____ Date __________

8.6 Practice B

In Exercises 1–4, find the frequency of the function.

1. $y = \cos 3x$
2. $y = -\cos 4x - 3$
3. $y = \sin \dfrac{\pi x}{2}$
4. $y = 4 \cos 0.4x - 3$

5. A sub-contra-octave A tuning fork (corresponds to the lowest note on a piano keyboard) vibrates with a frequency f of 27.5 hertz (cycles per second). You strike a sub-contra-octave A tuning fork with a force that produces a maximum pressure of 4 Pascals. Write and graph a sine model that gives the pressure P as a function of the time t (in seconds).

In Exercises 6 and 7, write a function for the sinusoid.

6.

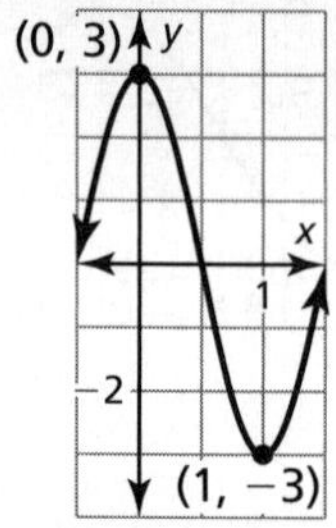

7.

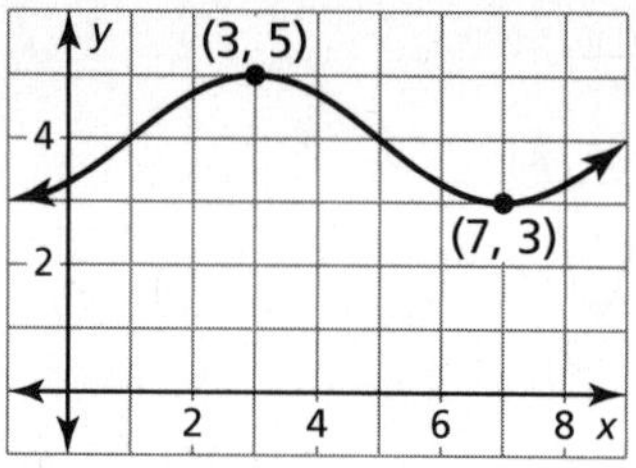

8. When you ride a Ferris wheel, your distance from the ground will vary with respect to the number of seconds that have elapsed since the wheel started. The table shows your height h (in meters) above the ground at time t as you ride the Ferris wheel.

t	0	1	2	3	4	5	6	7	8	9	10	11	12	15	20
h	1	2.3	5.8	10.2	13.7	15	13.7	10.2	5.8	2.3	1	2.3	5.8	15	1

a. Use sinusoidal regression to find a model that gives h as a function of t.

b. Predict your height above the ground after 42 seconds have elapsed.

Name______________________________ Date__________

8.6 Enrichment and Extension

Modeling with Trigonometric Functions

In Exercises 1–4, use the table which shows the monthly amounts of carbon dioxide in the atmosphere in parts per million (ppm) for three different years during a 40-year period.

Year	Jan	Feb	Mar	Apr	May	June	July	Aug	Sept	Oct	Nov	Dec
1965	319	320	320	322	321	321	320	318	317	318	319	320
1985	344	346	347	347	348	346	344	343	342	344	345	345
2005	376	377	378	380	379	379	377	375	374	374	375	377

1. Write a sinusoidal model that gives the amount of carbon dioxide y in the atmosphere as a function of the time x (in months) for the year 1965.

2. Write a sinusoidal model that gives the amount of carbon dioxide y in the atmosphere as a function of the time (in months) for the year 1985.

3. Write a sinusoidal model that gives the amount of carbon dioxide y in the atmosphere as a function of the time (in months) for the year 2005.

4. Explain why a sinusoidal model is appropriate in this situation. What do you notice when you compare the three graphs? What implications does this have for our environment?

Name ______________________________ Date __________

8.6 Puzzle Time

What Do You Get When You Cross A Monster With A Cat?

Write the letter of each answer in the box containing the exercise number.

Find the frequency of the function whose equation or graph is given.

1. $y = \sin 3x$
2. $y = 4\cos 3\pi x$
3. $y = 2\sin\dfrac{x}{3}$
4. $y = 3\cos\dfrac{x}{2}$
5. $y = 2\cos 4x$
6. $y = \sin 4\pi x$
7. $y = 2\sin 0.2x$
8. $y = \cos x - 4$
9. 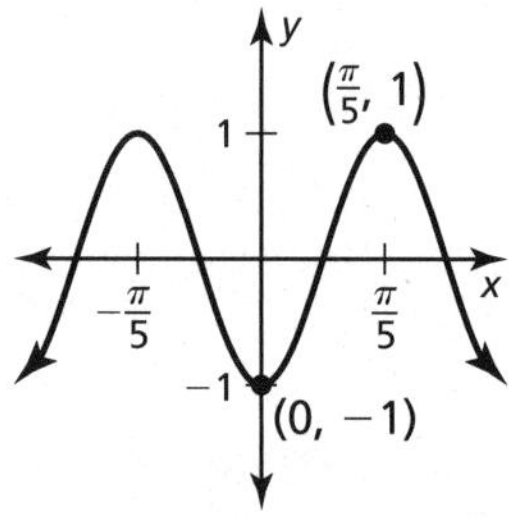

10. 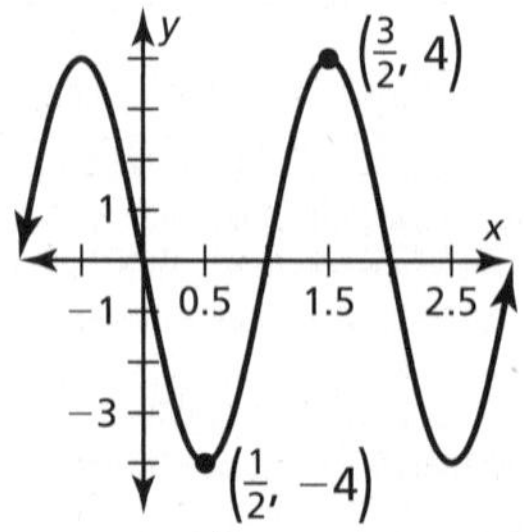

Answers

- **O.** $\dfrac{5}{2\pi}$
- **T.** $\dfrac{1}{10\pi}$
- **A.** $\dfrac{3}{2\pi}$
- **M.** $\dfrac{3}{2}$
- **T.** $\dfrac{2}{\pi}$
- **W.** $\dfrac{1}{4\pi}$
- **N.** $\dfrac{1}{2}$
- **A.** 2
- **E.** $\dfrac{1}{6\pi}$
- **I.** $\dfrac{1}{2\pi}$

1		2	3	4	5	6	7	8	9	10

Name____ Date__________

Chapter 8 Cumulative Review

In Exercises 1–8, order the expression by value from least to greatest.

1. $-|4|, |-6|, |7-8|, |-1+6|$

2. $|-4+3|, -|4+1|, |-8+3|, |9|$

3. $|2|, |-2+5|, |-6|, \dfrac{|-6|}{|8|}$

4. $|4-9|, |-5+5|, \dfrac{|5|}{|-7|}, |-8|$

5. $|8|+|-4|, |9-9|, |-6|-|4|, |-7 \bullet -6|$

6. $|9+6|, |4 \bullet -4|, |-8|+|4|, |-4+6|$

7. $|7-5|, |3^2|, |9|+|-4|, |-2 \bullet -9|$

8. $|-6 \bullet -6|, |7 \bullet 3|, |-8|+|-7|, |-4^2|$

In Exercises 9–18, find the midpoint between the two points.

9. $(17, -20), (3, 14)$

10. $(2, -14), (-20, -12)$

11. $(-11, 13), (11, -1)$

12. $(13, -10), (7, -2)$

13. $(-16, -11), (10, 7)$

14. $(-4, -9), (4, 9)$

15. $(-4, -20), (18, -12)$

16. $(-18, 7), (-10, 5)$

17. $(-19, 21), (19, -11)$

18. $(9, -7), (-13, 5)$

In Exercises 19–28, find the distance between the two points.

19. $(17, 14), (3, -6)$

20. $(7, -19), (1, 8)$

21. $(-8, 8), (13, 14)$

22. $(-16, 9), (9, 14)$

23. $(2, -15), (1, -18)$

24. $(11, 2), (8, 11)$

25. $(3, 2), (-18, -16)$

26. $(-3, -12), (17, -15)$

27. $(-2, 9), (-7, -1)$

28. $(18, 3), (0, -17)$

29. A sandbox is in the shape of a right triangular prism. Both side lengths are 4 feet, and it is 1.5 feet tall.

a. What is the length of the third side of the sandbox, or hypotenuse of the triangle? Round your answer to the nearest tenth of a foot.

b. What is the volume of the sandbox?

c. You want to put sand in the sandbox, but you only want the sand to be one foot deep. How much sand do you need?

d. Sand costs \$1.19 per cubic foot. Use your answer from part (c) to determine how much it will cost to fill the sandbox 1 foot deep. Round your answer to the nearest cent.

Name ______________________________ Date ________

Cumulative Review (continued)

In Exercises 30–35, find the missing side length of the triangle.

30.

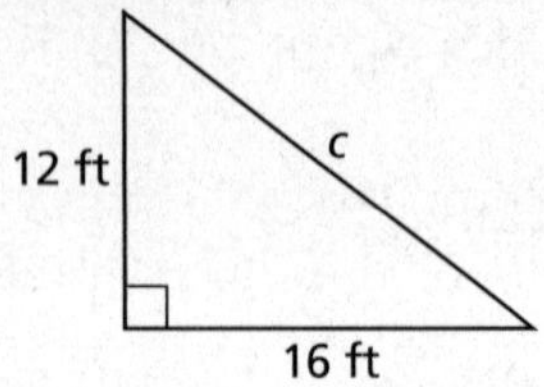

31.

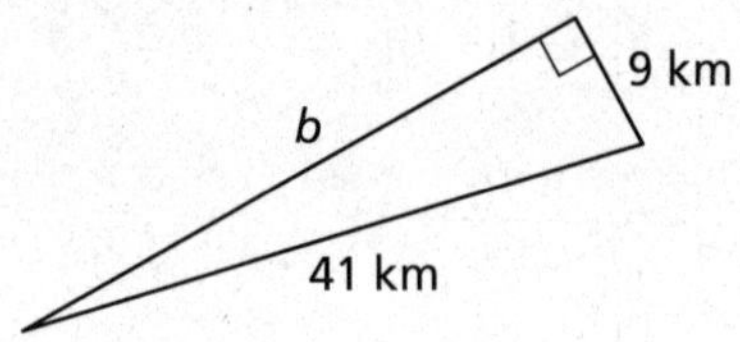

32.

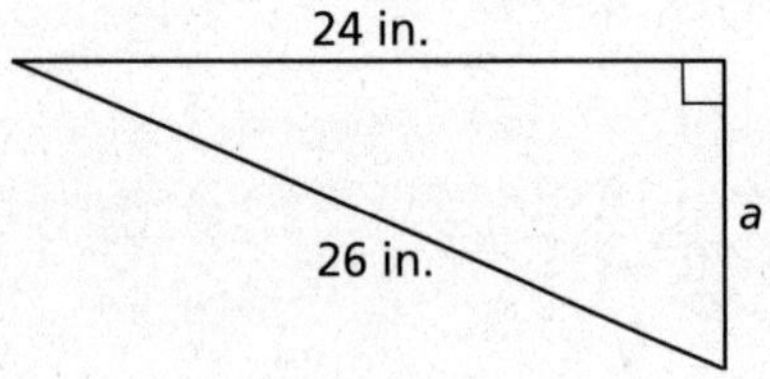

33.

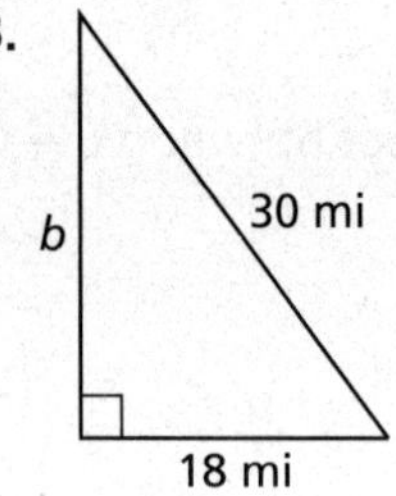

34.

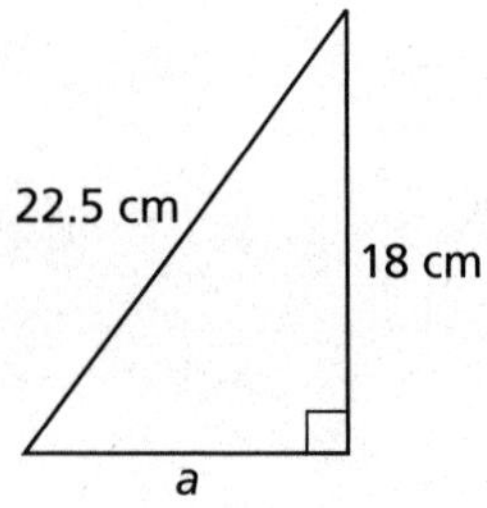

35.

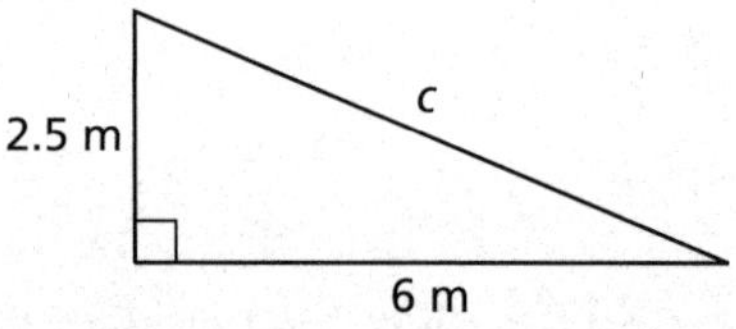

36. A mirror is in the shape of a right triangle and has side lengths of 6.5 inches and 7.5 inches. What is the area of the mirror? Round your answer to the nearest tenth.

37. A flower box is in the shape of a right triangular prism. Both side lengths are 3 feet, and it is 1 foot tall.

a. What is the length of the third side of the flower box, or hypotenuse of the triangle? Round your answer to the nearest tenth of a foot.

b. What is the volume of the flower box?

c. You want to put potting soil in the sandbox, but you only want the soil to be 0.75 foot deep. How much soil do you need? Round your answer to the nearest tenth.

d. Potting soil costs \$2.49 per cubic foot. Use your answer from part (c) to determine how much it will cost to fill the sandbox 0.75 foot deep. Round your answer to the nearest cent.

Name____ Date__________

Chapter 8

Cumulative Review (continued)

In Exercises 38–46, find the circumference of the circle with the given radius or diameter. Leave your answers in terms of π.

38. $r = 22$ inches

39. $r = 17$ kilometers

40. $d = 2$ miles

41. $d = 20$ centimeters

42. $r = 3$ feet

43. $d = 14$ millimeters

44. $r = 16$ yards

45. $d = 24$ meters

46. $r = 12$ centimeters

In Exercises 47–55, find the area of the circle with the given radius or diameter. Leave your answer in terms of π.

47. $d = 12$ miles

48. $r = 3$ inches

49. $d = 8$ kilometers

50. $r = 22$ yards

51. $r = 15$ feet

52. $r = 7$ centimeters

53. $d = 4$ meters

54. $d = 18$ miles

55. $r = 2$ yards

In Exercises 56–64, evaluate the expression for (a) $x = -3$ and (b) $x = 2$.

56. 3^x

57. 5^x

58. $2 \bullet 3^x$

59. $4 \bullet 2^x$

60. $3 \bullet 5^x$

61. $3 + 4^x$

62. $2 + 2^x$

63. $3^x - 4$

64. $2^x - 7$

In Exercises 65–79, simplify the expression.

65. $e^{-6} \bullet e^{-5}$

66. $e^9 \bullet e^5$

67. $e^6 \bullet e^{-3}$

68. $\dfrac{15e^2}{5e^{-3}}$

69. $\dfrac{-4e^7}{8e^5}$

70. $\dfrac{18e^9}{3e^2}$

71. $\left(2e^{-8x}\right)^3$

72. $\left(3e^{-9x}\right)^3$

73. $\left(-3e^{6x}\right)^2$

74. $\sqrt{16e^{12x}}$

75. $\sqrt{36e^{4x}}$

76. $\sqrt[3]{27e^{15x}}$

77. $e^x \bullet e^5 \bullet e^{6x}$

78. $e^7 \bullet e^{-3x} \bullet e^{x+2}$

79. $e^{2x} \bullet e^2 \bullet e^{x+2}$

80. The value of a pair of rollerblades y (in dollars) can be approximated by the model $y = 150(0.8)^t$, where t is the number of years since the pair of rollerblades was new.

a. Tell whether the model represents exponential growth or exponential decay.

b. Identify the annual percent increase or decrease in the value of the rollerblades.

c. Estimate when the value of the rollerblades will be $60.

Name ______________________________ Date __________

Chapter 8 Cumulative Review (continued)

In Exercises 81–89, evaluate the logarithm.

81. $\log_4 1024$ **82.** $\log_2 64$ **83.** $\log_{1/4} 1$

84. $\log_3 \frac{1}{243}$ **85.** $\log_7 \frac{1}{343}$ **86.** $\log_9 \frac{1}{729}$

87. $\log_5 0.04$ **88.** $\log_2 0.25$ **89.** $\log_4 0.0625$

In Exercises 90–101, expand the logarithmic expression.

90. $\log_2 5x$ **91.** $\log_7 8x$ **92.** $\log_9 5x$

93. $\log_4 8x^3$ **94.** $\log_2 4x^5$ **95.** $\log_2 3x^4$

96. $\ln \frac{2x}{y}$ **97.** $\ln \frac{x}{5y}$ **98.** $\ln \frac{4x}{3y}$

99. $\log_9 3\sqrt{x}$ **100.** $\log_8 \sqrt[3]{xy^4}$ **101.** $\log_4 2\sqrt{x^3}$

In Exercises 102–113, condense the logarithmic expression.

102. $\log_7 6 + \log_7 11$ **103.** $\ln 6 - \ln 3$ **104.** $\ln 7 + \ln 6$

105. $3 \log_8 x - 2 \log_8 y$ **106.** $9 \log_5 y + 4 \log_5 z$ **107.** $10 \log_9 w + 8 \log_9 x$

108. $7 \ln 5 + 6 \ln 4$ **109.** $7 \log 4 - 4 \log 5$ **110.** $2 \ln 7 + 7 \ln x + 9 \ln y$

111. $\log_4 5 + \log_4 x$ **112.** $3 \ln z - 4 \ln y$ **113.** $6 \log_3 x + 4 \log_3 y + 2 \log_3 5$

In Exercises 114–119, simplify the expression, if possible.

114. $\frac{4x}{3x^3 - x}$ **115.** $\frac{2x^2}{6x^4 + 8x}$

116. $\frac{3x^3}{9x - 21}$ **117.** $\frac{x^2 + 4x + 3}{x^2 + x - 6}$

118. $\frac{x^3 + 9x^2 + 27x + 27}{x^2 + 9x + 18}$ **119.** $\frac{x^3 - 6x^2 + 12x - 8}{x^2 + 6x - 16}$

In Exercises 120–128, evaluate the trigonometric function using a calculator. Round your answer to four decimal places.

120. $\cos 16^\circ$ **121.** $\tan 35^\circ$ **122.** $\csc 68^\circ$

123. $\sin 40^\circ$ **124.** $\cot 3^\circ$ **125.** $\sec 14^\circ$

126. $\cos 89^\circ$ **127.** $\tan 2^\circ$ **128.** $\csc 12^\circ$

Chapter 9

Name ______________________________ Date __________

Chapter 9 Trigonometric Identities and Formulas

Dear Family,

In this chapter, your student will learn about trigonometric identities. Listed below are just a few examples of how trigonometry is used in real life. Take time together as a family to research and discover how useful trigonometry is.

- Consider music. Whether your student enjoys rock and roll or country or anything in between, the music industry makes use of trigonometry. Research how sound travels and how people in the music industry use trigonometry.
- Consider architecture. Research to learn how trigonometry is used in the architectural world. After researching, take a trip to a city with many tall buildings to check out your findings.
- Consider digital imaging. Since the introduction of computers, images made through computer software have increasingly become more complex. Research to learn how computer programs make digital images.
- Consider the field of medicine. Has anyone you know ever had a computed tomography (CT) scan, magnetic resonance imaging (MRI) scan, or laser treatment? Research to learn CT and MRI scanners use trigonometry to detect tumors and how laser treatment works.
- Consider navigation. Research to learn how trigonometry is used in navigation. Why is trigonometry important to this field?

These are just a few examples of trigonometry applications. Can you find any other applications? Make a list. You may be surprised about how much trigonometry is all around you!

Nombre ______________________________ Fecha __________

Capítulo 9 Identidades trigonométricas y fórmulas

Estimada familia:

En este capítulo, su hijo aprenderá sobre identidades trigonométricas. A continuación, se enumeran unos pocos ejemplos de cómo se usa la trigonometría en la vida real. En familia, tómense un tiempo juntos para investigar y descubrir cuán útil es la trigonometría.

- Consideren la música. Independientemente de que a su hijo le guste el rock and roll o la música country o algo en el medio, la industria musical hace uso de la trigonometría. Investiguen cómo viaja el sonido y cómo las personas en la industria musical usan la trigonometría.
- Consideren la arquitectura. Investiguen para aprender cómo se usa la trigonometría en el mundo de la arquitectura. Después de investigar, vayan de viaje a una ciudad donde haya muchos edificios altos para comprobar sus hallazgos.
- Consideren la imagen digital. Desde la introducción de las computadoras, las imágenes creadas con software informático se han vuelto cada vez más complejas. Investiguen para aprender cómo se crean imágenes digitales mediante programas informáticos.
- Consideren el campo de la medicina. ¿Alguien que conocen alguna vez se hizo un estudio de tomografía computarizada (TC), imagen por resonancia magnética (IRM) o un tratamiento con láser? Investiguen para aprender cómo los escáneres de TC e IRM usan la trigonometría para detectar tumores y cómo funciona un tratamiento con láser.
- Consideren la navegación. Investiguen para aprender cómo se usa la trigonometría en la navegación. ¿Por qué la trigonometría es importante para este campo?

Estos son solo unos pocos ejemplos de los usos de la trigonometría. ¿Pueden encontrar algún otro uso? Hagan una lista. ¡Tal vez se sorprendan sobre lo mucho que la trigonometría los rodea!

9.1 Start Thinking

Use the figure and the basic definitions of trigonometric functions to write the equation of the circle in terms of sine and cosine.

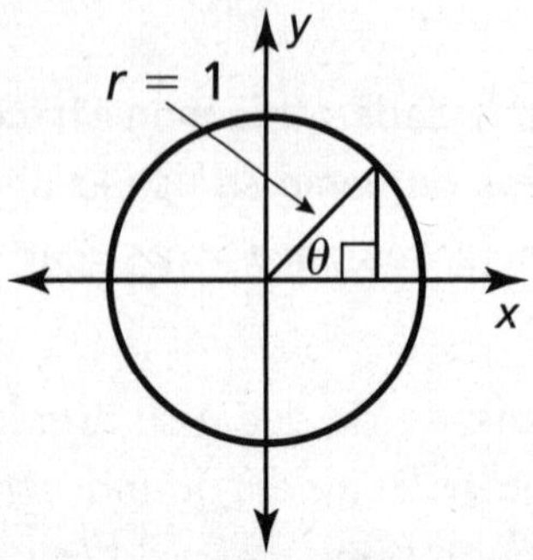

What is the length of the hypotenuse of the triangle? Write the tangent function in terms of sine and cosine.

9.1 Warm Up

Evaluate the six trigonometric functions for the angle θ.

1. 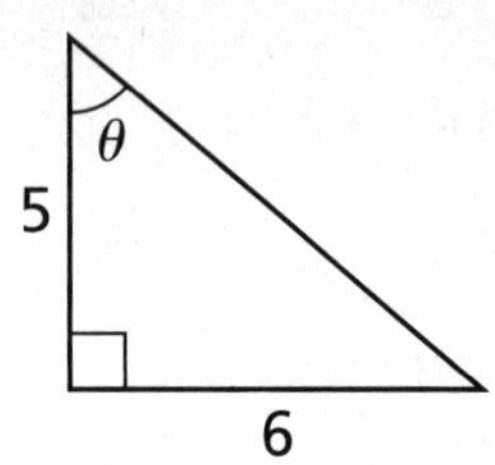

2. 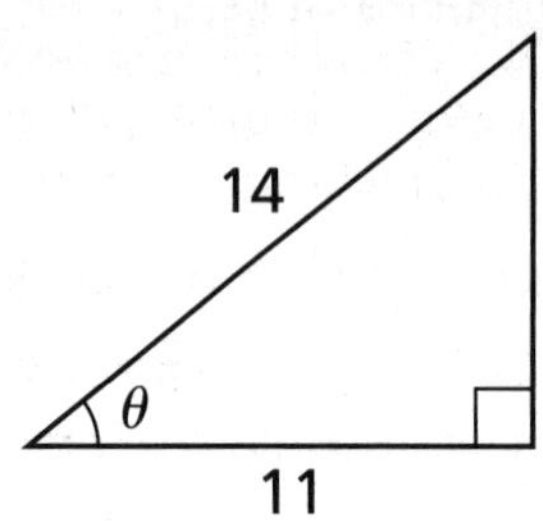

3. 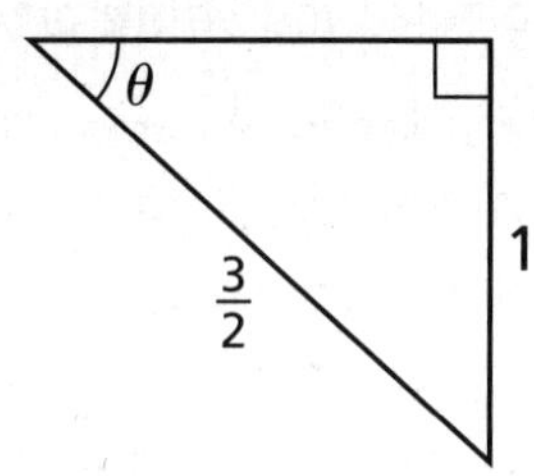

9.1 Cumulative Review Warm Up

Find the sum or difference.

1. $\dfrac{2x}{x+4} - \dfrac{x-1}{x+4}$

2. $\dfrac{x-1}{x^2-x-12} + \dfrac{3}{x-4}$

3. $\dfrac{-3x}{x^2+5x} + \dfrac{1-x}{x^2-25}$

4. $\dfrac{7}{2x^2} - \dfrac{3x+4}{4x-2}$

Name____ Date__________

9.1 Practice A

In Exercises 1–6, find the values of the other five trigonometric functions of θ.

1. $\sin\theta = \frac{2}{3}, 0 < \theta < \frac{\pi}{2}$

2. $\cos\theta = -\frac{3}{5}, \frac{\pi}{2} < \theta < \pi$

3. $\tan\theta = \frac{4}{7}, \pi < \theta < \frac{3\pi}{2}$

4. $\cot\theta = -\frac{5}{4}, \frac{\pi}{2} < \theta < \pi$

5. $\sec\theta = \frac{7}{3}, \frac{3\pi}{2} < \theta < 2\pi$

6. $\csc\theta = -\frac{6}{5}, \pi < \theta < \frac{3\pi}{2}$

In Exercises 7–12, simplify the expression.

7. $\cos x \tan x$

8. $\sin x\left(\csc^2 x - 1\right)$

9. $\frac{\tan(-\theta)}{\sin(-\theta)}$

10. $\frac{\sin^2\theta}{\tan^2\theta}$

11. $\frac{\tan\left(\frac{\pi}{2} - x\right)}{\csc x}$

12. $\frac{1 - \sin^2 x}{\sec x}$

13. Describe and correct the error in simplifying the expression.

$$\begin{aligned} \tan^2 x + \sec^2 x &= \tan^2 x + \left(\tan^2 x - 1\right) \\ &= \tan^2 x + \tan^2 x - 1 \\ &= 2\tan^2 x - 1 \end{aligned}$$

In Exercises 14–17, verify the identity.

14. $\tan x \cot x = 1$

15. $\sin\left(\frac{\pi}{2} - x\right)\sec x = 1$

16. $\frac{\sin^2 x + \left(1 - \cos^2 x\right)}{\tan x \cos x} = 2\sin x$

17. $\frac{\cos^2(-x)\csc x}{\cot x} = \cos x$

18. As the value of $\sin\theta$ decreases, what happens to the value of $\csc\theta$? Explain your reasoning.

Name ______________________________ Date __________

9.1 Practice B

In Exercises 1–6, find the values of the other five trigonometric functions of θ.

1. $\sin\theta = \frac{3}{8}, \frac{\pi}{2} < \theta < \pi$

2. $\cos\theta = -\frac{1}{3}, \pi < \theta < \frac{3\pi}{2}$

3. $\tan\theta = 3, 0 < \theta < \frac{\pi}{2}$

4. $\cot\theta = \frac{10}{3}, \pi < \theta < \frac{3\pi}{2}$

5. $\sec\theta = -\frac{9}{5}, \frac{\pi}{2} < \theta < \pi$

6. $\csc\theta = -\frac{7}{2}, \frac{3\pi}{2} < \theta < 2\pi$

In Exercises 7–12, simplify the expression.

7. $\sin x\left(\cot^2 x + 1\right)$

8. $\dfrac{\cot(-\theta)}{\cos(-\theta)}$

9. $\cos\left(\frac{\pi}{2} - \theta\right)\cot\theta$

10. $\dfrac{\tan^2\theta - \sec^2\theta}{\cos\left(\frac{\pi}{2} - \theta\right)\sec\theta}$

11. $\dfrac{\sin(-x)}{\tan(-x)\sec x} + \sin^2 x$

12. $\dfrac{\csc^2 x - \tan x \cot x}{\csc x - 1}$

13. Describe and correct the error in simplifying the expression.

$$\text{✗} \quad \csc\theta\cot\theta\cos\theta = \frac{1}{\cos\theta} \bullet \frac{\cos\theta}{\sin\theta} \bullet \cos\theta$$
$$= \frac{\cos\theta}{\sin\theta}$$
$$= \cot\theta$$

In Exercises 14–17, verify the identity.

14. $\tan\left(\frac{\pi}{2} - x\right)\cos^2\left(\frac{\pi}{2} - x\right) = \cos x \sin x$

15. $\dfrac{1 + \sin(-x)}{\cos x \tan x - 1} = -1$

16. $\dfrac{1 - \cos x}{\sin x} = \csc x - \cot x$

17. $\dfrac{1 + \sin\theta}{\cos\theta} + \dfrac{\cos\theta}{1 - \sin\theta} = 2\sec\theta + 2\tan\theta$

18. Use the sine and cosine functions to verify the identity $\tan x \cot x = 1$.

Name____ Date__________

9.1 Enrichment and Extension

Using Trigonometric Identities

You can use the fundamental trigonometric identities to write one trigonometric function in terms of another.

Example: Write $\tan\theta$ in terms of $\sin\theta$.

Solution:

$\sin^2\theta + \cos^2\theta = 1$	Write Pythagorean identity.
$\cos^2\theta = 1 - \sin^2\theta$	Subtract $\sin^2\theta$ from each side.
$\cos\theta = \pm\sqrt{1 - \sin^2\theta}$	Take the square root of each side.

Substitute $\pm\sqrt{1 - \sin^2\theta}$ for $\cos\theta$ in the tangent identity.

$$\tan\theta = \frac{\sin\theta}{\cos\theta}$$

$$= \frac{\sin\theta}{\pm\sqrt{1 - \sin^2\theta}}$$

In Exercises 1–3, write the function in terms of $\cos\theta$.

1. $\tan\theta$ **2.** $\cot\theta$ **3.** $\csc\theta$

In Exercises 4–6, write the function in terms of $\tan\theta$.

4. $\cot\theta$ **5.** $\sin\theta$ **6.** $\cos\theta$

In Exercises 7–8, write the function in terms of $\csc\theta$.

7. $\cos\theta$ **8.** $\tan\theta$ **9.** $\sec 0$

In Exercises 10–12, write the function in terms of $\sec\theta$.

10. $\tan\theta$ **11.** $\sin\theta$ **12.** $\cot\theta$

Name ______________________________ Date __________

9.1 Puzzle Time

What Do You Call A Little Blue Man Who Lives On The West Coast?

Write the letter of each answer in the box containing the exercise number.

Find the value of cos θ.

1. $\sin \theta = \frac{1}{4}, 0 < \theta < \frac{\pi}{2}$

2. $\tan \theta = -\frac{4}{5}, \frac{\pi}{2} < \theta < \pi$

Find the value of sin θ.

3. $\csc \theta = \frac{5}{4}, \frac{\pi}{2} < \theta < \pi$

4. $\cos \theta = -\frac{2}{3}, \pi < \theta < \frac{3\pi}{2}$

Find the value of tan θ.

5. $\sin \theta = -\frac{2}{3}, \frac{3\pi}{2} < \theta < 2\pi$

6. $\cot \theta = 4, 0 < \theta < \frac{\pi}{2}$

Find the value of csc θ.

7. $\sec \theta = -\frac{7}{3}, \frac{\pi}{2} < \theta < \pi$

8. $\tan \theta = -\frac{8}{5}, \frac{3\pi}{2} < \theta < 2\pi$

Answers

A. $-\frac{5\sqrt{41}}{41}$

U. $\frac{1}{4}$

P. $\frac{\sqrt{15}}{4}$

P. $\frac{4}{5}$

F. $-\frac{\sqrt{89}}{8}$

A. $-\frac{\sqrt{5}}{3}$

S. $-\frac{2\sqrt{5}}{5}$

R. $\frac{7\sqrt{10}}{20}$

1	2	3	4		5	6	7	8

9.2 Start Thinking

The formulas given below can be used to find the sine of angles that would normally require the use of a calculator. For each value of θ, determine two familiar angles whose sum or difference is equal to θ. Then use the appropriate formula to determine the value of $\sin\theta$.

$$\sin(a + b) = \sin a \cos b + \cos a \sin b$$

$$\sin(a - b) = \sin a \cos b - \cos a \sin b$$

1. $\theta = 75°$

2. $\theta = \dfrac{\pi}{12}$

3. $\theta = \dfrac{19\pi}{12}$

9.2 Warm Up

Evaluate the expression given that $\sin a = -\dfrac{3}{7}$ with $\pi < a < \dfrac{3\pi}{2}$ and $\cos b = \dfrac{5}{8}$ with $\dfrac{3\pi}{2} < b < 2\pi$.

1. $\sin b$

2. $\cos a$

3. $\tan b$

4. $\tan a$

5. $\csc b$

6. $\sec a$

9.2 Cumulative Review Warm Up

Use the given annual interest rate and method of compounding to find the amount *A* in an account earning compound interest after 8 years when the principal is \$5000.

1. $r = 2.25\%$, daily

2. $r = 1.125\%$, quarterly

3. $r = 1.84\%$, monthly

4. $r = 2.16\%$, continuously

Name ____ Date __________

9.2 Practice A

In Exercises 1–6, find the exact value of the expression.

1. $\tan 105^\circ$
2. $\tan(-75^\circ)$
3. $\sin \frac{\pi}{12}$
4. $\sin 255^\circ$
5. $\cos(-165^\circ)$
6. $\cos \frac{7\pi}{12}$

In Exercises 7–12, evaluate the expression given that $\sin a = \frac{3}{5}$ with $0 < a < \frac{\pi}{2}$ and $\cos b = -\frac{5}{13}$ with $\pi < b < \frac{3\pi}{2}$.

7. $\sin(a + b)$
8. $\sin(a - b)$
9. $\cos(a - b)$
10. $\cos(a + b)$
11. $\tan(a + b)$
12. $\tan(a - b)$

In Exercises 13–15, simplify the expression.

13. $\sin\left(x + \frac{\pi}{2}\right)$
14. $\tan(x - \pi)$
15. $\cos(x + \pi)$

16. Describe and correct the error in simplifying the expression.

$$\text{✗} \quad \cos\left(x + \frac{3\pi}{2}\right) = \cos x \cos \frac{3\pi}{2} + \sin x \sin \frac{3\pi}{2}$$
$$= (0)\cos x + (-1)\sin x$$
$$= -\sin x$$

In Exercises 17–20, solve the equation for $0 \le x < 2\pi$.

17. $\cos\left(x - \frac{\pi}{2}\right) = 0$
18. $\tan(x + \pi) = 1$
19. $\cos\left(x + \frac{\pi}{4}\right) + \cos\left(x - \frac{\pi}{4}\right) = 1$
20. $\sin\left(x + \frac{\pi}{6}\right) - \sin\left(x - \frac{\pi}{6}\right) = 0$

21. Verify that the tangent function has a period of π by deriving the identity $\tan(x - \pi) = \tan x$ using the difference formula for tangent.

Name________________________________ Date__________

9.2 Practice B

In Exercises 1–6, find the exact value of the expression.

1. $\tan 165°$
2. $\sin \dfrac{13\pi}{12}$
3. $\sin(-105°)$
4. $\cos 75°$
5. $\cos\left(-\dfrac{5\pi}{12}\right)$
6. $\tan \dfrac{25\pi}{12}$

In Exercises 7–12, evaluate the expression given that $\sin a = \dfrac{12}{13}$ with $0 < a < \dfrac{\pi}{2}$ and $\cos b = -\dfrac{8}{17}$ with $\pi < b < \dfrac{3\pi}{2}$.

7. $\sin(a + b)$
8. $\sin(a - b)$
9. $\cos(a - b)$
10. $\cos(a + b)$
11. $\tan(a + b)$
12. $\tan(a - b)$

In Exercises 13–15, simplify the expression.

13. $\tan(x + 3\pi)$
14. $\cos\left(x + \dfrac{3\pi}{2}\right)$
15. $\sin(x - \pi)$

16. Describe and correct the error in simplifying the expression.

$$\times \quad \sin\left(x + \frac{\pi}{2}\right) = \sin x \sin \frac{\pi}{2} + \cos x \cos \frac{\pi}{2}$$
$$= (1) \sin x + (0) \cos x$$
$$= \sin x$$

In Exercises 17–20, solve the equation for $0 \le x < 2\pi$.

17. $\cos\left(x - \dfrac{3\pi}{2}\right) = \dfrac{1}{2}$
18. $\cos\left(x + \dfrac{\pi}{3}\right) + \cos\left(x - \dfrac{\pi}{3}\right) = 0$
19. $\tan\left(x + \dfrac{\pi}{4}\right) - \tan\left(\dfrac{\pi}{4} - x\right) = 0$
20. $\sin(x - \pi) - \cos(x + \pi) = 0$

21. Verify the identity $\tan(a + b) = \dfrac{\sin(a + b)}{\cos(a + b)}$ by using the angle sum formula for tangent.

Name ____________________ Date ________

9.2 Enrichment and Extension

Sum and Difference Formulas

In Exercises 1–4, use your knowledge of the sum and difference formulas for sine, cosine, and tangent to decide whether the statement is true or false. If false, correct the statement.

1. $\sin\left(x - \dfrac{11\pi}{2}\right) = \cos x$

2. $\cos(n\pi + \theta) = (-1)^n \cos\theta$, where n is an integer.

3. $\sin(n\pi + \theta) = (-1)^{n+1} \sin\theta$, where n is an integer.

4. $\sin(n\pi - \theta) = (-1)^n \sin\theta$, where n is an integer.

In Exercises 5 and 6, write a formula for the trigonometric function of the sum.

5. $\cos(u + v + w)$

6. $\sin(u + v + w)$

In Exercises 7–10, use the sum and difference formulas to prove the even/odd identity.

7. $\sin(-u) = -\sin u$

8. $\cos(-u) = \cos u$

9. $\tan(-u) = -\tan u$

10. $\sec(-u) = \sec u$

11. Simplify $\tan(x + \pi)$ and interpret your answer graphically. What does your answer illustrate about the period of the tangent function?

Name ______________________ Date __________

9.2 Puzzle Time

What Do You Get When You Cross A Secretary With A CD Player?

Write the letter of each answer in the box containing the exercise number.

Find the exact value of the expression.

1. $\sin 195^\circ$
2. $\cos 165^\circ$
3. $\tan(-15^\circ)$
4. $\cos 75^\circ$
5. $\sin 105^\circ$
6. $\tan \dfrac{13\pi}{12}$

Simplify the expression.

7. $\sin\left(x + \dfrac{\pi}{2}\right)$
8. $\cos(x - \pi)$
9. $\tan(x - \pi)$
10. $\sin(x - 2\pi)$

Answers

Y. $-\cos x$

R. $\dfrac{\sqrt{6} - \sqrt{2}}{4}$

T. $\dfrac{-\sqrt{6} - \sqrt{2}}{4}$

S. $\dfrac{\sqrt{2} - \sqrt{6}}{4}$

E. $\sqrt{3} - 2$

O. $2 - \sqrt{3}$

E. $\sin x$

T. $\cos x$

E. $\dfrac{\sqrt{6} + \sqrt{2}}{4}$

P. $\tan x$

1	2	3	4	5	6	7	8	9	10

9.3 Start Thinking

Use the triangle in the diagram to find the value of each ratio. How do the three ratios relate? Do you think this relationship would be the same if the triangle was not a right triangle?

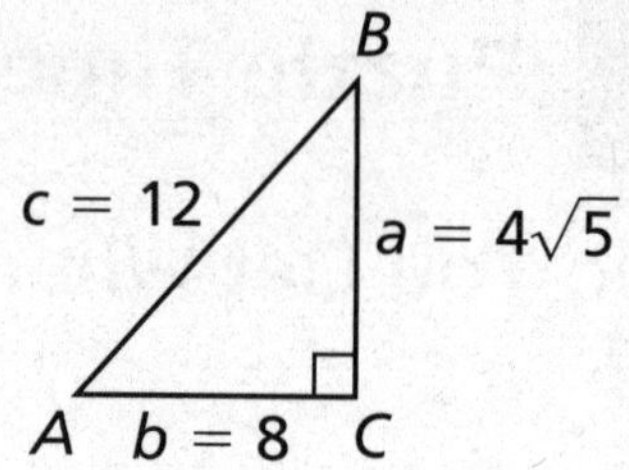

1. $\dfrac{a}{\sin A}$

2. $\dfrac{b}{\sin B}$

3. $\dfrac{c}{\sin C}$

9.3 Warm Up

Solve the proportion. Round your answer to the nearest tenth.

1. $\dfrac{a}{\sin 28^\circ} = \dfrac{21}{\sin 65^\circ}$

2. $\dfrac{15}{\sin 40^\circ} = \dfrac{c}{\sin 94^\circ}$

3. $\dfrac{b}{\sin 9^\circ} = \dfrac{63}{\sin 105^\circ}$

4. $\dfrac{54}{\sin B} = \dfrac{61}{\sin 73^\circ}$

5. $\dfrac{16}{\sin 81^\circ} = \dfrac{15}{\sin A}$

6. $\dfrac{110}{\sin C} = \dfrac{85}{\sin 36^\circ}$

9.3 Cumulative Review Warm Up

Find the measures of $\angle X$ and $\angle Z$.

1.

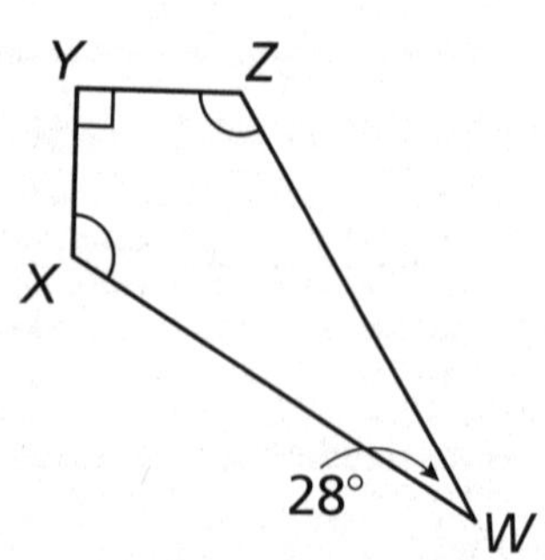

2.

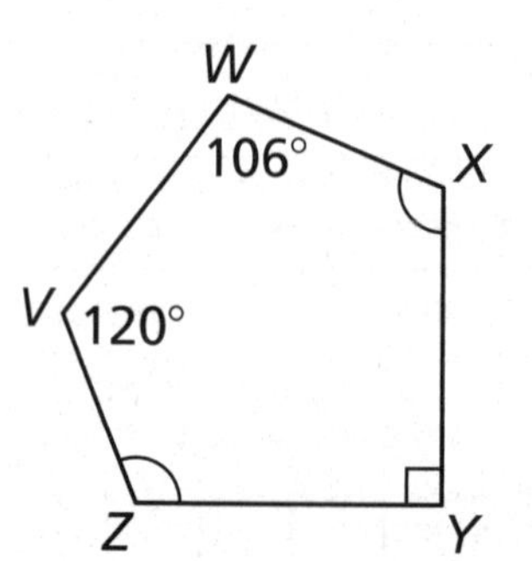

3.

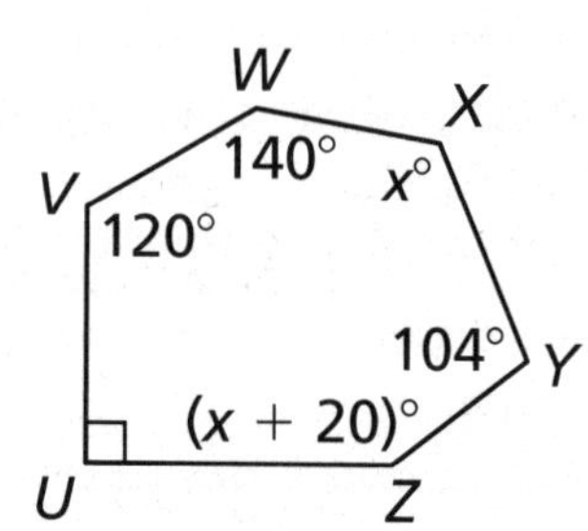

Name____ Date__________

9.3 Practice A

In Exercises 1–3, evaluate the trigonometric function using a calculator. Round your answer to four decimal places.

1. $\cos 115^\circ$ **2.** $\tan 95^\circ$ **3.** $\sin 148^\circ$

In Exercises 4 and 5, find the area of the triangle. Round your answer to the nearest tenth.

4.

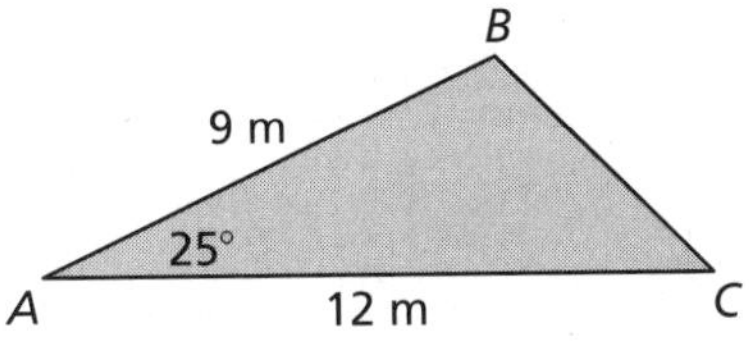

5.

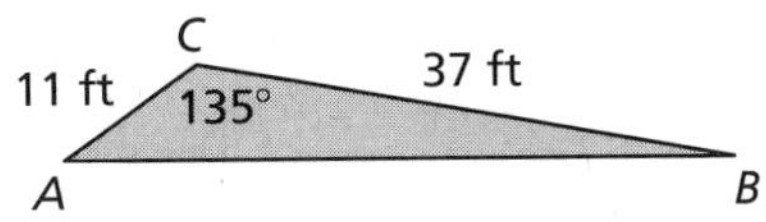

In Exercises 6–10, state the case (AAS, ASA, or SSA) applicable to the given measurements. Then decide whether the measurements determine *one triangle*, *two triangles*, or *no triangle*.

6. $A = 52^\circ, a = 31, b = 18$

7. $A = 32^\circ, B = 95^\circ, a = 26$

8. $C = 103^\circ, a = 21, c = 16$

9. $B = 37^\circ, b = 10, c = 12$

10. $A = 43^\circ, B = 35^\circ, c = 15$

In Exercises 11–13, solve the triangle. Round decimal answers to the nearest tenth.

11.

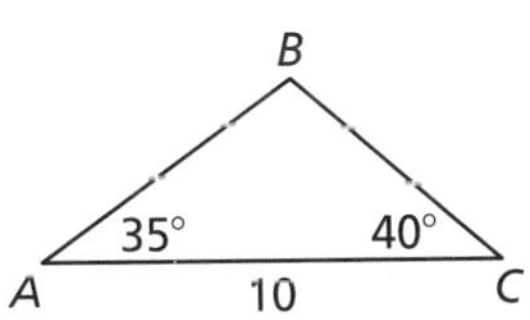

12.

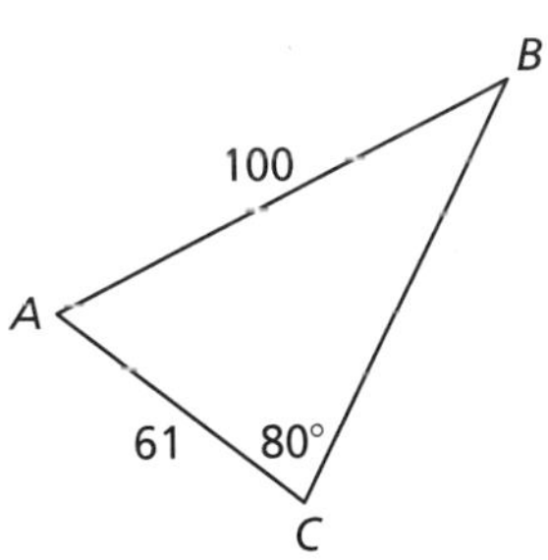

13.

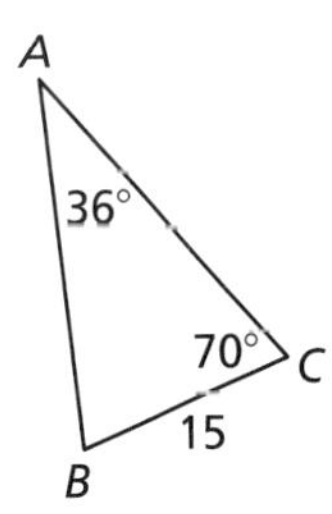

14. You are standing at a swing in a park looking at a slide. When you turn 123° clockwise, you see the slide. The swing and the slide are 19 feet apart. The slide and the water fountain are 27 feet apart. Estimate the distance from the swing to the water fountain.

Name ______________________________ Date __________

9.3 Practice B

In Exercises 1–3, evaluate the trigonometric function using a calculator. Round your answer to four decimal places.

1. tan 133°

2. cos 128°

3. sin 91°

In Exercises 4 and 5, find the area of the triangle. Round your answer to the nearest tenth.

4.

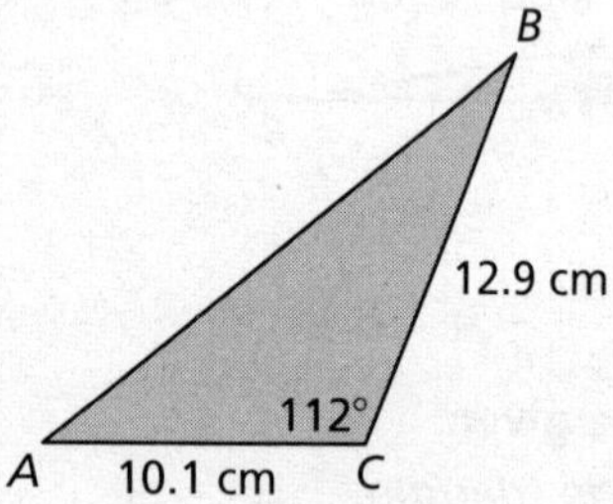

5.

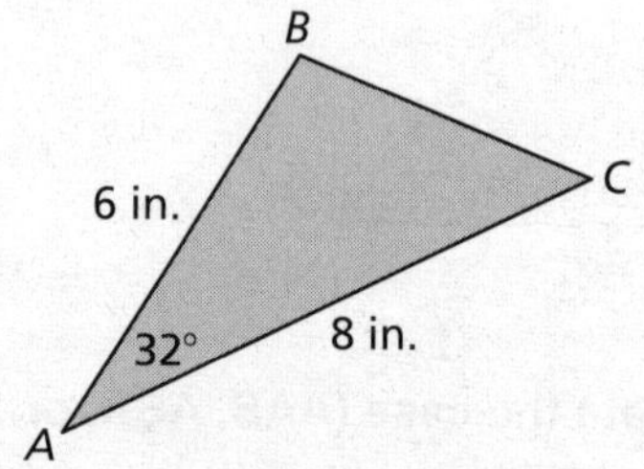

6. A parking lot has the shape of a parallelogram, as shown. Explain how you can find the area of the parking lot without using right triangles. Then find the area of the parking lot.

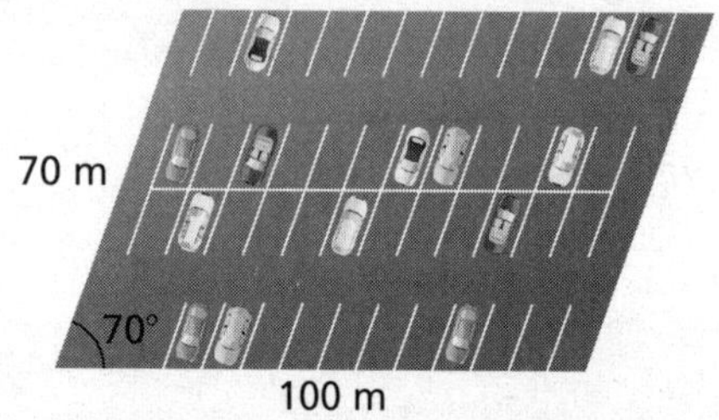

In Exercises 7–9, solve the triangle. Round decimal answers to the nearest tenth.

7.

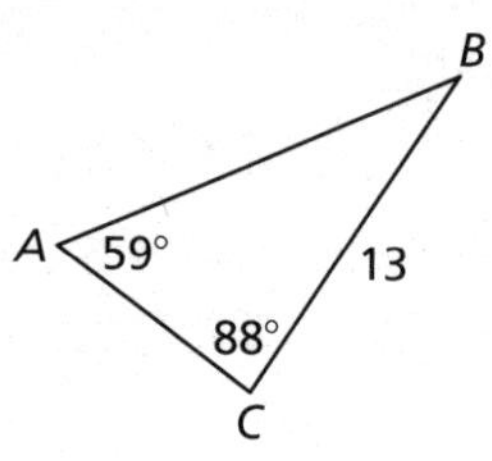

8.

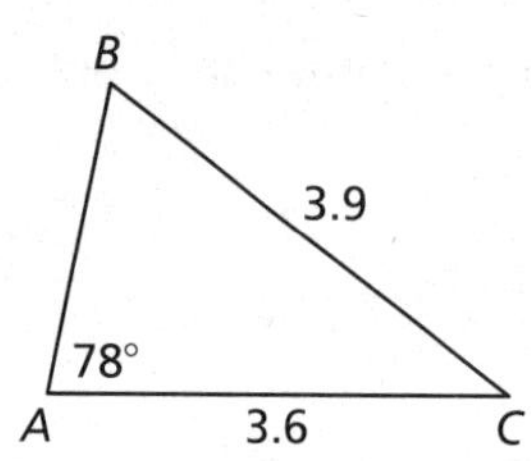

9.

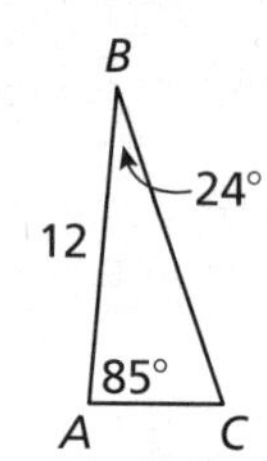

10. Find the values of x and y.

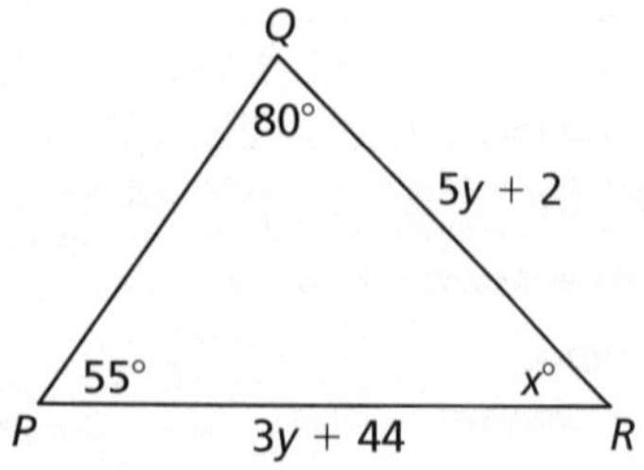

Name____ Date__________

9.3 Enrichment and Extension

Possible Triangles in the SSA Case

Previously, you were instructed to fix side b and angle A and sketch the possible positions of side a to figure out how many triangles can be formed in the SSA case. You can determine the number of possible triangles that can be formed using a given SSA case without making a sketch.

Example 1: Without making a sketch, decide whether the given measurements determine *one triangle*, *two triangles*, or *no triangle*.

$A = 51°, a = 3.5, b = 5$

Solution: When you use the Law of Sines to find the possible measures of B, you obtain $\sin B \approx 1.1102$. Because the sine ratio cannot exceed 1, there is no angle B for which $\sin B \approx 1.1102$. So, there is no triangle with the indicated measurements.

Example 2: Without making a sketch, decide whether the given measurements determine *one triangle*, *two triangles*, or *no triangle*.

$A = 61°, a = 55, b = 35$

Solution: When you use the Law of Sines to find the possible measures of B, you obtain $\sin B \approx 0.5565$. Because the sine ratio is positive in Quadrants I and II, there are two possible measures for B.

$B = \sin^{-1} 0.5566 \approx 33.8°$ or $B \approx 180° - 33.8° \approx 146.2°$

Notice that if $B = 146.2°$, then the sum of the measures of angles A and B is about $207.2°$. Because the sum of the angle measures of a triangle cannot exceed $180°$, the possible measure for B is $33.8°$. So, there is only one triangle with the indicated measurements.

In Exercises 1–10, without making a sketch, decide whether the given measurements determine *one triangle*, *two triangles*, or *no triangle*. Explain your reasoning.

1. $A = 40°, a = 20, b = 15$

2. $A = 30°, a = 10, b = 40$

3. $A = 112°, a = 42.1, b = 37$

4. $A = 150°, a = 10, b = 30$

5. $B = 50°, b = 30, c = 20$

6. $B = 162°, b = 6.1, c = 4$

7. $C = 37°, a = 12, c = 16.1$

8. $C = 12°, b = 28, c = 7$

9. $C = 49°, b = 125, c = 95$

10. $C = 80°, a = 52, c = 28$

Name ______________________ Date ________

9.3 Puzzle Time

Why Should You Always Walk A Mile In People's Shoes Before You Criticize Them?

A	B	C	D	E	F
G	H	I	J	K	L

Complete each exercise. Find the answer in the answer column. Write the word under the answer in the box containing the exercise letter.

Answer	Word
−0.8391	BECAUSE
347.5	SOCKS
no triangle	YOU'LL
−0.1736	TO
one triangle	AND
43.8	SHOES
not possible	ROAD
0.9063	BE
129.6	END
173.7	MILE

Evaluate the trigonometric function using a calculator. Round your answer to four decimal places.

A. tan 140°

B. sin 170°

C. cos 135°

D. sin 115°

Find the area of $\triangle ABC$ in square units. Round your answer to the nearest tenth.

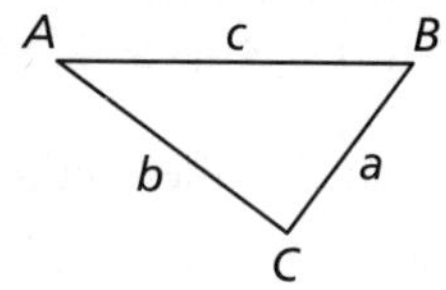

E. $A = 44°, b = 18, c = 21$

F. $B = 63°, a = 15, c = 26$

Decide whether the given measurements determine *one triangle*, *two triangles*, or *no triangle*.

G. $A = 40°, a = 7, b = 10$

H. $B = 75°, C = 39°, a = 10$

I. $C = 50°, b = 8, c = 3$

Find the indicated measure, if possible. Round decimal answers to the nearest tenth.

J. $A = 100°, a = 14, b = 9$; Find B.

K. $A = 40°, B = 82°, c = 14$; Find a.

L. $C = 98°, b = 15, c = 24$; Find A.

Answer	Word
−0.9063	THE
14.9	SO
two triangles	AWAY
0.1736	THEN
262.6	PATH
10.6	THEIR
−0.7071	YOU'LL
39.3	HAVE
131.3	A
174.0	NICE

9.4 Start Thinking

Draw $\overline{BD}$ perpendicular to $\overline{AC}$. Label the length of $\overline{BD}$ as h. How can you use h to find c? If you were given that $A = x°$, how could you use the sine function to rewrite h?

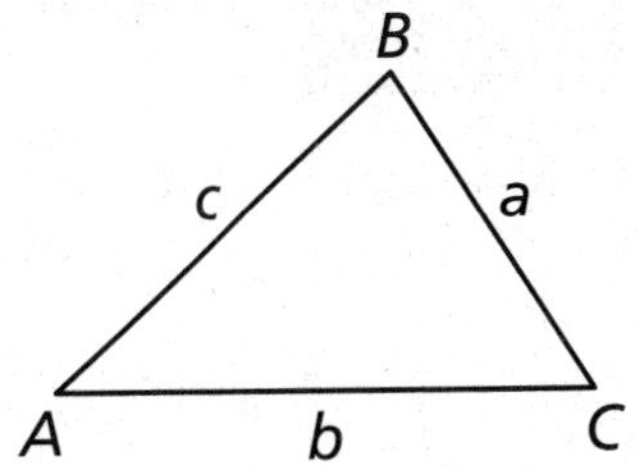

9.4 Warm Up

Use the Pythagorean Theorem to find the missing side length. Round decimal answers to the nearest tenth.

1.

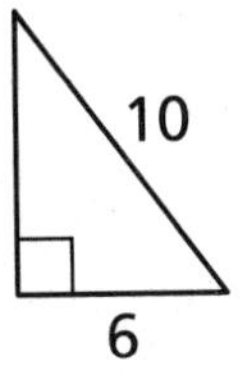

2.

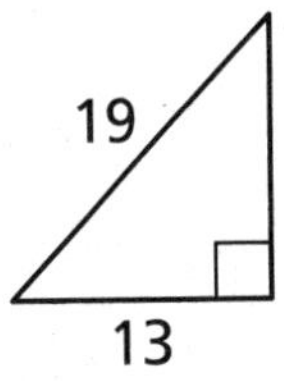

9.4 Cumulative Review Warm Up

Use a graphing calculator to find a polynomial function that best fits the data.

1. $(-5, -3), (-2, 14), (1, -5), (4, 48), (7, 191), (10, 712)$
2. $(0, -15), (1, -6), (2, -3), (3, 6), (4, 33), (5, 90)$
3. $(-3, -226), (-2, -38), (-1, 22), (0, 6), (1, -2), (2, 2),$ $(3, -48), (4, -198)$

Name ______________________________ Date ________

9.4 Practice A

In Exercises 1–3, solve $\triangle ABC$. Round decimal answers to the nearest tenth.

1.

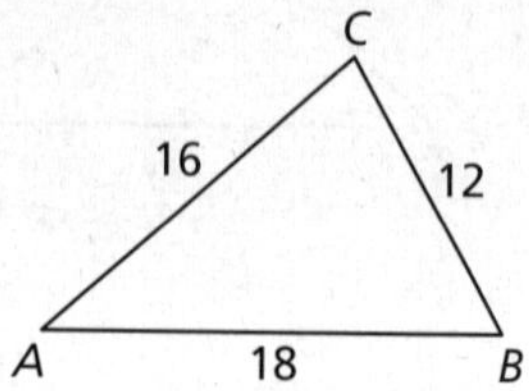

2.

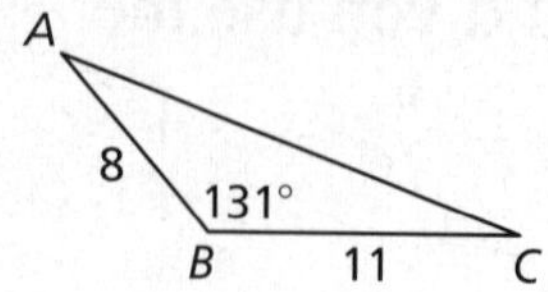

3.

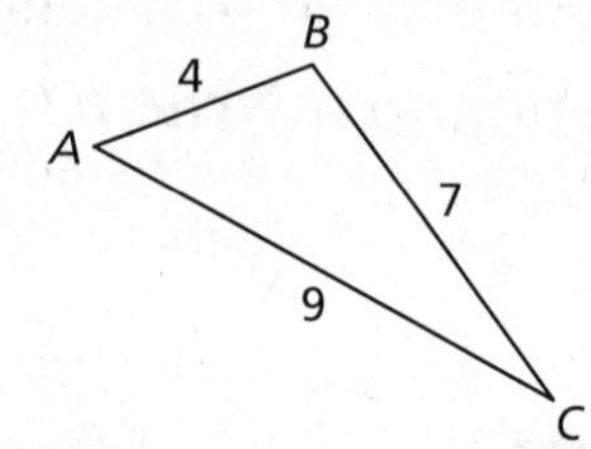

In Exercises 4 and 5, find the area of $\triangle ABC$. Round your answer to the nearest tenth.

4.

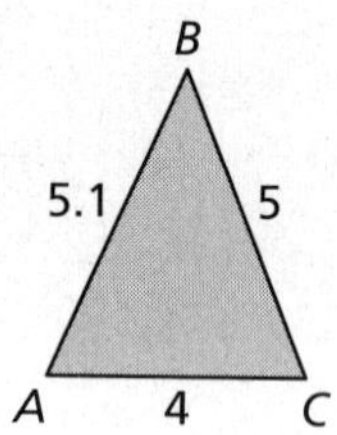

5.

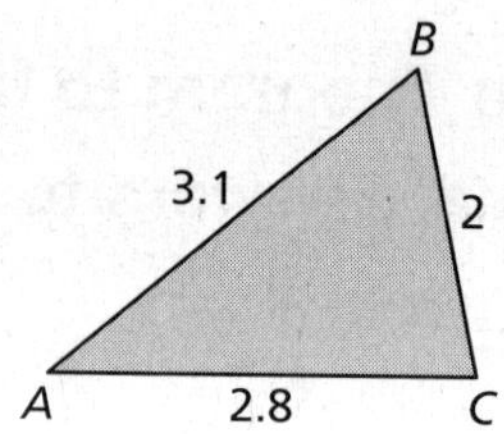

6. Place each triangle case into one of the three categories according to the first step in solving the triangle.

Law of Sines	Law of Cosines	Neither

AAA AAS ASA SSS SSA SAS

7. Determine the measure of angle A in the design of the streetlamp shown in the diagram.

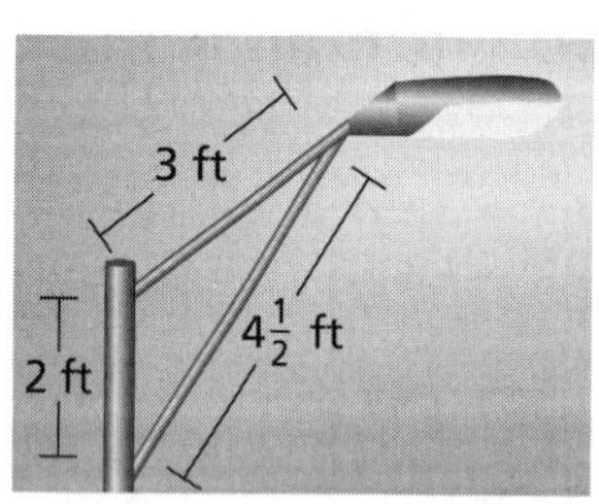

Name______________________________________ Date__________

9.4 Practice B

In Exercises 1–5, solve $\triangle ABC$. Round decimal answers to the nearest tenth.

1. 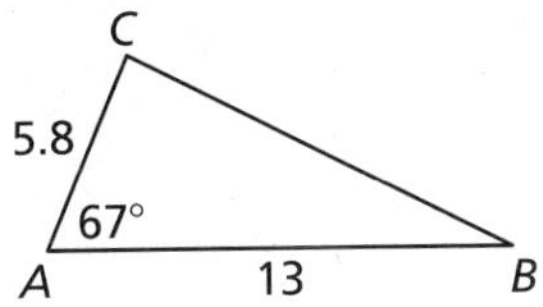

2. 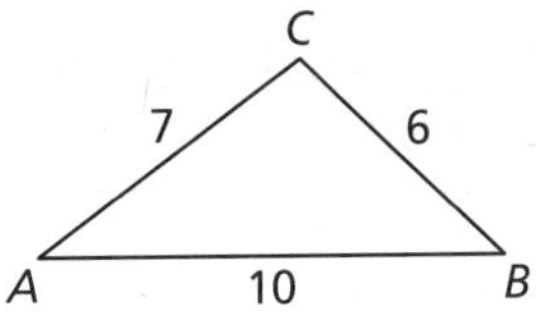

3. 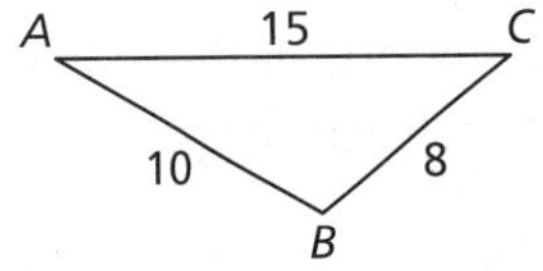

4. $a = 15, b = 18, c = 19$

5. $a = 51, B = 19°, c = 61$

In Exercises 6 and 7, find the area of $\triangle ABC$. Round your answer to the nearest tenth.

6. 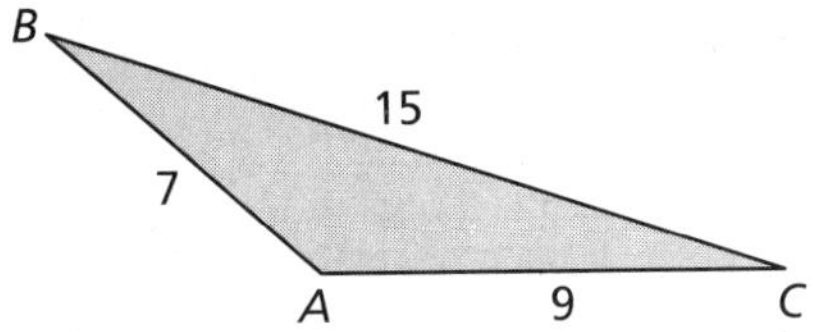

7. 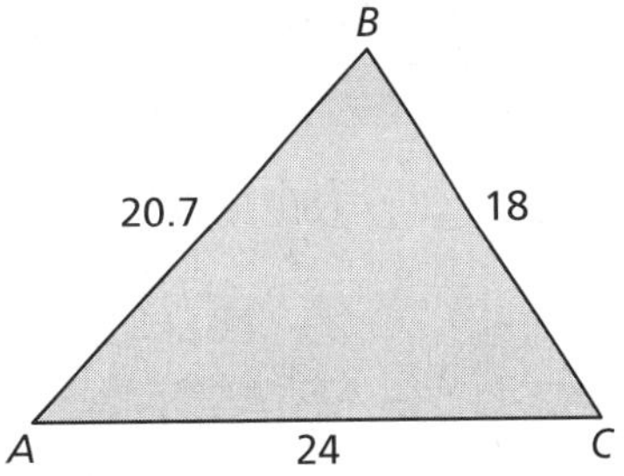

8. A bike frame has a top tube length of 20.75 inches, a seat tube length of 8.9 inches, and a seat tube angle of 71°.

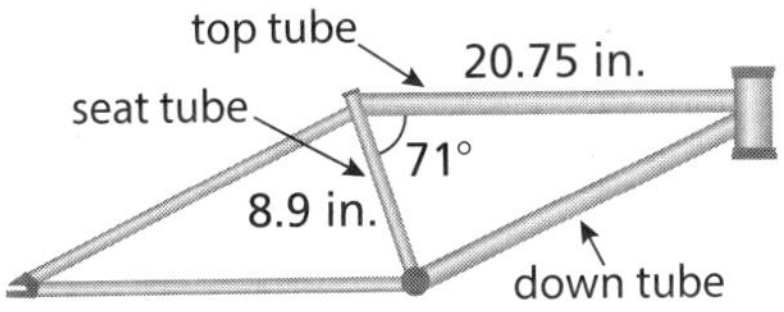

a. Find the approximate length of the down tube.

b. Find the angle between the seat tube and down tube.

Name ______________________________ Date __________

9.4 Enrichment and Extension

Law of Tangents

In addition to the Law of Sines and the Law of Cosines, the *Law of Tangents* can be used to solve a triangle.

The Law of Tangents can be written as

$$\frac{a+b}{a-b} = \frac{\tan\frac{A+B}{2}}{\tan\frac{A-B}{2}} \text{ or } \frac{a-b}{a+b} = \frac{\tan\frac{A-B}{2}}{\tan\frac{A+B}{2}},$$

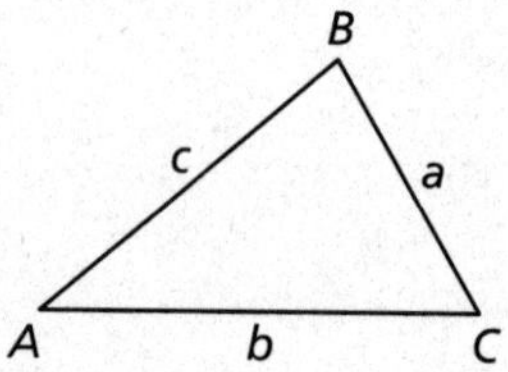

for ΔABC with sides of length a and b.

Example: Solve the triangle at the right using the Law of Tangents.

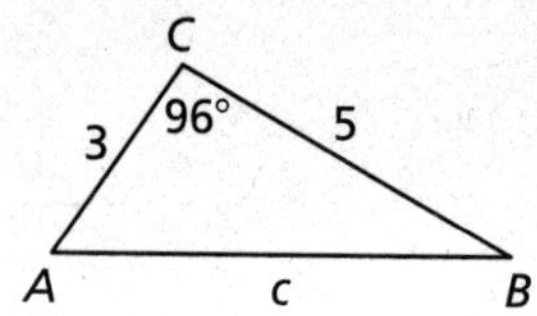

Solution: Because $A + B + C = 180°$,

$$A + B = 180° - C = 180° - 96° = 84°,$$

and $\frac{A+B}{2} = 42°$.

$$\frac{a-b}{a+b} = \frac{\tan\frac{A-B}{2}}{\tan\frac{A+B}{2}} = \frac{5-3}{5+3} = \frac{\tan\frac{A-B}{2}}{\tan 42°}$$ Substitute.

$$\frac{2}{8}\tan 42° = \tan\frac{A-B}{2}$$ Multiply each side by tan 42°.

$$0.2251 = \tan\frac{A-B}{2}$$ Use a calculator.

$$12.7° = \frac{A-B}{2}$$ Take the inverse tangent of each side.

$$25.4° = A - B$$ Multiply each side by 2.

Using the equations $A - B = 25.4°$ and $A + B = 84°$, solve the system to obtain $A = 54.7°$ and $B = 29.3°$. You can then find the remaining side c using the Law of Sines: $\frac{c}{\sin C} = \frac{a}{\sin A} \rightarrow c = \frac{a \sin C}{\sin A} = \frac{5 \sin 96°}{\sin 54.7°} \approx 6.109$

In Exercises 1–3, use the Law of Tangents to solve ΔABC. Round your answers to the nearest tenth.

1. $C = 80°, a = 52, b = 28$

2. $A = 62°, b = 9, c = 14$

3. $A = 66°, b = 13, c = 20$

Name____ Date__________

9.4 Puzzle Time

Why Did The Bicycle Fall Over?

Circle the letter of each correct answer in the boxes below. The circled letters will spell out the answer to the riddle.

Complete the sentence.

1. You can use the Law of ____________ to solve triangles when two sides and the included angle are known, or when all three sides are known.

2. In the SSS case, the ____________ angle is found first, so when you use the Law of Sines to find another angle measure, you will know that it is between 0° and 90°.

3. The area of any triangle with side lengths a, b, and c is $\text{Area} = \sqrt{s(s-a)(s-b)(s-c)}$, where $s =$ ____________.

Find the indicated measure. Round decimal answers to the nearest tenth.

4. $a = 11, b = 15, c = 21$; Find A.

5. $a = 6, b = 8, c = 12$; Find C.

6. $A = 48°, b = 3, c = 14$; Find a.

7. $C = 101°, a = 3, b = 7$; Find c.

8. $A = 50°, b = 15, c = 30$; Find B.

9. $C = 108°, a = 10, b = 7$; Find A.

10. $B = 10°, a = 40, c = 30$; Find A.

11. $a = 9, b = 3, c = 11$; Find B.

Find the area of $\triangle ABC$ in square units. Round your answer to the nearest tenth.

12.

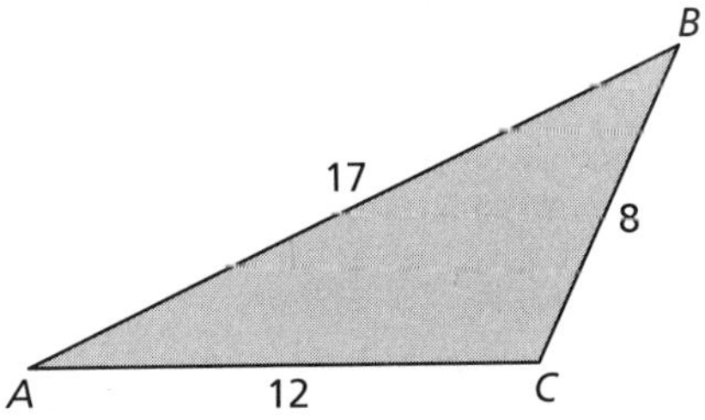

13.

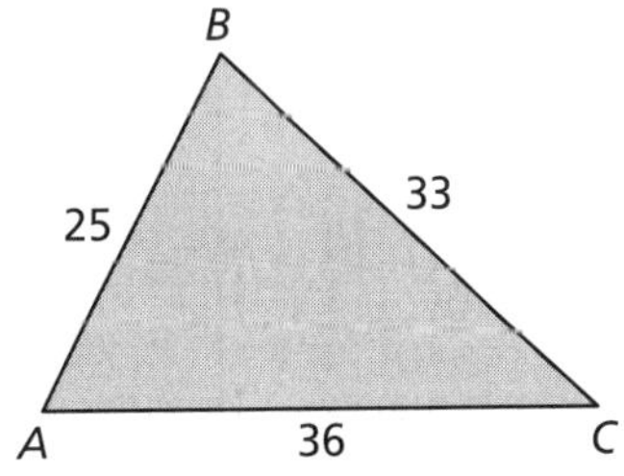

I	S	T	W	P	A	M	S	T
Cosines	10.1	largest	43.5	7.0	$\frac{1}{2}(a+b+c)$	smallest	117.3°	43.3°
W	**O**	**L**	**T**	**I**	**N**	**R**	**E**	**D**
399.0	30.1°	$\frac{1}{3}(a+b+c)$	12.2	12.9°	Sines	143.5°	29.4°	8.1

Name ______________________________ Date __________

Chapter 9 Cumulative Review

Solve the equation, if possible.

1. $x + 3x + 4(x + 1) = 6(x - 1) + x$

2. $4|6y - 3| + 6 = 2$

Solve the inequality, if possible.

3. $|7h + 14| \geq 28$

4. $5 \geq -2(x + 1) - 3$

5. $4|3x + 9| + 2 > 14$

6. You have just been informed that your cell phone plan will be increasing in price. Your new pricing plan has a flat fee of \$24.00 per month, plus a cost of \$0.11 per minute of usage. Based on your budget, you have determined that you can only afford to use a maximum of 263 minutes per month. Based on this information, what is the highest amount you will pay for your cell phone per month?

7. Three times the quantity of a number x plus 4 is at least 20. Write this sentence as an inequality.

Graph the linear equation or linear inequality.

8. $y = -\frac{2}{5}x - 2$

9. $y < -4x + 2$

10. $y \geq -x$

Write the equation of the line in slope-intercept form that passes through the given point and is parallel to the given line.

11. $(-2, 4); y = 2x + 9$

12. $(8, 0); y + 2 = -\frac{1}{4}(x + 9)$

13. $(-8, -12); 18x - 9y = 27$

Describe the domain of the function.

14. $y = \frac{2}{3}\sqrt{x}$

15. $y = \sqrt{x - 4}$

16. $f(x) = \frac{3}{11}\sqrt{-2x + 4}$

17. An Internet music store charges \$35 for monthly access, and you can download up to 56 songs every month. There is a \$1.20 charge for every song downloaded after 56 songs per month.

    a. Write an equation for the Internet music store modeling the total cost y for x songs over the monthly 56-song limit.

    b. How much does it cost to download 50 songs?

    c. How much does it cost to download 64 songs?

    d. The Internet music company decides to omit the monthly access fee, but charge \$0.75 per song. When is this a better deal, compared to the monthly access fee?

Name_____________________________ Date__________

Chapter 9 Cumulative Review (continued)

Simplify the expression. Write your answer using only positive exponents.

18. $\dfrac{5^2 x^{-2} y^{-9} z^5}{10^2 x^7 y^{-8} z^6}$

19. $\dfrac{343m^{-10}}{49m^{-13}}$

20. $\dfrac{x^4 y^5 z^{-51}}{x^5 y^4 z^{-52}}$

Find the sum.

21. $\displaystyle\sum_{i=1}^{4} 4i$

22. $\displaystyle\sum_{m=0}^{9} m^2$

23. $\displaystyle\sum_{k=4}^{9} (3k + 1)$

24. $\displaystyle\sum_{n=2}^{7} (-4n + 5)$

25. $\displaystyle\sum_{d=3}^{5} \frac{d}{d + 2}$

26. $\displaystyle\sum_{f=8}^{13} 2$

27. A boulder is launched from the top of a tall cliff. The distance d (in feet) between the boulder and the ground t seconds after it is launched is given by $d = -16t^2 - 96t + 144$. Approximately how long after the boulder is launched does it hit the ground? Give your answer to two decimal places.

Simplify the expression.

28. $e^4 \bullet e^6$

29. $e^{-2} \bullet e^{10}$

30. $\dfrac{12e^7}{2e^{-2}}$

31. $\dfrac{9e^8}{3e^{-10}}$

32. $\left(4e^{12x}\right)^3$

33. $\left(3e^{-2x}\right)^6$

34. A dance team is arranged in rows on a stage. The first row has three dancers, and each row after the first has one more dancer than the row before it.

a. Write a rule for the number of dancers in the nth row.

b. How many dancers are on the stage with six rows?

c. How many dancers are on the stage with seven rows?

d. How many more dancers are on stage where there are seven rows, compared to when there are six rows?

35. The function $f(t) = -16t^2 + s_0$ represents the approximate height (in feet) of an object falling t seconds after it is dropped from an initial height s_0 (in feet). A tomato is dropped from a height of 1296 feet.

a. After how many seconds does the tomato hit the ground?

b. Suppose the initial height is adjusted by k feet. How will this affect the answer in part (a)?

Name ______________________________ Date ________

Chapter 9

Cumulative Review (continued)

Find the vertex and the axis of symmetry of the graph of the function.

36. $f(x) = \frac{9}{2}(x + 7)^2$ **37.** $g(x) = 0.5(x - 9)^2$ **38.** $g(x) = 2(x - 1)^2 - 9$

Graph the function. Compare the graph to the graph of $f(x) = x^2$.

39. $f(x) = \frac{7}{9}(x - 2)^2$ **40.** $f(x) = 5(x + 6)^2 + 1$

Simplify the expression.

41. $\sqrt[3]{\frac{125x^7}{54y^4}}$ **42.** $\frac{8}{5 + \sqrt{7}}$ **43.** $-3\sqrt{3} - \sqrt{8} - 3\sqrt{3}$

Solve the equation by graphing.

44. $x^2 + 17x + 42 = 0$ **45.** $7x + 198 = x^2$

Solve the equation using square roots.

46. $2x^2 = 50$ **47.** $3x^2 = 3$ **48.** $2x^2 + 2 = 100$

Solve the equation by completing the square. Round your solutions to the nearest hundredth, if necessary.

49. $x^2 + 13x + 22 = 7$ **50.** $5y^2 - 21 = 10y$

51. You want to enclose a rectangular vegetable garden with 70 feet of fence, with one side of the garden being your garage. What should the three side lengths of the garden be to maximize its area?

Solve the equation using the Quadratic Formula. Round your solutions to the nearest tenth, if necessary.

52. $9x^2 - 7x - 4 = 0$ **53.** $y^2 + 2y - 1 = 2$

54. A rectangle has an area of 40 square inches and a perimeter of 26 inches. Find the dimensions of the rectangle.

Describe the domain of the function.

55. $y = 2\sqrt{x}$ **56.** $y = \sqrt{x + 9}$ **57.** $f(x) = 3\sqrt{-x - 3}$

58. $y = \frac{1}{3}\sqrt{5x}$ **59.** $f(x) = \sqrt{\frac{1}{3}x - 4}$ **60.** $g(x) = \frac{3}{4}\sqrt{-x + 5}$

Graph the function. Describe the range.

61. $f(x) = \sqrt{x} + 2$ **62.** $f(x) = \sqrt{x - 3}$ **63.** $h(x) = 2\sqrt{x}$

Name____ Date__________

Chapter 9 Cumulative Review (continued)

Graph the function. Compare the graph to the graph of $f(x) = \sqrt[3]{x}$.

64. $h(x) = \sqrt[3]{x - 3}$

65. $g(x) = \sqrt[3]{x} + 4$

66. $p(x) = 7\sqrt[3]{x}$

67. $d(x) = 4\sqrt[3]{x + 2} - 1$

68. $c(x) = \frac{1}{2}\sqrt[3]{x - 1} + 2$

69. $j(x) = 0.25\sqrt[3]{x + 2}$

Solve the equation. Check your solution.

70. $\sqrt{x} = 5$

71. $\sqrt{x} - 4 = 2$

72. $5 - \sqrt{y} = -2$

73. $\sqrt{w - 2} + 4 = 7$

74. $2\sqrt{r - 2} = 20$

75. $-3 = \sqrt{4x + 12} - 7$

76. $\sqrt{5x - 8} = \sqrt{x}$

77. $\sqrt{7x + 5} = \sqrt{4x + 17}$

Find the inverse of each relation.

78.

Input	−2	−1	0	1	2	3	4
Output	6	5	4	3	2	1	0

79.

Input	5	2	−1	−4	−7	−10	−13
Output	−9	−5	−1	3	7	11	15

Find the values of the other five trigonometric functions of θ.

80. $\sin\theta = -\frac{4}{5}, \pi < \theta < \frac{3\pi}{2}$

81. $\cot\theta = \frac{1}{6}, 0 < \theta < \frac{\pi}{2}$

82. $\cos\theta = -\frac{3}{8}, \frac{\pi}{2} < \theta < \pi$

83. $\sec\theta = 5, \frac{3\pi}{2} < \theta < 2\pi$

Solve $\triangle ABC$. Round decimal answers to the nearest tenth.

84. $B = 54°\ a = 12, b = 17$

85. $A = 132°, B = 22°, b = 7$

86. $a = 9, b = 11, c = 16$

87. $C = 78°, a = 21, b = 10$

Chapter 10

Name____ Date__________

Chapter 10 Data Analysis and Statistics

Dear Family,

What types of surveys have you participated in? Did you volunteer for the survey or did the surveyor contact you? The Internet has changed the way we do business with each other. Examples include buying and selling merchandise, researching contractors to hire, and restaurants to eat at. Within an online community, there is a reputation system in place such that the consumer can rate a product, service, packaging, and more. Other consumers use this information to decide whether or not to trust a store, buy a product, or make a purchase. Most places use a five-star rating system such that five is the best and one is the worst.

As a family, discuss the following questions:

- What are the benefits of the reputation system?
- What are the drawbacks of the reputation system?
- What stores have you rated?
- What products have you rated?
- Does the rating system help you when you are making a decision?
- What are some things to consider when comparing the ratings of different products or stores?
- What are some strategies business owners can do to boost their ratings?

People who leave feedback are a self-selected sample, so the ratings may be biased. In this chapter, you will learn about the different ways to conduct an experiment by choosing a sample of the population, collecting the data, and analyzing the data. From the data, you can make predictions about the population.

The next time you buy something online, remember to leave feedback for other consumers!

Nombre ____ Fecha ________

Capítulo 10 Análisis de datos y estadística

Estimada familia:

¿En qué tipos de encuestas han participado? ¿Se ofrecieron a responder la encuesta o el encuestador los contactó? Internet ha cambiado la manera en que hacemos negocios con los demás. Entre los ejemplos, se incluyen comprar y vender mercadería, investigar contratistas para contratar o restaurantes donde comer. Dentro de una comunidad en línea, hay un sistema de reputación para que el comprador pueda calificar un producto, servicio, embalaje y más. Otros compradores usan esta información para decidir si confían o no en una tienda, compran un producto o hacen una compra. La mayoría de los lugares usan un sistema de calificación de cinco estrellas, donde cinco es el mejor y uno es el peor.

En familia, comenten las siguientes preguntas:

- ¿Cuáles son los beneficios de un sistema de reputación?
- ¿Cuáles son las desventajas de un sistema de reputación?
- ¿Qué tiendas ustedes han calificado?
- ¿Qué productos ustedes han calificado?
- ¿El sistema de calificación los ayuda cuando están por tomar una decisión?
- ¿Cuáles son algunas cosas a considerar cuando comparan las calificaciones de diferentes productos o tiendas?
- ¿Cuáles son algunas estrategias que pueden implementar los dueños de los negocios para aumentar sus calificaciones?

Las personas que dejan comentarios son una muestra autoseleccionada, entonces las calificaciones pueden ser tendenciosas. En este capítulo, aprenderán sobre las diferentes maneras de llevar a cabo un experimento eligiendo una muestra de población, recopilando datos y analizando los datos. A partir de los datos, pueden hacer predicciones sobre la población.

La próxima vez que compren algo en línea, ¡recuerden dejar un comentario para otros compradores!

10.1 Start Thinking

The scores for a recent calculus test are shown in the table. Find the mean of the data set and complete the frequency table using the described intervals. Use your frequency table to create a histogram. Does your histogram appear to have a symmetric distribution?

Calculus Test Scores

54	70	78	84
56	72	79	84
60	72	79	87
62	75	80	90
68	76	83	91

Score	Frequency
50–59	
60–69	
70–79	
80–89	
90–99	

10.1 Warm Up

Find the mean and the median of the data set. Round to the nearest hundredth, when necessary.

1. 18, 22, 15, 12, 25, 13, 21, 16, 22, 17
2. 120, 140, 108, 115, 116, 135, 121
3. 1.8, 4.5, 3.2, 6.1, 4.7, 5.1, 4.3, 5.7

10.1 Cumulative Review Warm Up

Write a function $g(x)$ whose graph represents the transformation of the graph of $f(x) = |x|$.

1. 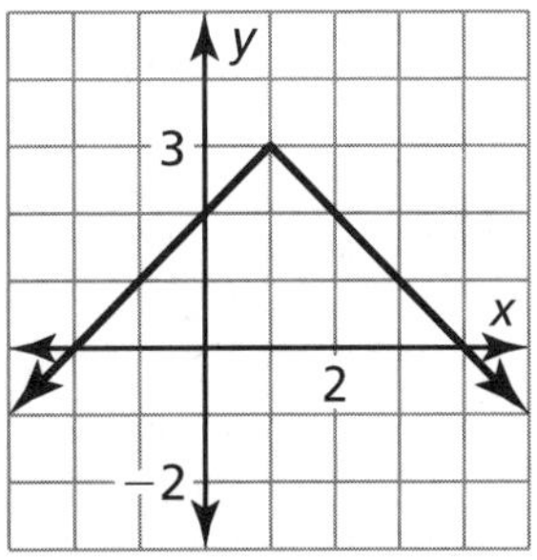

2. 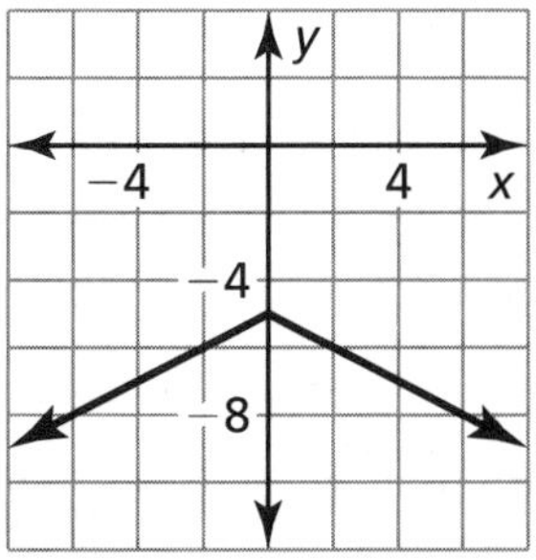

Name ______________________________ Date ________

10.1 Practice A

In Exercises 1 and 2, give the percent of the area under the normal curve represented by the shaded region.

1.

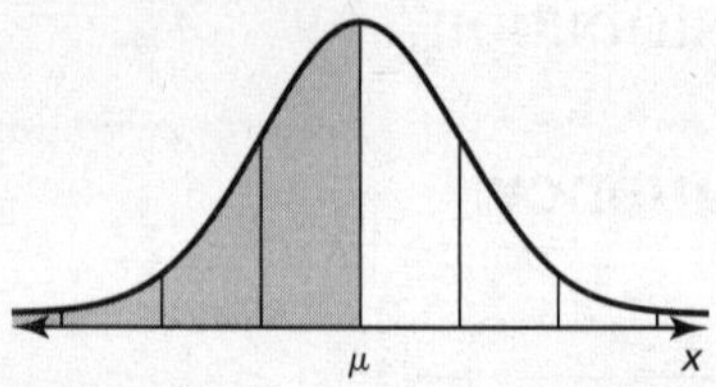

2.

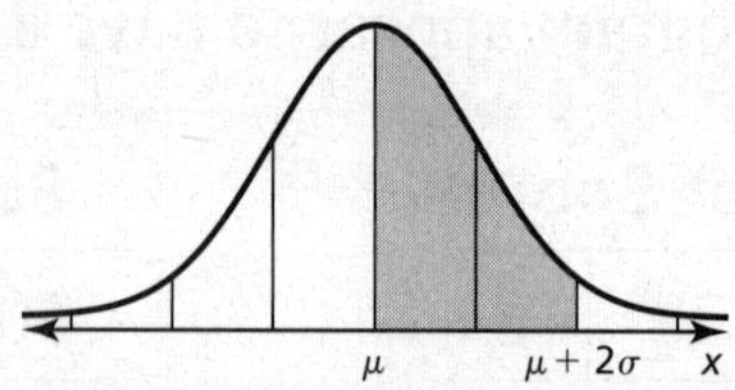

In Exercises 3–5, a normal distribution has mean μ and standard deviation σ. Find the indicated probability for a randomly selected x-value from the distribution.

3. $P(x \geq \mu + 3\sigma)$

4. $P(x \geq \mu - 3\sigma)$

5. $P(\mu - 2\sigma \leq x \leq \mu)$

In Exercises 6–8, a normal distribution has a mean of 28 and a standard deviation of 3. Find the probability that a randomly selected x-value from the distribution is in the given interval.

6. between 25 and 31

7. between 22 and 28

8. at least 31

9. The scores on an entrance mathematics exam are normally distributed with a mean of 76 and a standard deviation of 9.

a. About what percent of students have scores between 67 and 94?

b. About what percent of students have scores above 85?

10. A normal distribution has a mean of 18 and a standard deviation of 3. Describe and correct the error in finding the probability that a randomly selected x-value is in the given interval.

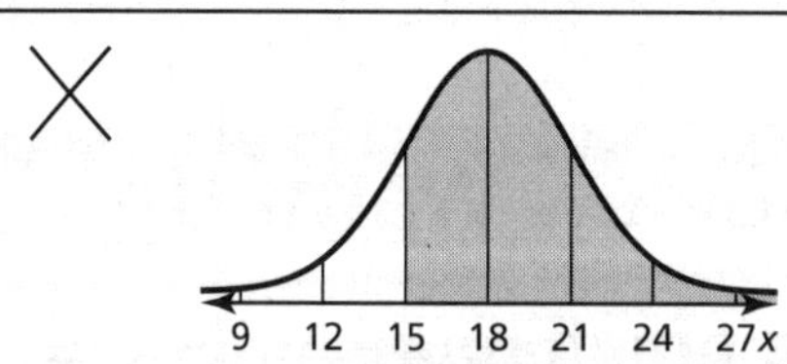

The probability that x is at least 15 is 0.8385.

Name_______________________________ Date__________

10.1 Practice B

In Exercises 1 and 2, give the percent of the area under the normal curve represented by the shaded region.

1.

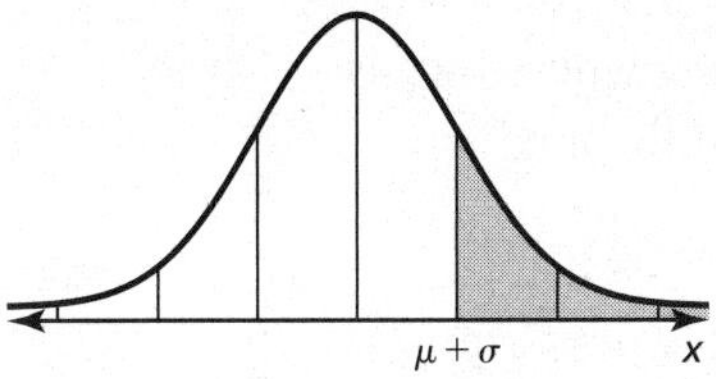

2.

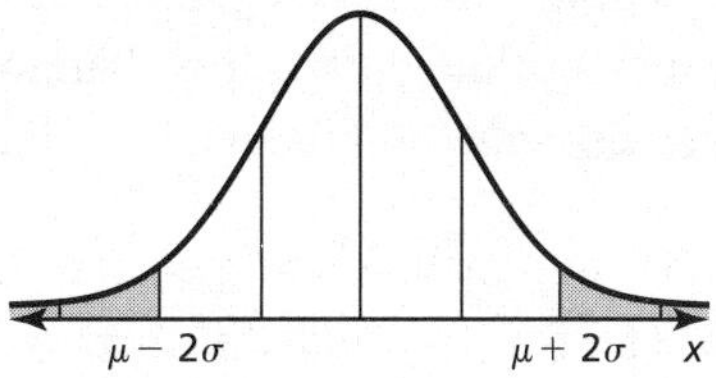

In Exercises 3–5, a normal distribution has mean μ and standard deviation σ. Find the indicated probability for a randomly selected *x*-value from the distribution.

3. $P(x \geq \mu - 2\sigma)$

4. $P(\mu - \sigma \leq x \leq \mu + 3\sigma)$

5. $P(\mu + \sigma \leq x \leq \mu + 2\sigma)$

In Exercises 6–8, a normal distribution has a mean of 28 and a standard deviation of 3. Find the probability that a randomly selected *x*-value from the distribution is in the given interval.

6. between 19 and 34

7. at most 31

8. at least 34

9. The times a restaurant takes to prepare its "quick lunch" specials are normally distributed with a mean of 3 minutes and a standard deviation of 0.5 minute.

a. About what percent of customers have their "quick lunch" between 2 minutes and 4 minutes?

b. About what percent of customers have their "quick lunch" in fewer than 2 minutes?

10. A normal distribution has a mean of 18 and a standard deviation of 3. Describe and correct the error in finding the probability that a randomly selected *x*-value is in the given interval.

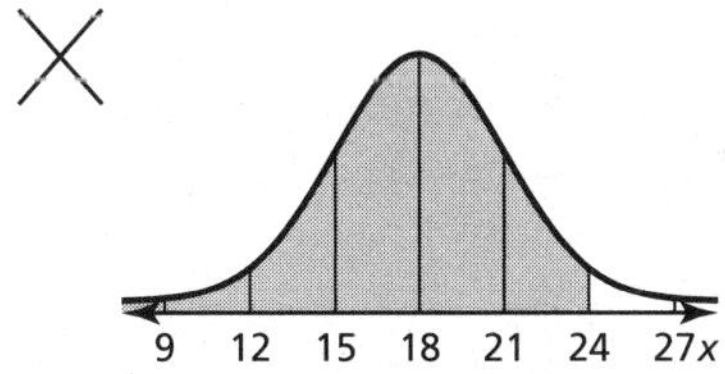

The probability that x is at most 24 is 0.475.

Name ____________________ Date __________

10.1 Enrichment and Extension

Pearson's Coefficient of Skewness

Skewness is a measure of symmetry. Negative skewness means that the distribution is skewed left, and the mean of the data is less than the median. Positive skewness means the distribution is skewed right, and the mean of the data is greater than the median.

Karl Pearson developed two methods to measure the skewness of a data set using descriptive statistics. The two formulas for Pearson's Coefficient of Skewness are

$S_{k1} = \dfrac{3(\overline{X} - \text{mode})}{s}$ and $S_{k2} = \dfrac{3(\overline{X} - \text{median})}{s}$, where $\overline{X}$ is the mean and s is the standard deviation.

There is a general rule to interpret the coefficient. If the value is between -1 and 1, then the distribution is approximately symmetric. If the value is less than -1 then the distribution is skewed left, and if the value is greater than 1, then the distribution is skewed right.

Use the second formula to find and interpret the skewness of the data.

1. The table shows the prices (in cents per pound) of bananas reported from 14 markets surveyed by the U.S. Department of Agriculture.

51	52	45	48	53	52	50
49	52	48	43	46	45	42

2. The table shows the changes in score (Friday's score–Monday's score) of 25 students who were given a vocabulary test on Friday and then retested when they returned to school on Monday.

6	0	−1	10	4
5	4	4	9	2
13	3	3	0	7
3	−2	4	9	9
9	6	−5	11	1

Name____ Date__________

10.1 Puzzle Time

What is Moby Dick's Favorite Dinner?

Write the letter of each answer in the box containing the exercise number.

A normal distribution has mean μ and standard deviation σ. Find the indicated probability for a randomy selected *x*-value from the distribution.

1. $P(x \geq \mu + \sigma)$
2. $P(x \geq \mu - 2\sigma)$
3. $P(x \leq \mu - 2\sigma)$
4. $P(\mu - 2\sigma \leq x \leq \mu + 2\sigma)$
5. $P(x \leq \mu + 3\sigma)$

A normal distribution has a mean of 24 and a standard deviation of 3. Find the probability that a randomly selected *x*-value from the distribution is in the given interval.

6. between 18 and 27
7. between 21 and 33
8. between 24 and 33
9. at least 27
10. at most 15
11. not greater than 24
12. not less than 18

Answers

S. 2.5%

A. 99.85%

F. 16%

H. 95%

S. 97.5%

N. 81.5%

S. 49.85%

I. 97.5%

I. 0.15%

H. 16%

D. 83.85%

P. 50%

1	2	3	4		5	6	7		8	9	10	11	12

10.2 Start Thinking

When three coins are tossed, what is the theoretical probability that all three will land heads up? Conduct an experiment to verify your answer. How many times did you have to toss the three coins before you got all heads? Compare your results with your classmates and discuss how the experimental probability relates to the theoretical probability. What would you expect to be the relationship between the two probabilities as you increase the number of times you toss the coins?

10.2 Warm Up

Describe the sample space for the experiment.

1. A card is drawn from a regular deck of 52 cards.
2. A spinner that contains four colors, red, blue, yellow and green, is spun twice.
3. Each member of your class is asked if they want to participate in a mud run.
4. For one full day, patients at a clinic are asked if the wait time is acceptable.

10.2 Cumulative Review Warm Up

Solve the system of linear equations.

1. $2x - y = 8$
   $x + 3y = 4$
2. $x - y = 8$
   $-2x + 2y = -16$
3. $3x + 5y = -1$
   $-2x + 3y = 7$
4. $\frac{1}{2}x - \frac{2}{3}y = \frac{5}{2}$
   $\frac{1}{4}x + \frac{4}{3}y = 0$
5. $y = -\frac{1}{2}x + 5$
   $y = \frac{7}{2}x - 11$
6. $y = x$
   $y = -x$

Name____ Date__________

10.2 Practice A

In Exercises 1 and 2, determine whether the data is collected from a population or a sample. Explain your reasoning.

1. the number of popcorn kernels in 25 bags of popcorn

2. the college degrees of every employee at the hospital

In Exercises 3 and 4, identify the population and sample. Describe the sample.

3. In a university, a survey of 1641 students found that 479 of them do not know the name of their college's mascot.

4. In the United States, a survey of 1000 households with at least one child found that 874 of them have at least two computers.

In Exercises 5 and 6, determine whether the numerical value is a parameter or a statistic. Explain your reasoning.

5. On a high school football team, 2% of the players are vegetarians.

6. The average amount of the surveyed utility bills is \$176.42.

7. You roll a six-sided die 5 times and get all ones. The probability of this happening is, $\left(\frac{1}{6}\right)^5 \approx 0.0001286$, so you suspect this die favors ones. The die maker claims the die does not favor ones. You simulate rolling the die 30 times by repeatedly drawing 100 samples of size 30. The histogram shows the results.

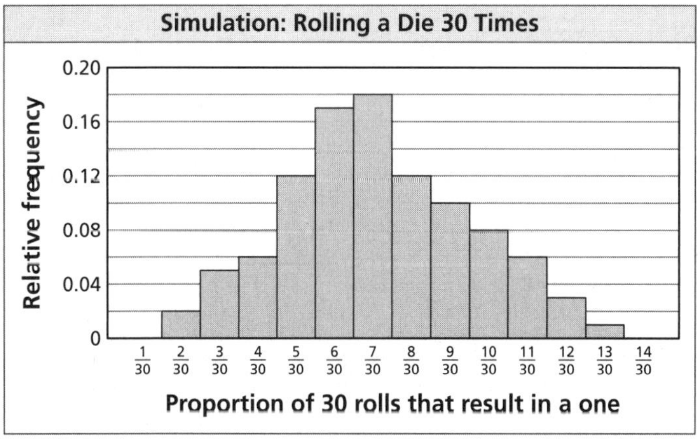

   a. What should you conclude when you roll the actual die 30 times and get 6 ones?

   b. What should you conclude when you roll the actual die 30 times and get 12 ones?

Name ______________________________ Date __________

10.2 Practice B

In Exercises 1 and 2, determine whether the data is collected from a population or a sample. Explain your reasoning.

1. the address of every student in the school

2. a survey of 80 people who access a website

In Exercises 3 and 4, identify the population and sample. Describe the sample.

3. In an office building, a survey of 648 employees found that 147 of them ride the subway to work each day.

4. In Florida, a survey of 2500 homeowners found that 1145 of them have switched their homeowner's insurance policy to a different company within the last 3 years.

In Exercises 5 and 6, determine whether the numerical value is a parameter or a statistic. Explain your reasoning.

5. Thirty-four percent of the surveyed hockey players first played hockey before their 10th birthday.

6. Eighty-two percent of all the tickets sold were for the Saturday matinee.

7. You roll a six-sided die 8 times and get either threes or fours. The probability of this happening is $\left(\frac{2}{6}\right)^8 \approx 0.0001524$, so you suspect this die favors threes and fours. The die maker claims the die does not favor threes or fours. You simulate rolling the die 30 times by repeatedly drawing 200 samples of size 30. The histogram shows the results.

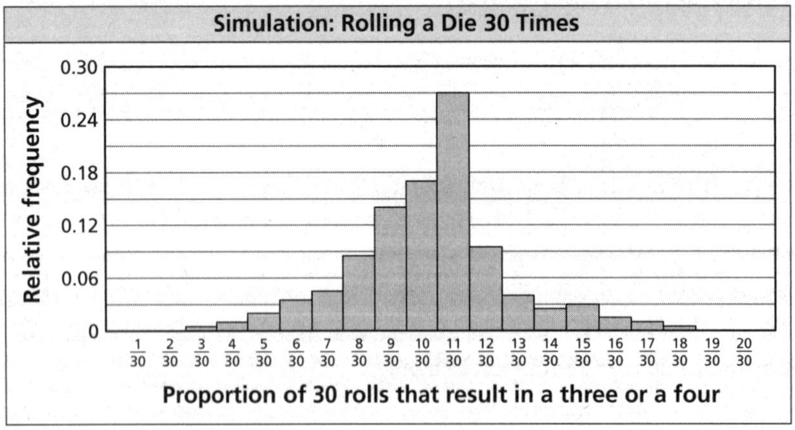

a. What should you conclude when you roll the actual die 30 times and get 5 threes and fours?

b. What should you conclude when you roll the actual die 30 times and get 10 threes and fours?

Name____ Date__________

10.2 Enrichment and Extension

Type I and Type II Errors

The *null hypothesis* states that an observed difference is due to chance alone. You commit a Type I error when you decide that the null hypothesis is false when the null hypothesis is true. You commit a Type II error when you decide that the null hypothesis is true when the null hypothesis is false. In experiments, the researcher often has to decide which error is worse.

Example: Researchers tested the effectiveness of a new drug for migraine headaches against a placebo using a cross-over design. The null hypothesis states that the new drug is not effective in reducing migraines. For the experiment, (a) state the Type I and Type II errors, and describe the consequences of each error, and (b) decide whether one error is more serious than the other. Explain.

Solution: a. Type I: Decide the new drug is effective in reducing migraines when it really is not effective; Consequence: Migraine patients use an ineffective pain medication and do not get the necessary pain relief; Type II: Decide the new drug is not effective when it really is effective; Consequence: Migraine patients do not get to use an effective pain medication.

b. Either could be argued, but a Type I error is probably more serious because people are not getting relief.

In Exercises 1–3, (a) state the Type 1 and Type II errors, and describe the consequences of each error, and (b) decide whether one error is more serious than the other. Explain.

1. A recent study considered whether the daily consumption of 1200 milligrams of garlic could reduce tick bites. This was tested against a placebo using a cross-over design. The null hypothesis states that garlic is not effective in reducing tick bites.

2. Production managers on an assembly line monitor the output to be sure that the level of defective products remains low. They periodically inspect a random sample of the items produced. If they find a significant increase in the proportion of items that must be rejected, they will halt the assembly process until the problem can be identified and repaired. The null hypothesis states that there are not too many defects.

3. A clean air standard requires that vehicle exhaust emissions do not exceed specified limits for various pollutants. Many states require cars be tested annually to be sure they meet these standards. Suppose state regulators double-check a random sample of cars that a repair shop has certified as acceptable. The null hypothesis states that the shop is meeting clean air standards.

Name ____ Date __________

10.2 Puzzle Time

Where Do You Find Canada?

Write the letter of each answer in the box containing the exercise number.

Determine whether the data are collected from a population or a sample.

1. a survey of 400 high school sophomores at a school with 842 sophomores
2. the number of people 65 or older in the United States

Identify the population and sample.

3. In the United States, a survey of 5000 high school seniors found that 3284 of them plan on attending a 4-year university following graduation.
4. In the United States, a survey of 2500 homeowners with a monthly mortgage payment found that 1280 of them have refinanced their mortgage within the last five years.
5. In a school district, a survey of 1000 middle school students found that 894 of them participate in an extracurricular activity.
6. A survey of 5000 elementary students found that 1974 of them have cell phones.

Answers

P. population: all elementary students, sample: the 5000 elementary students surveyed

M. population: all homeowners with mortgage payments in the United States, sample: the 2500 homeowners surveyed

O. sample

A. population: all high school seniors in the United States, sample: the 5000 high school seniors surveyed

N. population

A. population: all middle school students in the school district, sample: the 1000 middle school students surveyed

1	2		3		4	5	6

10.3 Start Thinking

The purpose of a statistical study is to collect data so that an informed and reliable decision can be made. In each of the scenarios described below, determine if a random survey of 50 students from your school would provide good information for making a reliable decision.

1. The school board wants to make a decision about allocating funds for a robotics team at your school.

2. The Student Government Association at your school wants to know if families would be interested in participating in a school carnival.

10.3 Warm Up

Explain why the sample might be biased.

1. Twenty pet owners are surveyed at a veterinary clinic to determine if pets should be permitted in hotel rooms.

2. One hundred shoppers are asked if the mall hours should be extended.

3. Fifty high school students are surveyed to determine if the amount of homework should be decreased.

4. Twenty members of an athletic club were surveyed to determine how many times the average American exercises each week.

10.3 Cumulative Review Warm Up

Graph the function.

1. $y = \dfrac{2}{x - 5}$

2. $y = \dfrac{3}{x + 2} - 1$

3. $y = \dfrac{2x - 1}{x + 4}$

Name ______________________________ Date __________

10.3 Practice A

In Exercises 1 and 2, identify the type of sample described.

1. A bank wants to know whether its drive-thru customers are satisfied with the service. Customers receive a code on their receipt that allows them to go online and fill out a survey.

2. A school wants to know whether high school teachers prefer before-school or after-school faculty meetings. Ten teachers with classrooms near the front office are surveyed.

In Exercises 3 and 4, identify the type of sample and explain why the sample is biased.

3. Every fifth customer who walks into a locally owned hardware store answers a survey that asks for opinions about lowering taxes for locally owned businesses to increase the likelihood that they can stay open for business.

4. A fitness center wants to find out whether its members would sign up for an afternoon yoga class. It surveys the first twenty members on the alphabetized membership list.

In Exercises 5 and 6, determine whether the sample is biased. Explain your reasoning.

5. Every fourth passenger who boards an airplane is asked whether they like the new method of seat assignment.

6. Your teacher is choosing 4 students to participate in a pizza-eating contest. The names of all 30 students are put in a hat and four names are randomly drawn from the hat.

In Exercises 7 and 8, explain why the survey question may be biased or otherwise introduce bias into the survey. Then describe a way to correct the flaw.

7. A guidance counselor asks high school students, "Do you miss school often?"

8. "The fish being caught in our lake are found to have lesions, which could be toxic when eaten. Do you think the city should address this health problem?"

Name____ Date__________

10.3 Practice B

In Exercises 1 and 2, identify the type of sample described.

1. An amphitheater wants to know if Saturday morning concerts would be attended. The amphitheater is divided into 30 different seating areas. It surveys all the members in five randomly selected seating areas.

2. The manager of an orange juice plant wants to test the quality of the plant's new extra-pulp orange juice. She does a taste test of every twentieth container of extra-pulp orange juice.

In Exercises 3 and 4, identify the type of sample and explain why the sample is biased.

3. A town council wants to know whether residents support lower taxes for those who own waterfront property. Thirty-four owners of waterfront property are surveyed.

4. A local radio station is considering switching from country music to classical music. Randomly throughout the day it announces the phone number and website where listeners can go to place their vote for either country music or classical music.

In Exercises 5 and 6, determine whether the sample is biased. Explain your reasoning.

5. A local senator is proposing a bill to lower the limit on the number of lobster that can be caught each day. Participants in a local fishing tournament are asked to complete a survey.

6. A triathlon organizer wants to know if a Saturday morning triathlon is preferred over a Sunday morning triathlon. The organizer randomly surveys five triathletes in one of the age groups.

In Exercises 7 and 8, explain why the survey question may be biased or otherwise introduce bias into the survey. Then describe a way to correct the flaw.

7. "Our plant has not been making a profit. Do you agree that the company should close this plant?"

8. "Do you agree that the mayor is receiving a salary that is far beyond what is reasonable?"

Name ______________________________ Date __________

10.3 Enrichment and Extension

Collecting Data

1. A researcher is considering three methods of evaluating two different types of fertilizer for corn. Classify each method as a survey, an experiment, or an observational study. Then identify which method would be most reliable. Explain your reasoning.

Method 1	Method 2	Method 3
Randomly divide the rows of corn in the same test field into two groups. Put fertilizer on one group of rows, and record the yield for those rows and for the rows without the fertilizer. Draw conclusions based on your findings.	Choose some of the rows of corn at random from a field. Determine which type of fertilizer has been applied in the past, and what the yield is for each row. Extrapolate the findings to the whole field.	Monitor the yield of all rows of corn in various test fields being fertilized with a product chosen by the farmer. Record the yield for those rows.

2. A researcher is considering three methods of evaluating the expected charge-length of two new batteries that can be used in electric cars. Classify each method as a survey, an experiment, or an observational study. Then identify which method would be most reliable. Explain your reasoning.

Method 1	Method 2	Method 3
Randomly divide some electric cars into two groups. Have each group use a different battery, and record how long the batteries last in each group before they need to be recharged. Confirm or reject your hypothesis.	Monitor the batteries in a large number of electric cars. Record how long the batteries last before they need to be recharged.	Choose a group of owners of electric cars at random. Ask which types of batteries they have used in the past, and how quickly the batteries need to be recharged. Estimate parameters for the entire population based on your data.

Name______________________________ Date__________

10.3 Puzzle Time

Why Did The Watch Go On Vacation?

Write the letter of each answer in the box containing the exercise number.

Identify the type of sample described.

1. You list all the counties in Pennsylvania and choose every third county.
2. You distribute questionnaires in the cafeteria and use only the questionnaires that are returned.
3. You ask all of the players on your baseball team whether they plan to attend the school dance.
4. You randomly select four students from each table in the cafeteria.
5. You divide all the students at a university into commuting students and campus-dwelling students, and then you randomly select all the students from the commuting group.

Identify the method of data collection each situation describes.

6. A student researcher records whether people at a restaurant leave a tip for their server.
7. A scientist provides natural light to 50 plants and fluorescent light to 50 plants in a study. The scientist then compares the plant growths after one week to determine which light is better.
8. A teacher asks the students in her fifth-period class how many siblings they have in their immediate family so that she can estimate the number of siblings for all her students.

Answers

N. experiment

I. observational

U. convenience

T. systematic

O. self-selected

N. stratified

W. cluster

D. survey

1	2		3	4	5	6	7	8

10.4 Start Thinking

Discuss how you might determine if a spinner or a coin are "fair" (equally likely to come up with each possibility).

10.4 Warm Up

Identify the method of data collection the situation describes.

1. A researcher places 20 plants in natural sunlight and 20 plants in artificial light. The researcher then compares the growth of the plants after 10 weeks and determines which light is better.
2. A researcher records whether people at a gym clean the machines after they use them.

10.4 Cumulative Review Warm Up

Simplify the expression.

1. $\dfrac{48^{1/3}}{16^{1/3}}$
2. $\left(3^{1/2} \bullet 6^{1/2}\right)^4$
3. $\sqrt{27}\sqrt[3]{27}$
4. $\dfrac{\sqrt{12}}{\sqrt[3]{12}}$
5. $\dfrac{\sqrt[5]{64}\sqrt[5]{4}}{\sqrt[5]{8}}$
6. $\dfrac{25^{3/2}25^{-1}}{5^{1/2}}$

Name____ Date__________

10.4 Practice A

In Exercises 1 and 2, determine whether the study is a randomized comparative experiment. If it is, describe the treatment, the treatment group, and the control group. If it is not, explain why not and discuss whether the conclusions drawn from the study are valid.

1. To test the effect of using a computer for testing, 300 students are randomly divided into two groups. One group is tested using paper and pencil, and one group is tested using the computer. After analyzing the test results, it was found that the average test scores in both groups were not significantly different.

2. At a health fair, people can choose to enroll in a six-month healthy eating plan. Sixty people who chose the "eat six vegetables a day" plan were monitored for 6 months, as were 60 people who chose the “drink 12 glasses of water a day” plan. At the end of the 6 months, people who chose the "drink 12 glasses of water a day" plan had 20% lower cholesterol than people in the other group.

In Exercises 3 and 4, explain whether the research topic is best investigated through an experiment or an observational study. Then describe the design of the experiment or observational study.

3. A cycling team wants to know whether incorporating yoga into the workout routine improves racing times.

4. A researcher wants to compare the effects of a new experimental cancer drug with a cancer drug that has been used for at least 10 years.

5. A researcher wants to test whether stretching after exercising decreases the number of injuries due to muscle damage. Identify a potential problem, if any, with each experimental design. Then describe how you can improve it.

    a. The researcher selects 400 people who exercise on a regular basis. The people are divided into two groups based on age. Within each age group, the people are randomly assigned to stretch after exercising or to not stretch. The people’s occurrence of injuries due to muscle damage is monitored. The stretching after exercise significantly decreases the number of injuries due to muscle damage.

    b. The researcher randomly selects 150 people who exercise on a regular basis. Half the people stretch after exercising, and the number of their injuries due to muscle damage is monitored. The other half do not stretch after exercising, and the number of their injuries due to muscle damage is monitored. The number of injuries due to muscle damage significantly decreases for those who stretch after exercising.

Name ______________________________ Date __________

10.4 Practice B

In Exercises 1 and 2, determine whether the study is a randomized comparative experiment. If it is, describe the treatment, the treatment group, and the control group. If it is not, explain why not and discuss whether the conclusions drawn from the study are valid.

1. A pool cleaning service is offering a new chlorine solution to its customers. Of 90 customers, the 45 customers who chose to switch to the new chlorine solution were monitored for a year, as were the 45 customers who did not switch. At the end of the year, the customers who switched to the new chlorine solution were 30% more satisfied with the condition of their pools while those who did not switch experienced no significant change.

2. A recycling company is testing the use of recyclables containers rather than recyclables bins, in the hopes of increasing the amount of recycling. It randomly divided 150 customers into two groups. One group received the new recyclables containers, and the other group continued using their recyclables bins. After 6 months, the customers with the new recyclables containers recycled 25% more pounds of recyclables than customers with the recyclables bins.

In Exercises 3 and 4, explain whether the research topic is best investigated through an experiment or an observational study. Then describe the design of the experiment or observational study.

3. An organization wants to know whether donating 20% of one's income to charities affects one's satisfaction with his or her job.

4. A rancher wants to know whether a new feed affects the quality of the milk produced by cows.

5. A researcher wants to test whether drinking diet soda increases sugar cravings. Identify a potential problem, if any, with each experimental design. Then describe how you can improve it.

a. The researcher randomly selects 200 people. Half of the people drink diet soda, and the occurrence of sugar cravings is monitored. The other half of the people do not drink diet soda, and their occurrence of sugar cravings is monitored. The occurrence of sugar cravings for the people who drink diet soda is significantly higher than the occurrence of cravings for those who do not.

b. The researcher selects 300 people. The people are divided into two groups based on exercise habits. Within each group, the people are randomly assigned to drink diet soda or to not drink diet soda. The people's sugar cravings are monitored. There is no significant difference in the occurrence of sugar cravings between the two groups.

Name____ Date__________

10.4 Enrichment and Extension

Experimental Design

The *response variable* is the variable that is being measured after the treatment is applied in an experiment. In a *blind or blinded experiment*, information that might lead to biased results is concealed from the researcher, the subject, or both until after the experiment. Bias may be intentional or unconscious. When both researcher and subject are blinded, the trial is a *double-blind trial.*

In Exercises 1–4, (a) identify the treatment, (b) identify the response variable, (c) determine the number of treatment groups, and (d) decide whether the experiment could be blind or double blind. Explain.

1. A mathematics education researcher was interested in determining the effects of class size (small, medium, and large) and the use of a traditional statistics textbook versus a new pilot textbook. The researcher conducted her own experiment, assigning each combination of class size and type of textbook to two classes at each of five chosen schools. The average final grade for each class was then recorded at the end of the year.

2. A botanist was interested in determining the effects of scheduled watering (three days a week or daily) and the use of fertilizer (traditional, and organic) in increasing the heat rating of jalapeño peppers. The botanist conducted his own experiment, assigning each combination of watering schedule and type of fertilizer to plots that had similar soil and full sun. The average final heat rating for the plot was then recorded at the end of the growing season.

3. A new dog food, specially designed for dogs with kidney problems, has been developed. A veterinarian wants to test this new food against another dog food currently on the market to determine whether it improves a dog's health. Thirty dogs with kidney problems were recruited to participate in the study. They were fed either the "new" or "old" food for 6 months and the improvement in kidney health was rated.

4. Does talking while walking slow you down? In a study reported in the journal Physical Therapy, the cadence was measured for subjects who were walking (using no device, a standard walker, or a rolling walker) and who were required (or not required) to respond to a signal while walking.

5. Researchers have developed a new insulin inhaler to replace daily insulin shots needed by patients with diabetes. Design an experiment to test the effects of this new insulin treatment on a volunteer group of 100 diabetic patients.

Name ____ Date __________

10.4 Puzzle Time

What Kind Of Hair Does The Ocean Have?

Write the letter of each answer in the box containing the exercise number.

Identify a potental problem with the experimental design.

1. A school district's administrators want to decide whether to continue their current after-school tutoring program next year. Six students who were recommended by their teachers to participate in the program were randomly assigned to either receive tutoring or not. The three students who received tutoring were monitored for one year, as were the other three students who did not receive tutoring. At the end of the year, the tutored students did not perform significantly better than the other students, so the administrators discontinued the program.

2. The local humane society wants to test the effectiveness of a new flea medicine. In the experiment, 30 flea-infested cats receive the medicine and another 30 flea-infested dogs receive a placebo. After one week, a significantly larger number of cats have no fleas.

Answers

V. observational study

W. The sample size is not large enough

Y. experiment

A. The groups are not similar.

Determine whether the research topic is best investigated through an experiment or an observational study.

3. The manager of the produce department at a grocery store wants to know which type of apple more customers prefer.

4. The owner of a pet store wants to know whether distilled water or natural spring water leads to longer life in the fish he sells.

1	2	3	4

10.5 Start Thinking

You conduct a sample survey to determine what percent of high school students in your town have a part-time job. Based on the students that you know, you predict that 25% of high school students have a part-time job. You survey 40 randomly chosen students and find that 14 of them have a part-time job. Based on your sample survey, what percent of the students have a part-time job? Do you think that you should change your prediction based on your sample survey? Explain your reasoning.

10.5 Warm Up

Sample surveys are conducted to determine the monthly cell phone bill and the number of sick or personal days taken in a calendar year. Calculate the mean, $\bar{x}$, of the data shown in the table. Round your answer to the nearest tenth, if necessary.

1. Monthly cell phone bill

148	210	160	190	124
165	189	194	280	215
210	160	115	220	240
228	140	169	187	218
240	215	192	115	122

2. Sick or personal days taken annually

2	4	1	0	8
3	12	6	7	5
7	4	15	4	2
1	2	9	5	7
0	3	6	5	2

10.5 Cumulative Review Warm Up

Use the Binomial Theorem to write the binomial expansion.

1. $(x - 3)^5$

2. $(2x + 5)^4$

3. $(1 - 2x)^7$

4. $(x + y^2)^4$

5. $(2a - 3b)^6$

6. $(x^3 - y^4)^8$

Name ______________________________ Date ________

10.5 Practice A

1. The numbers of words per text message in a random sample of 30 text messages are shown in the table.

Number of Words Per Text Message									
3	10	9	12	21	17	4	6	18	2
10	3	9	12	5	24	19	4	1	7
5	11	14	6	2	4	14	6	9	12

a. Estimate the population mean μ.

b. Estimate the population proportion ρ of text messages that contain more than 12 words.

c. Estimate the population proportion ρ of text messages that contain less than five words.

2. A survey asks a random sample of U.S. college students how many hours they spend surfing the Internet each day. The survey reveals that the sample mean is 2.7 hours per day. How confident are you that the average time spent surfing the Internet each day of all U. S. college students is exactly 2.7 hours per day? Explain your reasoning.

3. The number of households with pets is increasing. A national polling company claims that 43% of U.S. households have at least one pet. You survey a random sample of 50 households.

a. What can you conclude about the accuracy of the claim that the population proportion is 0.43 when 29 households have at least one pet?

b. What can you conclude about the accuracy of the claim that the population proportion is 0.43 when 22 households have at least one pet?

c. Assume that the true proportion is 0.43. Estimate the variation among sample proportions for samples of size 50.

Name________________________________ Date__________

10.5 Practice B

1. The numbers of bait fish caught in a random sample of 40 cast net throws are shown in the table.

Number of Bait Fish Per Cast Net Throw									
4	8	6	15	1	6	8	0	1	14
9	6	8	7	3	10	4	11	2	4
15	2	0	5	1	2	7	5	6	6
9	11	8	9	5	1	4	7	2	1

   a. Estimate the population mean μ.

   b. Estimate the population proportion ρ of cast net throws that produce at least eight bait fish.

   c. Estimate the population proportion ρ of cast net throws that produce fewer than three bait fish.

2. A survey asks a random sample of U.S. voters how many times they have gone to the polls unknowledgeable about who they are voting for. The survey reveals that the sample mean is 5.8 times. How confident are you that the average number of times all U. S. voters have gone to the polls unknowledgeable is exactly 5.8 times? Explain your reasoning.

3. A national polling company claims that 45% of U.S. drivers do not adhere to the speed limits in construction areas. You survey a random sample of 50 households.

   a. What can you conclude about the accuracy of the claim that the population proportion is 0.45 when 15 drivers do not adhere to the speed limits in construction areas?

   b. What can you conclude about the accuracy of the claim that the population proportion is 0.45 when 23 drivers do not adhere to the speed limits in construction areas?

   c. Assume that the true proportion is 0.45. Estimate the variation among sample proportions for samples of size 50.

Name ______________________________ Date __________

10.5 Enrichment and Extension

Making Inferences from Sample Surveys

You want to determine whether you should start a new club at your school. In Exercises 1–8, answer the question(s) to design an unbiased survey.

√	Characteristic
	1. Write the survey question.
	2. Is the survey question clear? Explain.
	3. Describe the population you will sample.
	4. Describe the method you will use to sample the population.
	5. Is your sample biased? Explain how you know.
	6. Is the sample representative of the entire population? Explain.
	7. Do a trial run of the survey with several students in your class. Re-evaluate your survey question. Does the question need modifying? Explain.
	8. Conduct the survey. Find the margin of error of your sample to help you predict the opinion of the entire school. Based on the results of the survey, make a prediction about the opinion of the entire population of the school. Did your results become more accurate the more people you surveyed? Explain.

Name____ Date__________

10.5 Puzzle Time

What Do You Call A Bird That Has Been Eaten By A Cat?

Write the letter of each answer in the box containing the exercise number.

1. The table shows the numbers of lunch-hour customers to a local fish market for a random sample of 16 days. Estimate the population mean μ.

Number of Customers			
15	16	27	8
12	21	26	24
18	32	38	34
24	14	42	25

2. The table shows the total numbers of free throws made by a high school basketball team for a random sample of eight games. Estimate the population mean μ.

Total Number of Free Throws Made			
24	19	7	14
13	11	12	4

Find the margin of error for a survey that has the given sample size. Round your answer to the nearest tenth of a percent.

3. 100 **4.** 2400 **5.** 32

6. 289 **7.** 750 **8.** 170

Answers

L. ± 0.06

W. ± 0.1

O. ± 0.04

W. ± 0.08

A. ± 0.02

A. 23.5

L. ± 0.18

S. 13.25

1		2	3	4	5	6	7	8

10.6 Start Thinking

A marketing expert is consulted to help a company increase sales of their new sports drink. The expert advises them to consider repackaging the sports drink by changing the size of the bottle as well as the label. Because of the cost of such changes, the company wants to test the marketing expert's hypothesis before making a large investment in the change. How could the hypothesis be tested?

10.6 Warm Up

The table shows results of two different randomized comparative experiments. Find the mean value of the control group, the mean value of the treatment group, and the difference between the two means.

1. **Height (inches)**

Control Group	14	15.2	14.2	15	15.5	15	14.8	15.5	14.1	14.6
Treatment Group	14.3	15.3	14.8	15	15.4	15.4	14.8	15.8	14.5	14.9

2. **Weight (pounds)**

Control Group	58	60	61	65	57	55	62	61	60	57
Treatment Group	51	58	55	61	50	55	60	58	55	51

10.6 Cumulative Review Warm Up

Simplify the expression.

1. $\sin x \csc x$

2. $\dfrac{1 - \cos^2(-x)}{\cos^2 x}$

3. $\dfrac{\sin x}{1 + \cos x} + \dfrac{1 + \cos x}{\sin x}$

4. $\dfrac{\cos\left(\dfrac{\pi}{2} - x\right)}{\tan x}$

5. $\cos x\left(1 + \tan^2 x\right)$

6. $\dfrac{\sin x \cos x - \cos\left(\dfrac{\pi}{2} - x\right)}{\sin x}$

Name____ Date__________

10.6 Practice A

1. A randomized comparative experiment tests whether pet therapy affects the mood scores of hospital patients. The mood scores range from 1 to 10, with scores close to 10 indicating a good mood. The control group has eight patients and the treatment group, which receives the pet therapy, has eight patients. The table shows the results.

	Mood Score							
Control Group	1	4	2	3	8	5	4	2
Treatment Group	3	9	4	7	3	6	2	9

   a. Find the mean score of the control group.

   b. Find the mean score of the treatment group.

   c. Find the experimental difference of the means.

   d. Display the data in a double dot plot.

   e. What can you conclude?

   f. Resample the data using a simulation.

   g. Find the mean score of the new control group.

   h. Find the mean score of the new treatment group.

   i. Find the experimental difference of the new means.

Name ______________________________ Date __________

10.6 Practice B

1. A randomized comparative experiment tests whether photo therapy affects the size of the area of skin irritations. The control group has eight patients and the treatment group, which receives the photo therapy, has eight patients. The table shows the results.

	Area of Skin Irritation (square centimeters)							
Control Group	4.2	1.3	5.1	3.4	2.7	4.1	3.6	2.4
Treatment Group	1.2	2.5	4.1	3.7	2.1	1.5	2.4	1.9

a. Find the mean area of the control group.

b. Find the mean area of the treatment group.

c. Find the experimental difference of the means.

d. Display the data in a double dot plot.

e. What can you conclude?

f. Resample the data using a simulation.

g. Find the mean score of the new control group.

h. Find the mean score of the new treatment group.

i. Find the experimental difference of the new means.

Name____ Date__________

10.6 Enrichment and Extension

Bootstrap and Jackknife Resampling

Another way to resample is to combine the measurements from both groups, and repeatedly create new "control" and "treatment" groups at random from the measurements *with repeats*. This resampling strategy is called *bootstrap*. This method is more accurate than resampling without replacement.

A third way to resample is to remove an extreme measurement, combine the remaining measurements from both groups, and repeatedly create new "control" and "treatment" groups at random from the measurements without repeats. This resampling strategy is called *jackknife*. This method is useful when the data contain outliers.

In Exercises 1–3, (a) resample the data using the bootstrap method, (b) resample the data using the jackknife method, and (c) find the experimental differences of the means using both methods.

1. A randomized comparative experiment tests whether a soil supplement affects the total yield of cherry tomato plants.

	Total Yield of Tomato Plants (kilograms)									
Control Group	1.2	1.3	0.9	1.4	2.0	1.2	0.7	1.9	1.4	1.7
Treatment Group	1.4	0.9	1.5	1.8	1.6	1.8	2.4	1.9	1.9	1.7

2. A randomized comparative experiment tests whether music therapy affects the depression score of college students.

	Depression Score			
Control Group	49	45	43	47
Treatment Group	39	40	39	30

Control Group	46	45	47	46
Treatment Group	41	40	42	43

3. A randomized comparative experiment tests whether low-level laser therapy affects the waist circumference of adults.

	Circumference (inches)			
Control Group	34.6	35.4	33	34.6
Treatment Group	31.4	33	32.4	32.6

Control Group	35.2	35.2	38.2	35
Treatment Group	33.4	33.4	34.8	33

Name ______________________________ Date __________

10.6 Puzzle Time

How Do You Get An Astronaut's Baby To Fall Asleep?

Write the letter of each answer in the box containing the exercise number.

Complete the exercises.

A randomized comparative experiment tests whether time spent outdoors affects the calculus exam scores of college students during finals week. The control group has five students and the treatment group, which receives time outdoors, has five students. The table shows the results.

	Calculus Exam Score				
Control Group	80	64	95	92	71
Treatment Group	83	98	88	79	94

1. Find the mean mileage of the control group.
2. Find the mean mileage of the treatment group.
3. Find the experimental difference of the means.

A randomized comparative experiment tests whether tire tread thickness affects the gas mileage of a vehicle. The control group, which receives tires that are 1 year old, has 10 vehicles and the treatment group, which receives brand new tires, has 10 vehicles. All vehicles are identical except for their tires. The table shows the results.

	Vehicle Gas Mileage (mi/gal)									
Control Group	18	15	19	21	20	17	20	21	17	20
Treatment Group	19	24	21	22	18	23	20	25	21	23

4. Find the mean mileage of the control group.
5. Find the mean mileage of the treatment group.
6. Find the experimental difference of the means.

Answers

K. 18.8
R. 80.4
T. 2.8
C. 8
O. 88.4
E. 21.6

1	2	3	4	5	6

Name____ Date__________

Chapter 10 Cumulative Review

In Exercises 1–6, find the mean, median, and mode of the data set. Round your answer to the nearest tenth, when necessary.

1. 14, 1, 14, 10, 15, 18

2. 46, 47, 37, 49, 45, 46

3. 64, 73, 80, 63, 68, 80, 60

4. 40, 43, 45, 47, 47, 44, 43

5. 3, 17, 20, 8, 10, 19, 16

6. 15, 19, 20, 15, 14, 14, 11

In Exercises 7–12, find and interpret the standard deviation of the data set. Round your answer to the nearest tenth, when necessary.

7. 30, 29, 30, 20, 26, 22

8. 6, 7, 2, 9, 6, 3

9. 69, 53, 50, 51, 60, 54, 54

10. 18, 16, 17, 19, 14, 18, 17

11. 85, 82, 81, 87, 86, 85, 89, 85

12. 63, 62, 60, 60, 69, 70, 62, 65

In Exercises 13–18, tell whether *x* and *y* show *direct variation*, *inverse variation*, or *neither*.

13. $xy = 4$

14. $\frac{y}{7} = x$

15. $y = x + 4$

16. $x = y - 3$

17. $\frac{x}{3} = y$

18. $y = -4x$

In Exercises 19–24, the variables *x* and *y* vary inversely. Use the given value to write an equation relating *x* and *y*. Then find *y* when $x = 3$.

19. $x = -1, y = -7$

20. $x = -10, y = 4$

21. $x = -7, y = 1$

22. $x = 5, y = -2$

23. $x = \frac{8}{9}, y = 7$

24. $x = 2, y = \frac{3}{4}$

25. The results for your test scores are as follows: 73, 95, 76, 99, 85, 92, 78, 72, 71, 87, 98, 81, 88, 92, 93, 79, 75, 95, 77, 71, 90, 77, 100, 75, 94, 89, 84. What is the mean, median, and mode of the test scores? Round your answers to the nearest tenth, where necessary.

In Exercises 26–31, simplify the expression.

26. $\frac{3x}{6x^2 - 27x^5}$

27. $\frac{6x + 2x^4}{3x}$

28. $\frac{x^2 - 3x - 4}{x^2 + 6x + 5}$

29. $\frac{x^2 + 8x - 9}{x^2 + 3x - 4}$

30. $\frac{x^2 - 16x + 63}{x^2 - x - 42}$

31. $\frac{x^2 - 13x + 40}{x^2 - 5x - 24}$

Name ______________________________ Date __________

Chapter 10 Cumulative Review (continued)

In Exercises 32–37, find the product.

32. $\frac{2xy^4}{x^3y} \bullet \frac{y}{6x^2}$

33. $\frac{5x^3y}{xy^2} \bullet \frac{y^4}{30x^2y}$

34. $\frac{x(x-4)}{x+8} \bullet \frac{(x+8)(x-2)}{x^2}$

35. $\frac{x^2-x-6}{x^2-16x+60} \bullet \frac{x^2-7x+6}{x^2-4x+3}$

36. $\frac{x^2-10x+16}{x^2+x-72} \bullet \frac{x^2+17x+72}{x^2+15x+56}$

37. $\frac{x^2+16x+55}{x^2-3x-40} \bullet \frac{x^2-5x-24}{x^2+x-110}$

In Exercises 38–43, find the sum or difference.

38. $\frac{12}{7x} + \frac{8}{7x}$

39. $\frac{8}{x-2} + \frac{3x}{x-2}$

40. $\frac{x}{3x+4} - \frac{8x}{3x+4}$

41. $\frac{x+8}{x^2+3x-40} - \frac{x+2}{x+8}$

42. $\frac{3x+7}{x-1} - \frac{x-9}{x^2-12x+11}$

43 $\frac{x+11}{x^2-11x-12} + \frac{x+3}{x-12} - \frac{4}{x+1}$

44. The sides of a rectangle are $(2x+7)$ inches and $(3x+9)$ inches.

a. Write an equation for the perimeter P of the rectangle.

b. Write an equation for the area A of the rectangle.

c. The value of x is 2. What is the perimeter and area of the rectangle?

In Exercises 45–48, evaluate the six trigonometric functions of the angle θ.

45.

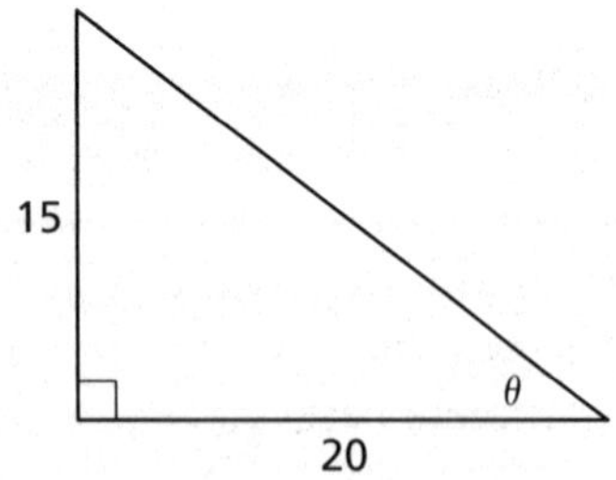

46.

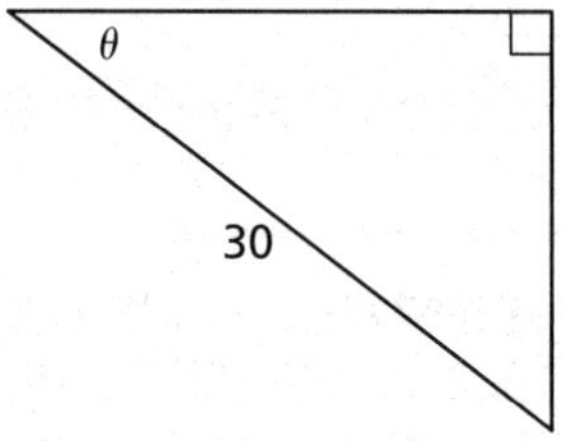

47.

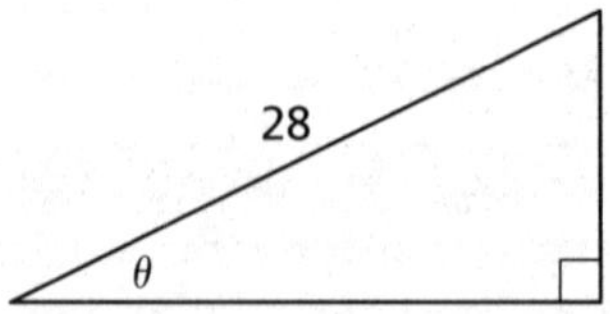

48.

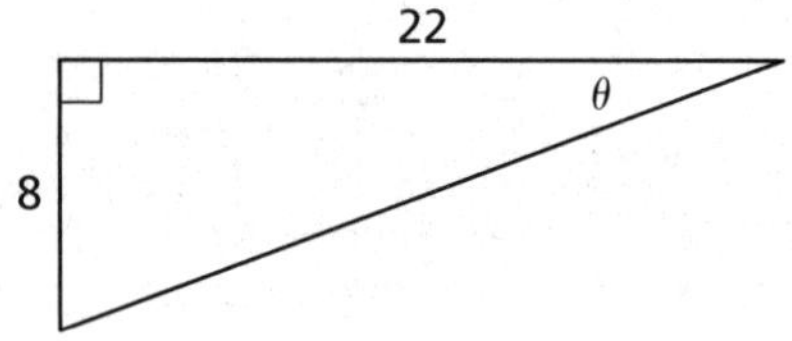

Name______________________________ Date__________

Chapter 10

Cumulative Review (continued)

In Exercises 49–52, find the value of *x* for the right triangle.

49.

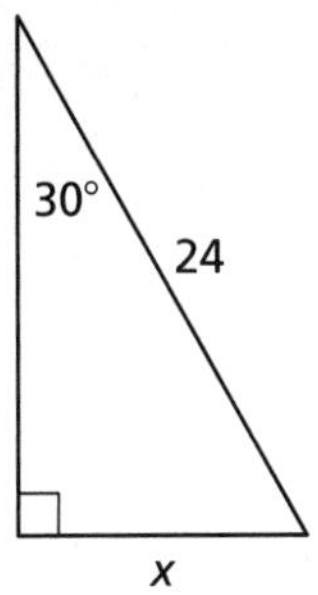

50.

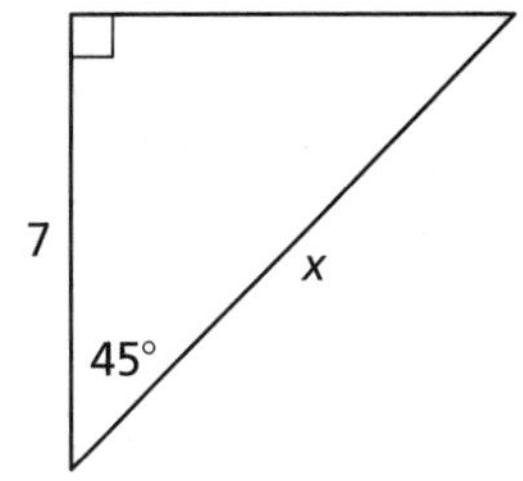

51.

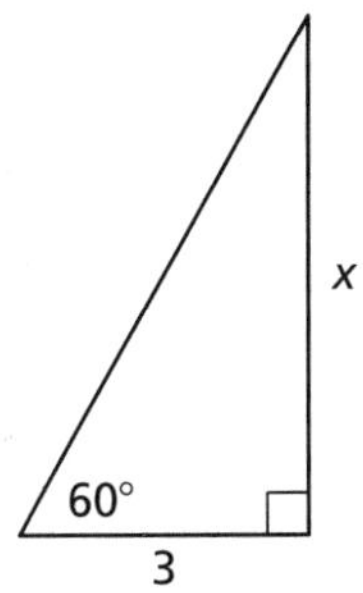

52.

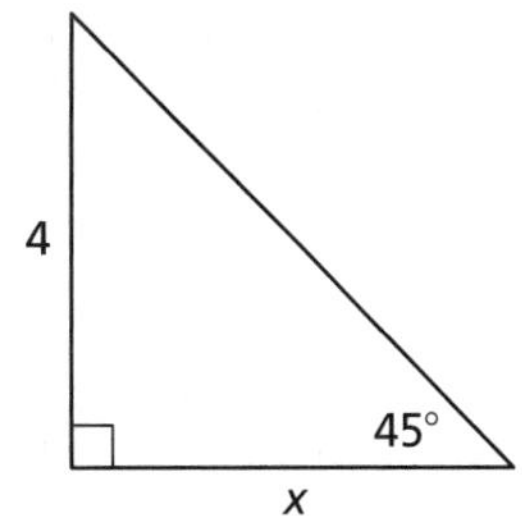

In Exercises 53–58, find one positive angle and one negative angle that are coterminal with the given angle.

53. $88°$

54. $-305°$

55. $-\frac{3\pi}{4}$

56. $\frac{5\pi}{6}$

57. $420°$

58. $-\frac{4\pi}{3}$

In Exercises 59–64, convert the degree measure to radians or the radian measure to degrees.

59. $110°$

60. $310°$

61. $-\frac{4\pi}{3}$

62. $\frac{5\pi}{6}$

63. $-60°$

64. $-\frac{3\pi}{4}$

Name ______________________________ Date __________

Chapter 10 Cumulative Review (continued)

In Exercises 65–70, use the unit circle to evaluate the six trigonometric functions of θ.

65. $180°$

66. $495°$

67. $-\frac{\pi}{3}$

68. $\frac{\pi}{6}$

69. $-120°$

70. $\frac{7\pi}{4}$

In Exercises 71–82, graph the function.

71. $g(x) = 5 \sin x$

72. $g(x) = 2 \cos x$

73. $g(x) = \cos 3x$

74. $g(x) = \sin 4x$

75. $g(x) = \frac{1}{2} \sin 2x$

76. $g(x) = \frac{1}{4} \cos 2x$

77. $g(x) = \frac{1}{2} \sin 4\pi x$

78. $g(x) = 3 \cos 2\pi x$

79. $g(x) = \sin x - 1$

80. $g(x) = \cos x + 3$

81. $g(x) = 3 \sin 2x + 1$

82. $g(x) = \frac{1}{2} \cos x - 5$

In Exercises 83–86, a normal distribution has mean of 26 and standard deviation of 3. Find the indicated probability for a randomly selected *x*-value from the distribution.

83. between 23 and 26

84. at least 17

85. at most 32

86. between 20 and 35

In Exercises 87 and 88, identify the population and the sample. Describe the sample.

87. In a school district, a survey of 1230 high school students found that 1144 of them felt that the school should keep the music program.

88. In a small town, a survey of 1000 citizens found that 872 of them want to replace the street signs.

In Exercises 89 and 90, identify the type of sample and explain why the sample is biased.

89. A school newspaper is doing a story about study habits and wants to know how many hours, on average, students spend doing homework each night. The student reporter asks students in an AP Calculus class.

90. Every fifth person that walks into a gym answers a survey about the importance of person fitness.

Credits

35, 40, 45, 50, 55, 60, 65, 77, 82, 87, 92, 97, 102, 107, 112, 117, 129, 134, 139, 144, 149, 154, 166, 171, 176, 181, 186, 191, 203, 208, 213, 218, 223, 235, 240, 245, 250, 255, 267, 272, 277, 282, 287, 292, 304, 309, 331, 336, 341, 346, 351, 356 Jokelopedia, 3rd Edition. Copyright ©2000 by Somerville House, 2006 by Key Porter and Workman Publishing, 2013 by Workman Publishing, Inc. Used by permission of Workman Publishing Co., New York. All Rights Reserved.

Answers

Chapter 1

1.1 Start Thinking

Monroe: about 113 people per mi^2; Rockland: about 1793 people per mi^2; Suffolk: about 1637 people per mi^2; Westchester: 2202 people per mi^2; Monroe, Suffolk, Rockland, Westchester

1.1 Warm Up

1. about 254.47 cm^2 **2.** about 804.25 in.^2

3. 7 mi

1.1 Cumulative Review Warm Up

1. $x \leq -2$ or $x \geq 2$

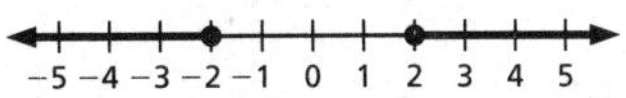

2. $-1.5 < n < 1.5$

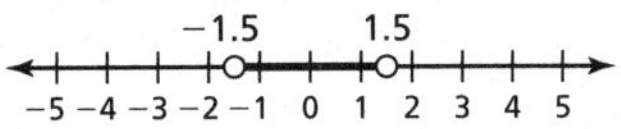

3. $t < -\frac{10}{3}$ or $t > 2$

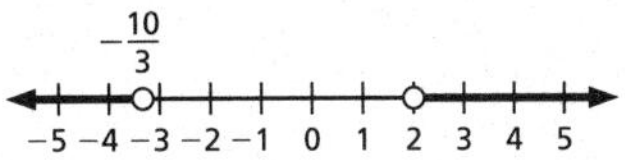

4. no solution

1.1 Practice A

1. about 11 people per mi^2

2. about 10,610 people/mi^2

3. about 883,573 people

4. about 12 mi

5. a. $A = 32w - w^2$

b. $w = 16$ ft and $\ell = 16$ ft; The area is maximized when the swimming pool is in the shape of a square.

6. The surface area increased by a factor of 9.

7. The surface area decreased by a factor of 64.

1.1 Practice B

1. about 0.31 otter per acre; about 1.66 raccoons per acre; about 0.08 fox per acre; about 0.03 bobcat per acre

2. about 756 people

3. about 5.5 mi **4.** about 10.5 mi

5. The surface area increased by a factor of about $\frac{150.8}{61.26} \approx 2.46$.

6. The surface area increased by a factor of $\frac{832}{608} \approx 1.37$.

7. a. no; The softball needs $\frac{16}{9}$, or $\left(\frac{4}{3}\right)^2$, times the amount of leather.

b. about 1.7 in.

1.1 Enrichment and Extension

1. a. $A = 400\ell - \frac{8}{3}\ell^2$

b. $\ell = 75$ ft; $w = 100$ ft

2. a. $V = 27s - \frac{1}{4}s^3$

b. $s = 6$ in.; $h = 3$ in.

3. a. $A = \left(\frac{24}{\ell} + 2\right)(\ell + 3)$

b. $\ell = 6$ in.; $w = 4$ in.

4. a. $V = 20\pi r^2 + \frac{320\pi}{r}$

b. $r = 2$ in.; $h = 5$ in.

1.1 Puzzle Time

COAT OF PAINT

1.2 Start Thinking

Water has a greater density than ice; a cubic foot of water; A greater density means that it has more mass for the same volume.

1.2 Warm Up

1. 10.16 cm **2.** 13.97 cm

3. 50 in. **4.** about 18.11 in.

1.2 Cumulative Review Warm Up

1. $x = 3$ **2.** $x = 10$ **3.** $x = 4$

4. $x = 5$ **5.** $x = 8$ **6.** $x = 2$

1.2 Practice A

1. about 2497 g

Answers

2. Density is $\dfrac{\text{mass}}{\text{volume}}$ not (mass)(volume);

   density $= \dfrac{14.5}{21} \approx 0.69\ \text{g/in.}^3$

3. The volume is $\frac{3}{4}$ times the original volume.

4. The volume is 8, or 2^3, times the original volume.

5. $3h$; $r\sqrt{3}$; *Sample answer:* The original volume is $V = \frac{1}{3}\pi r^2 h$ and the new volume is $V = \pi r^2 h$.

6. $\frac{98}{3}$, or $32\frac{2}{3}$ m

7. **a.** about 12,566 kg
   b. about 1.2 m

1.2 Practice B

1. 3700 g

2. **a.** about 59 kg
   b. 21,125 cm^3; no; The block loses 15,875 cm^3 in the second week, 11,375 cm^3 in the third week, 7625 cm^3 in the fourth week, and 4625 cm^3 in the fifth week.

3. The volume is $\frac{9}{4}$, or $\left(\frac{3}{2}\right)^2$, times the original volume.

4. The volume is 27, or 3^3, the original volume.

5. $\dfrac{1}{2}h$; $r\dfrac{\sqrt{2}}{2}$; *Sample answer:* The original volume is $V = \dfrac{1}{3}\pi r^2 h$ and the new volume is $V = \dfrac{1}{6}\pi r^2 h.$

6. **a.** 1.0 kg
   b. 0.5 kg
   c. 1.5 kg
   d. 12,000 kg/m^3

1.2 Enrichment and Extension

1. 3.2 cm
2. about 2.4 in.
3. about 0.48 g/cm^3
4. no; The density of the statuette is about 2.2 grams per cubic centimeter, which is not the density of bronze.

1.2 Puzzle Time

A HOLE

1.3 Start Thinking

1. *Sample answer:* circle and rectangle; the intersection of a cylinder and a plane

1.3 Warm Up

1.
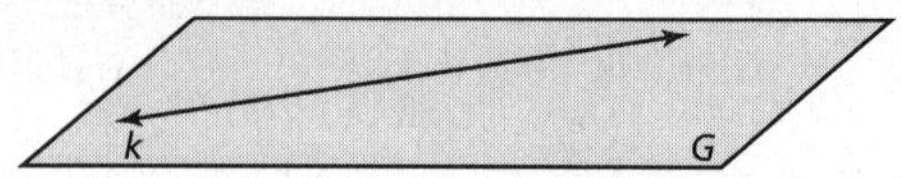

2.
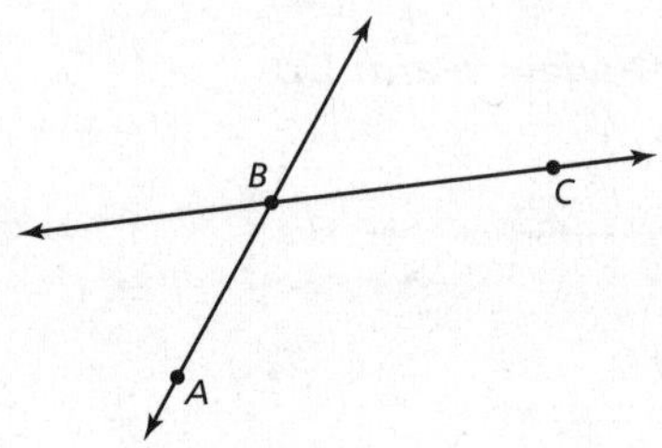

3.
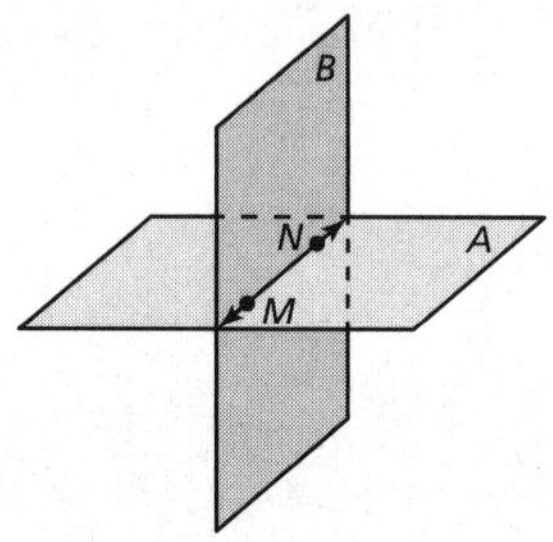

4.
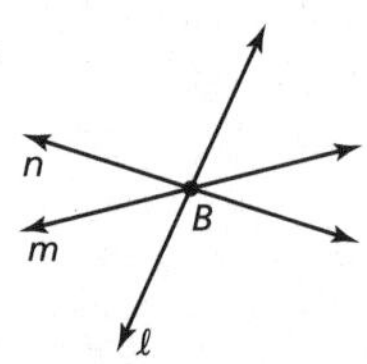

5.
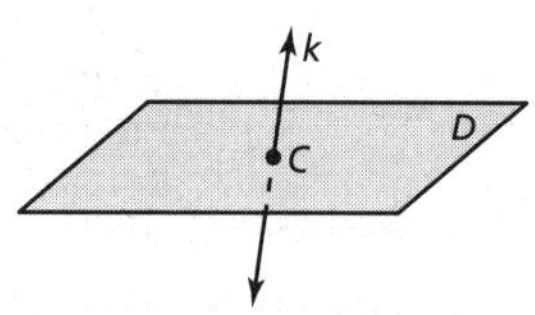

6.

1.3 Cumulative Review Warm Up

1. 8
2. 15
3. 8
4. 18.5
5. 21
6. 86.5

Answers

1.3 Practice A

1. circle **2.** triangle **3.** rectangle

4.

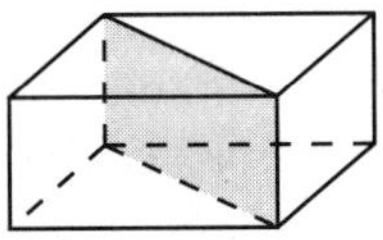

rectangle

5.

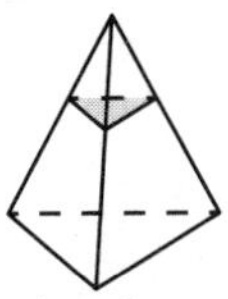

triangle

6.

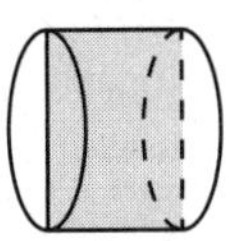

rectangle

7. *Sample answer:*

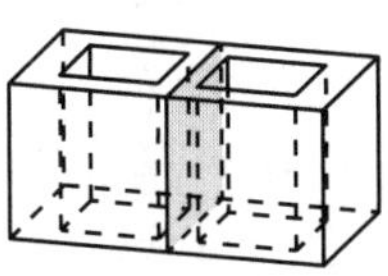

yes; yes; When the plane is parallel to the shorter sides and intersects the figure halfway between them, the cross section is one rectangle. When the plane is parallel to the longer sides and intersects the figure halfway between them, the cross section consists of three rectangles.

8.

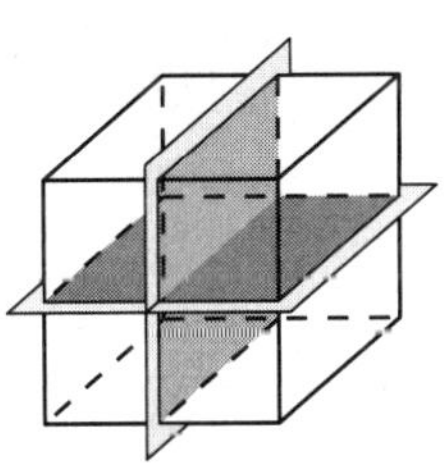

yes; *Sample answer*: parallel to a horizontal face and parallel to a vertical face

9. a. $4 + 4\sqrt{5} \approx 12.94$ in.

b. $4\sqrt{5} \approx 8.94$ in.^{2}

1.3 Practice B

1. triangle **2.** trapezoid **3.** semicircle

4.

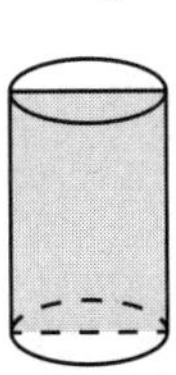

rectangle

5.

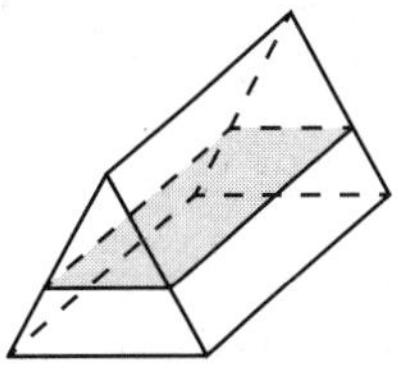

rectangle

6.

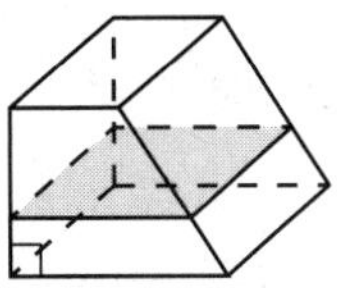

rectangle

7. a. trapezoid, triangle

b. trapezoid: perimeter $= 12 + 2\sqrt{13} \approx 19.21$ in., area $= 18$ in.^{2}

triangle: perimeter $= 4 + 2\sqrt{13} \approx 11.21$ in., area $= 6$ in.^{2}

c. no; *Sample answer*: It is flatter than a circle, more like an oval.

1.3 Enrichment and Extension

1. $\left(18\pi h - \pi h^2\right)$ in.^{2}; *Sample answer:* Any horizontal cross section of solid A is a circle. By the Pythagorean Theorem, the radius is $\sqrt{18h - h^2}$ inches.

2. $\left(18\pi h - \pi h^2\right)$ in.^{2}; *Sample answer:* Any horizontal cross section of container B is a circular washer with an outer radius of 9 inches and an inner radius of $(9 - h)$ inches. The surface area of the water can be found by subtracting the area of the inner circle from the area of the outer circle.

3. *Sample answer:* The cross-sectional areas of the two solids are the same.

4. They are the same.

5. no; *Sample answer:* Solid A and solid B each have a volume of about 1526.8 cubic inches and solid C has a volume of 1944 cubic inches.

1.3 Puzzle Time

THE LETTER E

Answers

1.4 Start Thinking

Rotate the triangle about line ℓ.

1.4 Warm Up

1. about 314.16 ft^3 **2.** about 31.4 cm^3

3. about 56.5 in.^3 **4.** about 12.6 mm^3

5. about 113.1 ft^3 **6.** about 904.8 in.^3

1.4 Cumulative Review Warm Up

1. yes; 5 **2.** no **3.** yes; 10

4. yes; −6 **5.** no **6.** yes; −17

1.4 Practice A

1.

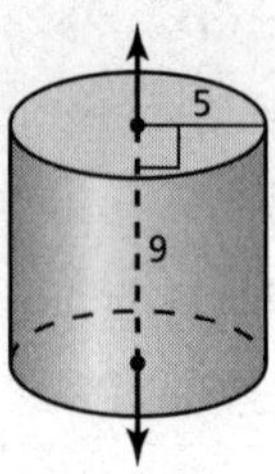

cylinder with height 9 and base radius 5

2.

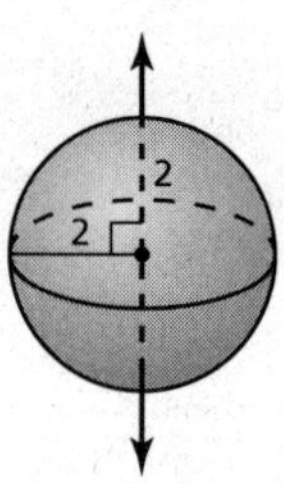

sphere with radius 2

3.

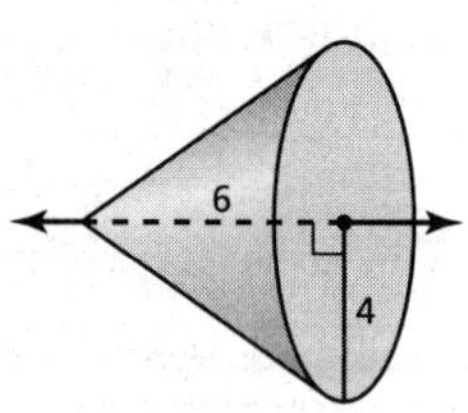

cone with height 6 and base radius 4

4.

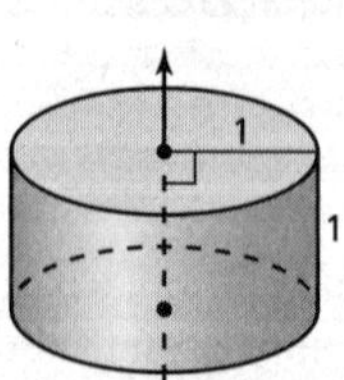

cylinder with radius 1 and height 1

5.

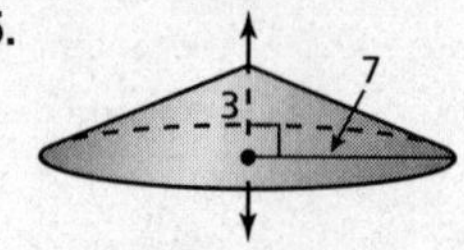

cone with radius 7 and height 3

6.

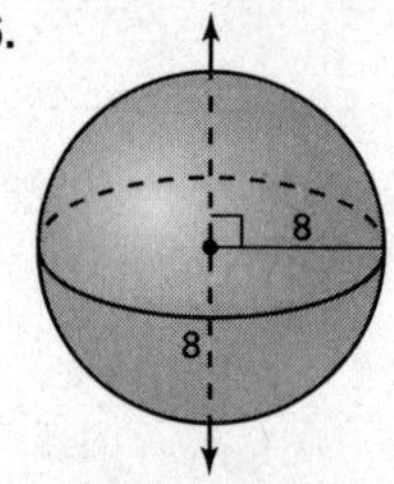

sphere with radius 8

7.

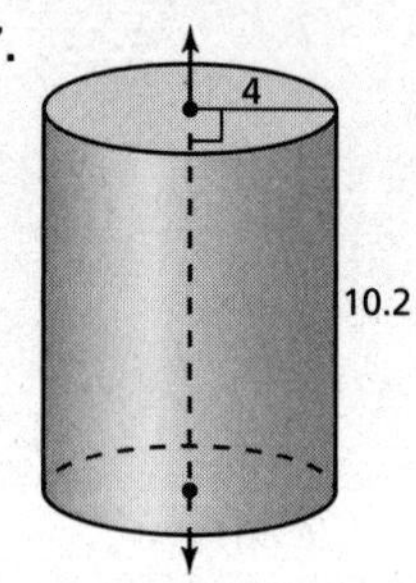

cylinder with radius 4 and height 10.2; about 356.88 square units, about 512.71 cubic units

8.

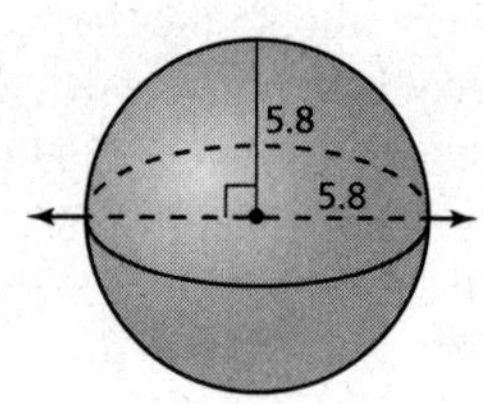

sphere with radius 5.8; about 422.73 square units, about 817.28 cubic units

9.

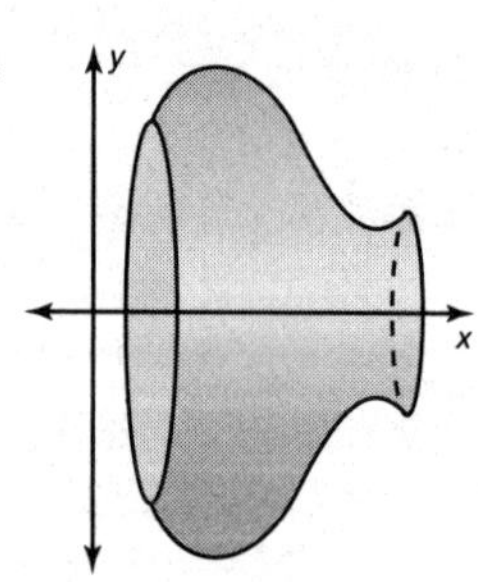

Answers

1.4 Practice B

1.

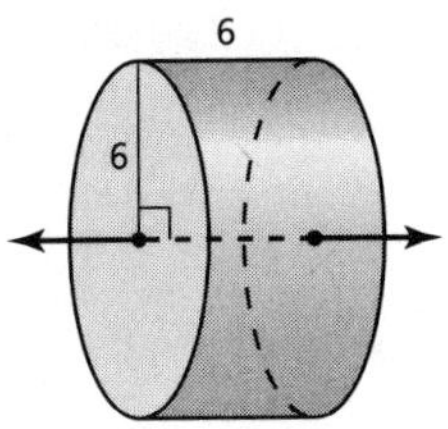

cylinder with height 6 and base radius 6

2.

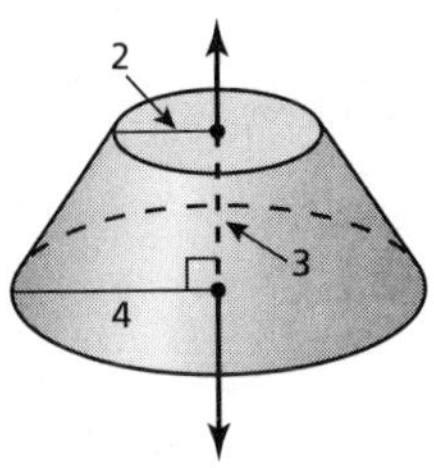

Sample answer: part of a cone with height 3 and two different circular bases with radii 2 and 4

3.

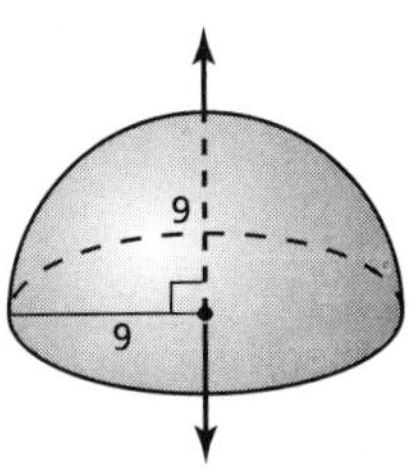

hemisphere with radius 9

4.

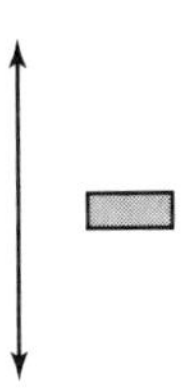

5.

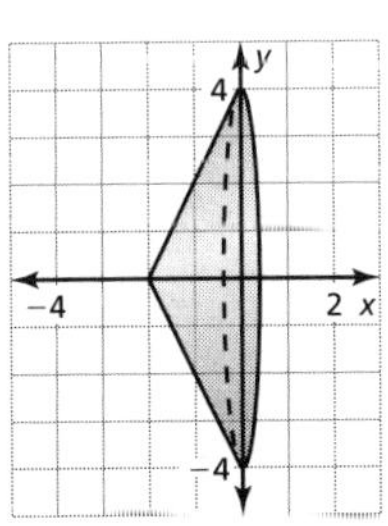

cone with radius 4 and height 2; about 33.5 square units

6.

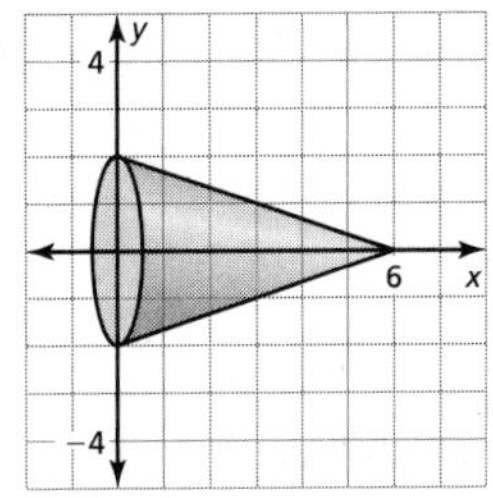

cone with radius 2 and height 6; about 25.1 square units

7.

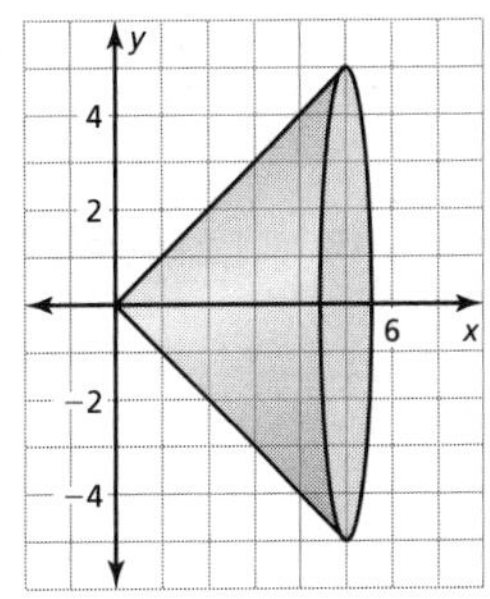

cone with radius 5 and height 5; about 25.1 square units

8.

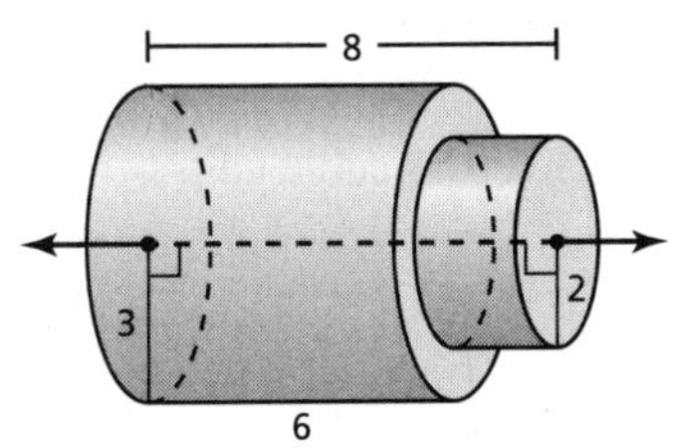

Sample answer: two adjacent cylinders that share the same axis of symmetry

9. yes; It is an isosceles right triangle, so both legs are congruent. Either rotation results in a cone with height 2 and base radius 2.

1.4 Enrichment and Extension

1. a.

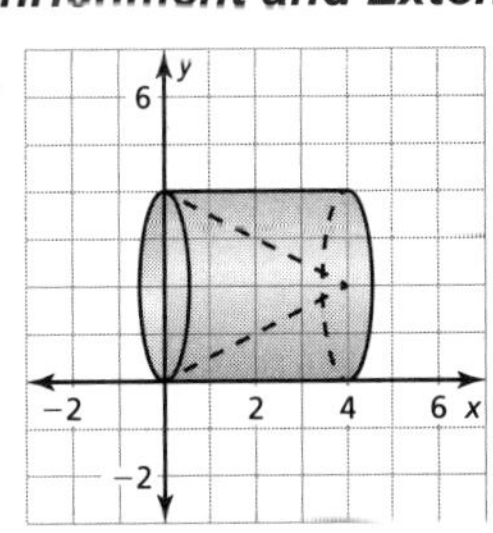

a cylinder with right cone removed; about 33.51 cubic units.

b.

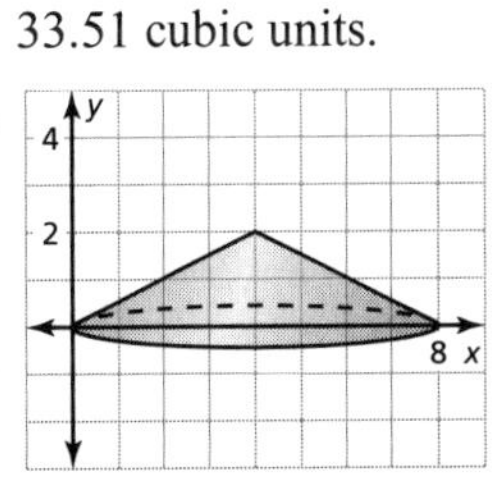

a right cone; about 33.51 cubic units

Answers

2. a.

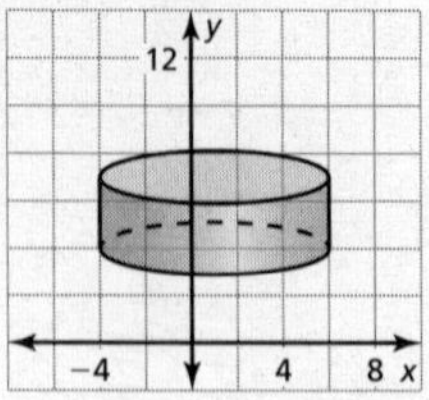

a cylinder; about 235.62 cubic units

b.

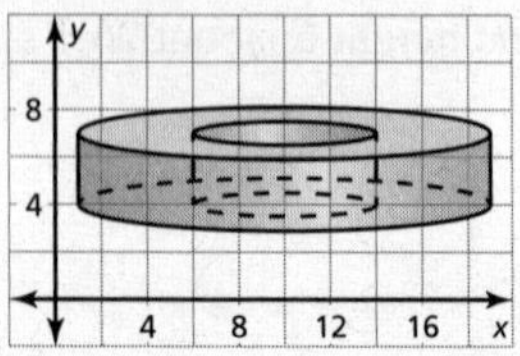

a cylinder with a smaller cylinder removed; about 612.61 cubic units

3. a.

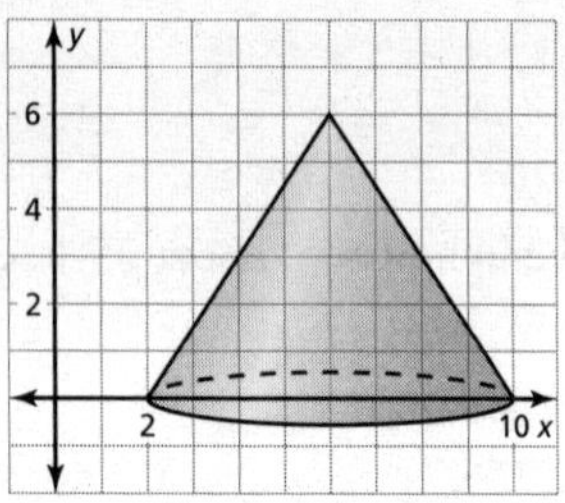

a right cone; about 100.53 cubic units

b.

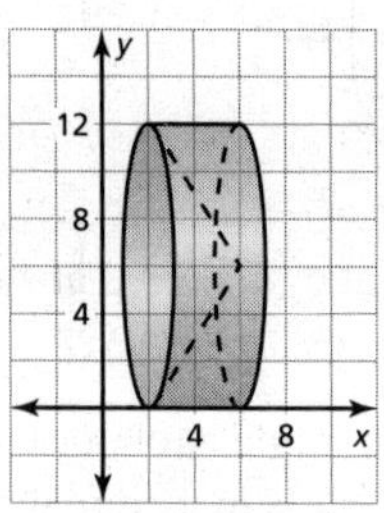

a cylinder with a right cone removed; about 301.59 cubic units

4. a.

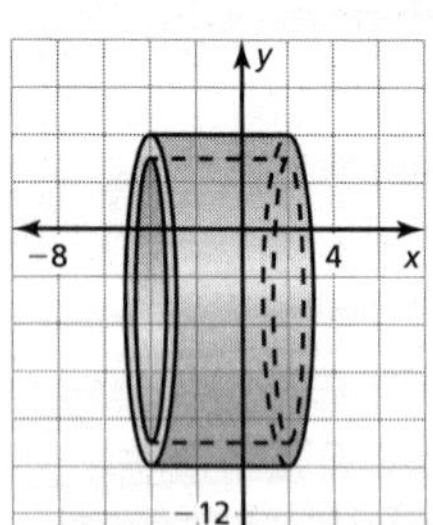

a cylinder with a smaller cylinder removed; about 245.04 cubic units.

b.

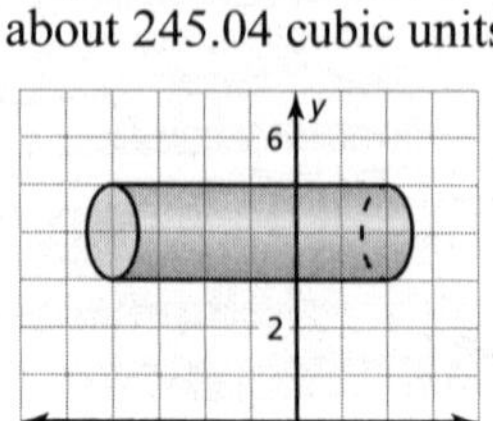

a cylinder; about 18.85 cubic units

1.4 Puzzle Time

IT ALWAYS WENT BACK FOUR SECONDS

Cumulative Review

1. solution **2.** solution

3. not a solution **4.** solution

5. not a solution **6.** solution

7. not a solution **8.** solution

9. not a solution **10.** solution

11. solution **12.** solution

13. $10.5 \leq t \leq 12$

14. a. $21 \leq m \leq 25$ **b.** 1.5 gal **c.** 4 gal

15. $z = 1$ **16.** $c = 4$ **17.** $y = -6$

18. $h = -10$ **19.** $p = 9$ **20.** $b = -49$

21. $r = -\frac{1}{8}$ **22.** $s = 7$

23. $f = -5$ and $f = 13$ **24.** $k = -3$ and $k = \frac{25}{3}$

25. $g = -\frac{9}{2}$ and $g = \frac{5}{2}$ **26.** $v = -\frac{3}{4}$ and $v = \frac{5}{4}$

27. $w = 1$ and $w = 9$ **28.** $d = -\frac{1}{3}$ and $d = 1$

29. $y = -3x + 2$ **30.** $y = 9x - 8$

31. $y = 2x - 1$ **32.** $y = -\frac{1}{2}x + \frac{5}{2}$

33. $y = -6x + 12$ **34.** $y = 5x - 30$

35. $\dfrac{108}{x}$ **36.** $55 - x$

37. $32 + x$ **38.** $40x$

39. a. $2.50 + 2.25m$ **b.** \$22.75 **c.** \$27.25 **d.** \$34.75

40. a. $2.12p$ **b.** \$10.60 **c.** \$7.28
d. 9.43 lbs, or about 9 lbs

41. Distributive Property

42. Additive Inverse Property

43. Commutative Property of Multiplication

44. Associative Property of Multiplication

Answers

45. Commutative Property of Addition

46. Associative Property of Addition

47. Additive Identity Property

48. Multiplicative Identity Property

49.

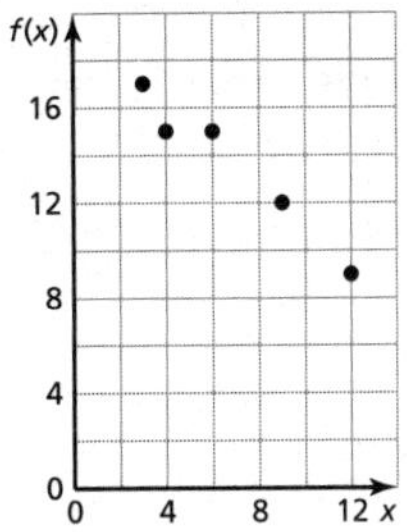

50.

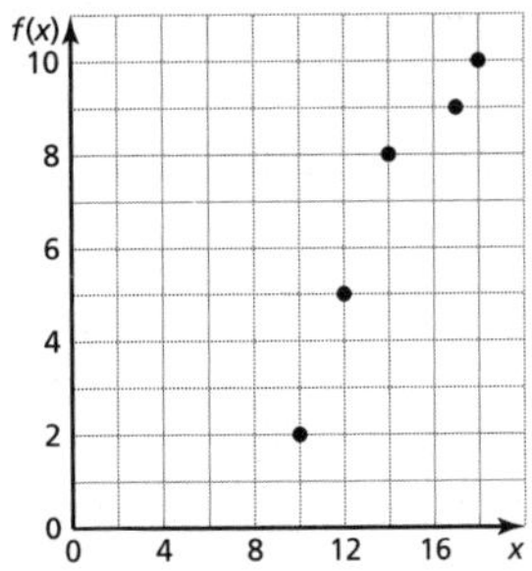

51.

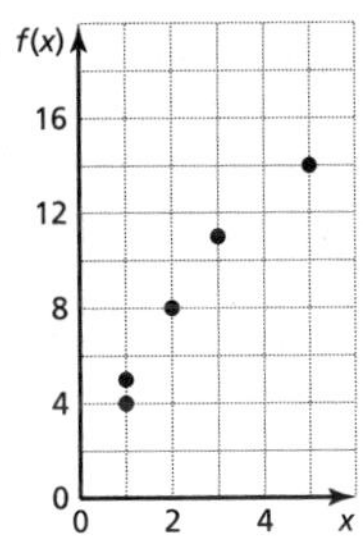

52.

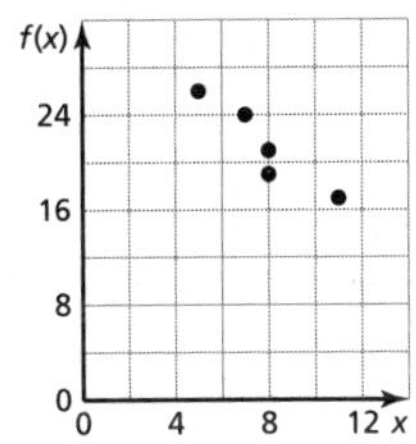

53. $(7, 3)$ **54.** $(4, 9)$ **55.** $(-1, -7)$

56. no solution **57.** $(-2, 7)$ **58.** no solution

59. no solution **60.** $(7, -2)$

61. infinitely many solutions

62. $(1, 7)$

63. infinitely many solutions

64. $(-2, -9)$ **65.** $9a^2 + 12a + 4$

66. $b^2 - 14b + 49$ **67.** $25c^2 + 90c + 81$

68. $9d^2 - 48d + 64$ **69.** $16m^2 - 32m + 16$

70. $4n^2 + 4n + 1$ **71.** $25p^2 - 60p + 36$

72. $q^2 - 4q + 4$

73. **a.** 904.8 cm^3 **b.** 1436.8 cm^3 **c.** 150.8 cm^3

74. $y = 2x + 3$ **75.** $y = x - 4$

76. $y = -\frac{1}{4}x + 1$ **77.** $y = -\frac{3}{4}x - 5$

78. $y = 4x$ **79.** $y = \frac{4}{5}$

80. $y = -\frac{2}{7}x - \frac{1}{2}$ **81.** $y = \frac{2}{3}x + 2$

82. 294 people **83.** about 5.59 mi

84. The surface area is $2\frac{4}{9}$ times the original surface area.

85. The surface area is $2\frac{11}{29}$ times the original surface area.

86.

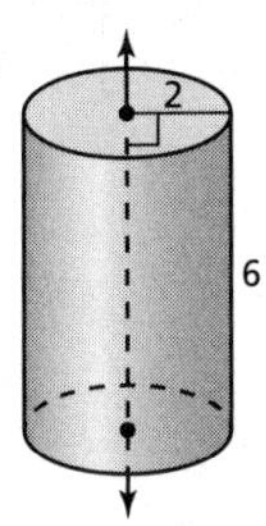

cylinder with height 6 and radius 2

87.

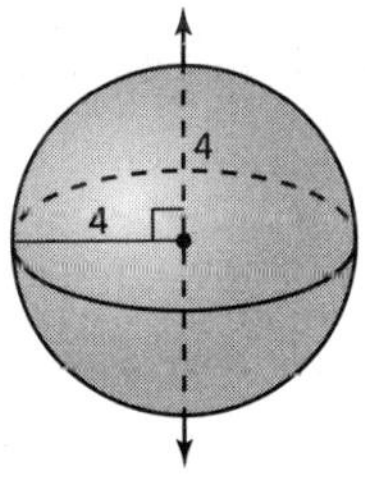

sphere with radius 4

88.

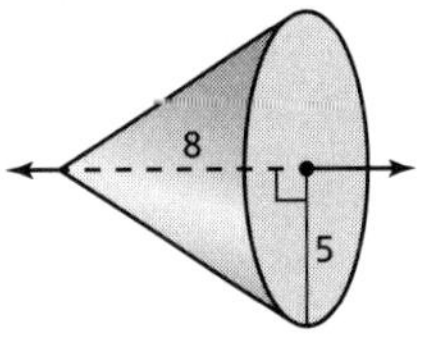

cone with height 8 and radius 5

Answers

Chapter 2

2.1 Start Thinking

As the string V gets wider, the points on the string move closer to the x-axis. This activity mimics a vertical shrink of a parabola.

2.1 Warm Up

1.

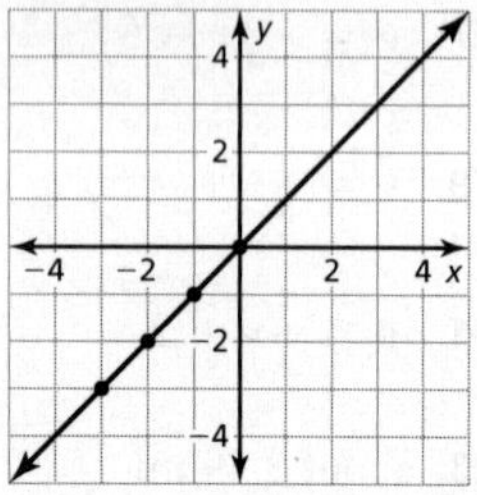

2.

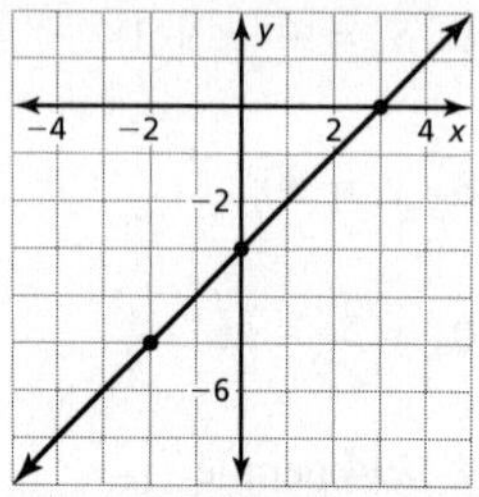

3.

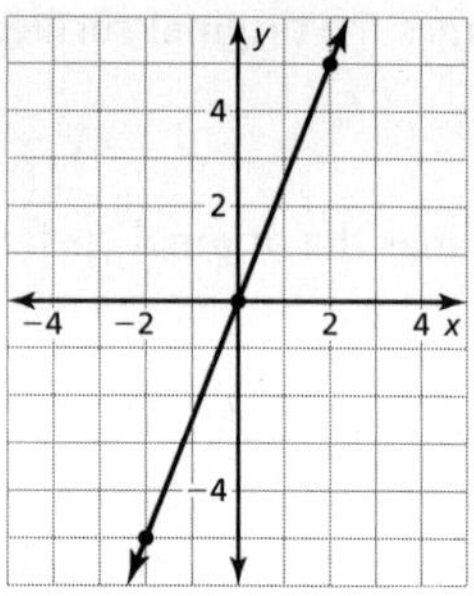

4.

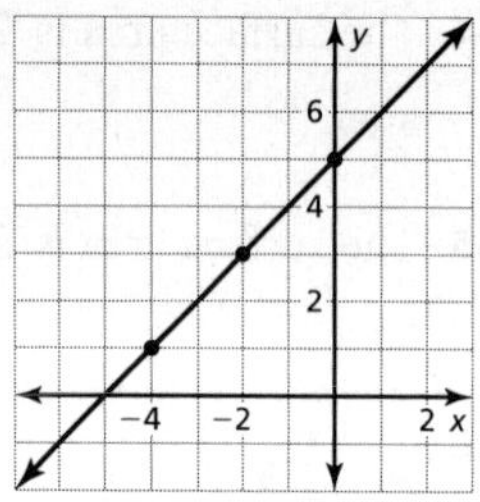

2.1 Cumulative Review Warm Up

1. $-\frac{1}{2}$ **2.** 1 **3.** $\frac{1}{2}$ **4.** 25

2.1 Practice A

1. quadratic; The graph of f is a vertical shrink by a factor of $\frac{1}{3}$ followed by a translation 1 unit down of the graph of the parent quadratic function.
2. constant; The graph of f is a translation 1 unit up of the graph of the parent constant function.
3. linear
4. 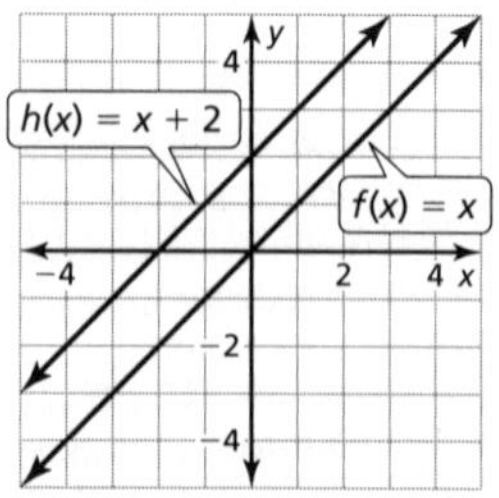

   Sample answer: The graph of h is a translation 2 units up of the graph of the parent linear function.

5. 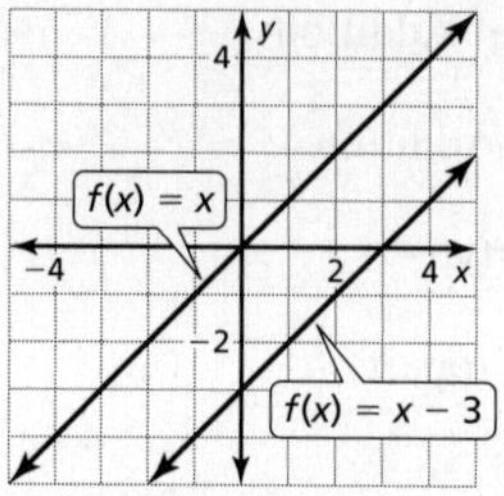

   Sample answer: The graph of f is a translation 3 units right of the parent linear function.

6. 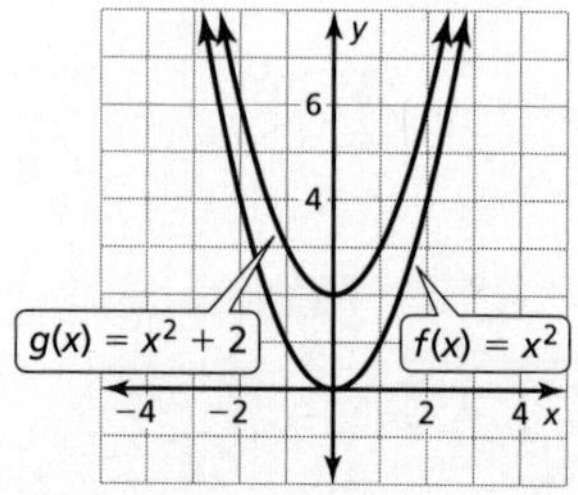

   The graph of g is a translation 2 units up of the parent quadratic function.

7. 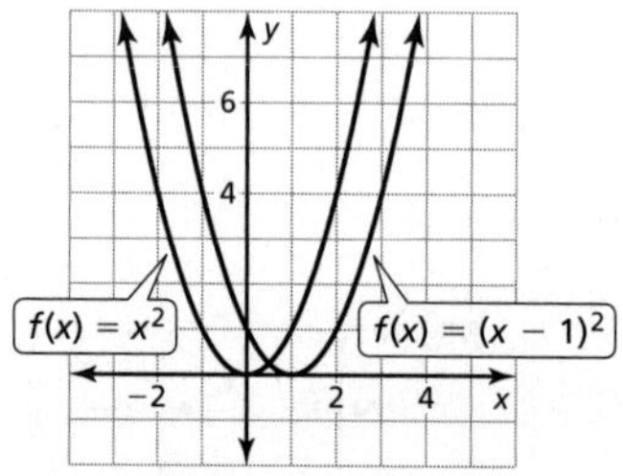

   The graph of f is a translation 1 unit right of the graph of the parent quadratic function.

8. 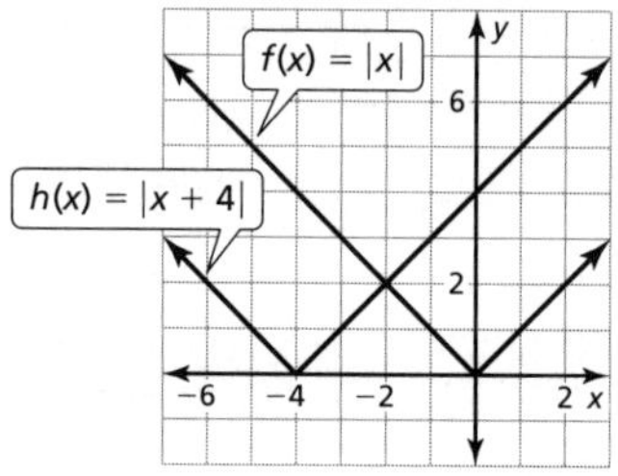

   The graph of h is a translation 4 units left of the graph of the parent function.

9. 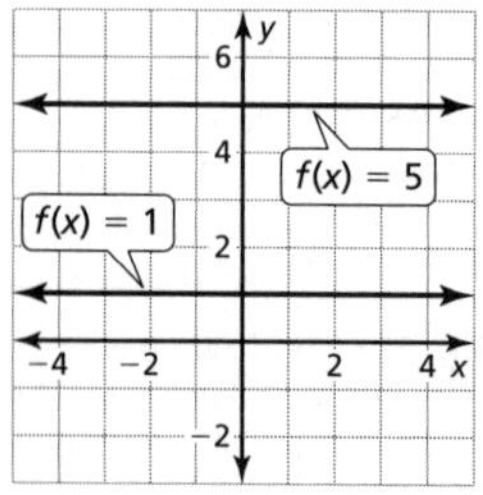

   The graph of f is a translation 4 units up of the graph of the parent constant function.

Answers

10. 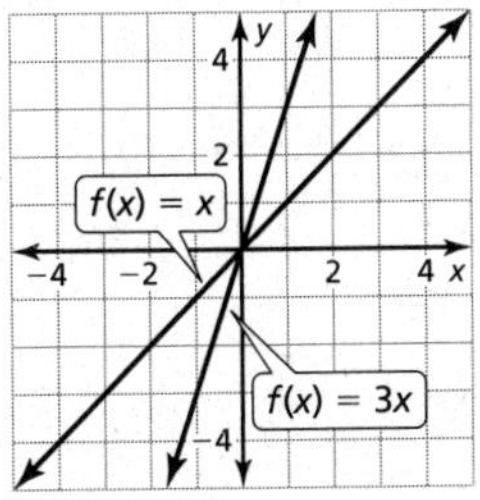

Sample answer: The graph of f is a vertical stretch by a factor of 3 of the graph of the parent linear function.

11. 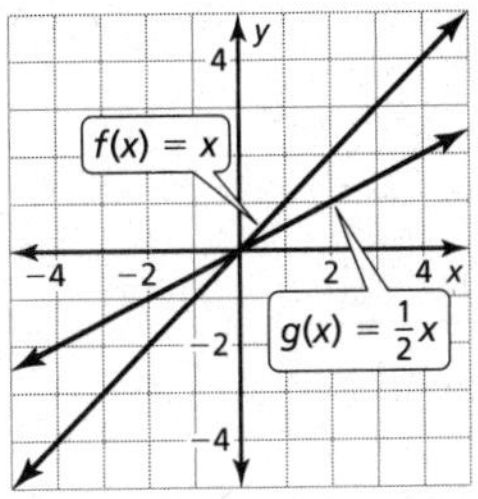

Sample answer: The graph of g is a vertical shrink by a factor of $\frac{1}{2}$ of the parent linear function.

12. 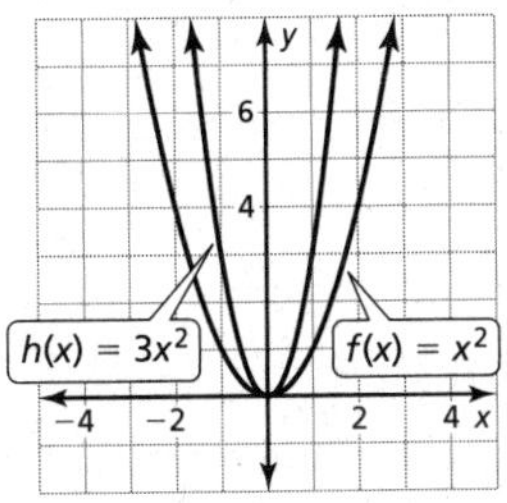

The graph of h is a vertical stretch by a factor of 3 of the graph of the parent quadratic function.

13. 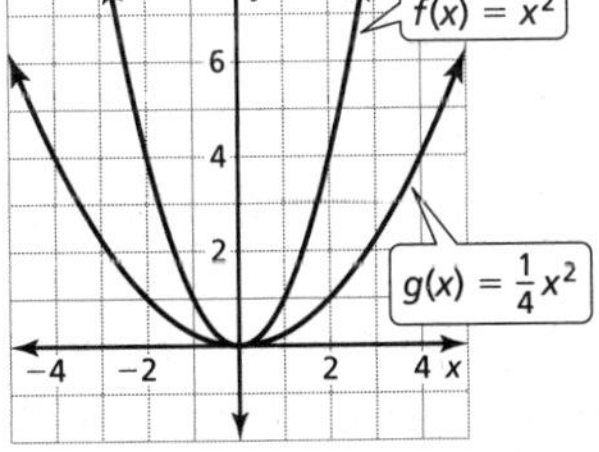

Sample answer: The graph of g is a vertical shrink by a factor of $\frac{1}{4}$ of the graph of the parent quadratic function.

14. 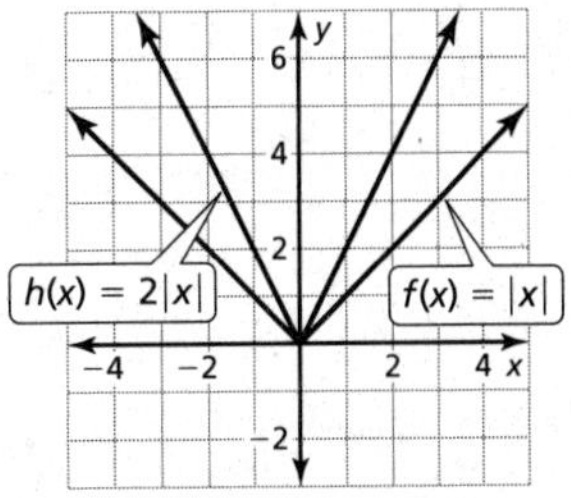

The graph of h is a vertical stretch by a factor of 2 of the graph of the parent absolute value function.

15. 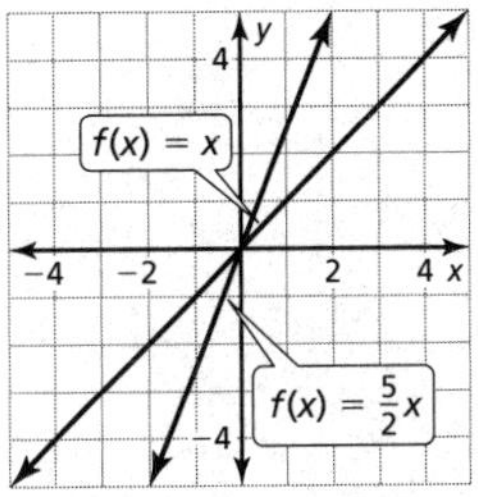

Sample answer: The graph of f is a vertical stretch by a factor of $\frac{5}{2}$ of the graph of the parent linear function.

16. 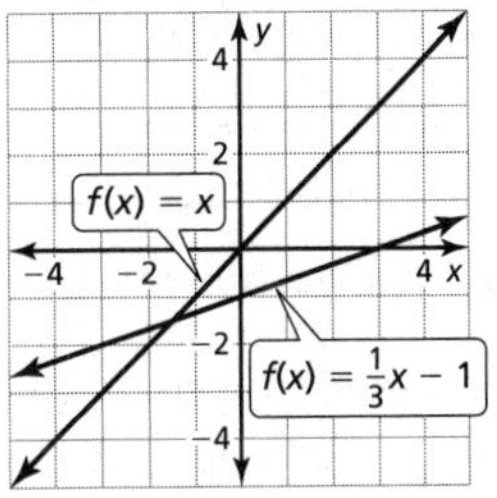

The graph of f is a vertical shrink by a factor of $\frac{1}{3}$ followed by a translation 1 unit down of the graph of the parent linear function.

17. 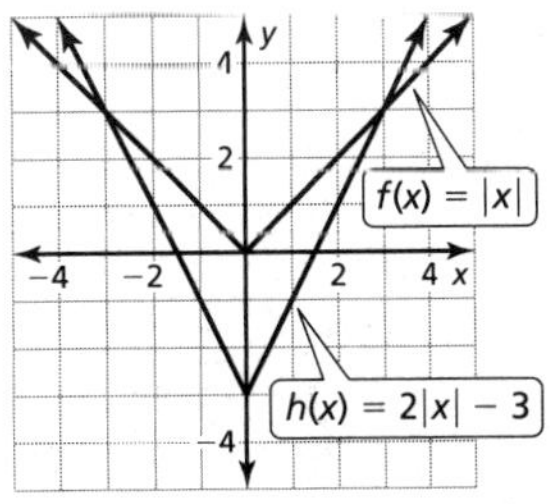

The graph of h is a vertical stretch by a factor of 2 followed by a translation 3 units down of the graph of the parent absolute value function.

Answers

18. 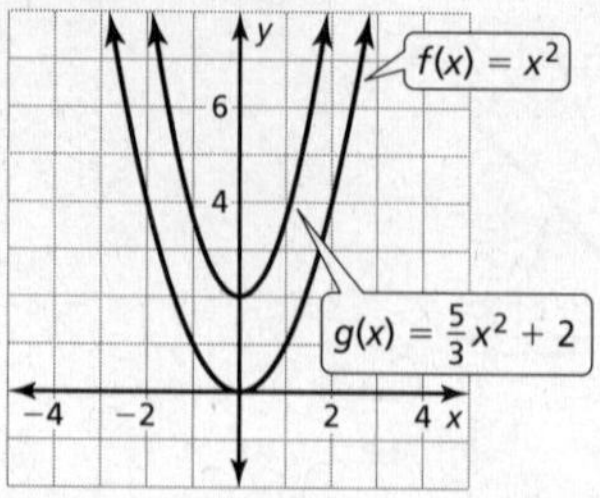

The graph of g is a vertical stretch by a factor of $\frac{5}{3}$ followed by a translation 2 units up of the graph of the parent quadratic function.

19. *Sample answer:*

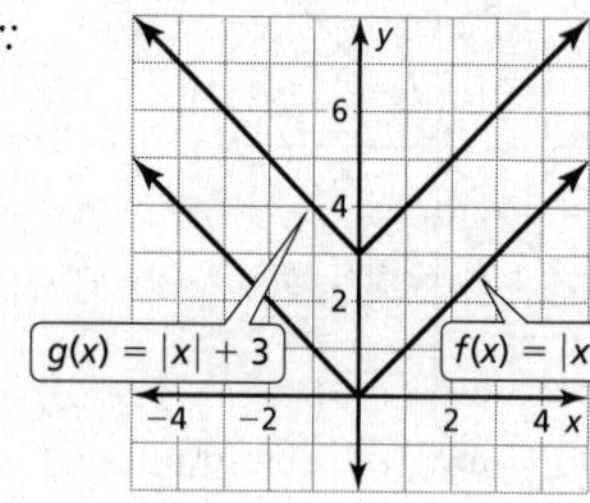

The graph of g is a translation 3 units up of the graph of the parent absolute value function.

2.1 Practice B

1. absolute value; The graph of f is a vertical shrink by a factor of $\frac{2}{5}$ followed by a translation 3 units right of the graph of the parent absolute value function.

2. linear; The graph of f is a vertical stretch by a factor of 2 followed by a translation 1 unit up of the graph of the parent linear function.

3. 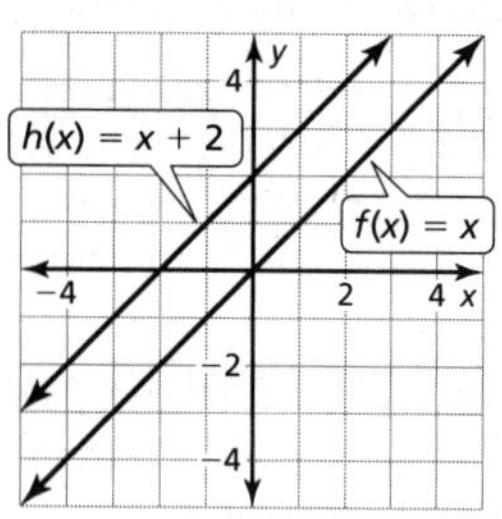

Sample answer: The graph of h is a translation 2 units up of the graph of the parent linear function.

4. 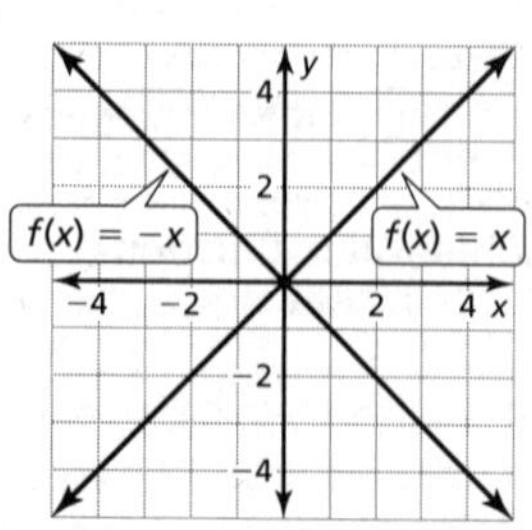

Sample answer: The graph of f is a reflection in the x-axis of the graph of the parent linear function.

5. 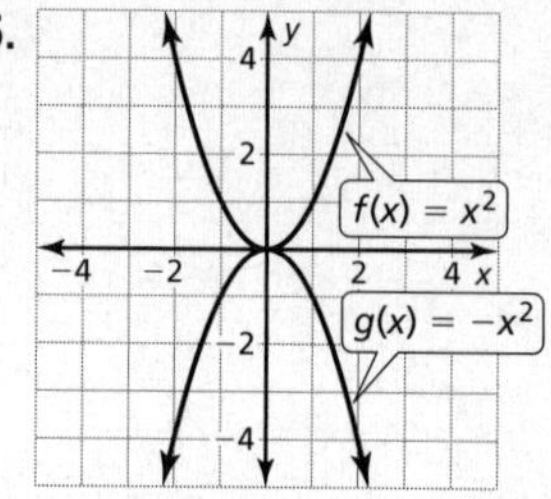

The graph of g is a reflection in the x-axis of the graph of the parent quadratic function.

6. 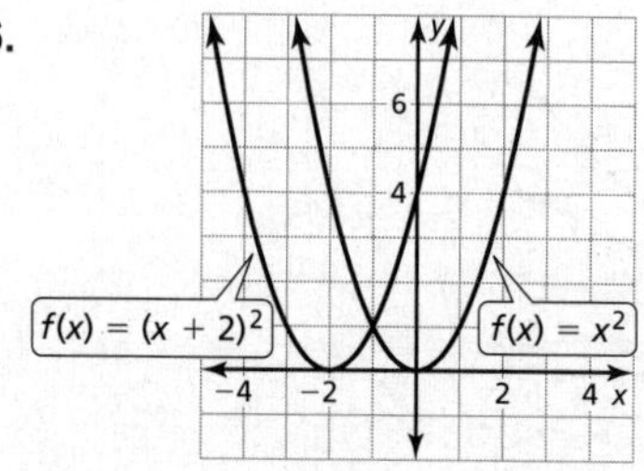

The graph of f is a translation 2 units left of the graph of the parent quadratic function.

7. 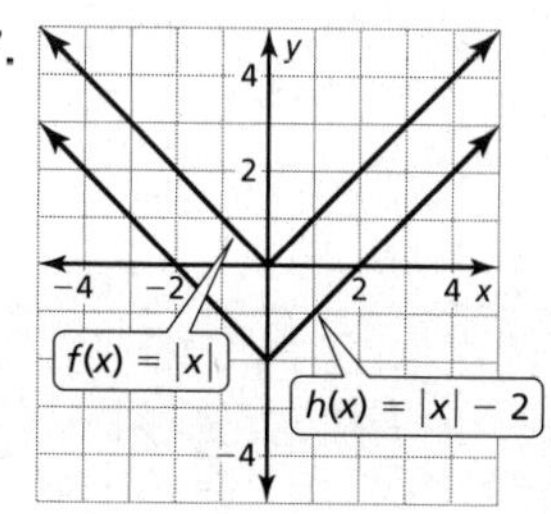

The graph of h is a translation 2 units down of the graph of the parent absolute value function.

8. 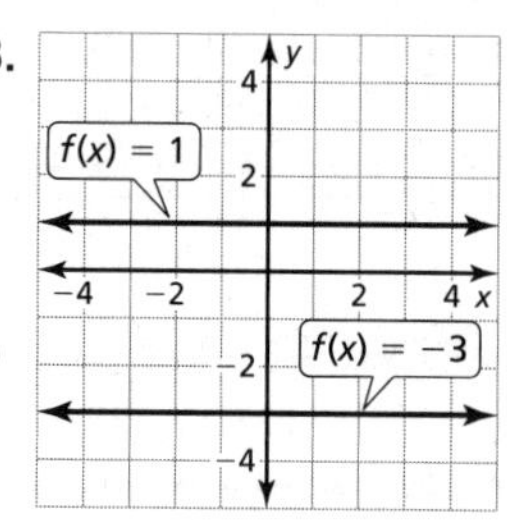

The graph of f is a translation 4 units down of the parent constant function.

Answers

9. 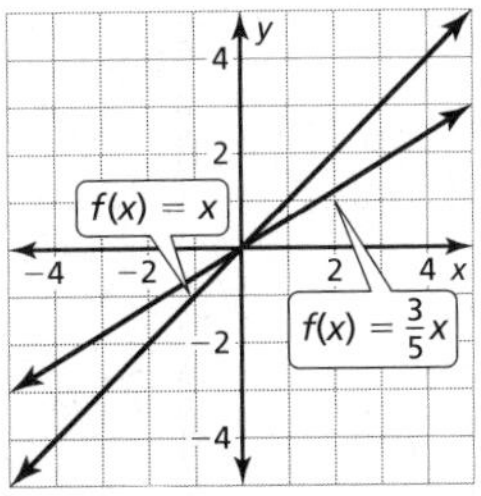

Sample answer: The graph of f is a vertical shrink by a factor of $\frac{3}{5}$ of the graph of the parent linear function.

10. 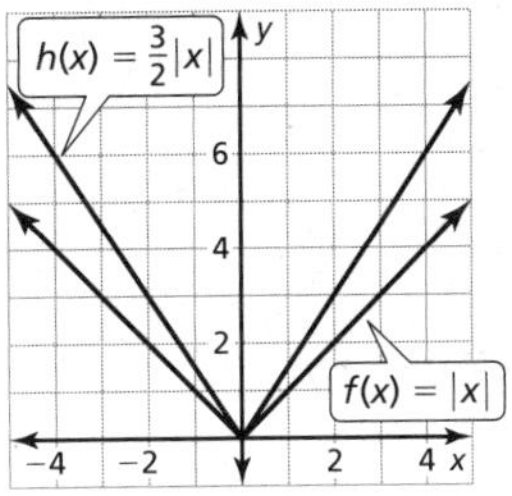

Sample answer: The graph of h is a vertical stretch by a factor of $\frac{3}{2}$ of the graph of the parent absolute value function.

11. 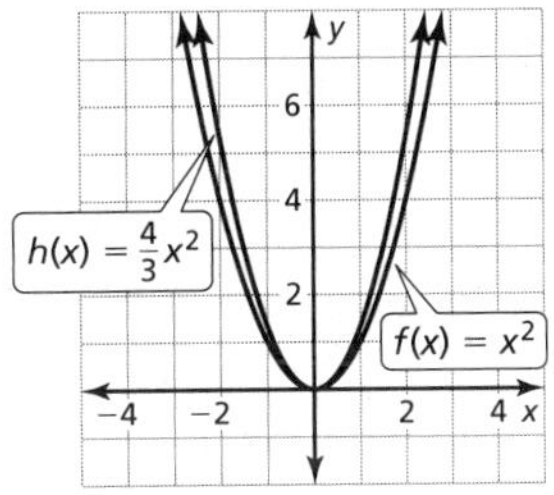

The graph of h is a vertical stretch by a factor of $\frac{4}{3}$ of the graph of the parent quadratic function.

12. 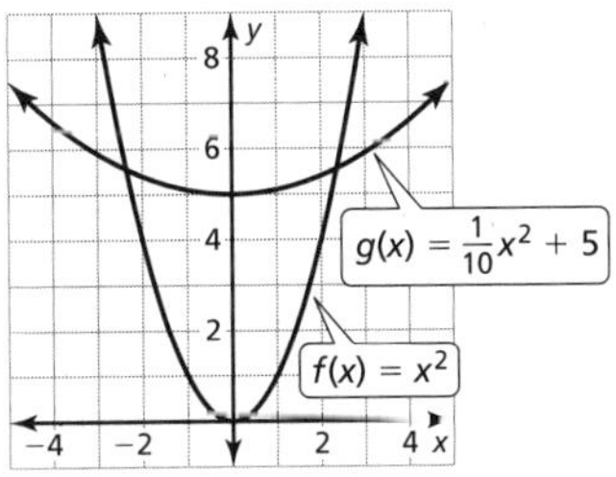

The graph of g is a vertical shrink by a factor of $\frac{1}{10}$ followed by a translation 5 units up of the graph of the parent quadratic function.

13. 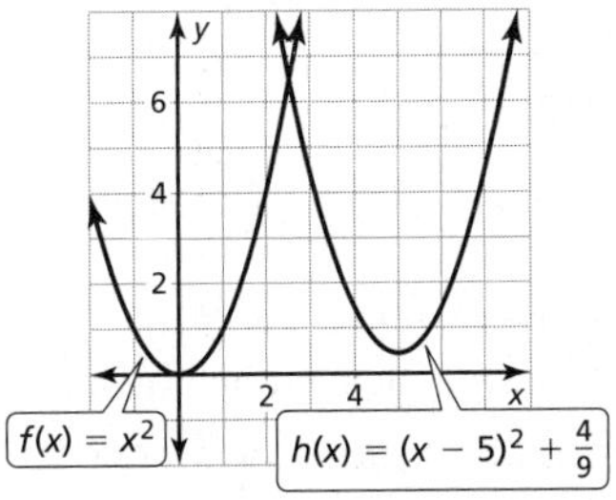

The graph of h is a translation 5 units right and $\frac{4}{9}$ units up of the graph of the parent quadratic function.

14. 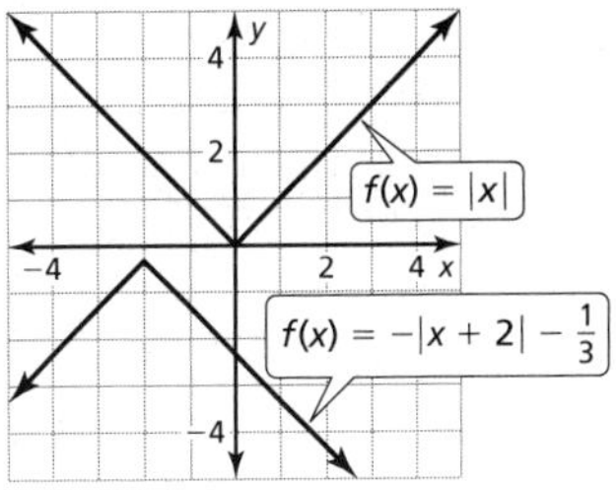

The graph of f is a reflection in the x-axis, followed by a translation 2 units left and $\frac{1}{3}$ units down of the graph of the parent absolute value function.

15. absolute value; domain: all real numbers, range: $y \geq 3$

16. linear; domain: all real numbers, range: all real numbers

17. quadratic; domain: all real numbers, range: $y \geq -3$

18. **a.** quadratic function

b. 0; t is the number of seconds after the ball is thrown, so when the ball is thrown $t = 0$.

c. 6 ft; $f(0) = 6$

2.1 Enrichment and Extension

1. 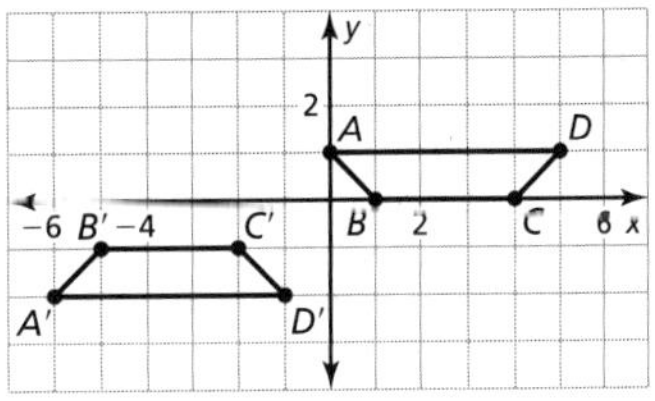

Sample answer: Trapezoid $A'B'C'D'$ a reflection in the x-axis, followed by translation 1 unit down and 6 units to the left of trapezoid $ABCD$.

Answers

2. **a.** quadratic function

   b. after $5\sqrt{2} \approx 7.07$ years

   c. domain: $x \geq 0$, range: $0 \leq y \leq 15{,}000$

3. *Sample answer:* domain: all real numbers; range: all real numbers; vertical stretch by a factor of $\frac{1}{2}$; reflection in y-axis; translation 5 units up

4. *Sample answer:* domain: all real numbers; range: $y \leq -3$; vertical stretch by a factor of 4; reflection in x-axis; translation 3 units down

5. *Sample answer:* domain: all real numbers; range: $y \geq 5$; horizontal stretch by a factor of 3; reflection in y-axis; translation 5 units up and 3 units to the left

2.1 Puzzle Time

BECAUSE PEOPLE ALWAYS SAY IF IT IS NOT BROKEN DO NOT FIX IT

2.2 Start Thinking

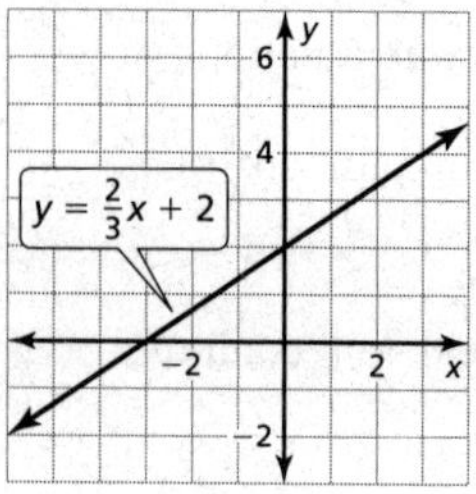

The equation becomes $y = \frac{2}{3}x + 3$; The equation becomes $y = \frac{2}{3}x + 1$; When 1 is added, by definition, the y-intercept moves up one unit. The slope is the same, so each point is moved up one unit. When -1 is added, the y-intercept moves down one unit, along with every other point on the line.

2.2 Warm Up

1. 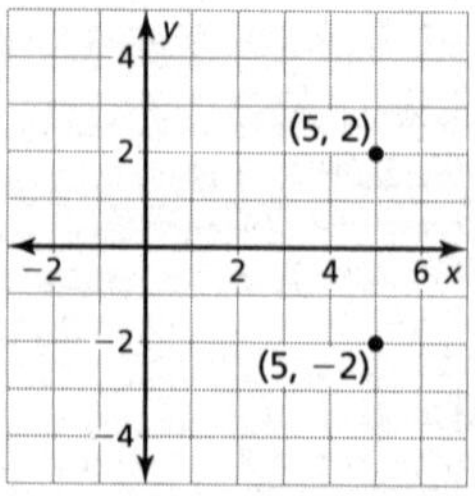

2. 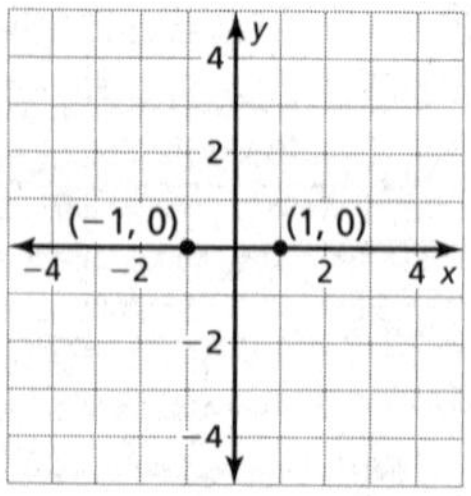

3. 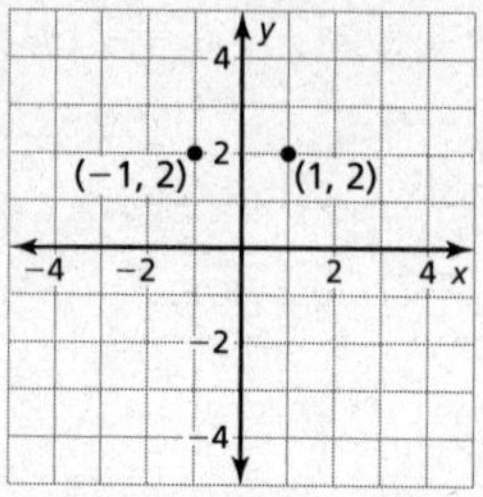

4. 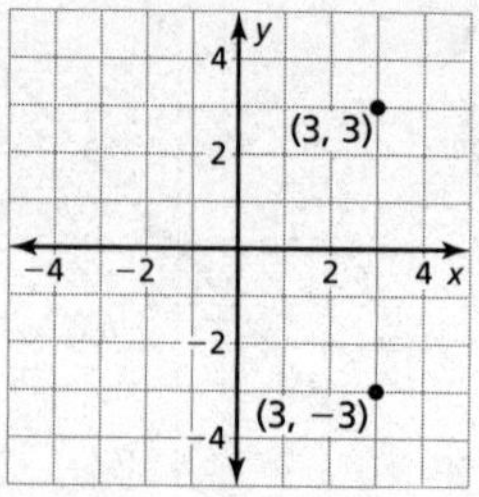

5. 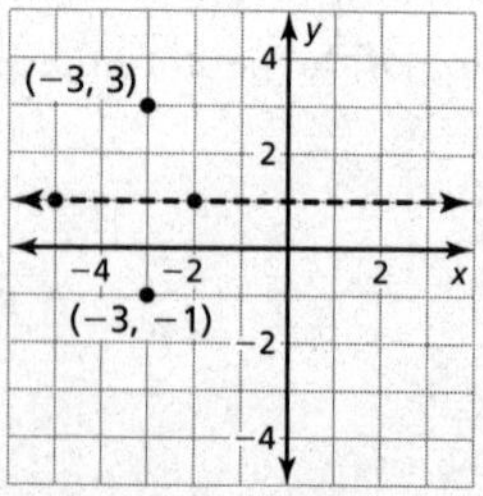

6. 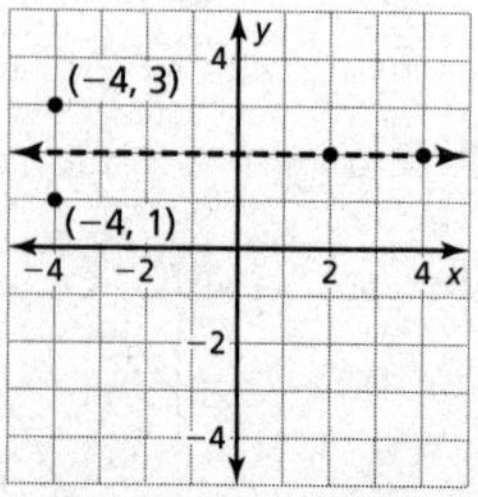

2.2 Cumulative Review Warm Up

1. one

2. one

3. zero

4. one

5. two

6. zero

2.2 Practice A

1. $g(x) = x + 3$
2. $g(x) = x - 3$
3. $g(x) = |3x + 2| + 1$
4. $g(x) = 4x - 2$
5. $g(x) = 3x - 7$
6. $g(x) = -\frac{1}{3}x + 2$
7. $g(x) = |-4x| - 6$
8. $g(x) = |-3x - 5| + 3$
9. $g(x) = 4x + 12$
10. $g(x) = \frac{4}{3}x + 1$
11. $g(x) = |9x| + 2$
12. $g(x) = \left|\frac{1}{3}x + 1\right|$
13. $g(x) = \frac{1}{3}x - 4$
14. $g(x) = |2x + 3|$

2.2 Practice B

1. $g(x) = 5x - 27$
2. $g(x) = 3x + 10$
3. $g(x) = 3 - |x|$
4. $g(x) = |2x| + 1$
5. $g(x) = x + 3$
6. $g(x) = -\frac{2}{3}x + 4$

Answers

7. $g(x) = -5 + |-x + 8|$
8. $g(x) = |-4x - 1| + 2$
9. $g(x) = 3 - \frac{1}{2}x$
10. $g(x) = x + \frac{5}{3}$
11. $g(x) = |9x| + 2$
12. $g(x) = -4|x - 2| + 8$
13. $g(x) = \frac{1}{4}x + \frac{5}{4}$
14. $g(x) = -|x + 2|$

2.2 Enrichment and Extension

1. $g(x) = 2x - 8; x = 4$
2. $g(x) = -2x - 1; x = -\frac{1}{2}$
3. $g(x) = 6x - 6; x = 1$
4. $g(x) = 6x + 4; x = -\frac{2}{3}$
5. $g(x) = 4x - 14; x = \frac{7}{2}$
6. $g(x) = -2x - 6; x = -3$
7. $g(x) = |x - 1| - 2; x = 3, x = -1$
8. $g(x) = |x + 3| + 1$; no x-intercepts
9. $g(x) = -|x + 5|; x = -5$
10. $g(x) = 2|x + 1| - 6; x = 2, x = -4$
11. $g(x) = -|4x - 38|; x = \frac{19}{2}$
12. $g(x) = -\frac{1}{2}|x - 3| + 2; x = 7, x = -1$

2.2 Puzzle time

JAMES MONROE

2.3 Start Thinking

You can model this situation with the equation $y = 14.99x - 8.50x$, where x represents the number of units sold and y represents the total profit. You are looking for the point that has a y-value of 150,000. By substituting 150,000 for y in the equation and solving for x, you obtain $x = 23{,}113$.

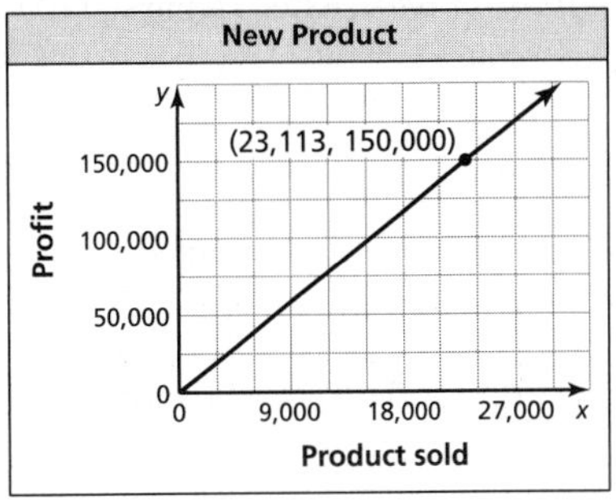

2.3 Warm Up

1. $D = \{0 \le x \le 82\}$
   $R = \{0 \le y \le 12{,}300{,}000\}$
2. $D = \{0 \le x \le 5\}$
   $R = \{0 \le y \le 4220\}$
3. $D = \{0 \le x \le 2500\}$
   $R = \{0 \le y \le 13{,}125\}$

2.3 Cumulative Review Warm Up

1. 5.1
2. 9.3
3. 10.3
4. \$130,533.30

2.3 Practice A

1. $y = \frac{3}{50}x$; The sales tax rate is $\frac{3}{50} = 6\%$.
2. $y = -\frac{1}{2}x + 10$; An amount of $\frac{1}{2}$ ounce of soap is used each day.
3. Soapy Car Wash; 6 extras
4. not linear
5. yes; $y = \frac{1}{2}x$; $y = 7.5$; This means 7.5 cars are washed in 15 minutes.
6. yes; A correlation coefficient close to -1 is a strong, negative correlation.

Answers

2.3 Practice B

1. $y = 5x + 15$; A child gains about 5 pounds per year.

2. $y = -\frac{11}{4}x + 20$; One loaf of bread uses $\frac{11}{4}$ cups of flour.

3. yes; $y = 22x + 10$; $y = 340$; After 15 days, 340 tickets were sold.

4. not linear

5. $y = 0.42x + 2.67$; $r = 0.65$; The points lie somewhat close to the line.

6. $y = -x + 6$; $r = -0.85$; The points lie close to the line.

2.3 Enrichment and Extension

1. $y = 4$
2. $y = \frac{2}{3}x - 2$
3. $y = 7$
4. $y = \frac{1}{2}x - 3$
5. -9
6. 14
7. $-\frac{3}{2}$
8. -18
9. 3
10. $-\frac{4}{3}$
11. $y = \frac{5}{4}x - 2488$; \$30,750
12. $y = \frac{7}{4}x - 3490$; \$36,250; \$5500

2.3 Puzzle Time

ROADTRIP

2.4 Start Thinking

The point of intersection of the three models would be when all three models create the same profit for the same cost.

2.4 Warm Up

1. $y = \frac{3}{2}x - \frac{3}{2}$
2. $y = -\frac{1}{3}x + \frac{7}{2}$
3. $y = -\frac{3}{2}x$
4. $y = -\frac{7}{2}x + \frac{13}{2}$
5. $y = \frac{1}{4}x - \frac{5}{4}$
6. $y = \frac{7}{8}x - \frac{19}{8}$

2.4 Cumulative Review Warm Up

1. yes
2. no
3. no
4. yes
5. no
6. yes

2.4 Practice A

1. $(1, -3, -7)$
2. $(2, 1, 5)$
3. Each term of the second equation should be multiplied by 5.

$$\begin{array}{r} 5x + 3y - z = 15 \\ -5x + 10y + 15z = 50 \\ \hline 13y + 14z = 65 \end{array}$$

4. no solution
5. Any ordered triple of the form $(x, 0, x - 2)$ is a solution of the system.
6. Roses cost \$3 each, carnations cost \$1 each, and tulips cost \$3 each.
7. $(4, -3, -4)$
8. $(7, 2, 1)$
9. a. $\ell + m + n = 90$

$$m = 5n - 5$$
$$\ell = \frac{5}{4}m - n$$
$$\ell = 41, m = 40, n = 9$$

   b. yes; $9^2 + 40^2 = 41^2$

2.4 Practice B

1. $(-1, 1, 3)$
2. $(2, 1, 4)$
3. Each term of the first equation should be multiplied by 3.

$$\begin{array}{r} 15x + 9y - 3z = 45 \\ 3x - 4y + 3z = 8 \\ \hline 18x + 5y = 53 \end{array}$$

4. no solution
5. Any ordered triple of the form $(x, 2 + 2x, -x + 1)$ is a solution.
6. $(5, 4, -3)$
7. no solution
8. no; The solution of the system is 16.4 nickels, 8.2 dimes, and 5.4 quarters. You cannot have part of a coin.
9. a. $\left(\frac{1}{5}, -\frac{7}{5}\right)$

   b. yes; There is one solution, so the lines intersect at a point.

Answers

2.4 Enrichment and Extension

1. $(3, -2, 2, 1)$ **2.** $(0, -3, -1, 4)$ **3.** $(5, -2, 3, 4)$

2.4 Puzzle Time

THE TALLEST LIVING HORSE WHO MEASURED SIX FEET TEN INCHES

2.5 Start Thinking

Sample answer:

$f(x) = x^2$

x	-2	-1	0	1	2
$f(x)$	4	1	0	1	4

$f(x) = (x - 1)^2$

x	-2	-1	0	1	2
$f(x)$	9	4	1	0	1

$f(x) = 2x^2$

x	-2	-1	0	1	2
$f(x)$	8	2	0	2	8

$f(x) = x^2 + 1$

x	-2	-1	0	1	2
$f(x)$	5	2	1	2	5

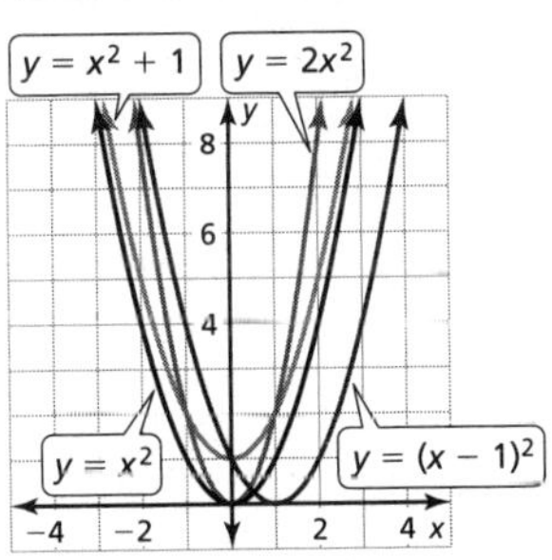

The function $f(x) = (x - 1)^2$ is a horizontal translation 1 unit right $(h = 1)$. The function $f(x) = 2x^2$ is a vertical stretch by a factor of 2 $(a = 2)$. The function $f(x) = x^2 + 1$ is a vertical translation 1 unit up $(k = 1)$.

2.5 Warm Up

1. $6x^2 - 16x + 8$ **2.** $20x^2 + 13x + 2$

3. $8x^2 - 10xy - 3y^2$ **4.** $12a^2 + 3a$

5. $20x^2 - 3x - 2$ **6.** $15y^2 + 22y + 8$

2.5 Cumulative Review Warm Up

1. $g(x) = x + 3$ **2.** $g(x) = x - 2$

3. $g(x) = |5x - 2| - 4$

2.5 Practice A

1. The graph of g is a translation 2 units down of the graph of f.

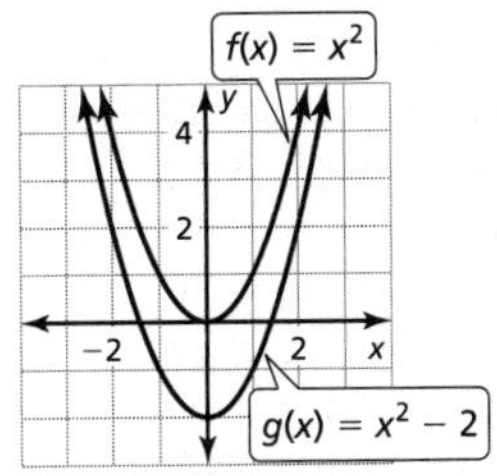

2. The graph of g is a translation 1 unit up of the graph of f.

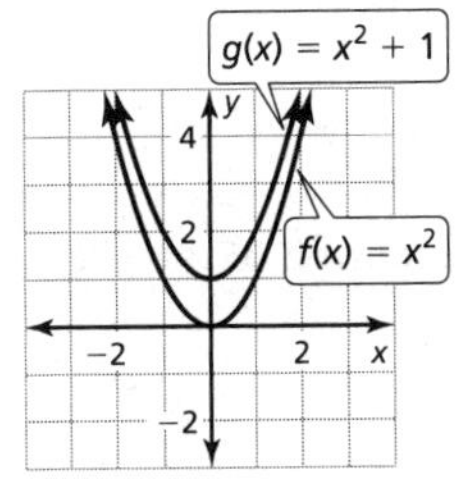

3. The graph of g is a translation 1 unit left of the graph of f.

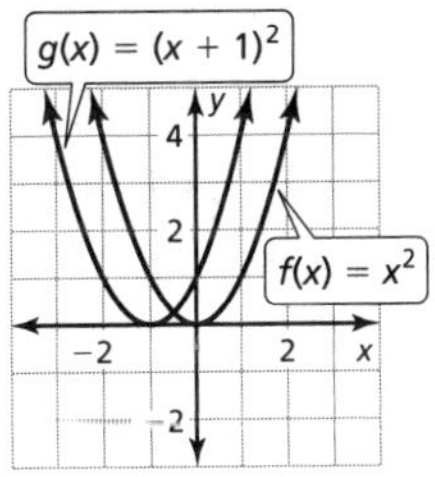

4. The graph of g is a translation 2 units right of the graph of f.

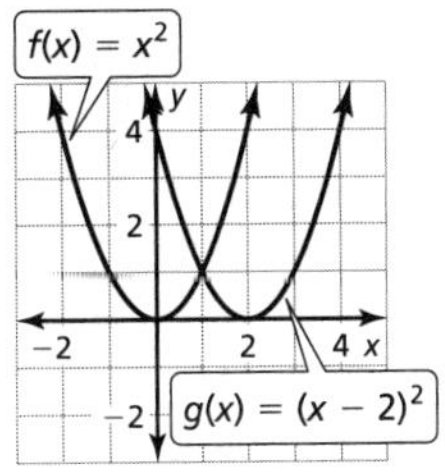

Answers

5. The graph of g is a translation 5 units right of the graph of f.

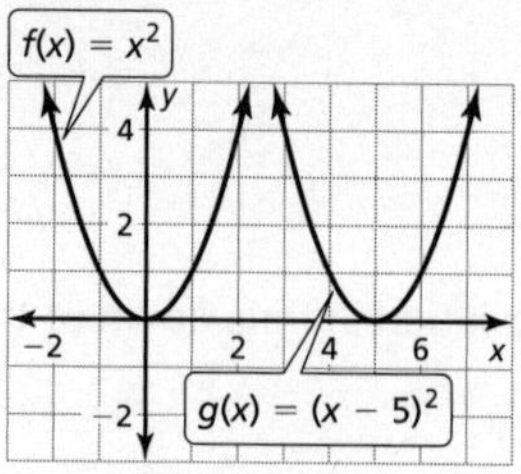

6. The graph of g is a translation 2 units left and 1 unit down of the graph of f.

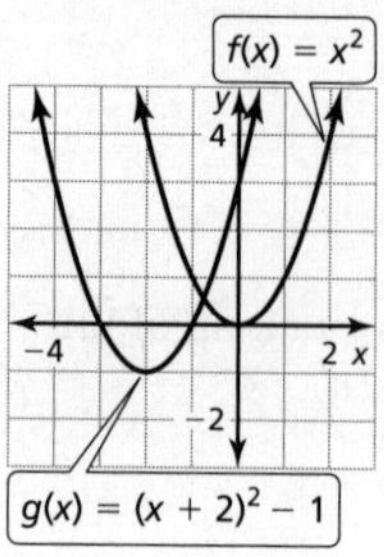

7. The graph of g is a reflection in the x-axis followed by a vertical stretch by a factor of 2 of the graph of f.

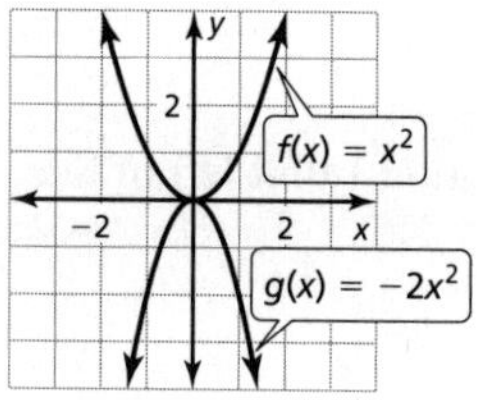

8. The graph of g is a reflection in the y-axis followed by a horizontal shrink of the graph of f by a factor of $\frac{1}{2}$.

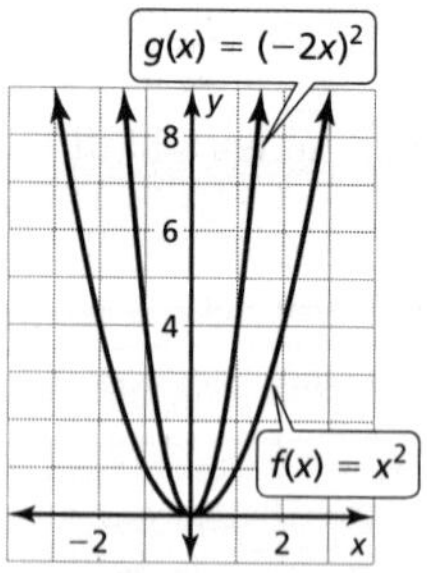

9. The graph of g is a vertical shrink by a factor of $\frac{1}{4}$ of the graph of f.

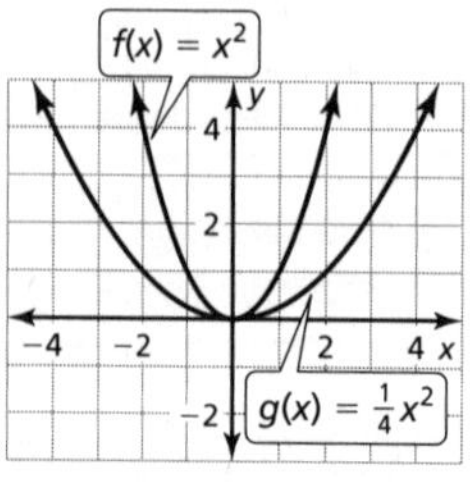

10. When $0 < a < 1$ in the function $g(x) = a \bullet f(x)$, the transformation is a vertical shrink, not stretch; The graph of g is a reflection in the x-axis followed by a vertical shrink by a factor of $\frac{1}{3}$ of the graph of the parent quadratic function.

11. The graph is a vertical stretch by a factor of 2, followed by a translation 3 units left and 2 units up of the parent quadratic function; $(-3, 2)$

12. The graph is a reflection in the x-axis, followed by a vertical stretch by a factor of 5 and a translation 1 unit down of the parent quadratic function; $(0, -1)$

13. $g(x) = -3x^2 - 3$; $(0, -3)$

14. $g(x) = -x^2 - 7$; $(0, -7)$

15. a. $a = 2, h = 3, k = -4$; $g(x) = (2x - 3)^2 - 4$

    b. $a = 4, h = 3, k = -4$; $g(x) = 4(x - 3)^2 - 4$

2.5 Practice B

1. The graph of g is a translation 3 units up of the graph of f.

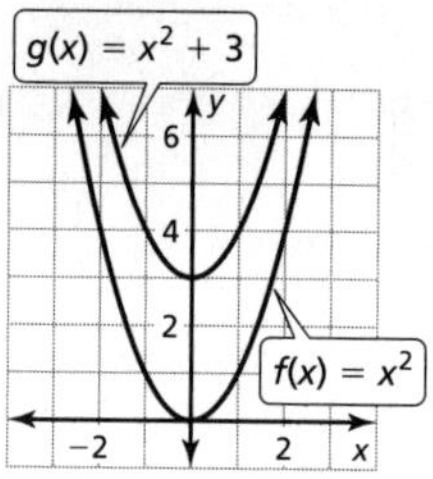

2. The graph of g is a translation 5 units left of the graph of f.

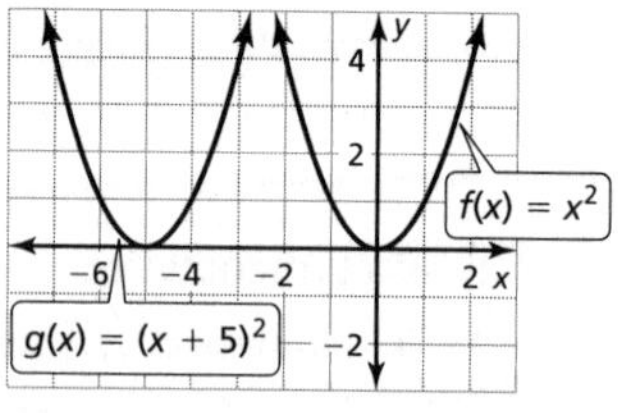

3. The graph of g is a translation 6 units left and 4 units down of the graph of f.

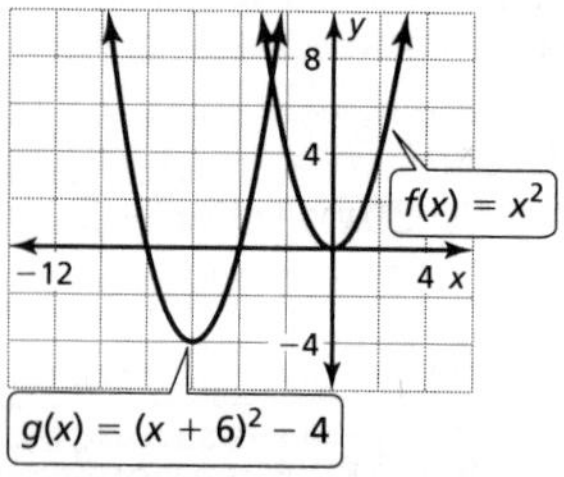

Answers

4. The graph of g is a translation 1 unit right and 5 units up of the graph of f.

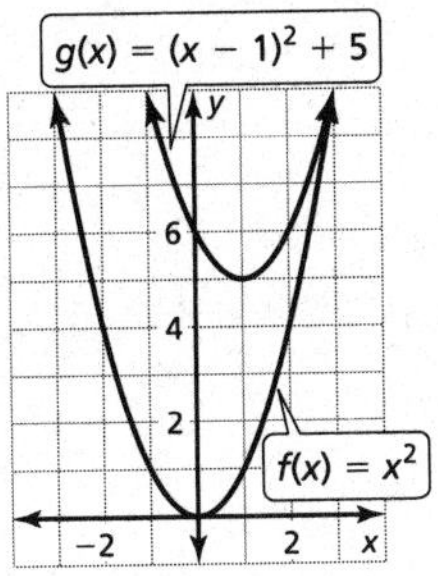

5. The graph of g is a translation 4 units right and 3 units up of the graph of f.

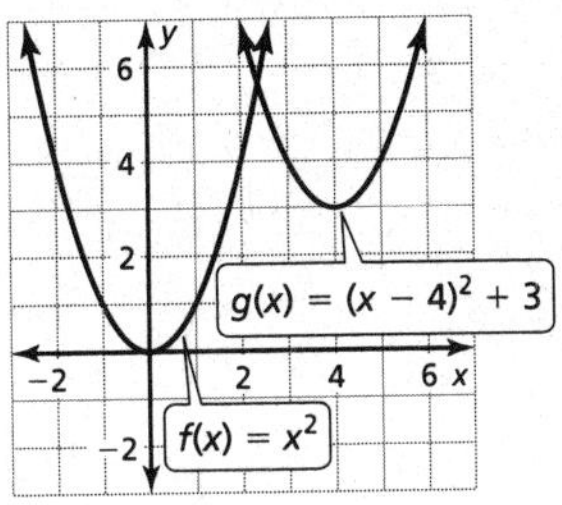

6. The graph of g is a translation 8 units left and 2 units down of the graph of f.

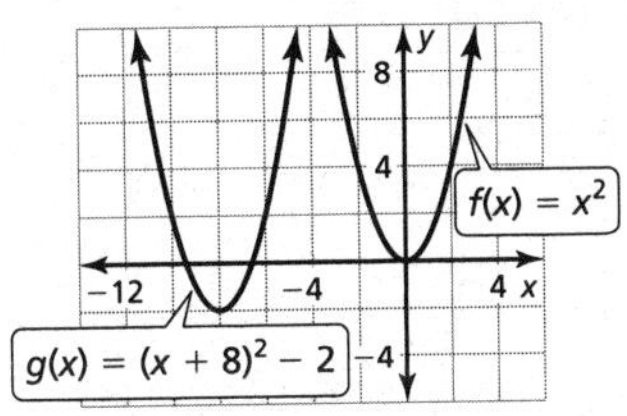

7. The graph of g is a reflection in the x-axis, followed by a horizontal stretch by a factor of 2 of the graph of f.

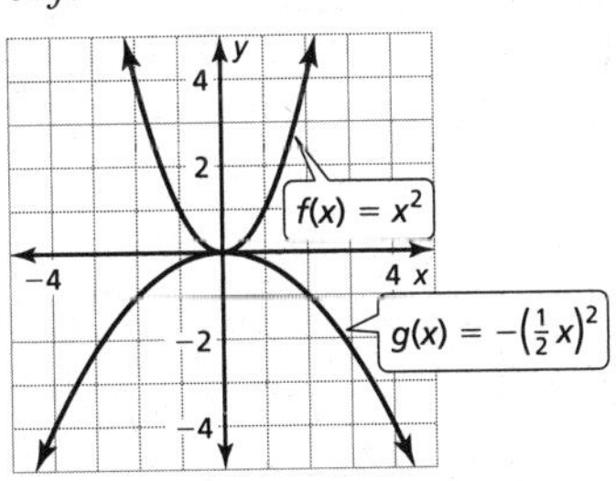

8. The graph of g is a vertical shrink by a factor of $\frac{1}{3}$, followed by a translation 2 units up of the graph of f.

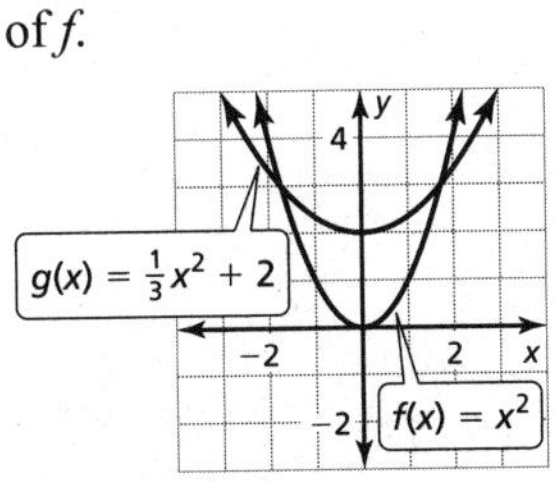

9. The graph of g is a vertical shrink by a factor of $\frac{1}{3}$, followed by a translation 1 unit left of the graph of f.

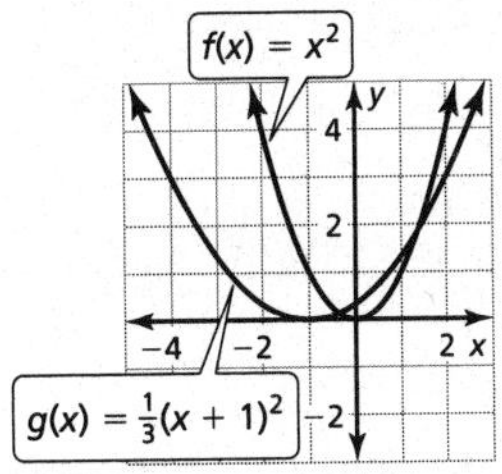

10. The graph is a reflection in the x-axis, followed by a vertical stretch by a factor of 3 and a translation 6 units left and 4 units down of the parent quadratic function; $(-6, -4)$

11. The graph is a vertical shrink by a factor of $\frac{1}{3}$, followed by a translation 2 units right and 1 unit up of the parent quadratic function; $(2, 1)$

12. $g(x) = \dfrac{(x + 2)^2}{2}$; $(-2, 0)$

13. $g(x) = -(3x + 4)^2 - 4$; $\left(-\frac{4}{3}, -4\right)$

14.

$h(x) = f(x) + 3$	Add 3 to the output.
$= 4x^2 - 3x + 3$	Substitute $f(x)$ and simplify.
$g(x) = h(-x)$	Multiply the input by -1.
$= 4x^2 + 3x + 3$	Substitute $-x$ into $h(x)$ and simplify.

2.5 Enrichment and Extension

$y = -3x^2$, $y = 3(x + 1)^2 + 1$, $y = 3(x - 1)^2 + 1$, $y = -3(x + 1)^2 - 1$, $y = 3(x - 4)^2 - 2$, $y = -3(x - 2)^2 + 2$, $y = -3(x + 2)^2 + 2$

2.5 Puzzle Time

EL SALVADOR

Answers

2.6 Start Thinking

x	-2	-1	0	1	2
$f(x)$	2	1	0	1	2

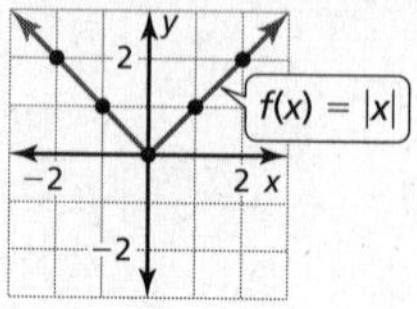

V shape; yes; yes; The line of symmetry is the y-axis.

2.6 Warm Up

1. $P'(5, 3)$

2. $P'(-5, -3)$

3. $P'(-5, -15)$

4. $P'(3, 3)$

2.6 Cumulative Review Warm Up

1. linear; $y = 11x$; $y = 220$; After jogging for 20 minutes, 220 calories were burned.

2. not linear

2.6 Practice A

1.

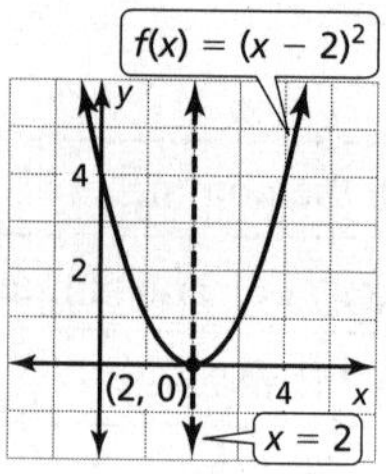

2.

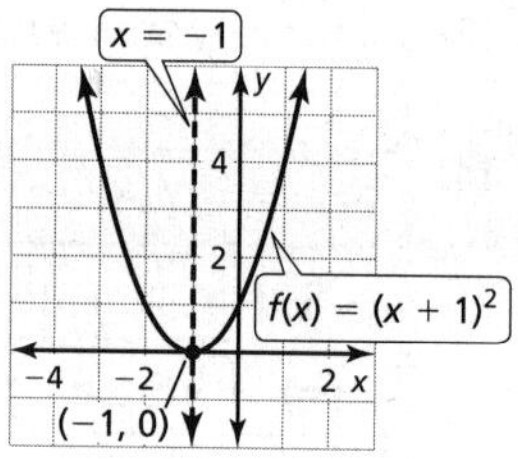

3.

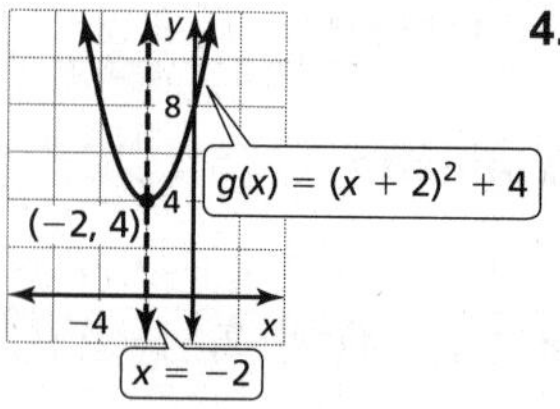

4.

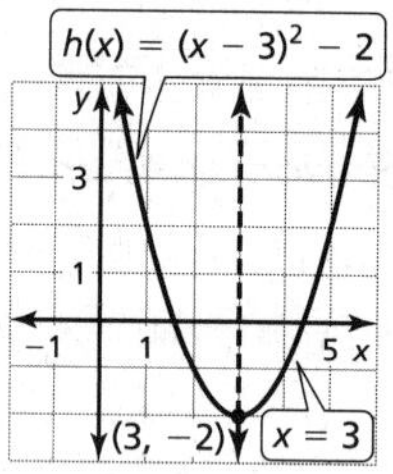

5.

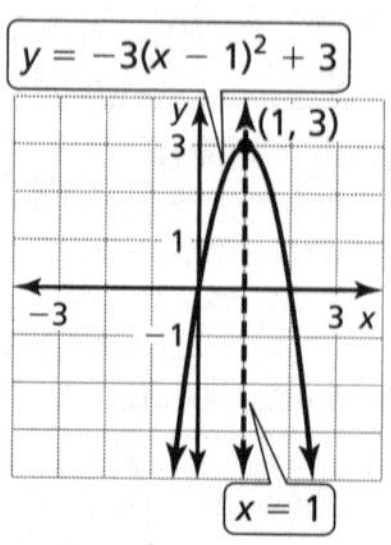

6.

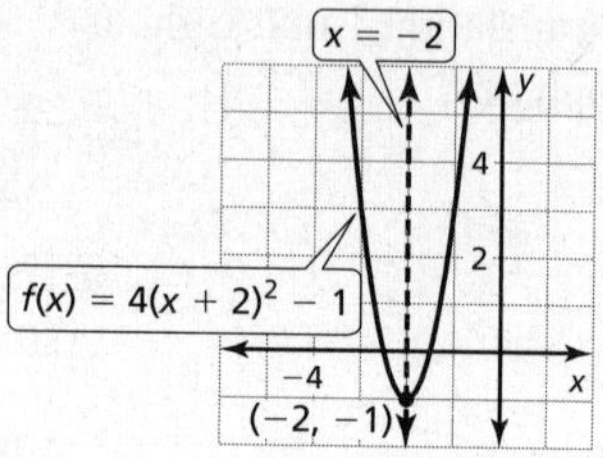

7.

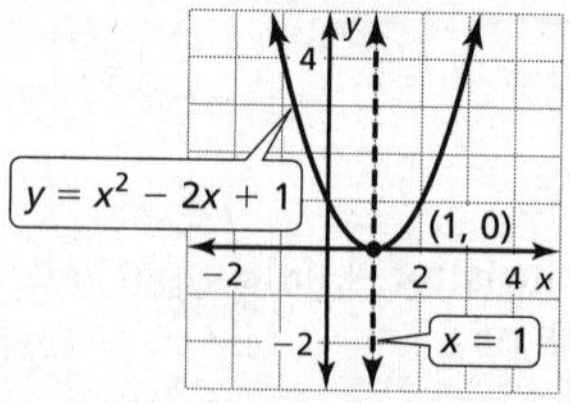

8.

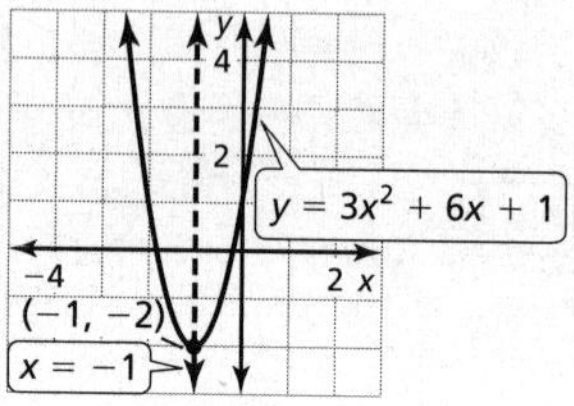

9.

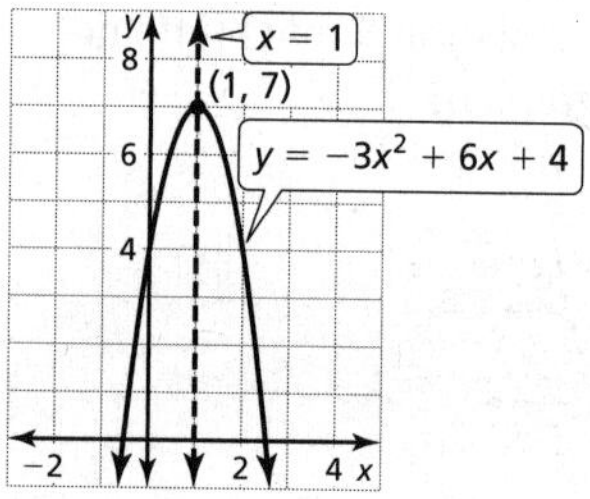

10.

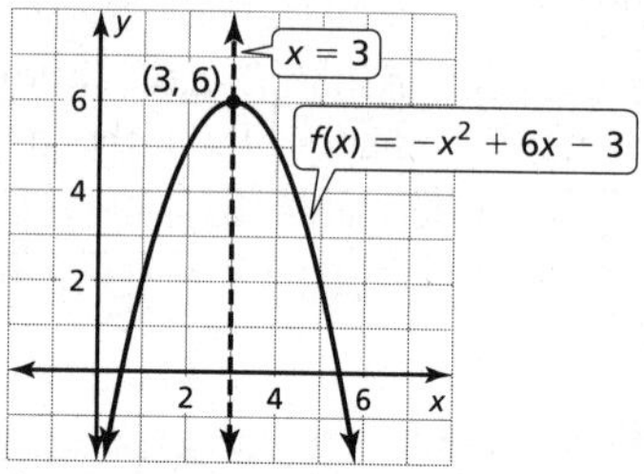

11.

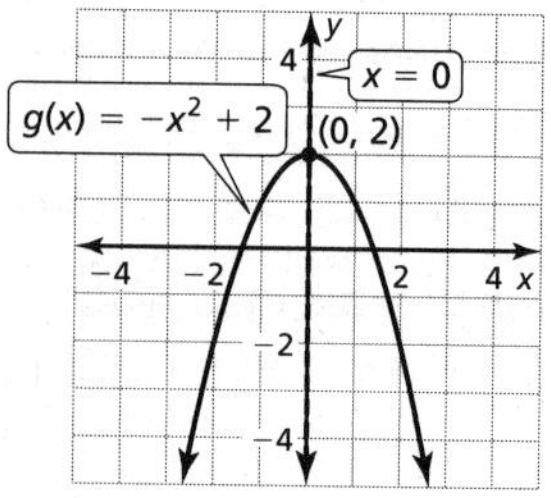

Answers

12. 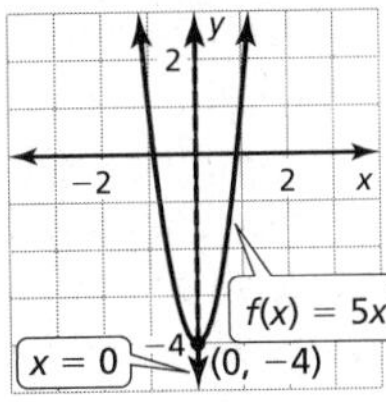

13. Both graphs have the same axis of symmetry, $x = -2$.

14. C; It has the largest leading coefficient, $a = 3$.

15. minimum: 2; domain: all real numbers, range: $y \geq 2$; increasing to the right of $x = 0$; decreasing to the left of $x = 0$

16. minimum: -3; domain: all real numbers, range: $y \geq -3$; increasing to the right of $x = 0$; decreasing to the left of $x = 0$

17. maximum: 3; domain: all real numbers, range: $y \leq 3$; increasing to the left of $x = 2$; decreasing to the right of $x = 2$

18. maximum: 11; domain: all real numbers, range: $y \leq 11$; increasing to the left of $x = 1$; decreasing to the right of $x = 1$

19. **a.** noon **b.** 75 customers

2.6 Practice B

1. 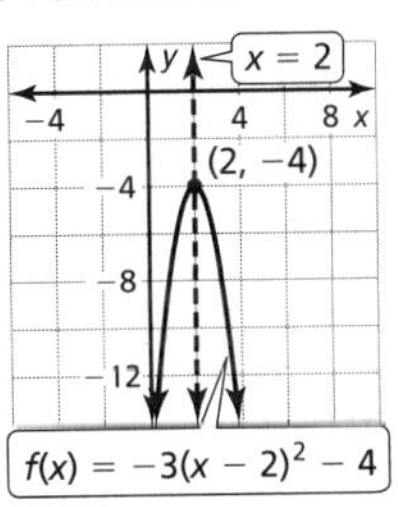

2. 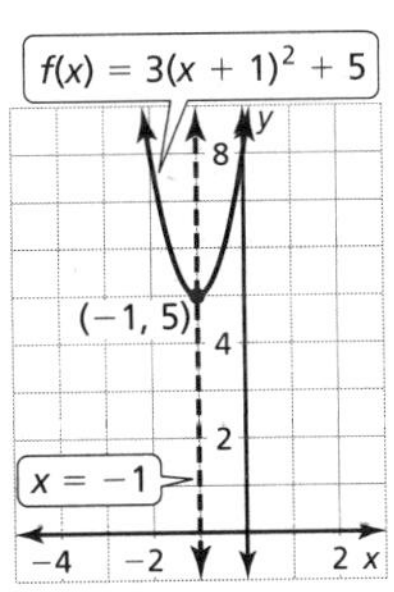

3. 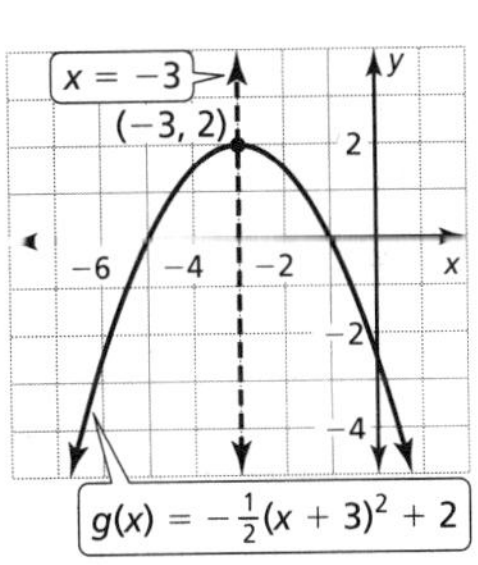

4. 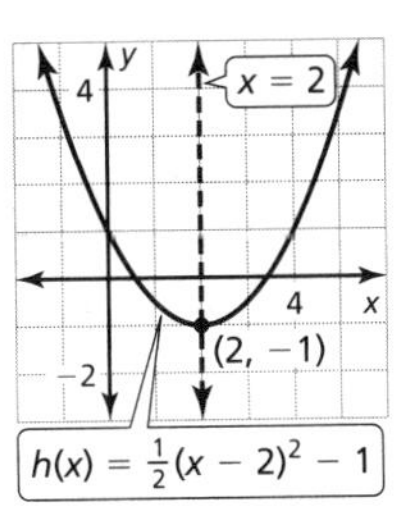

5. 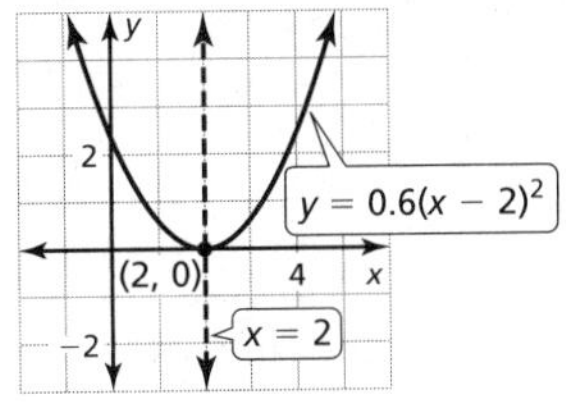

6. 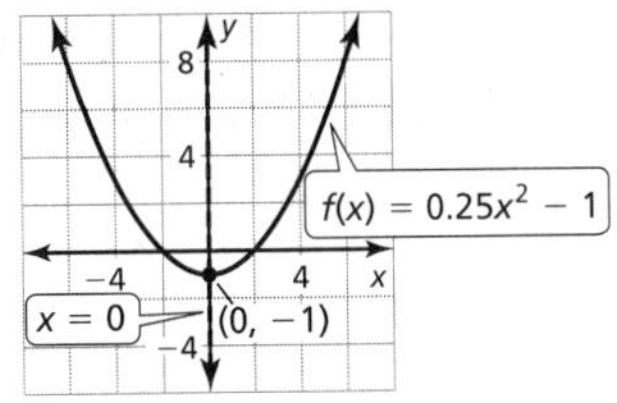

7. 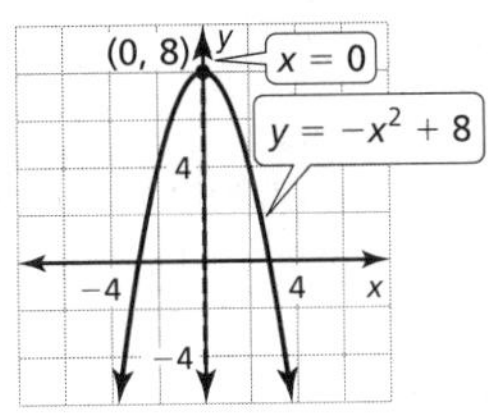

8. 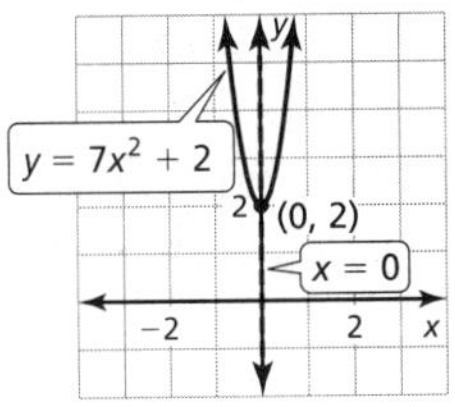

9. 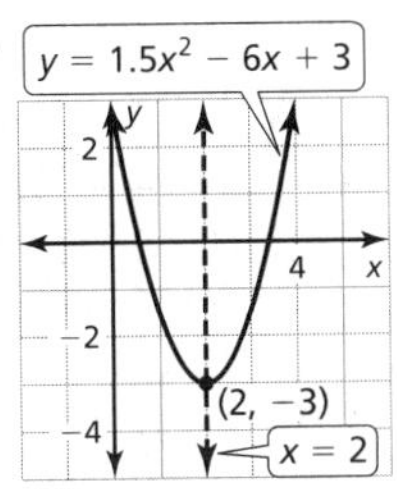

10. 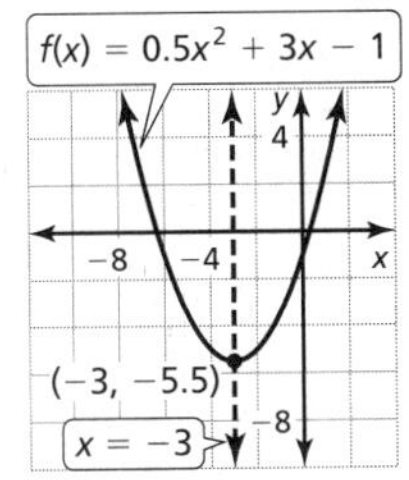

11. 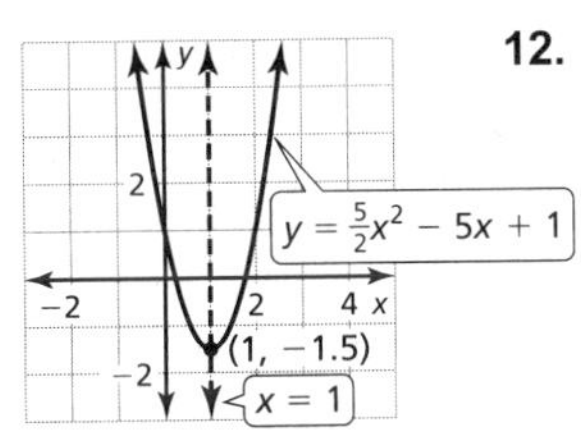

12. 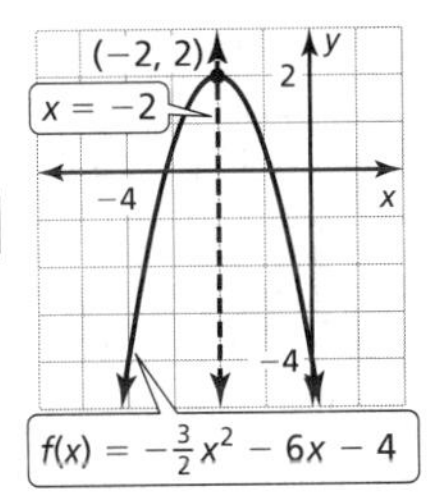

13. lowest; The y-values on either side of $x = 3$ are greater than 3.

14. A; Both have an axis of symmetry of $x = 2$.

15. minimum: 12; domain: all real numbers, range: $y \geq 12$; increasing to the right of $x = 0$; decreasing to the left of $x = 0$

16. maximum: 9; domain: all real numbers, range: $y \leq 9$; increasing to the left of $x = -3$; decreasing to the right of $x = -3$

Answers

17. maximum: 6; domain: all real numbers, range: $y \leq 6$; increasing to the left of $x = -3$; decreasing to the right of $x = -3$

18. minimum: 2.5; domain: all real numbers, range: $y \geq 2.5$; increasing to the right of $x = -3$; decreasing to the left of $x = -3$

19. a. The maximum height occurs $\frac{1}{6}$ mile from the base of the bridge.

b. The maximum height is $\frac{1}{12}$ mile.

20. The graph of the equation is a parabola with a vertex of $(2, 18)$.

21. The graph of the equation is a circle with a radius of $4\sqrt{3}$ and center $(2, -6)$.

2.6 Enrichment and Extension

1.

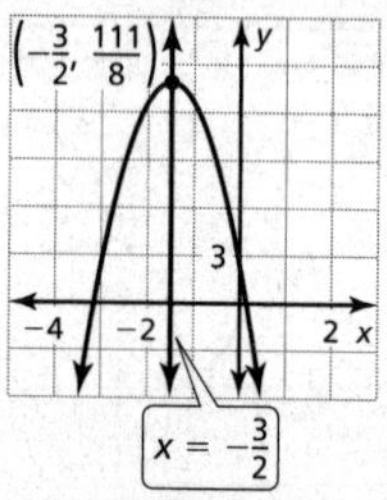

$x = -\frac{3}{2}$

2.

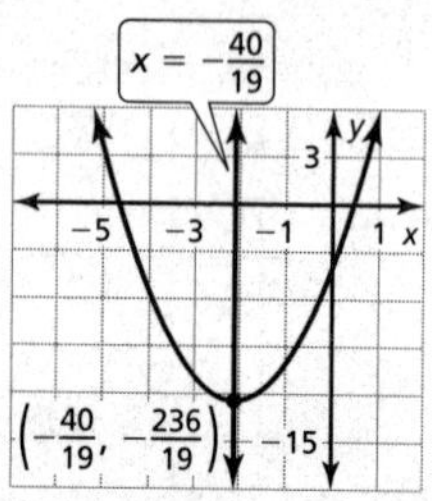

$x = -\frac{40}{19}$

3.

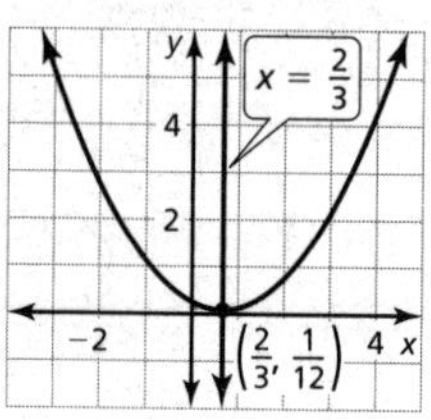

$x = \frac{2}{3}$

4.

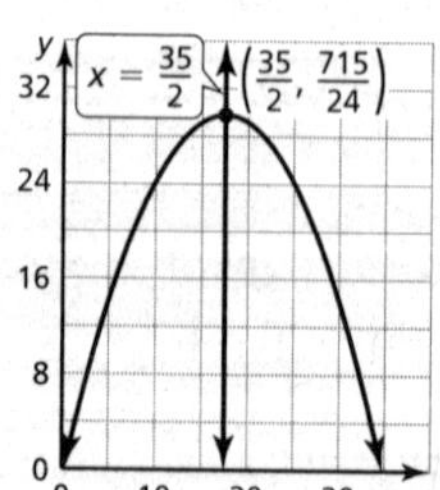

$x = \frac{35}{2}$

5. The x-coordinate of the vertex of the quadratic function is the solution of the related linear equation; The coefficient of the x^2-term is one-half of the coefficient of the x-term of the related linear equation; The coefficient of the x-term of the quadratic function is the same as the constant term of the related linear equation.

6. a. $8x - 5 = 0; \left(\frac{5}{8}, \frac{167}{16}\right)$

b. $-24x + 6 = 0; \left(\frac{1}{4}, -\frac{17}{4}\right)$

c. $-6x - 1 = 0; \left(-\frac{1}{6}, \frac{85}{12}\right)$

7.
$$\begin{aligned} f(x) &= ax^2 + bx + c \\ &= a\left(x^2 + \frac{b}{a}x + \frac{b^2}{4a^2}\right) - \frac{b^2}{4a} + c \\ &= a\left(x + \frac{b}{2a}\right)^2 - \frac{b^2}{4a} + c \\ &= a(x - h)^2 + k \end{aligned}$$

So, the vertex (h, k) is $\left(-\frac{b}{2a}, -\frac{b^2}{4a} + c\right)$.

8. $\frac{p^2}{8}$ ft^2

2.6 Puzzle Time

THIS BRITISH ROWER WAS THE FIRST WOMAN TO ROW ACROSS THREE OCEANS.

2.7 Start Thinking

Sample answer:

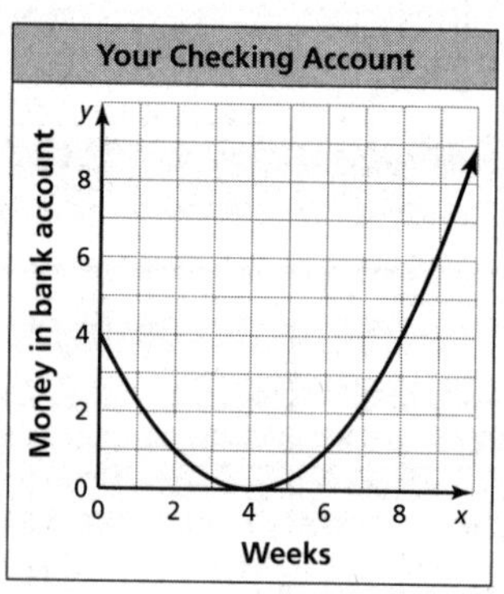

$y = 110.25$; After 25 weeks, you have \$110.25 in your bank account.

2.7 Warm Up

1. $y = -\frac{1}{6}(x - 6)$

2. $y - 3 = \frac{1}{2}(x - 1)$

3. $y + 1 = -2(x - 4)$

4. $y + 3 = 3(x - 3)$

Answers

5. $y + 18 = -\frac{1}{4}(x - 4)$ **6.** $y + 1 = -3(x - 6)$

2.7 Cumulative Review Warm Up

1.

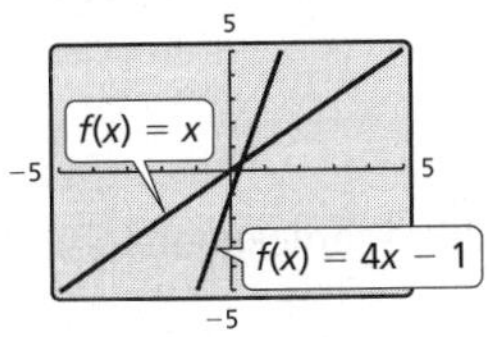

The graph is a vertical stretch by a factor of 4, followed by a translation 1 unit down of its parent function.

2.

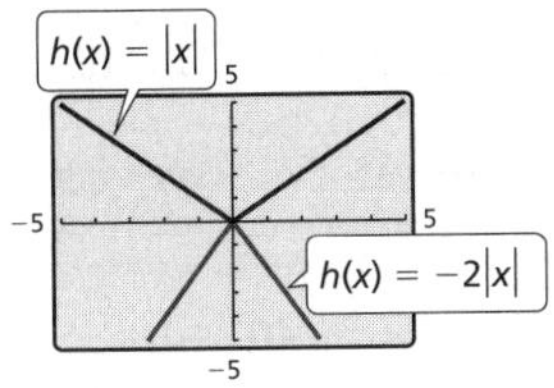

The graph is a vertical stretch by a factor of 2, followed by a reflection in the x-axis of its parent function.

3.

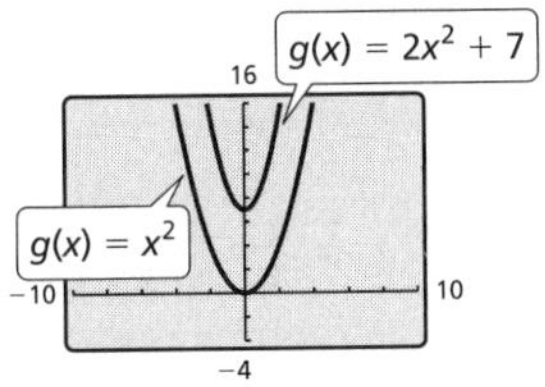

The graph is a vertical stretch by a factor of 2, followed by a translation 7 units up of its parent function.

4.

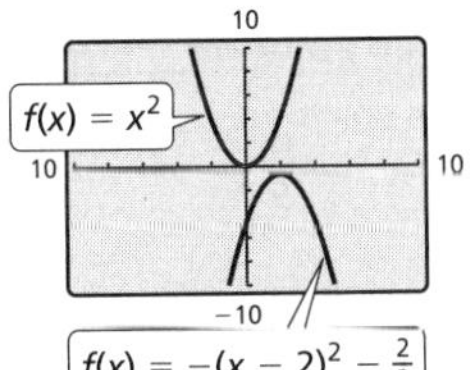

The graph is a reflection in the x-axis, followed by a translation 2 units right and $\frac{2}{3}$ unit down.

2.7 Practice A

1. $y = \frac{7}{16}(x - 2)^2 - 3$ **2.** $y = -\frac{1}{18}(x - 3)^2 - 8$

3. $y = -9(x + 1)^2 + 4$ **4.** $y = \frac{8}{5}(x - 10)(x - 6)$

5. $y = \frac{3}{16}(x - 2)(x - 8)$

6. $y = -\frac{2}{7}(x + 14)(x + 2)$

7. a. $y = -\frac{5}{9}(x - 1)^2 + 5$

b. $y = -\frac{5}{9}x^2 + \frac{10}{9}x + \frac{40}{9}$

c. $y = -\frac{5}{9}(x + 2)(x - 4)$

d. $y = -\frac{5}{9}x^2 + \frac{10}{9}x + \frac{40}{9}$

e. yes; intercept form; Two intercepts were given.

8. $6 = -0.069x^2 + 1.25x + 5$;
$17.277 - 0.839 \approx 16$ sec; about 21.5 sec

2.7 Practice B

1. $y = -\frac{1}{9}(x - 1)^2 - 6$ **2.** $-\frac{9}{49}(x + 2)^2 + 5$

3. $y = \frac{1}{3}(x + 1)^2 - 1$

4. $y = -\frac{5}{3}(x - 12)(x - 8)$

5. $y = \frac{1}{16}(x + 7)(x + 1)$

6. $y = -\frac{4}{81}(x + 9)(x - 9)$

7. The two given sets of coordinates were not substituted into the correct places.

$$y = a(x - h)^2 + k$$
$$-7 = a(1 - 3)^2 - 5$$
$$-7 = 4a - 5$$
$$-2 = 4a$$
$$-\frac{1}{2} = a$$

The equation is $y = -\frac{1}{2}(x - 3) - 5$.

8. a. The maximum area of 2500 square feet occurs when the length is 50 feet.

b. $A(x) = -x^2 + 100x$; $A(2) = 196$ ft^2

c. 0 to 50 ft: $50\frac{\text{ft}^2}{\text{ft}}$; 50 to 100 ft: $-50\frac{\text{ft}^2}{\text{ft}}$

2.7 Enrichment and Extension

1. linear; $y = -1.8x + 212.10$

Answers

2. quadratic
   a. $y = -4.9x^2 + 19.6x + 58.8$
   b. $y = -4.9(x - 2)^2 + 78.4$
   c. *Sample answer:* The graph of the function is a reflection in the x-axis, followed by a vertical stretch by a factor of 4.9 and a translation 2 units to the right and 78.4 units up of its parent function.
3. quadratic
   a. $y = -3x^2 + 30x + 12$
   b. $y = -3(x - 5)^2 + 87$
   c. *Sample answer:* The graph of the function is a reflection in the x-axis, followed by a vertical stretch by a factor of 3 and a translation 5 units to the right and 87 units up of its parent function.
4. linear; $y = 0.433x + 14.7$
5. quadratic
   a. $y = -5x^2 + 14x + 3$
   b. $y = -5(x - 1.4)^2 + 12.8$
   c. *Sample answer:* The graph of the function is a reflection in the x-axis, followed by a vertical stretch by a factor of 5 and a translation 1.4 units to the right and 12.8 units up of its parent function.

2.7 Puzzle Time

PINK FLAMINGO

Cumulative Review

1. $x = -18$ **2.** $x = -15$ **3.** $x = -12$

4. $x = -26$ **5.** $x = 75$ **6.** $x = -2$

7. $x = 3$ **8.** $x = 12$ **9.** $x = 21$

10. $x = 9$ **11.** $x = 8$ **12.** $x = 6$

13. $x = 26$ **14.** $x = -7$ **15.** $x = 8$

16. $x = -45$

17. **a.** 89%
b. 95%
c. 55 correct answers
d. 60 correct answers

18. **a.** 89%
b. 75%
c. 24 correct answers
d. 26 correct answers
e. 22 correct answers

19. 3 cups **20.** 4 tablespoons

21. 5.83 **22.** 7.07 **23.** 2.83

24. 10.44 **25.** 18.87 **26.** 7.07

27. 12.37 **28.** 7.21 **29.** 10.44

30. 19.24 **31.** 10.00 **32.** 10.00 **33.** 12.08

34. 17.03 **35.** 12.04 **36.** $\frac{35}{3}$ **37.** $-\frac{9}{4}$

38. $5\frac{4}{7}$ or $\frac{39}{7}$ **39.** -2 **40.** -10

41. 8 **42.** 5 **43.** -3 **44.** 2

45. $x = \frac{1}{8}y - 3$ **46.** $x = 2y - 6$

47. $x = -\frac{1}{5}y + 7$ **48.** $x = 4y + 7$

49. $x = 2y + \frac{8}{3}$ **50.** $x = \frac{3}{2}y - \frac{5}{4}$

51. $x = 2y + 10$ **52.** $x = -\frac{1}{4}y - 5$

53. $x = 3y - 8$ **54.** $x = \frac{1}{4}$

55. no solution **56.** no solution

57. no solution **58.** $x = \frac{28}{9}$

59. $x = 6$ **60.** $x = \frac{81}{2}$

61. $x = \frac{100}{3}$ **62.** $x = 1$

63. $x = \frac{100}{9}$ **64.** $x = 8$

65. no solution **66.** 5 h

67. 2.25 h **68.** about 52.5 h

69. $y = -\frac{1}{8}x - \frac{5}{8}$ **70.** $y = 4x - 41$

71. $y = \frac{1}{2}x + 4$ **72.** $y = -\frac{7}{2}x - \frac{19}{2}$

Answers

73. $y = -\frac{13}{10}x + \frac{22}{5}$

74. $y = -\frac{3}{7}x + \frac{20}{7}$

75. $y = \frac{3}{5}x + \frac{37}{5}$

76. $y = \frac{19}{8}x - \frac{21}{4}$

77. $y = -4x + 17$

78. $y = -\frac{5}{4}x - 2$

79. $y = \frac{11}{8}x - \frac{23}{2}$

80. $y = -\frac{3}{4}x + \frac{21}{4}$

81. $y = -\frac{3}{14}x + \frac{41}{14}$

82. $y = \frac{1}{7}x - \frac{27}{7}$

83. $y = -\frac{5}{4}x + \frac{21}{4}$

84. $(x + 3)(x - 2)$

85. $(x + 6)(x - 7)$

86. $(x - 11)(x + 7)$

87. $(x + 1)(x + 12)$

88. $(x - 6)(x + 4)$

89. $(x + 5)(x - 1)$

90. $(x + 6)(x + 5)$

91. $(x - 12)(x - 8)$

92. $(2x + 7)(x - 5)$

93. $(3x + 10)(x - 4)$

94. $(2x - 1)(x + 10)$

95. $(3x - 2)(x - 10)$

96. $(4x - 12)(x + 12)$

97. $(5x - 3)(x + 2)$

98. $(2x - 4)(3x + 11)$

99. $y = -x^2 - 4x + 32$

100. $y = 4x^2 + 4x - 48$

101. $y = -2x^2 - 2x + 4$

102. $y = -6x^2 + 6x + 72$

103. $y = -5x^2 + 55x - 120$

104. $y = 3x^2 - 9x - 84$

105. $y = 6x^2 - 36x + 54$

106. $y = -7x^2 - 14x - 7$

107. $y = 2x^2 - 16x + 32$

108. $y = x^2 - 8x + 17$

109. $y = x^2 + 10x + 18$

110. $y = x^2 - 16x + 73$

111. $y = 3x^2 + 6x + 7$

112. $y = -7x^2 - 70x - 178$

113. $y = -9x^2 + 54x - 74$

114. 6 in.

115. 42 ft^2

116. 18 in.

117. 5184 $in.^2$

118. a. about 64.47 cm^3; The volume is approximately equal to the volume of the sphere minus the volume of the cylindrical hole.

b. no; yes; For the surface area, the lateral area of the cylinder is added, while the areas of the bases of the cylinder are subtracted. Because the lateral surface area is greater than the total area of the bases, the surface area of the wooden ball increases after the hole is made. The volume decreases because part of the sphere is removed.

119. $x = 2 \pm 3i\sqrt{6}$

120. $x = -1 \pm \sqrt{15}$

121. $x = \dfrac{-7 \pm \sqrt{139}}{2}$

122. $x = \dfrac{5 \pm i\sqrt{47}}{6}$

123. $x = \dfrac{7 \pm \sqrt{129}}{-8}$

124. $x = \dfrac{2 \pm 2i\sqrt{39}}{5}$

125. $x = \dfrac{7 \pm i\sqrt{47}}{12}$

126. $x = \dfrac{-13 \pm i\sqrt{167}}{-8}$

127. $x = \dfrac{-3 \pm i\sqrt{55}}{4}$

128. $x = \dfrac{-5 \pm i\sqrt{119}}{-4}$

129. $x = \dfrac{-8 \pm i\sqrt{38}}{-3}$

130. $x = \dfrac{15 \pm i\sqrt{223}}{-8}$

131. $x = \dfrac{9 \pm \sqrt{103}}{2}$

132. $x = \dfrac{3 \pm i\sqrt{111}}{12}$

133.

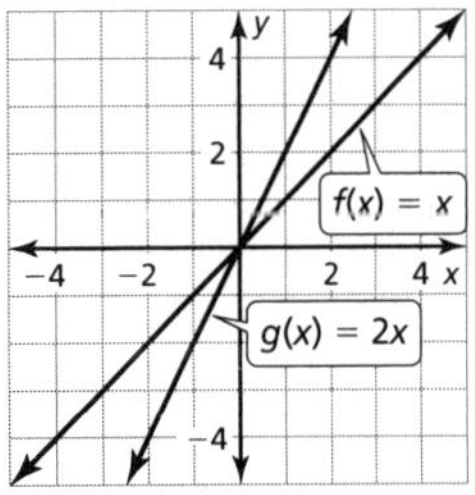

The graph of $g(x)$ is a vertical stretch by a factor of 2 of the graph of $f(x)$.

Answers

134.

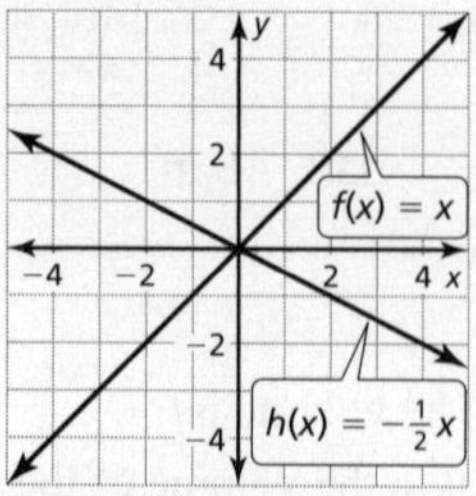

The graph of $h(x)$ is a vertical shrink by a factor of $\frac{1}{2}$ and a reflection in the x-axis of the graph of $f(x)$.

135.

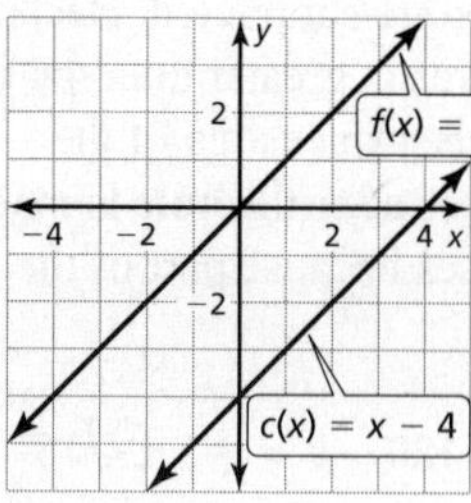

The graph of $c(x)$ is a vertical translation 4 units down of the graph of $f(x)$.

136.

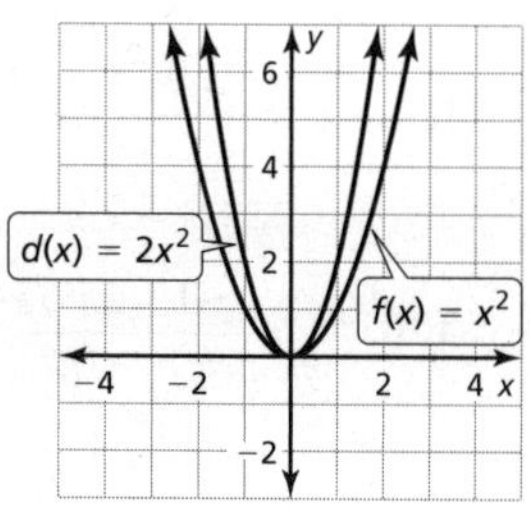

The graph of $d(x)$ is a vertical stretch by a factor of 2 of the graph of $f(x)$.

137.

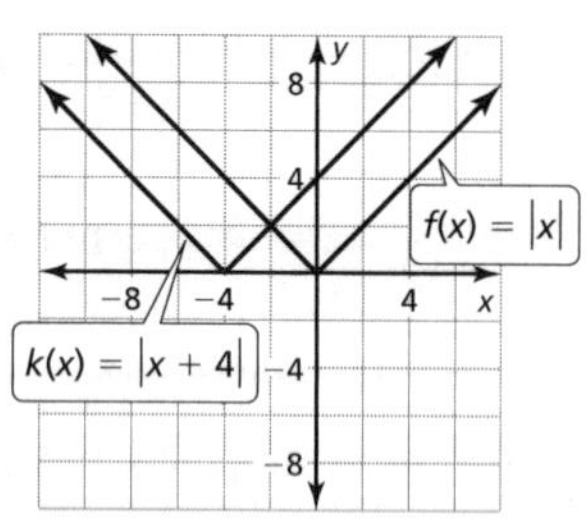

The graph of $k(x)$ is a horizontal translation 4 units left of the graph of $f(x)$.

138. The graph of $m(x)$ is a translation 3 units right and 2 units up of the graph of $f(x)$.

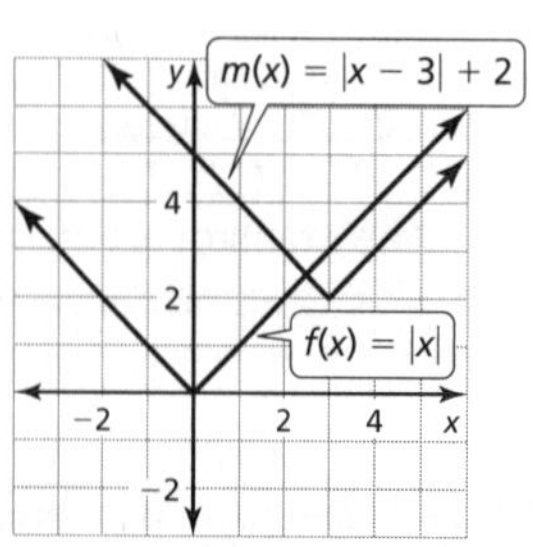

139. $g(x) = -2x - 3$

140. $g(x) = \frac{4}{7}x + 1$

141. $g(x) = |x + 2| + 4$

142. $g(x) = \frac{1}{2}|x - 6| - 7$

143. $g(x) = -|x - 3| - 5$

144. $g(x) = -|x + 1| + 2$

145. $g(x) = -3x - 1$

146. $g(x) = -\frac{3}{4}x + 5$

147. The graph of f is a vertical stretch by a factor of 2 followed by a translation 4 units left and 3 units up of the graph of the parent quadratic function; $(-4, 3)$

148. The graph of f is a vertical stretch by a factor of 4 followed by a reflection in the x-axis and a translation 6 units up of the graph of the parent quadratic function; $(0, 6)$

149. The graph of f is a vertical shrink by a factor of $\frac{1}{4}$ followed by a translation 2 units right of the graph of the parent quadratic function; $(2, 0)$

Chapter 3

3.1 Start Thinking

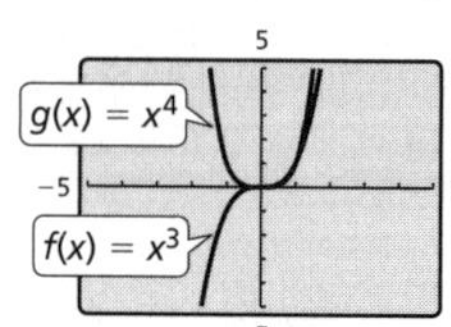

The graph of $f(x) = x^3$ is a curvy line that is moving upward from left to right as x increases. The graph of $g(x) = x^4$ is similar to a parabola that opens up with its vertex at the origin. Both graphs have positive y-values when x is positive. When x is negative, the y-values of $f(x)$ are negative, and the y-values of $g(x)$ are positive; The exponent is even; yes; $(0, 0)$ and $(1, 1)$

Answers

3.1 Warm Up

1. −20 **2.** 39 **3.** 10

4. 40 **5.** −45.8 **6.** 33.96

3.1 Cumulative Review Warm Up

1. down **2.** up **3.** down **4.** up

3.1 Practice A

1. polynomial function;
$f(x) = 5x^3 + 4x^2 - 3x - 7$, degree is 3, cubic, leading coefficient is 5

2. not a polynomial function

3. polynomial function;
$g(x) = x^4 - 4x^3 - \frac{1}{3}x^2 + 2x + 10$, degree is 4, quartic, leading coefficient is 1

4. polynomial function; $f(x) = 8x^2 - \sqrt{3}x + 2$, degree is 2, quadratic, leading coefficient is 8

5. −2 **6.** 6848 **7.** 15,651

8. $g(x) \to +\infty$ as $x \to +\infty$ and $g(x) \to +\infty$ as $x \to -\infty$.

9. $h(x) \to -\infty$ as $x \to +\infty$ and $h(x) \to +\infty$ as $x \to -\infty$.

10.
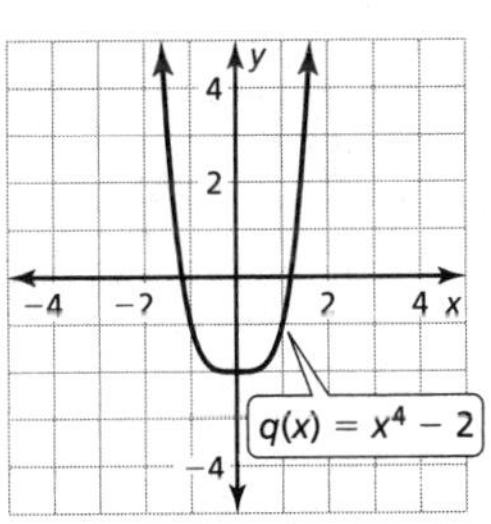

11.
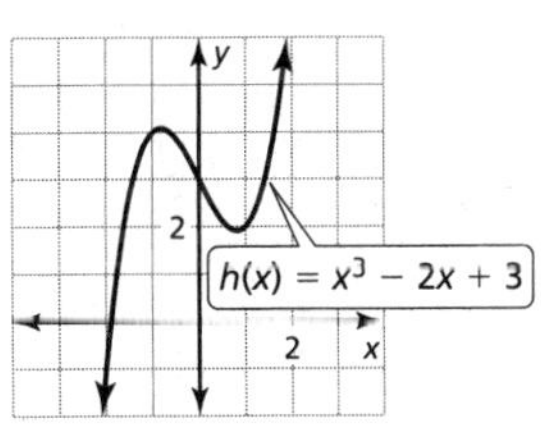

12.
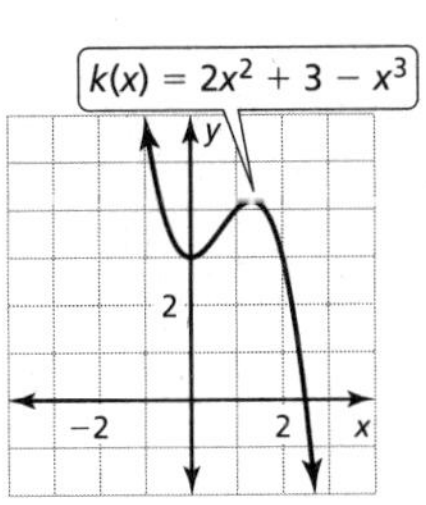

13.
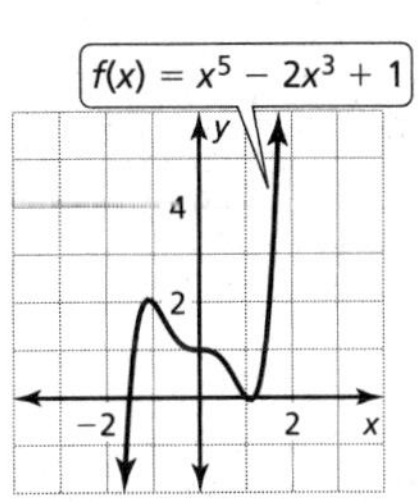

14. **a.** f is increasing when $x > 3$. f is decreasing when $x < 3$.
b. f is positive when $x < 2$ and $x > 4$.
c. f is negative when $2 < x < 4$.

15. **a.** f is increasing when $x < -1.2$ and $x > 1.2$. f is decreasing when $-1.2 < x < 1.2$.
b. f is positive when $-2 < x < 0$ and $x > 2$.
c. f is negative when $x < -2$ and $0 < x < 2$.

16. The degree is even and the leading coefficient is negative.

3.1 Practice B

1. not a polynomial function

2. polynomial function; $f(x) = 11x^2 + 12x - \sqrt{7}$, degree is 2, quadratic, leading coefficient is 11

3. polynomial function;
$g(x) = 2x^4 - \sqrt{14}x^3 - \frac{1}{3}x^2 + 2x - \frac{5}{3}$, degree is 4, quartic, leading coefficient is 2

4. not a polynomial function

5. 1841 **6.** $-\frac{47}{9}$ **7.** $-\frac{5}{8}$

8. $g(x) \to -\infty$ as $x \to +\infty$ and $g(x) \to -\infty$ as $x \to -\infty$.

9. $h(x) \to +\infty$ as $x \to +\infty$ and $h(x) \to -\infty$ as $x \to -\infty$.

10.
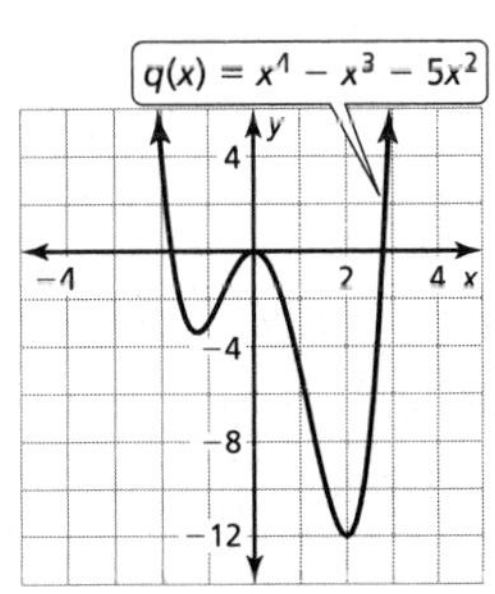

11.
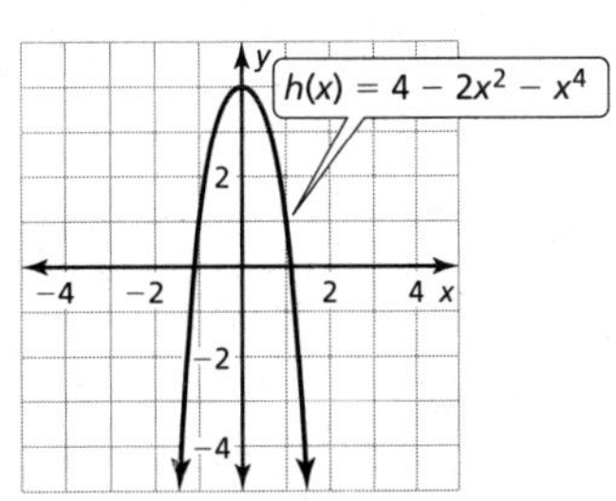

12.

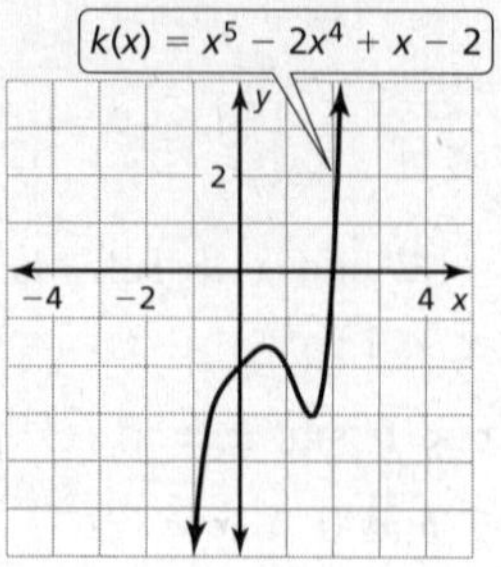

13.

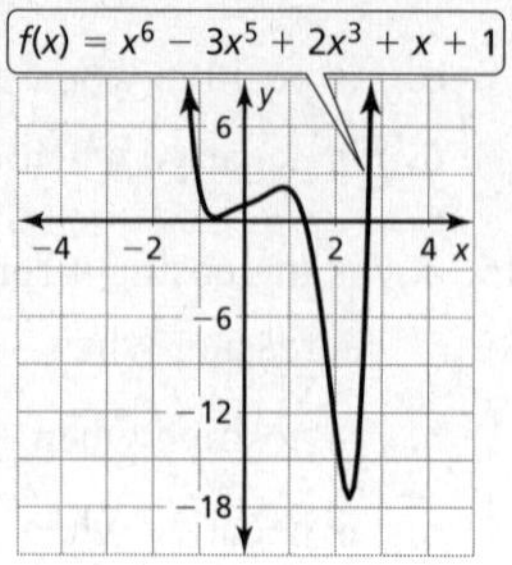

14.

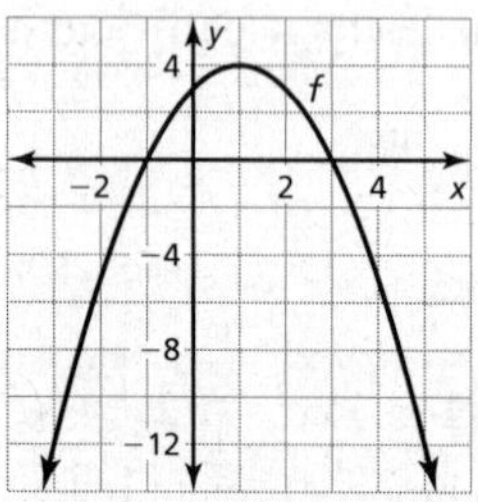

The degree is even and the leading coefficient is negative.

15.

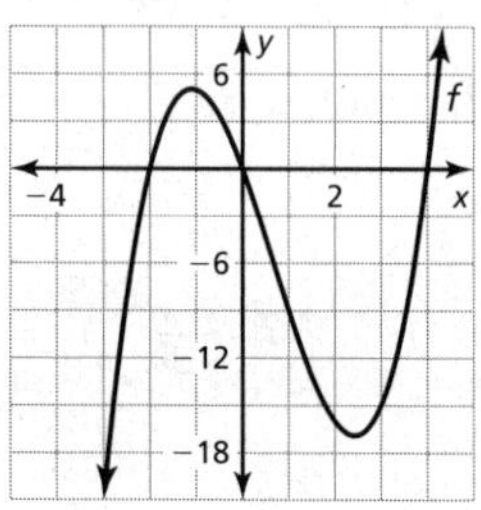

The degree is odd and the leading coefficient is positive.

16. *Sample answer*: $-2 \le x \le 8; -5 \le y \le 50$

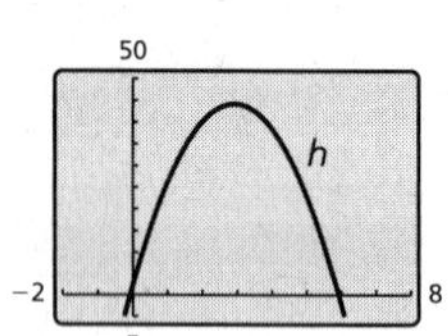

3.1 Enrichment and Extension

1. $y \approx 1.17x^4 - 14.33x^3 + 60.83x^2 - 102.67x + 56$

2. $y \approx 0.12x^4 - 2.89x^3 + 23.51x^2 - 76.96x + 84$

3. $y \approx -0.29x^4 + 1.07x^3 + 1.96x^2 - 5.93x - 0.60$

4. $y \approx 1.39 \times 10^{-4}x^4 - 0.03x^3 + 2.78x^2 - 88.57x + 969.47$

3.1 Puzzle Time

COBWEBS

3.2 Start Thinking

Sample answer:

$(x + 1)(x - 1) = x^2 - x + x - 1 = x^2 - 1;$

$(x + 3)(x - 3) = x^2 - 3x + 3x - 9 = x^2 - 9;$

yes; In each example, the middle terms cancel out, leaving only two terms. The first term is the square of the first term in each binomial. The second term is the square of the second term in each binomial;

$(x + 1)(x + 1) = x^2 + x + x + 1 = x^2 + 2x + 1;$

$(x - 3)(x - 3) = x^2 - 3x - 3x + 9 = x^2 - 6x + 9;$

no; The signs are the same inside the binomials, so the middle terms no longer cancel.

3.2 Warm Up

1. 7
2. $-x + 7$
3. $28m + 21$
4. $10r$
5. $-3z^2 - 2z$
6. $-2m - 4p$

3.2 Cumulative Review Warm Up

1. $\frac{25}{4}; \left(x + \frac{5}{2}\right)^2$
2. $9; (z + 3)^2$
3. $36; (w - 6)^2$
4. $\frac{625}{4}; \left(x - \frac{25}{2}\right)^2$
5. $16; (x - 4)^2$
6. $\frac{729}{4}; \left(s + \frac{27}{2}\right)^2$

3.2 Practice A

1. $4x^2 + 7x - 9$
2. $7x^5 + 5x^4 + 3x^2 - 3x - 5$
3. $5x^4 + 2x^3 - 4x^2 - 9$
4. $-4x^3 + 4x^2 - 4x + 2$
5. $8x^4 + x^3 - 3x^2 - 4x + 6$
6. $7x^5 - 6x^4 + 13x^3 - 3x^2 + 12x + 8$
7. $7x^2 + 9x - 8$
8. $15x^4 + 35x^3 + 30x^2$

Answers

9. $-20x^7 + 18x^6 + 14x^5 - 8x^4$

10. $-24x^3 + 25x^2 - 9x + 2$

11. $-3x^3 - 20x^2 - 21x - 54$

12. The negative was distributed incorrectly;
$-3x^2(4x^2 - 5x + 7) = -12x^4 + 15x^3 - 21x^2$

13. $x^3 - 13x + 12$

14. $x^3 - 13x^2 + 24x + 108$

15. $4x^3 + 8x^2 - 15x - 9$

16. $12x^3 - 25x^2 - 87x - 20$

17. $x^2 - 64$

18. $y^2 + 8y + 16$

19. $4p^2 - 12p + 9$

20. $k^4 + 12k^3 + 54k^2 + 108k + 81$

3.2 Practice B

1. $8x^7 + 15x^6 - 2x^5 + x^3 - 6x + 2$

2. $14x^4 - 7x^3 - 4x + 5$

3. $-3x^5 + 3x^4 - 8x^2 + 10x + 9$

4. $9x^4 + 5x^3 - 6x^2 - 7x + 11$

5. $x^4 - 10x^3 + 13x^2 + 48x + 12$

6. $-10x^4 - 19x^3 + 7x^2 + 14x - 4$

7. $4x^4 - 11x^3 + 20x^2 - 18x + 12$

8. $3x^6 - 6x^5 + x^4 + 3x^3 - 40x^2 \quad 25x$

9. The exponents were multiplied instead of added;
$4x^2(3x^4 - 2x^3 + 7) = 12x^6 - 8x^5 + 28x^2$

10. $6x^3 - 14x^2 - 14x + 6$

11. $8x^3 - 26x^2 - 67x - 15$

12. $-16x^3 + 12x^2 + 28x - 15$

13. $8x^3 - 30x^2 + 13x + 30$

14. $9x^2 - 25$

15. $36t^2 + 84t + 49$

16. $p^2q^2 + 4pq + 4$

17. a. *Sample answer*: $(3x - 1)(x + 6)^2$ ft^3

b. $3x^3 + 35x^2 + 96x - 36$ ft^3

18. $32m^4 - 400m^4 + 2000m^3 - 5000m^2 + 6250m - 3125$

19. $27s^3 + 27s^2t + 9st^2 + t^3$

3.2 Enrichment and Extension

1. 1 path; 3 paths; 3 paths; 1 path

2. *Sample answer:* The figure is symmetrical.

3. yes; *Sample answer:* The number of paths that a chip can travel to a given peg is the same as the corresponding entry in Pascal's Triangle.

4. H: 1 path, I: 9 paths, J: 36 paths, K: 84 paths, L: 126 paths, M: 126 paths, N: 84 paths, O: 36 paths, P: 9 paths, Q: 1 path

5. *Sample answer:* There are more paths to the center bins than to the other bins.

3.2 Puzzle Time

A WALKIE TALKIE

3.3 Start Thinking

$(x + 2)(x - 1)$; Inverse operations undo one another, so if two binomials are multiplied to make a product, you can divide the product by one binomial to obtain the other binomial; no; Factoring will only work as division if there is no remainder. It is possible to divide polynomials that are not factorable.

3.3 Warm Up

1. $13(t + 3y)$

2. $3k(k - 1)$

3. $ab(5ab - a + 11b)$

4. $(x + 5)(x - 5)$

5. $(n - 11)(n - 2)$

6. $3(x + 7)(x + 3)$

3.3 Cumulative Review Warm Up

1. $g(x) = 3x - 2$

2. $g(x) = \frac{1}{4}x + 1$

3. $g(x) = \left|\frac{x}{3} + \frac{1}{3}\right|$

Answers

3.3 Practice A

1. $x + 6 + \dfrac{42}{x - 5}$
2. $2x + 3 + \dfrac{5}{x - 2}$
3. $x + 1 + \dfrac{3}{x^2 - 9}$
4. $6x - 1 + \dfrac{2}{x^2 + 2}$
5. $x + 9 + \dfrac{28}{x - 3}$
6. $3x - 8 - \dfrac{12}{x - 1}$
7. $2x - 5 + \dfrac{15}{x + 2}$
8. $x^2 - 3x + 7 - \dfrac{15}{x + 3}$
9. $x + 5 + \dfrac{50}{x - 5}$
10. $5x + 2 + \dfrac{4}{x - 1}$
11. $k = -3$;

$$\begin{array}{r|rrrr} -3 & 1 & 2 & 0 & 7 \\ & & -3 & 3 & -9 \\ \hline & 1 & -1 & 3 & -2 \end{array}$$

$$\frac{x^3 + 2x^2 + 7}{x + 3} = x^2 - x + 3 - \frac{2}{x + 3}$$

12. 28
13. 41
14. 8
15. 18
16. $x^3 + x^2 - 3x + 3$; Multiply the result by $x + 1$.

3.3 Practice B

1. $x + 3 + \dfrac{6}{x^2 - 4}$
2. $4x^2 - 2x + 9 - \dfrac{17x}{x^2 + x - 4}$
3. $2x^2 + 10x + 14 + \dfrac{90x}{x^2 - 5x - 2}$
4. $4x - 7 - \dfrac{7}{x - 2}$
5. $x^2 - 3x + \dfrac{12}{x + 3}$
6. $x + 4 + \dfrac{32}{x - 4}$
7. $2x^2 - 7x + 7 - \dfrac{4}{x + 1}$
8. $x^3 + x^2 - 10x + 29 - \dfrac{102}{x + 4}$
9. $x^3 - 4x^2 + 24x - 140 + \dfrac{820}{x + 6}$
10. The powers in the quotient are too large by 1. The remainder (-2) was not divided by $(x + 3)$;

$$\begin{array}{r|rrrr} -3 & 1 & 2 & 0 & 7 \\ & & -3 & 3 & -9 \\ \hline & 1 & -1 & 3 & -2 \end{array}$$

$$\frac{x^3 + 2x^2 + 7}{x + 3} = x^2 - x + 3 - \frac{2}{x + 3}$$

11. 7
12. -10
13. 305
14. -95
15. $k = 4$

3.3 Enrichment and Extension

1. $y = -3x - 3$
2. $y = x + 5$
3. $y = 2x + 1$
4. $y = x + 6$
5. $y = -2x - 4$
6. $y = ax + b - ad$

3.3 Puzzle Time

IT WAS ALWAYS WILLING TO LEND AN EAR

3.4 Start Thinking

yes; You can group the terms with coefficients of 3 and 21 together, or you can group the terms with coefficients of -5 and 40 together; yes; You can group $3x$ and $21xy$ together, and you can group -5 and $40y$ together.

3.4 Warm Up

1. $3y$
2. $2rs$
3. $3x$
4. yz
5. ab
6. xy

3.4 Cumulative Review Warm Up

1.

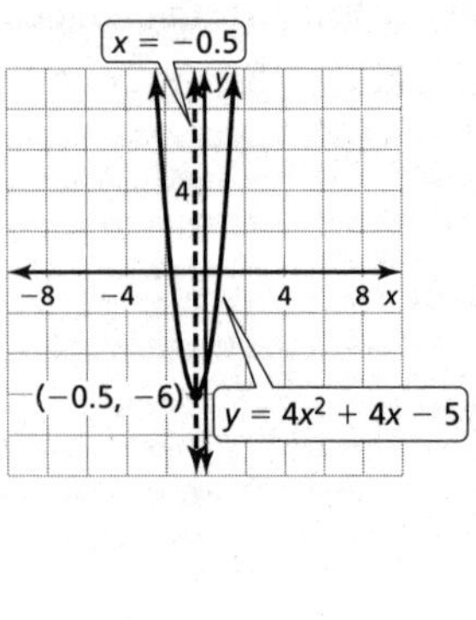

Answers

2.

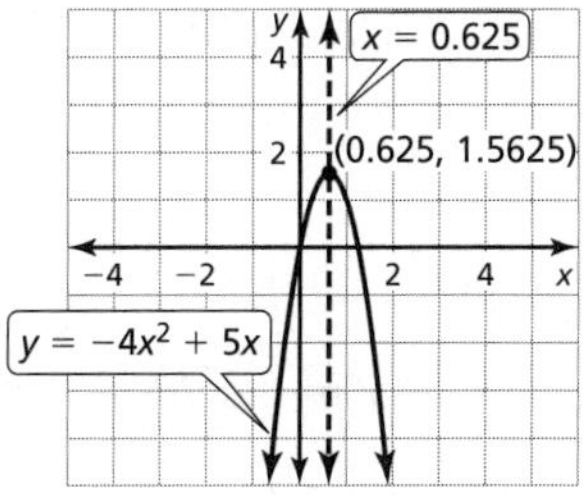

3.

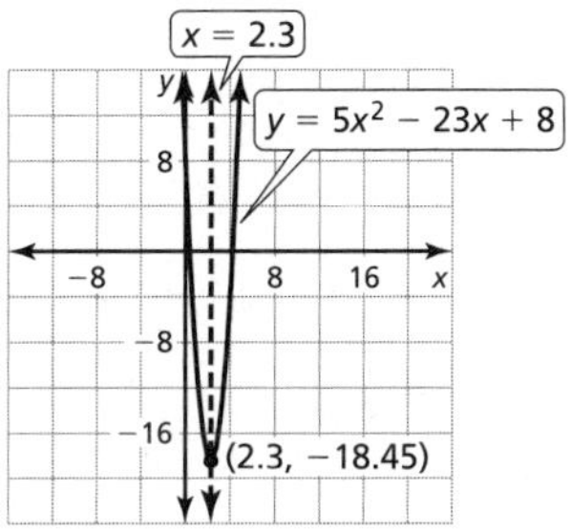

4.

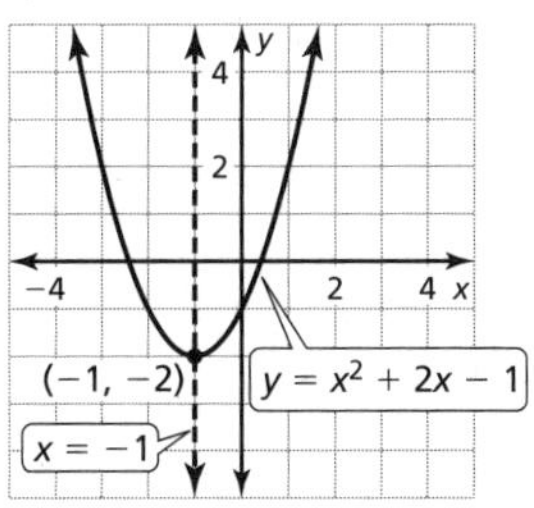

5.

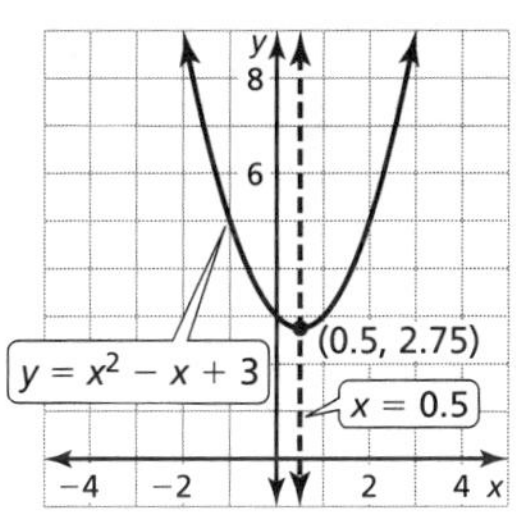

6.

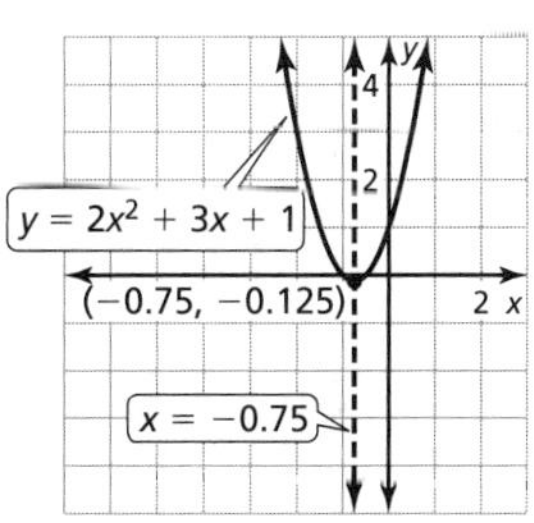

3.4 Practice A

1. $x(x-4)(x+3)$ **2.** $9p^5(p+2)(p-2)$

3. $3n^4(n-8)(n-3)$ **4.** $2k^2(k+11)(k-11)$

5. $w^2(2w+3)(w-5)$ **6.** $q^4(3q+4)(q-7)$

7. $(x+3)(x^2-3x+9)$

8. $(y+10)(y^2-10y+100)$

9. $(w-5)(w^2+5w+25)$

10. $(y-3)(y^2+4)$ **11.** $(q-2)(q^2+9)$

12. $(d+5)(2d^2+3)$ **13.** $(x-6)(x-3)(x+3)$

14. $(6p^2-5)(6p^2+5)$ **15.** $(n^2+4)(n^2+7)$

16. $(y^2+4)(y+2)(y-2)$

17. no **18.** yes **19.** yes **20.** no

21. *Sample answer*: $f(x) = x^3 + 5x^2 - 6x$; $(x-1)$;
$x(x-1)(x+6) = x^3 + 5x^2 - 6x$

22. $k = 9$; $f(x) = 3x^3 - 17x - 9x + 18$;
$3x^3 - 17x - 9x + 18 = (x-6)(3x^2 + x - 3)$

23. a. $(a^2+b^2)(5c-3d)$

b. $(x^n+3)^2$

3.4 Practice B

1. $5t^3(t+8)(t-8)$ **2.** $2p^4(p-7)(p-6)$

3. $3x^2(x+12)(x-12)$ **4.** $a^4(5a+9)(a-5)$

5. $j^7(2j-3)(6j-5)$ **6.** $q^8(3q+4)(5q+6)$

7. $2p^6(p-2)(p^2+2p+4)$

8. $25k^5(k+4)(k^2-4k+16)$

9. $2w^4(3w-2)(9w^2+6w+4)$

10. $(x-7)(x^2+5)$

11. $(m-2)(m+4)(m-4)$

12. $(w-3)(3w+2)(3w-2)$

13. $(s+4)(5s+1)(5s-1)$

14. $(9g^2 + 25)(3g + 5)(3g - 5)$

15. $2t^2(t^3 + 5)(t^3 - 2)$ **16.** $5v^2(v^4 - 3)(v^4 - 2)$

17. yes **18.** no

19. no **20.** no

21. *Sample answer:* $f(x) = 2x^3 + 7x^2 - 4x$; $(x + 4)$; $x(x + 4)(2x - 1) = 2x^3 + 7x^2 - 4x$

22. a. $(x - (-4))^2 + y^2 = 3^2$; $(h, k) = (-4, 0)$, $r = 3$

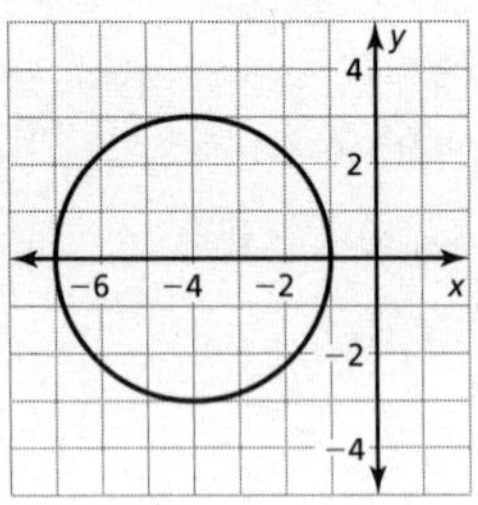

b. $(x - 5)^2 + y^2 = 2^2$; $(h, k) = (5, 0)$, $r = 2$

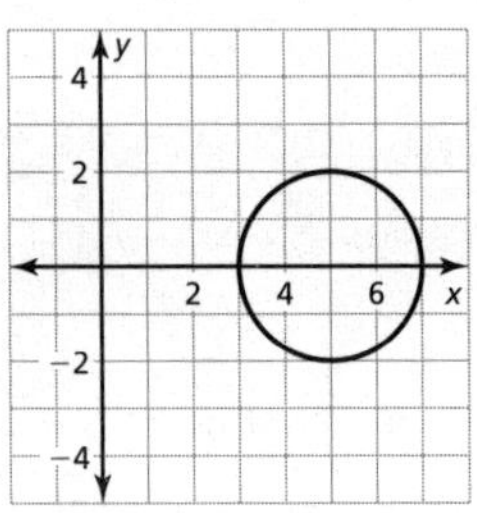

c. $(x - 2)^2 + (y - (-3))^2 = 4^2$; $(h, k) = (2, -3)$, $r = 4$

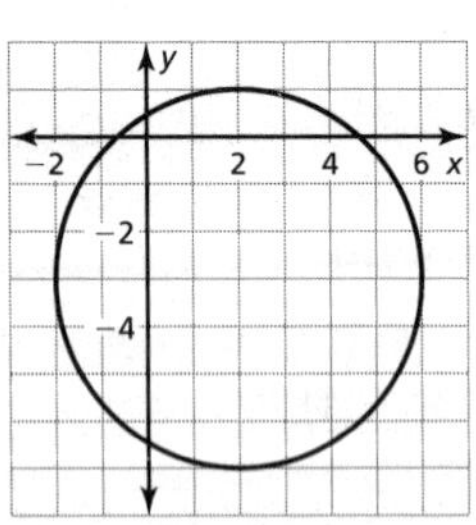

3.4 Enrichment and Extension

1. 27 **2.** −18 **3.** 3

4. −4 **5.** −1 **6.** 6

7. $(x^5 - y^5) = (x - y)(x^4 + x^3y + x^2y^2 + xy^3 + y^4)$

8. $(a^7 + b^7) = (a + b)(a^6 - a^5b + a^4b^2 - a^3b^3 + a^2b^4 - ab^5 + b^6)$

9. $(a^{14} - b^{14}) = (a^7 + b^7)(a^7 - b^7) = (a + b)(a - b)(a^6 - a^5b + a^4b^2 - a^3b^3 + a^2b^4 - ab^5 + b^6)(a^6 + a^5b + a^4b^2 + a^3b^3 + a^2b^4 + ab^5 + b^6)$

10. $(x^{10} - y^{10}) = (x^5 + y^5)(x^5 - y^5) = (x + y)(x - y)(x^4 - x^3y + x^2y^2 - xy^3 + y^4)(x^4 + x^3y + x^2y^2 + xy^3 + y^4)$

3.4 Puzzle Time

QUARTERBACK

3.5 Start Thinking

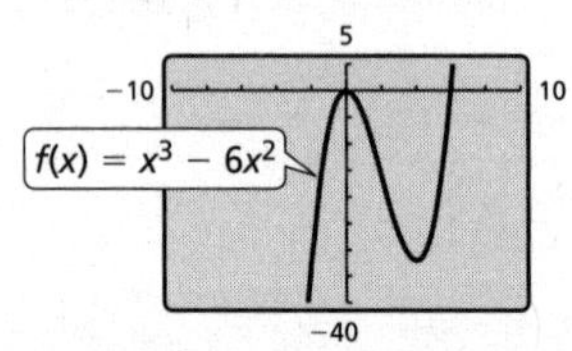

$(0, 0)$ and $(6, 0)$; The function simplifies to $0 = 0$; These points have x-values which yield a y-value of zero, meaning the graph crosses the x-axis. These are the only points that can be inserted into the function $f(x) = x^3 - 6x^2$ to get this result.

3.5 Warm Up

1. $t = -\frac{1}{4}$ **2.** $x = -5$

3. $r = -60$ **4.** $z = 20$

5. $m = -2$ **6.** $b = 20$

3.5 Cumulative Review Warm Up

1. $x < -\sqrt{5}$ or $x > \sqrt{5}$

2. $-9 \le x \le -3$ **3.** $x < -5$ or $x > -1$

4. $-\sqrt{5} < x < \sqrt{5}$ **5.** $x < -3$ or $x > 1$

6. $1 < x < 8$

Answers

3.5 Practice A

1. $q = -5, q = 0, q = 6$

2. $k = -3, k = 0$

3. $y = 0, y = 1$

4. $n = -3, n = -2, n = 3$

5. $p = 0, p = \pm\sqrt{7}$

6. $u = 0, u = \pm\sqrt{2}$

7. $x = -4, x = 0, x = 3$

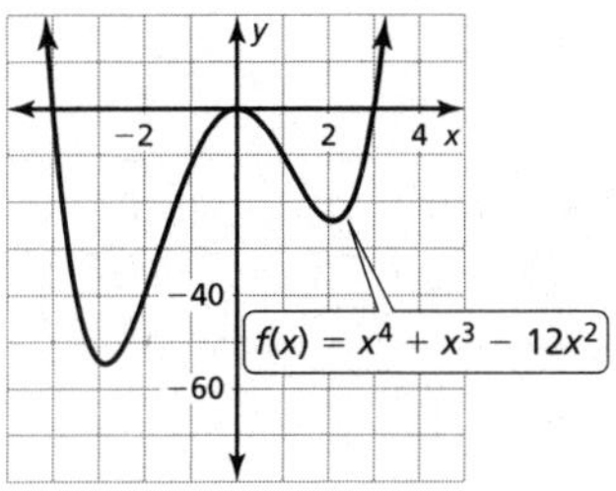

8. $x = -2, x = 2$

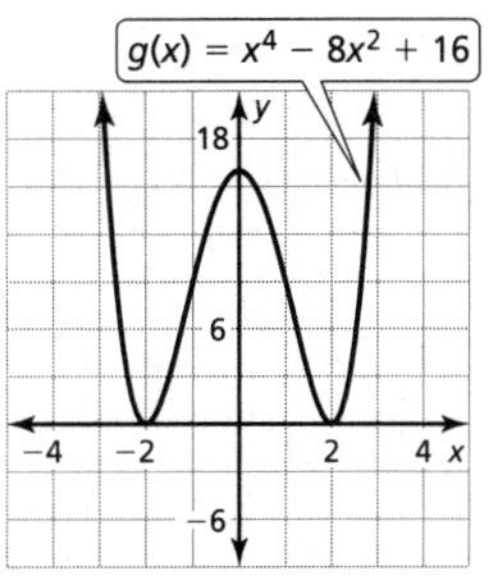

9. $x = -3, x = 0, x = 5$

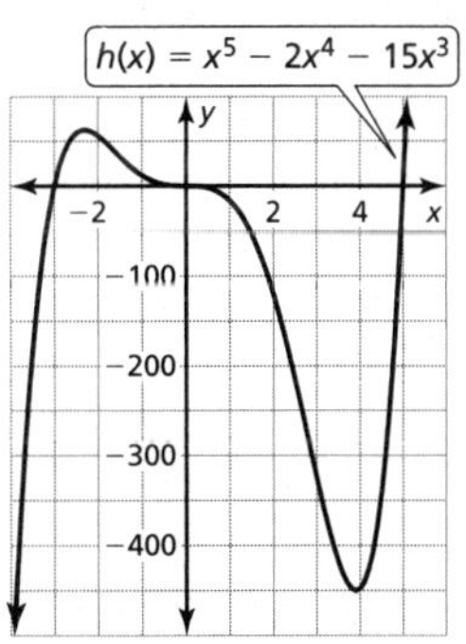

10. $x = -4, x = -1, x = 0$

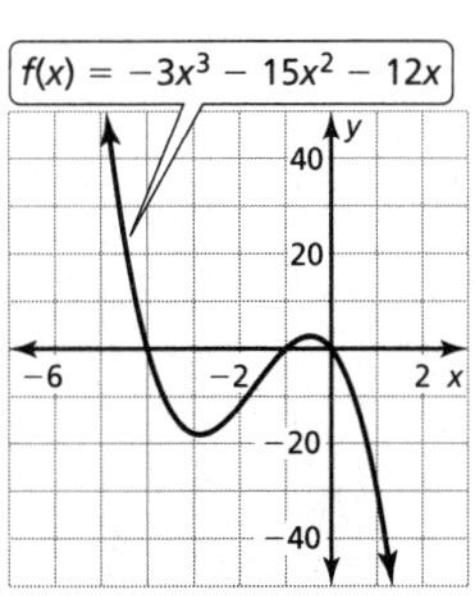

11. C

12. The factors of 18 include ± 1 and ± 18;

$f(x) = x^3 + 3x^2 - 8x - 18$;

Possible zeros: $\pm 1, \pm 2, \pm 3, \pm 6, \pm 9, \pm 18$

13. $x = 3, x = -3$

14. $x = -1, x = 2, x = -3$

15. *Sample answer*: $f(x) = 4x^3 + 4x^2 - 9x - 9$;

$f\left(\frac{3}{2}\right) = 0;\ f\left(-\frac{3}{2}\right) = 0$

16. **a.** $k = -18$ **b.** $k = -31$

3.5 Practice B

1. $x = -\frac{3}{2}, x = 0$

2. $h = 0, h = \pm\sqrt{2}$

3. $q = \pm\frac{1}{2}$

4. $w = \pm 3$

5. $p = -2, p = \pm 5$

6. $y = \pm 3, y = 8$

7. $x = -2, x = 0, x = 6$

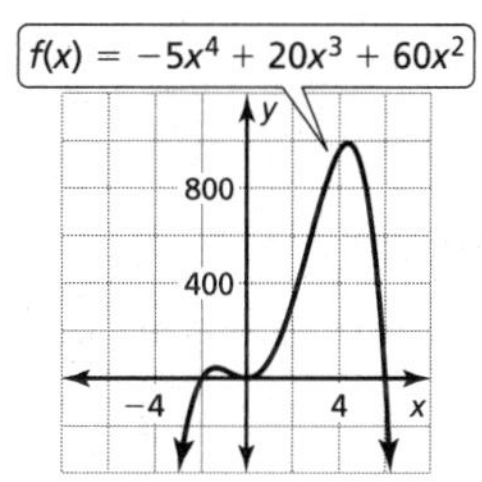

8. $x = -6, x = 0, x = 5$

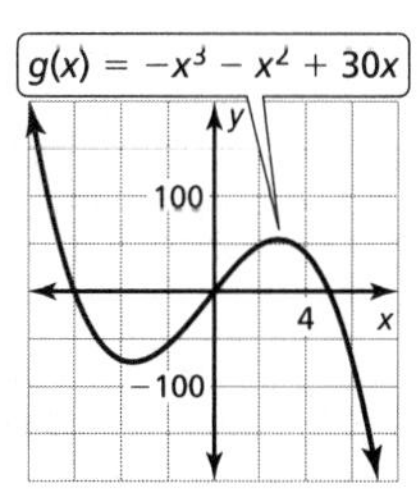

9. $x = -2, x = -1, x = 2$

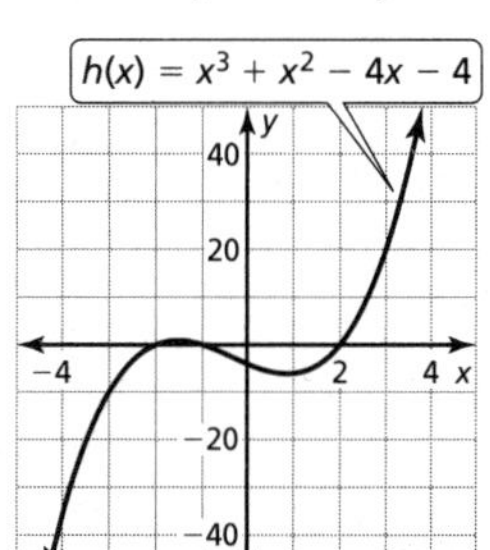

10. $x = -3, x = 3, x = 4$

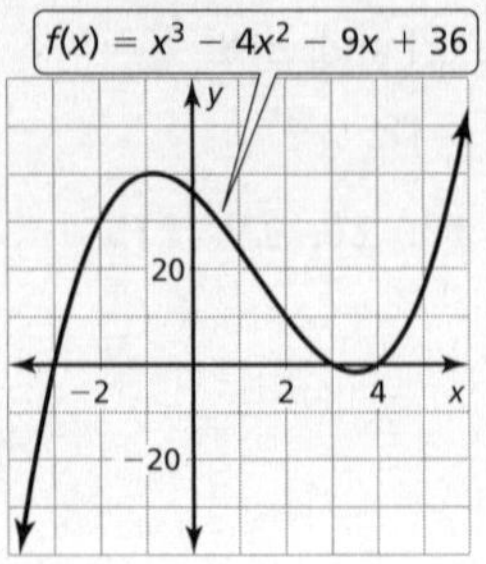

11. B

12. The factors of 2 include ± 2;

$f(x) = 2x^3 + 5x^2 - 2x - 6$;

Possible zeros: $\pm 1, \pm 2, \pm 3, \pm 6, \pm\frac{1}{2}, \pm\frac{3}{2}$

13. $x = \frac{3}{2}$

14. $x = -2, x = 3, x = 4$

15. *Sample answer*:

$f(x) = 25x^3 - 50x^2 - 49x + 98$;

$f\left(\frac{7}{5}\right) = 0$; $f\left(-\frac{7}{5}\right) = 0$

16. height is 11 cm, side length is 9 cm

3.5 Enrichment and Extension

1. a.

x	−3	−2	−1	0	1	2	3
$f(x)$	−19	−5	−1	−1	1	11	35

[0, 1]

b.

x	−3	−2	−1	0	1	2	3
$f(x)$	−180	−34	6	12	8	−30	−174

[−2, −1], [1, 2]

c.

x	−3	−2	−1	0	1	2	3
$f(x)$	193	8	−7	4	−7	8	193

[−2, −1], [−1, 0], [0, 1], [1, 2]

d.

x	−3	−2	−1	0	1	2	3
$f(x)$	−3	41	21	−3	29	177	501

[−3, −2], [−1, 0], [0, 1]

e.

x	−3	−2	−1	0	1	2	3
$f(x)$	−227	−18	1	−2	−3	−2	97

[−2, −1], [−1, 0], [2, 3]

f.

x	−3	−2	−1	0	1	2	3
$f(x)$	1110	115	−2	3	−2	115	1110

[−2, −1], [−1, 0], [0, 1], [1, 2]

2. a. ± 1; no; When you evaluate the function at either of these values, you do not obtain 0.

b.

x	−3	−2	−1	0	1	2	3
$f(x)$	−11	−1	1	1	5	19	49

[−2, −1]

c.

x	−2	−1.9	−1.8	−1.7
$f(x)$	−1	−0.539	−0.152	0.167

x	−1.6	−1.5	−1.4	−1.3
$f(x)$	0.424	0.625	0.776	0.883

x	−1.2	−1.1	−1
$f(x)$	0.952	0.989	1

[−1.8, −1.7]

d. −1.75488

3.5 Puzzle Time

IT WAS A BREEZE WITH ONLY A FEW FOGGY PATCHES

Answers

3.6 Start Thinking

Sample answer:

Function	Number of x-intercepts
$f(x) = x + 4$	1
$g(x) = x^2 - 5$	2
$h(x) = x^3 + 3x^2 - x - 1$	3
$j(x) = x^4 - x^3 - 4x^2 + 1$	4

The degree of the function and the number of x-intercepts are the same; no; Sometimes, there are solutions to polynomial functions that are imaginary numbers, which are not shown on the graph of the function.

3.6 Warm Up

1. 4 **2.** 3 **3.** 5

4. 6 **5.** 3 **6.** 5

3.6 Cumulative Review Warm Up

1.

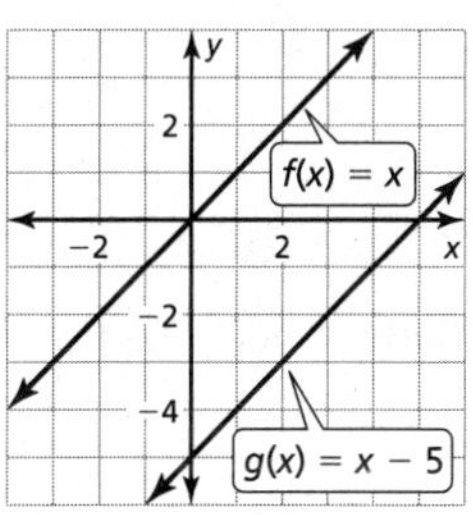

The graph of $g(x) = x - 5$ is a vertical translation 5 units down of the parent linear function.

2.

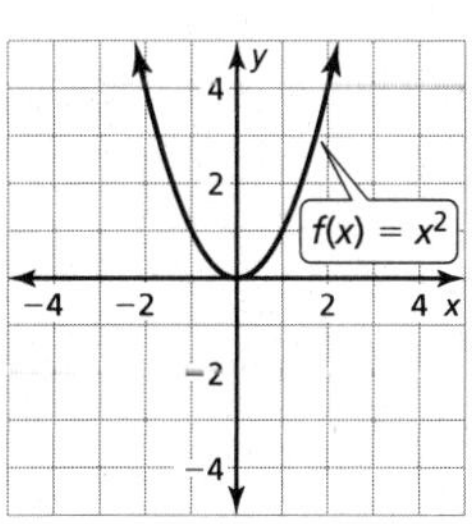

The graph of $f(x) = x^2$ is the parent quadratic function, so there was no transformation.

3.

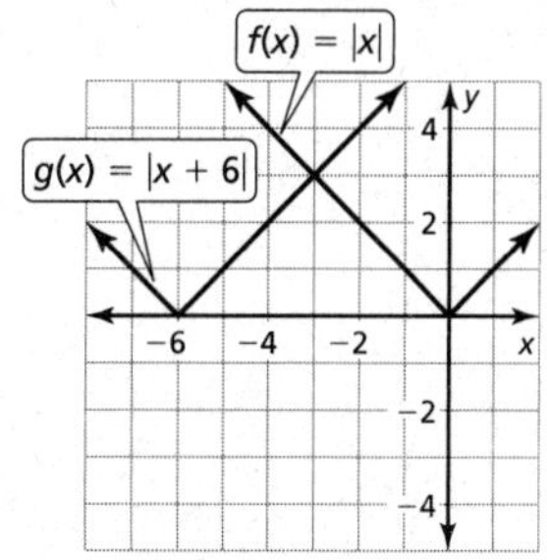

The graph of $g(x) = |x + 6|$ is a horizontal translation 6 units left of the parent absolute value function.

4.

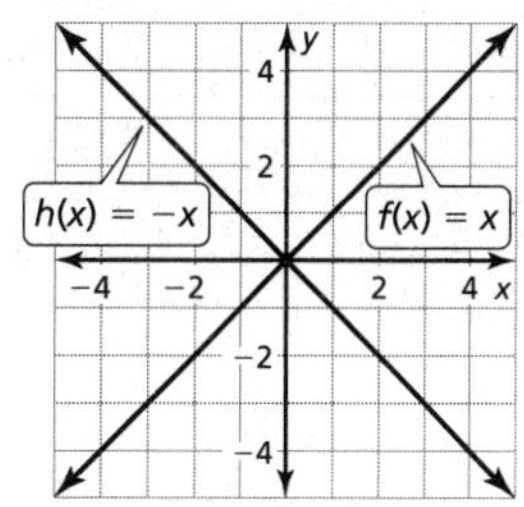

Sample answer: The graph of $h(x) = -x$ is a reflection in the x-axis of the parent linear function.

3.6 Practice A

1. $3; -5, \dfrac{-5 + 5i\sqrt{3}}{2}, \dfrac{-5 - 5i\sqrt{3}}{2}$

2. $5; 0, 0, -\dfrac{1}{3}, \dfrac{1 + i\sqrt{3}}{6}, \dfrac{1 - i\sqrt{3}}{6}$

3. $5; -1, -1, \dfrac{1 + i\sqrt{3}}{2}, \dfrac{1 - i\sqrt{3}}{2}$

4. $3; 2, i\sqrt{5}, -i\sqrt{5}$

5. $-3, -2i, 2i, 3$ **6.** $-3, -1, 1, 2$

7. $0, 1, -3i, 3i$ **8.** $-3, -\sqrt{2}, \sqrt{2}, 3$

9. 4; The graph shows 1 real zero, so the remaining zeros must be imaginary.

10. 0; There are 4 zeros for this function. The graph crosses the x-axis twice and touches the x-axis once at the repeated zero, leaving 0 imaginary zeros.

Answers

11. $f(x) = x^3 + x^2 - 10x + 8$

12. $f(x) = x^3 - 4x^2 + x + 6$

13. $f(x) = x^3 - 2x^2 - 3x + 6$

14. *Sample answer*: $f(x) = x^5 - 2x^3 - 2x^2 - 3x - 2$; Because i is a zero, $-i$ is also a zero. The graph touches the x-axis at -1 (has a multiplicity of 2) and the graph crosses the x-axis at 2.

15.

Positive real zeros	Negative real zeros	Imaginary zeros	Total zeros
2	1	0	3
0	1	2	3

16.

Positive real zeros	Negative real zeros	Imaginary zeros	Total zeros
1	1	2	4

17.

Positive real zeros	Negative real zeros	Imaginary zeros	Total zeros
3	2	0	5
3	0	2	5
1	2	2	5
1	0	4	5

3.6 Practice B

1. $3; -\frac{3}{2}, \frac{3 + 3i\sqrt{3}}{4}, \frac{3 - 3i\sqrt{3}}{4}$

2. $5; 0, 0, 2, -1 + i\sqrt{3}, -1 - i\sqrt{3}$

3. $8; -1, -1, 1, 1, i, -i$

4. $5; -2, -3, 3, 1 + i\sqrt{3}, 1 - i\sqrt{3}$

5. $-1, 1, 2, 2$

6. $3, 3, 3, 3$

7. $-2, -2, -1, -1, 2$

8. $-3, -2, -2, 2, 3$

9. 2; *Sample answer*: There are four zeros for this function. The graph crosses the x-axis twice, leaving two imaginary zeros.

10. 2; *Sample answer*: There are three zeros for this function. The graph crosses the x-axis only once, leaving two imaginary zeros.

11. $f(x) = x^3 - 8x^2 + 22x - 20$

12. $f(x) = x^4 - 2x^3 + 6x^2 - 8x + 8$

13. $f(x) = x^3 - 3x^2 - 7x + 21$

14. Complex zeros come in pairs, so the remaining zero cannot be complex.

15. C

3.6 Enrichment and Extension

1.
$$\begin{array}{r|rrrr} 1 & 6 & -4 & 3 & -5 \\ & & 6 & 2 & 5 \\ \hline & 6 & 2 & 5 & 0 \end{array}$$

Because $a_n = 6 > 0$, $k = 1 > 0$, and all numbers in the last row are positive, f has no real zero greater than 1.

2.
$$\begin{array}{r|rrrr} -4 & 1 & 3 & -2 & 8 \\ & & -4 & 4 & -8 \\ \hline & 1 & -1 & 2 & 0 \end{array}$$

Because $a_n = 1 > 0$, $k = -4 < 0$, and the numbers in the last row alternate in sign, f has no real zero less than -4.

3.
$$\begin{array}{r|rrrrrr} 1 & 2 & -1 & 2 & -2 & 4 & -4 \\ & & 2 & 1 & 3 & 1 & 5 \\ \hline & 2 & 1 & 3 & 1 & 5 & 1 \end{array}$$

Because $a_n = 2 > 0$, $k = 1 > 0$, and all numbers in the last row are positive, f has no real zero greater than 1.

4.
$$\begin{array}{r|rrrrr} 2 & 1 & -1 & 3 & -8 & 8 \\ & & 2 & 2 & 10 & 4 \\ \hline & 1 & 1 & 5 & 2 & 12 \end{array}$$

Because $a_n = 1 > 0$, $k = 2 > 0$, and all numbers in the last row are positive, f has no real zero greater than 2.

5. $$\begin{array}{r|rrrrrr} -1 & 1 & 0 & 2 & -2 & 5 & 5 \\ & & -1 & 1 & -3 & 5 & -10 \\ \hline & 1 & -1 & 3 & -5 & 10 & -5 \end{array}$$

Because $a_n = 1 > 0$, $k = -1 < 0$, and the numbers in the last row alternate in sign, f has no real zeros less than -1.

6. $$\begin{array}{r|rrrrrr} -3 & 1 & 0 & -3 & 0 & 1 & 2 \\ & & -3 & 9 & -18 & 54 & -165 \\ \hline & 1 & -3 & 6 & -18 & 55 & -163 \end{array}$$

Because $a_n = 1 > 0$, $k = -3 < 0$, and the numbers in the last row alternate in sign, f has no real zeros less than -3.

7. $$\begin{array}{r|rrrr} -1 & 1 & -4 & 0 & 1 \\ & & -1 & 5 & -5 \\ \hline & 1 & -5 & 5 & -4 \end{array}$$

Because $a_n = 1 > 0$, $k = -1 < 0$, and the numbers in the last row alternate in sign, f has no real zeros less than -1.

8. $$\begin{array}{r|rrrrr} 5 & 1 & -4 & 0 & -16 & -15 \\ & & 5 & 5 & 25 & 45 \\ \hline & 1 & 1 & 5 & 9 & 30 \end{array}$$

Because $a_n = 1 > 0$, $k = 5 > 0$, and all numbers in the last row are positive, f has no real zeros greater than 5.

9. $$\begin{array}{r|rrrrr} -2 & 1 & 1 & -1 & 0 & 3 \\ & & -2 & 2 & -2 & 4 \\ \hline & 1 & -1 & 1 & -2 & 7 \end{array}$$

Because $a_n = 1 > 0$, $k = -2 < 0$, and the numbers in the last row alternate in sign, f has no real zeros less than -2.

10. $$\begin{array}{r|rrrrr} 3 & 2 & 0 & 0 & -8 & 3 \\ & & 6 & 18 & 54 & 138 \\ \hline & 2 & 6 & 18 & 46 & 141 \end{array}$$

Because $a_n = 2 > 0$, $k = 3 > 0$, and all numbers in the last row are positive, f has no real zeros greater than 3.

3.6 Puzzle Time

A PUP TENT

3.7 Start Thinking

Function	Transformation	Function	Transformation
$g(x) = (x - 2)^5$	Translation 2 units right	$g(x) = -x^5$	Reflection in the x-axis
$g(x) = (x + 2)^5$	Translation 2 units left	$g(x) = (2x)^5$	Horizontal shrink by a factor of $\frac{1}{2}$
$g(x) = x^5 - 2$	Translation 2 units down	$g(x) = \left(\frac{1}{2}x\right)^5$	Horizontal stretch by a factor of 2
$g(x) = x^5 + 2$	Translation 2 units up	$g(x) = 2x^5$	Vertical stretch by a factor of 2

The transformations of $g(x) = x^5$ behave in the same manner as other parent function transformations. Numbers added or subtracted inside parentheses translate the graph left or right, and numbers added or subtracted outside the parentheses translate the graph up or down. Numbers multiplied inside the parentheses horizontally stretch or shrink the graph, and numbers multiplied outside the parentheses vertically stretch or shrink the graph.

3.7 Warm Up

1. *Sample answer:* The graph of g is a vertical stretch by a factor of 5 of the graph of the parent function.
2. The graph of h is a reflection in the x-axis, followed by a vertical shrink by a factor of $\frac{1}{3}$ of the graph of the parent function.
3. *Sample answer:* The graph of g is a vertical stretch by a factor of 2 of the graph of the parent function.
4. The graph of f is a reflection in the x-axis, followed by a vertical shrink by a factor of $\frac{1}{2}$ of the graph of the parent function.

Answers

3.7 Cumulative Review Warm Up

1. The graph of $g(x) = x^2 + 5$ is a vertical translation 5 units up of the graph of the parent quadratic function.

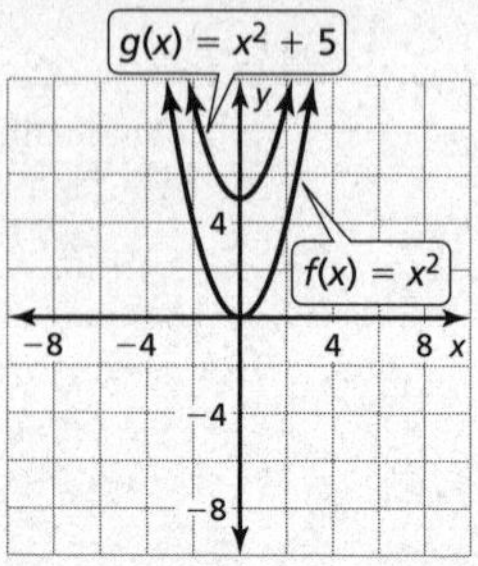

2. The graph of $g(x) = (x - 1)^2$ is a horizontal translation 1 unit right of the graph of the parent quadratic function.

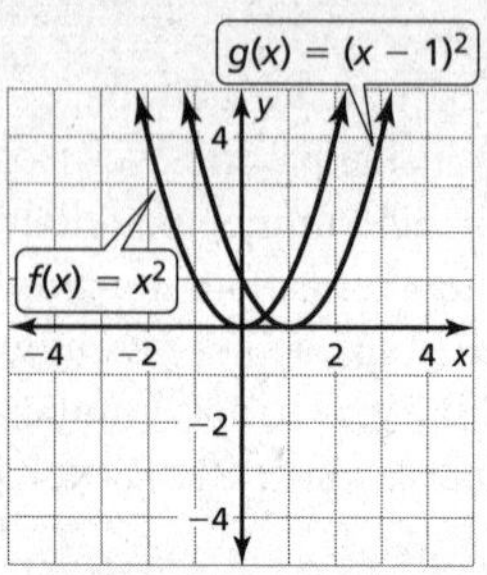

3. The graph of $g(x) = (x + 2)^2$ is a horizontal translation 2 units left of the graph of the parent quadratic function.

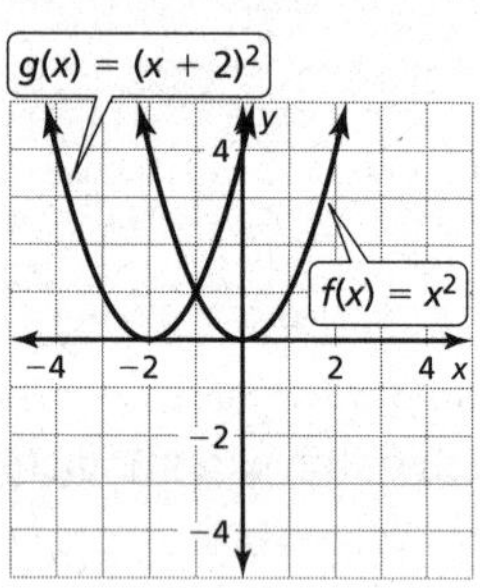

4. The graph of $g(x) = (x - 5)^2 + 3$ is a horizontal translation 5 units right, followed by a vertical translation 3 units up of the graph of the parent quadratic function.

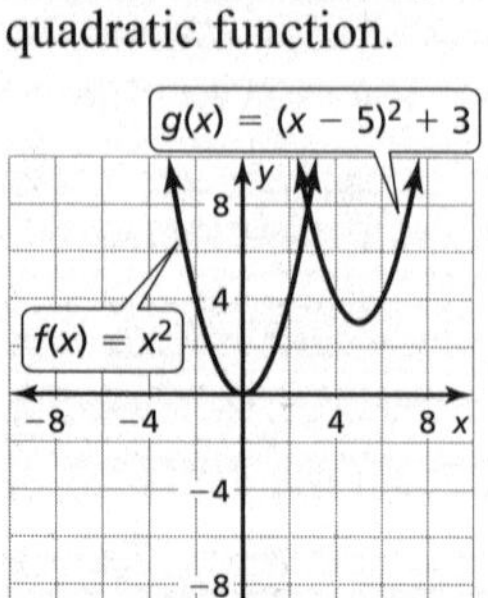

3.7 Practice A

1. The graph of *g* is a vertical translation 2 units down of the graph of the parent function *f*.

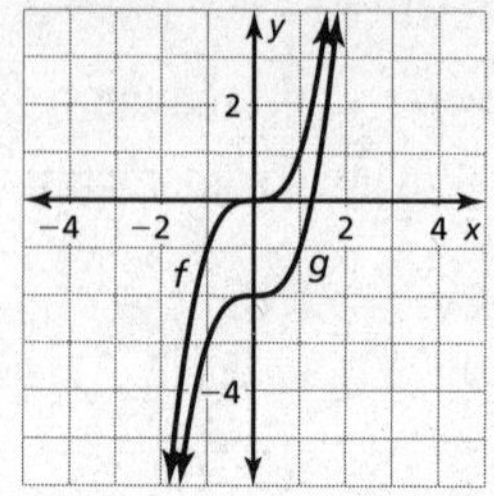

2. The graph of *g* is a horizontal translation 3 units left of the graph of the parent function *f*.

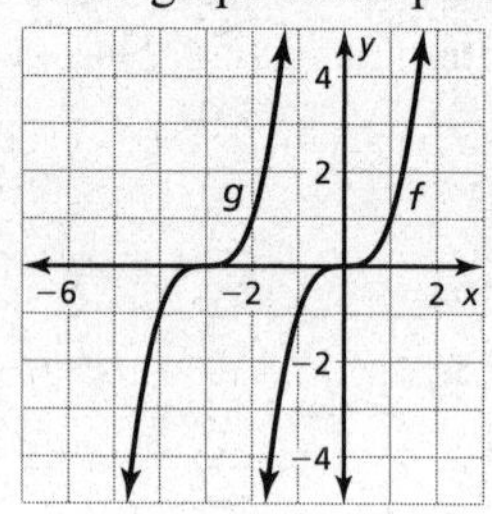

3. The graph of *g* is a reflection in *x*-axis, followed by a vertical stretch by a factor of 5 of the graph of the parent function *f*.

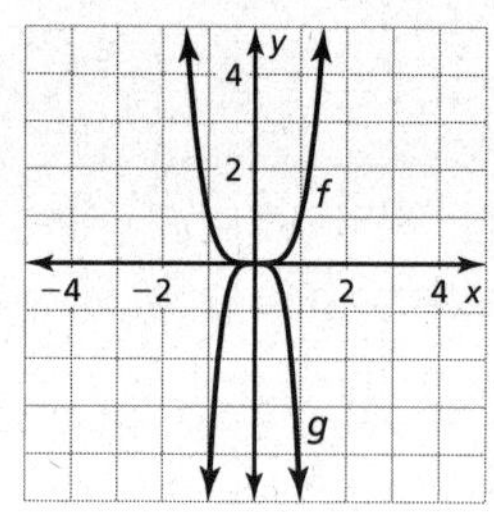

4. The graph of *g* is a vertical stretch by a factor of 4, followed by a vertical translation 3 units down of the graph of the parent function *f*.

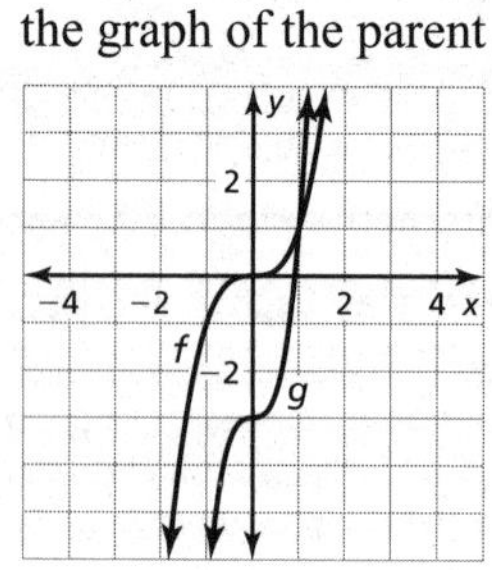

Answers

5. The graph of g is a vertical shrink by a factor of $\frac{2}{3}$, followed by a vertical translation 5 units down of the graph of the parent function f.

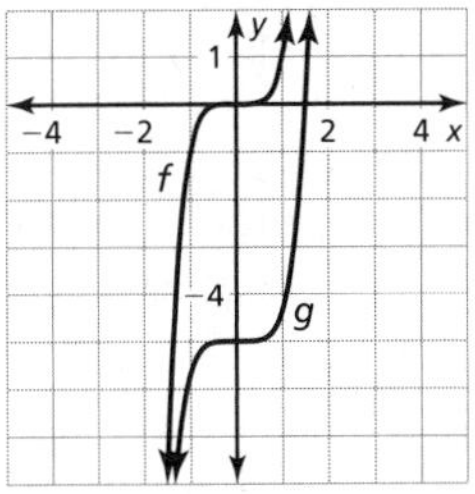

6. The graph of g is a vertical shrink by a factor of $\frac{1}{2}$, followed by a horizontal translation 2 units right of the graph of the parent function f.

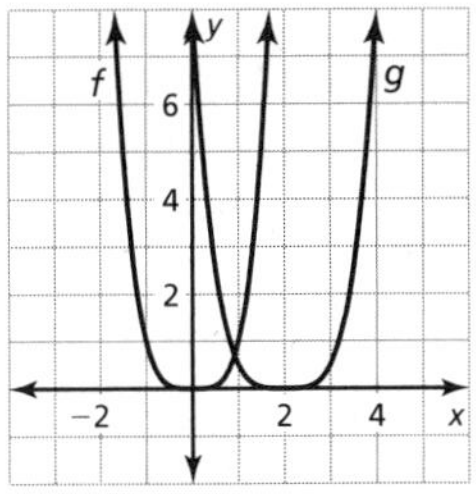

7. $g(x) = (x - 1)^3 + 2$

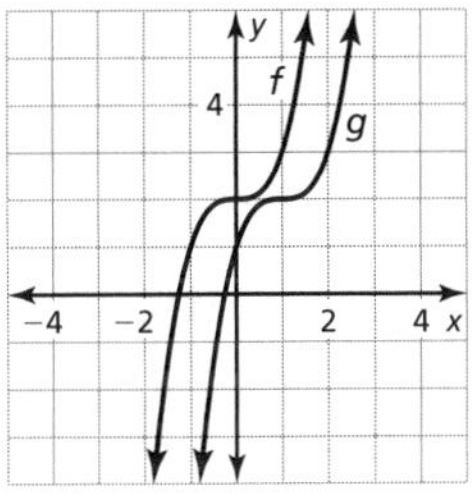

The graph of g is a horizontal translation 1 unit right of the graph of the parent function f.

8. $g(x) = 2x^4 - 6x + 2$

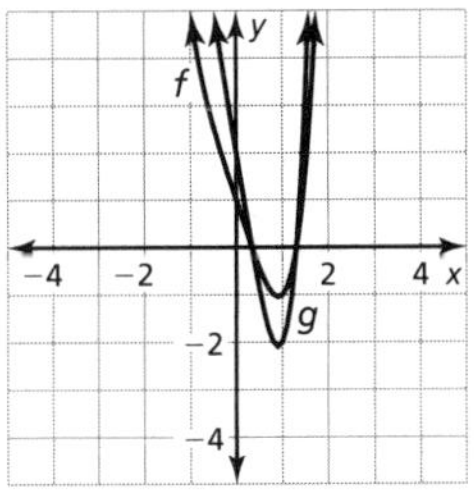

The graph of g is a vertical stretch by a factor of 2 of the graph of the parent function f.

9. The parent function was translated 2 units down instead of 2 units up.

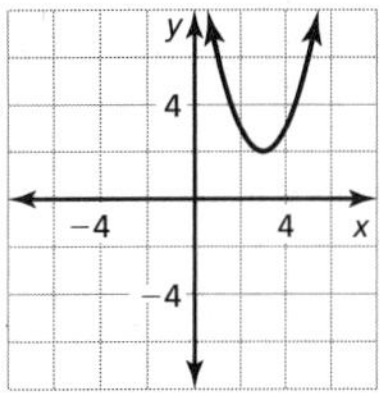

10. $g(x) = (-x - 2)^3 + 5$

11. $g(x) = \frac{1}{3}x^4 - x - \frac{5}{3}$

12. a. $W(x) = 27x^3 + 12x + 3$

b. $Z(x) = 46{,}656x^3 + 144x + 3$

3.7 Practice B

1. The graph of g is a horizontal translation 3 units right, followed by a vertical translation 2 units down of the graph of the parent function f.

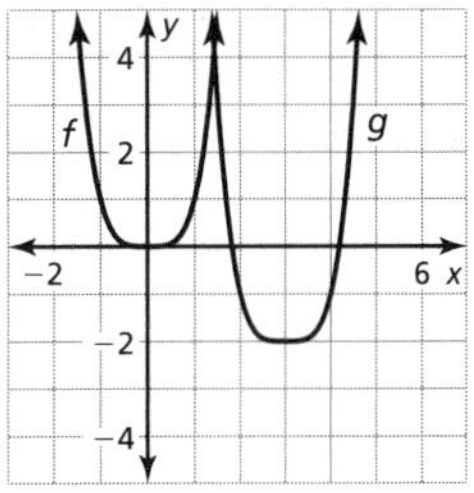

2. The graph of g is a horizontal translation 1 unit right, followed by a vertical translation 4 units up of the graph of the parent function f.

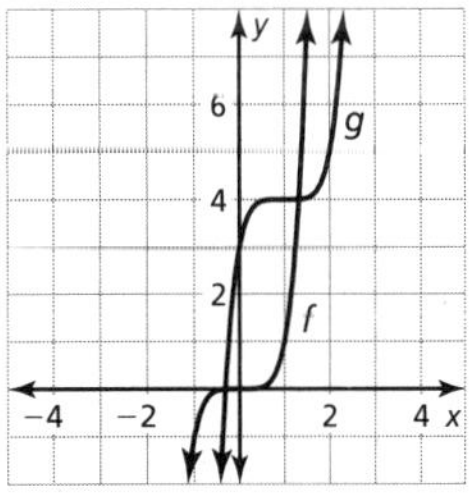

3. The graph of g is a reflection in the x-axis, followed by a vertical stretch by a factor of 3 of the graph of the parent function f.

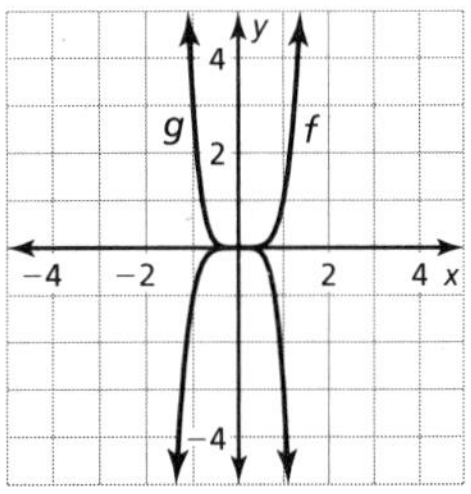

Answers

4. The graph of g is a vertical stretch by a factor of 3, followed by a vertical translation 2 units up of the graph of the parent function f.

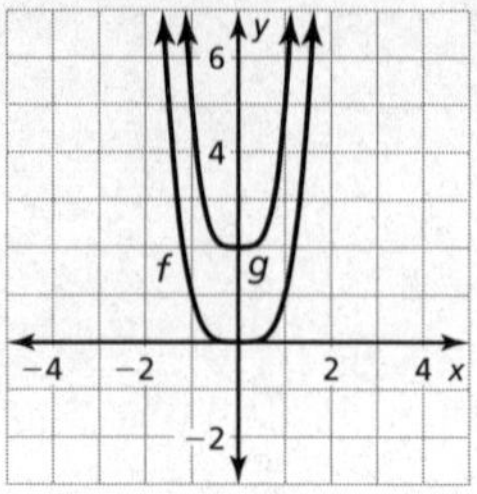

5. The graph of g is a vertical shrink by a factor of $\frac{1}{3}$, followed by a vertical translation 3 units down of the graph of the parent function f.

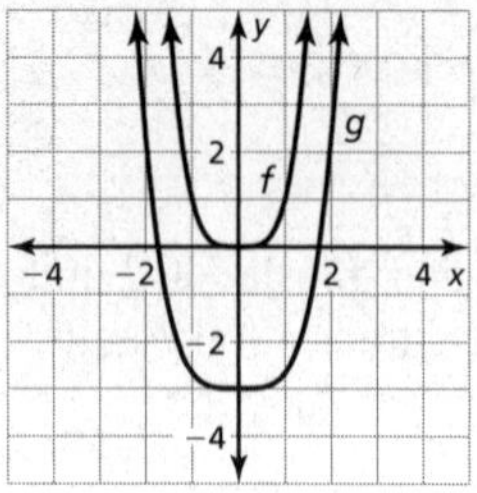

6. The graph of g is a vertical shrink by a factor of $\frac{2}{3}$, followed by a horizontal translation 3 units left of the graph of the parent function f.

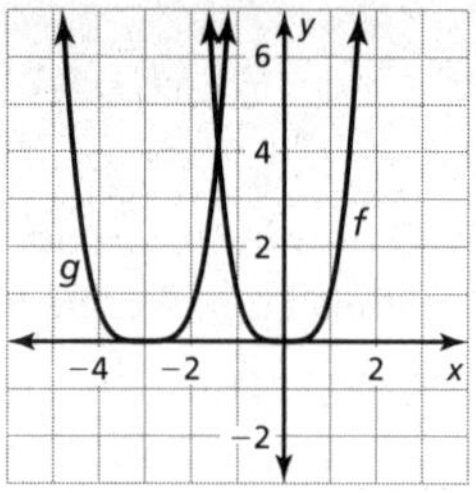

7. $g(x) = -\frac{1}{4}x^3 + x^2 - \frac{1}{2}$

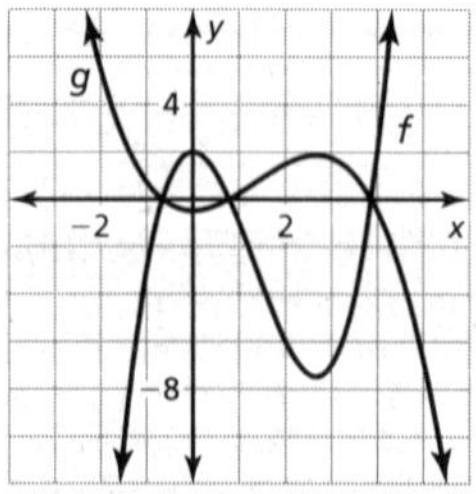

The graph of g is a reflection in the x-axis, followed by a vertical shrink by a factor of $\frac{1}{4}$ of the graph of the parent function f.

8. $g(x) = x^4 - x + 3$

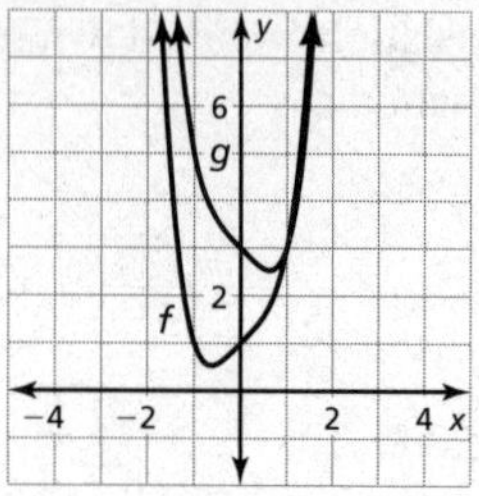

The graph of g is a reflection in the y-axis, followed by a vertical translation 2 units up of the graph of the parent function f.

9. The graph of g is a vertical stretch by a factor of 4, not a vertical shrink by a factor of $\frac{1}{4}$, of the graph of the parent function $f(x) = x^4$; The graph of g is a vertical stretch by a factor of 4, followed by a translation 3 units up of the graph of f.

10. $g(x) = -\frac{1}{27}x^3 + \frac{1}{3}x^2 - 5$

11. $g(x) = -\frac{3}{2}x^5 + \frac{1}{2}x^3 + \frac{5}{2}x^2 + \frac{3}{2}$

12. a. $W(x) = \frac{1}{864}x^3 + 9; W(6) = 9.25 \text{ ft}^3;$

When x is 6 inches, the volume of the box is 9.25 cubic feet.

b. $Z(x) = \frac{1}{23{,}328}x^3 + 9$

3.7 Enrichment and Extension

$y = (x+3)^3 - 6$, $y = (2x)^3$, $y = (x-3)^3 + 2$, $y = 2(x-3)^3$, $y = (x-5)^3$, $y = (x-5)^3 - 2$, $y = (x-8)^3 + 1$

3.7 Puzzle Time

FRYDAY

Answers

3.8 Start Thinking

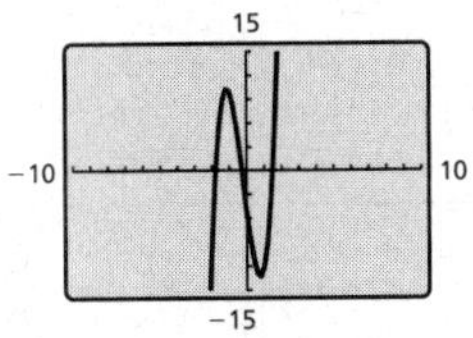

The shape of the graph of the function is a rounded "N"; There is one zero where $-2 < x < -1$, another zero where $-1 < x < 0$, and a third zero where $1 < x < 2$; no; The y-values are changing signs, but the table does not show an x-value when $y = 0$.

3.8 Warm Up

1. $(5, -33)$
2. $(3, 12)$
3. $(-2, -19)$
4. $\left(\frac{4}{3}, 5\frac{1}{3}\right)$
5. $(8, 45)$
6. $(2, 10)$

3.8 Cumulative Review Warm Up

1. $f(x) = (x + 3)^2 - 29; (-3, -29)$
2. $g(x) = (x + 2)^2 - 2; (-2, -2)$
3. $g(x) = (x - 5)^2 - 63; (5, -63)$
4. $h(x) = (x - 10)^2 - 191; (10, -191)$
5. $h(x) = (x - 1)^2 + 48; (1, 48)$
6. $f(x) = (x - 3)^2 - 6; (3, -6)$

3.8 Practice A

1. 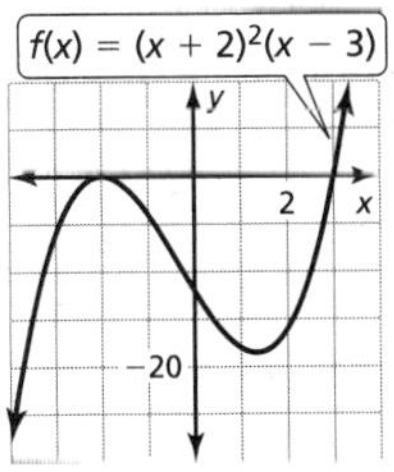

2. 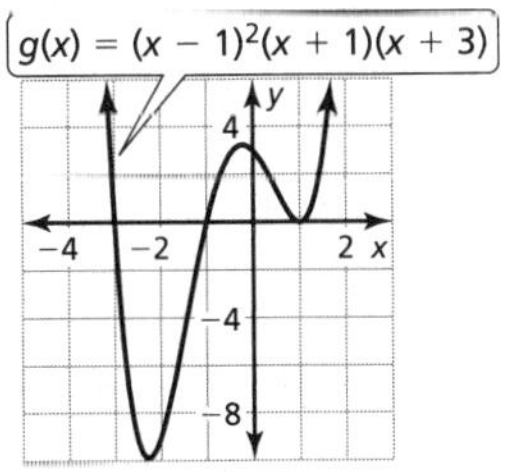

3. 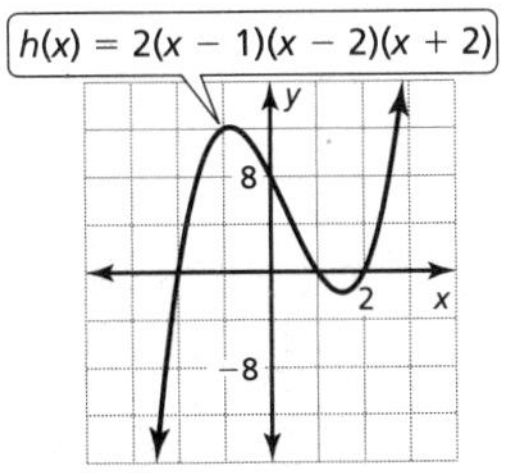

4. 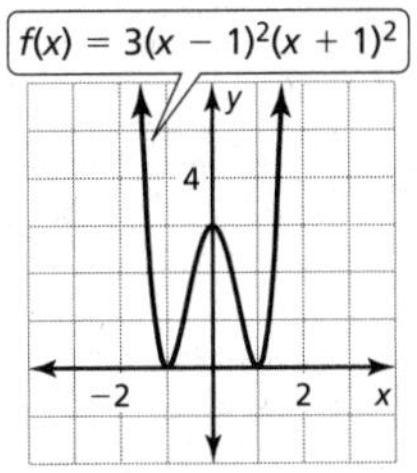

5. The function was graphed as if the zero $x = -3$ had a multiplicity of 2 instead of the zero $x = 1$.

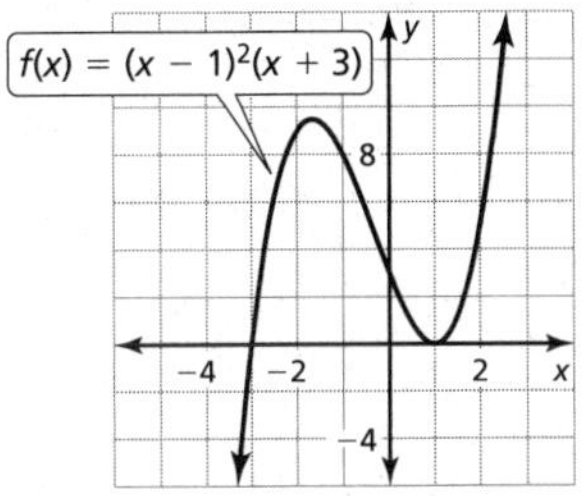

6. −3, −2, 2
7. −7, −1, 1
8. 5
9. $-3, \frac{3}{2}, 3$
10. 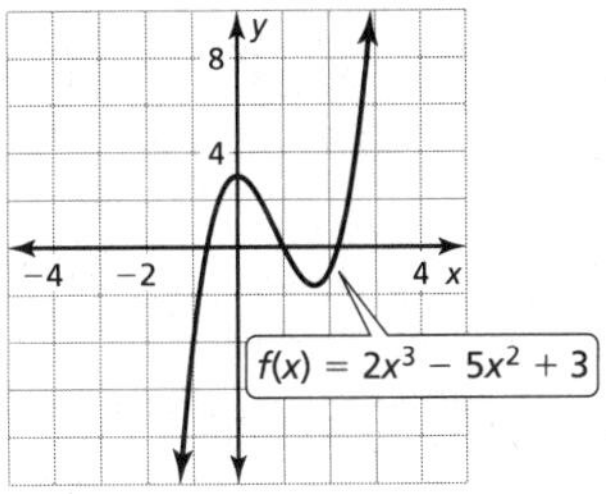

x-intercepts: $-0.69, 1, 2.19$; local maximum: $(0, 3)$; local minimum: $(1.67, -1.63)$; increasing: $x < 0$, $x > 1.67$; decreasing: $0 < x < 1.67$

11. 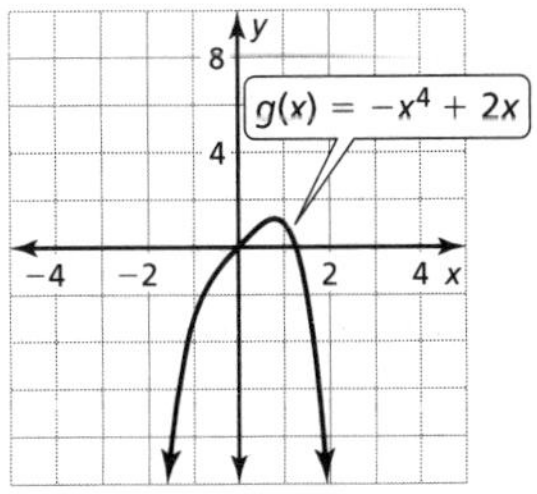

x-intercepts: $0, 1.26$; local maximum: $(0.790, 1.19)$; local minimum: none; increasing: $x < 0.79$; decreasing: $x > 0.79$

Answers

12. 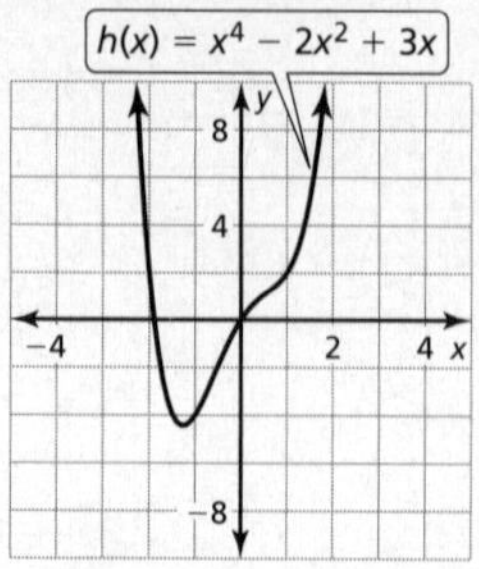

x-intercepts: -1.89, 0; local maximum: none; local minimum: $(-1.26, -4.43)$; increasing: $x > -1.26$; decreasing: $x < -1.26$

13. 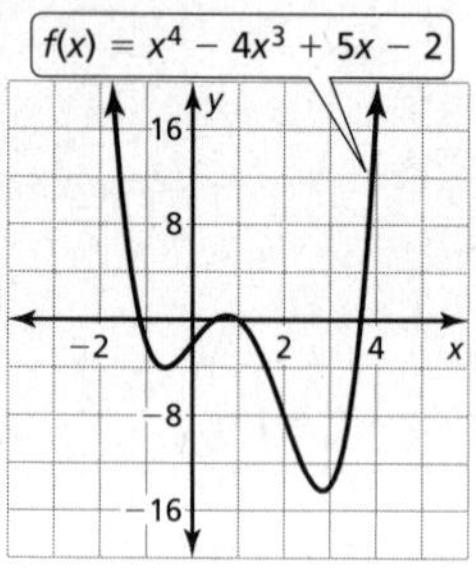

x-intercepts: -1.15, 0.48, 1, 3.67; local maximum: $(0.74, 0.38)$; local minimum: $(-0.59, 4.01)$, $(2.85, -14.37)$; increasing: $-0.59 < x < 0.74$, $x > 2.85$; decreasing: $x < -0.59$, $0.74 < x < 2.85$

14. $(-0.82, 1.09)$, $(0.82, -1.09)$; The point $(-0.82, 1.09)$ is a local maximum. The point $(0.82, -1.09)$ is a local minimum; The real zeros are -1.41, 0, and 1.41; The minimum degree is 3.

15. $(1.50, 3.69)$; The point $(1.50, 3.69)$ is a local maximum; The real zeros are -0.89 and 2.19; The minimum degree is 4.

3.8 Practice B

1. 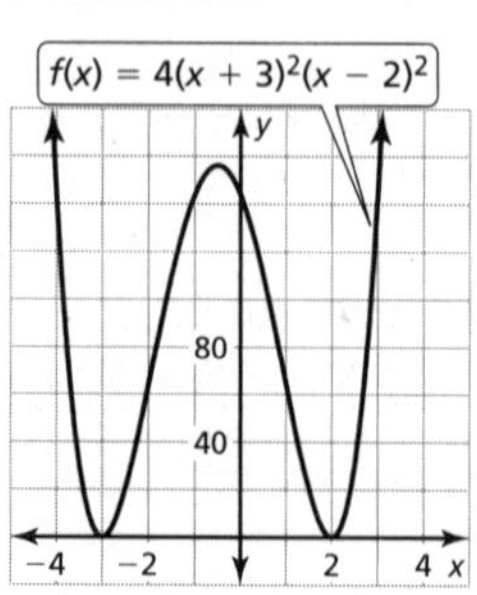

2. 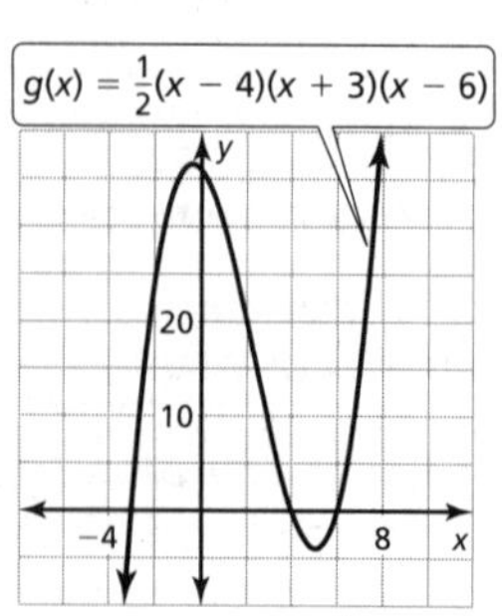

3. 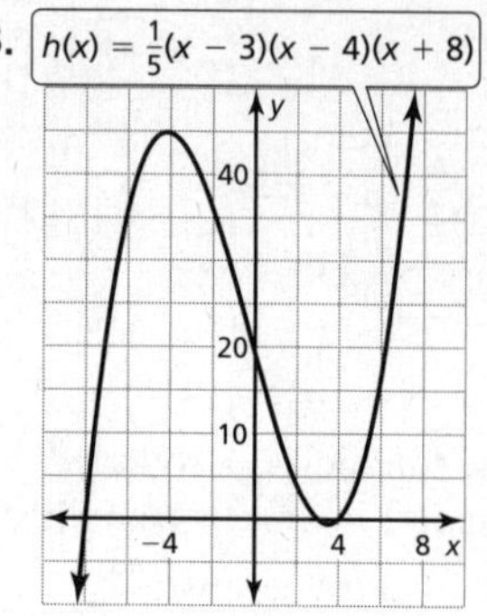

4. 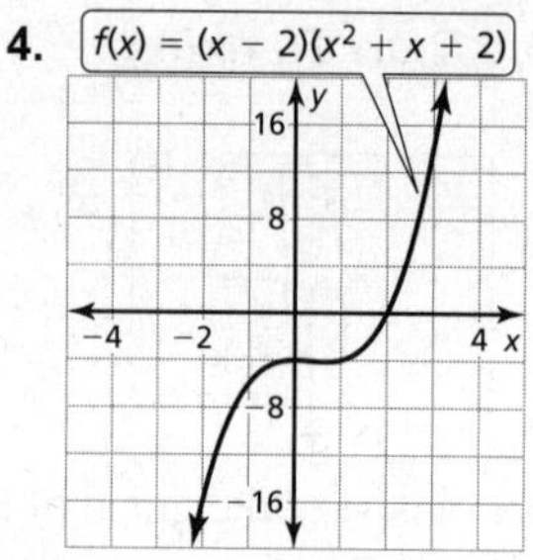

5. The function was graphed as if the zero $x = 0$ had a multiplicity of 1 instead of a multiplicity of 2.

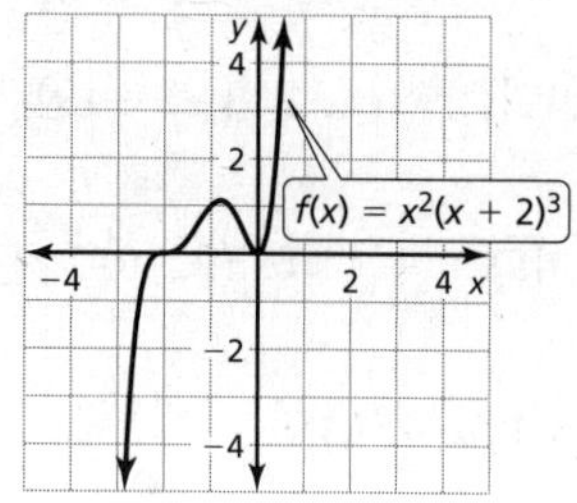

6. $\frac{1}{2}$

7. $-\frac{5}{2}, -2, 1$

8. $-3, \frac{3}{4}, 3$

9. $-\frac{3}{2}$

10. 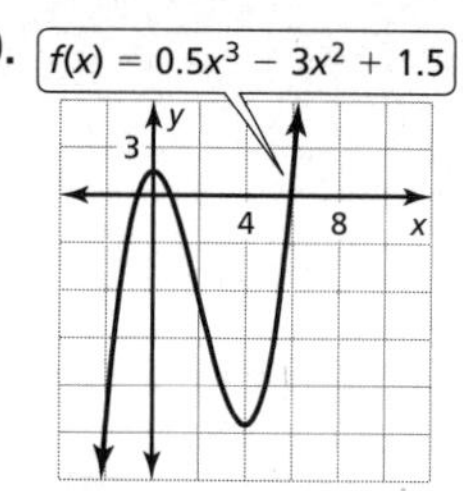

x-intercepts: -0.67, 0.76, 5.91; local maximum: $(0, 1.5)$; local minimum: $(4, -14.5)$; increasing: $x < 0, x > 4$; decreasing: $0 < x < 4$

11. 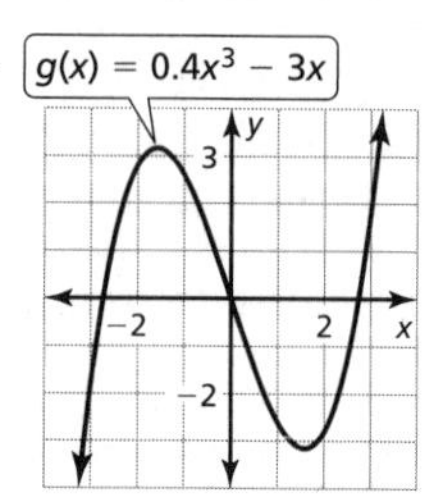

x-intercepts: -2.74, 0, 2.74; local maximum: $(-1.58, 3.16)$; local minimum: $(1.58, -3.16)$; increasing: $x < 1.58, x > 1.58$; decreasing: $-1.58 < x < 1.58$

Answers

12.

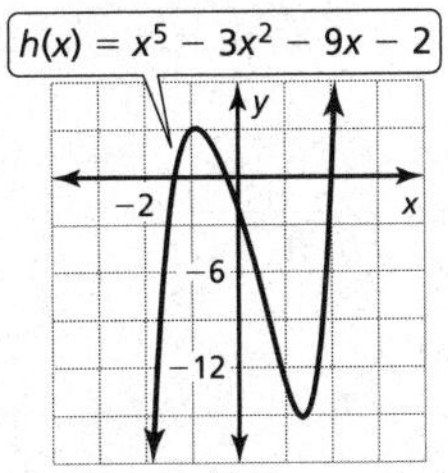

x-intercepts: $-1.36, -0.24, 2$; local maximum: $(-0.92, 3.08)$; local minimum: $(1.36, -15.14)$; increasing: $x < -0.92, 1.36 < x$; decreasing: $-0.92 < x < 1.36$

13.

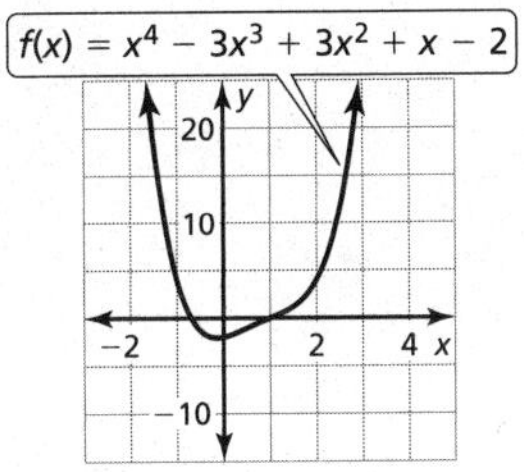

x-intercepts: $-0.7, 1$; local maximum: none; local minimum: $(-0.14, -2.07)$; increasing: $-0.14 < x$; decreasing: $x < -0.14$

14. **a.** about 1.6 in.

b. about 67.6 in.^{3}

c. length $\approx$ 8.8 in., width $\approx$ 4.8 in., height $\approx$ 1.6 in.

3.8 Enrichment and Extension

1. $x \approx 3.16$ **2.** $x \approx 1.52$ **3.** $x \approx -1.31$

4. $x \approx -2.99$ **5.** $x \approx 1.92$

3.8 Puzzle Time

SHIFTY

3.9 Start Thinking

Sample answer:

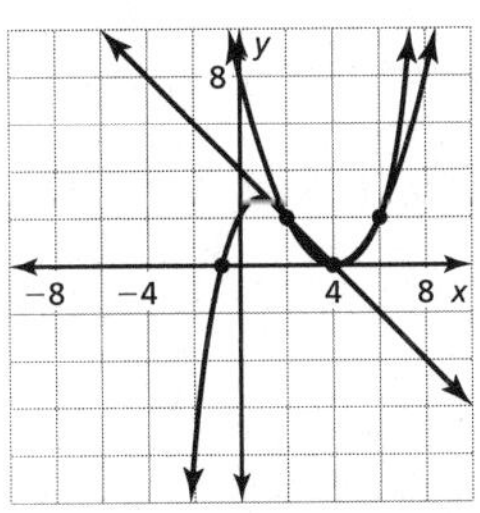

quadratic; parabola; cubic

3.9 Warm Up

1. 4 **2.** $-\dfrac{1}{3}$ **3.** 4 **4.** 1

3.9 Cumulative Review Warm Up

1. $(1, 6)$ **2.** $(-3, -11)$

3.9 Practice A

1. $f(x) = x^3 - 3x^2 - x + 3$

2. $f(x) = -\dfrac{1}{2}x^3 + \dfrac{3}{2}x^2 + 3x - 4$

3. 3; $f(x) = \dfrac{1}{6}x^3 + \dfrac{5}{6}x$

4. 2; $f(x) = \dfrac{1}{8}x^2 + \dfrac{13}{4}x + 8$

5. 3; $f(x) = \dfrac{1}{3}x^3 - 5x^2 + \dfrac{8}{3}x + 32$

6. a. $f(x) = \dfrac{1}{2}x^2 + \dfrac{7}{2}x - 2$

b. yes; The cumulative number of customers will continue to increase.

3.9 Practice B

1. $f(x) = \dfrac{1}{2}x^3 - 3x^2 + \dfrac{3}{2}x + 5$

2. $f(x) = \dfrac{2}{3}x^3 - \dfrac{14}{3}x - 4$

3. 2; $f(x) = \dfrac{5}{2}x^2 - \dfrac{23}{2}x - 1$

4. 3; $f(x) = \dfrac{1}{2}x^3 - \dfrac{31}{2}x + 39$

5. 4; $f(x) = \dfrac{1}{6}x^4 - \dfrac{7}{3}x^3 + \dfrac{95}{6}x^2 - \dfrac{149}{3}x + 56$

6. a. $f(x) = \dfrac{1}{6}x^3 - \dfrac{1}{2}x^2 + \dfrac{17}{6}x - 2$

b. no; The wave height will decrease eventually. After 14 seconds, the height of the wave will not be 397 inches.

Answers

3.9 Enrichment and Extension

1. $f(x) = \frac{7}{6}x^3 - x^2 - \frac{7}{6}x + 2$
2. $f(x) = -3x^3 + 20x^2 - 47x + 28$
3. $f(x) = -3x^4 + x^3 + 7x^2 + x + 1$
4. $f(x) = \frac{17}{12}x^4 - \frac{5}{6}x^3 - \frac{53}{12}x^2 + \frac{11}{6}x + 4$
5. $f(x) = 2x^5 - 3x^4 - 6x^3 + 10x^2 + 6x - 5$
6. $f(x) = -x^5 + 3$
7. $f(x) = 3x^6 - 4$

3.9 Puzzle Time

ANCHOR

Cumulative Review

1. $4x^2 + 28x + 12$
2. $-10x^2 - 25x + 50$
3. $x^2 + 16x + 48$
4. $-8x^2 + 160x - 792$
5. $3x^2 + 6x - 9$
6. $16x^2 - 12x - 28$
7. $-96x^2 - 444x + 462$
8. $2x^2 + 20x - 10$
9. $7x^2 + 23x - 17$
10. $54x^2 + 9x + 31$
11. $-6x^2 - 47x - 28$
12. $4x^2 + 21x - 87$
13. brother = 7 min, sister = 12 min
14. **a.** width = 14 ft, length = 19 ft
    b. \$1056
    c. 266 ft^2
15. $(x + 6)(x + 4)$
16. $(x - 11)(x + 3)$
17. $(x + 12)(x - 7)$
18. $(x + 11)(x + 5)$
19. $(x - 7)(x - 2)$
20. $(x - 10)(x - 5)$
21. $(5x + 7)(x + 8)$
22. $(x - 6)(9x + 1)$
23. $(2x - 5)(x + 4)$
24. $(3x + 4)(x + 3)$
25. $(2x - 3)(x + 1)$
26. $(2x + 7)(x - 4)$
27. **a.** 252 ft^2
    b. \$829.08
28. **a.** $p = 2(x + 4) + 2(2x - 23)$ or $p = 6x - 38$
    b. $a = (x + 4)(2x - 23)$ or $a = 2x^2 - 15x - 92$
    c. length = 23 ft, width = 15 ft
    d. 345 ft^2
29. $x = -2$ and $x = 3$
30. $x = -1$ and $x = 9$
31. $x = -7$ and $x = -5$
32. $x = 3$
33. $x = -11$ and $x = 4$
34. $x = 1$ and $x = 7$
35. $x = -\frac{9}{2}$ and $x = -\frac{4}{3}$
36. $x = -\frac{7}{2}$ and $x = 3$
37. $x = -\frac{1}{5}$ and $x = 4$
38. $x = -\frac{1}{2}$ and $x = \frac{9}{4}$
39. $x = \frac{7 \pm \sqrt{17}}{2}$
40. $x = \frac{-13 \pm \sqrt{173}}{2}$
41. $x = 7 \pm \sqrt{47}$
42. $x = \frac{-9 \pm \sqrt{97}}{2}$
43. $x = \frac{12 \pm \sqrt{123}}{3}$
44. $x = \frac{-9 \pm \sqrt{105}}{4}$
45. $x = \frac{9 \pm 4\sqrt{5}}{2}$
46. $x = \frac{21 \pm \sqrt{461}}{5}$
47. $\frac{9 + i\sqrt{3}}{2}$ and $\frac{9 - i\sqrt{3}}{2}$
48. -4 and 5
49. $\frac{3 + \sqrt{13}}{2}$ and $\frac{3 - \sqrt{13}}{2}$
50. $\frac{5 + \sqrt{101}}{2}$ and $\frac{5 - \sqrt{101}}{2}$

Answers

51. $\dfrac{5 + 3i\sqrt{15}}{8}$ and $\dfrac{5 - 3i\sqrt{15}}{8}$

52. $\dfrac{7 + \sqrt{301}}{6}$ and $\dfrac{7 - \sqrt{301}}{6}$

53. $\dfrac{11 + i\sqrt{39}}{10}$ and $\dfrac{11 - i\sqrt{39}}{10}$

54. $\dfrac{9 + \sqrt{137}}{4}$ and $\dfrac{9 - \sqrt{137}}{4}$

55. a. 1 student
    b. 4 students

56. a. 4 students
    b. 2 students

57. The graph of g is a translation 4 units right of the graph of f.

58. The graph of g is a translation 8 units up of the graph of f.

59. The graph of g is a translation 3 units right and 5 units up of the graph of f.

60. The graph of g is a translation 1 unit left and 4 units down of the graph of f.

61. The graph of g is a reflection in the x-axis, followed by a translation 3 units down of the graph of f.

62. The graph of g is a reflection in the x-axis, followed by a translation 1 unit right and 5 units up of the graph of f.

63. The graph of g is a reflection in the x-axis, followed by a vertical stretch by a factor of 5 and a translation 2 units down of the graph of f.

64. The graph of g is a vertical shrink by a factor of $\frac{1}{3}$, followed by a translation 5 units left and 3 units down of the graph of f.

65. $g(x) = -x + 3$

66. $g(x) = \frac{1}{4}|x + 3| + 1$

67. $g(x) = 4\sqrt{x - 2} + 16$

68. $g(x) = -(x + 2)^2 + 8$

69. a. 41 ft
    b. 44 ft
    c. 2 ft; The ball was initially hit from 2 feet above the ground.
    d. maximum
    e. (4.75, 47.125); 4.75 sec

70. $7x^2 + 12x - 8$

71. $x^2 + 20x - 15$

72. $33x - 24$

73. $-x^2 + 25x - 35$

74. $27x^3 - 7x^2 + 41x + 9$

75. $-x^5 + 4x^4 - 16x^2 - 7x - 6$

76. $8x^3 + 10x^2 + 22x + 19$

77. $-10x^4 + 35x^3 - 19x^2 + 8x - 20$

78. $4x^5 - 40x^4 + 44x^3$

79. $-10x^7 + 5x^6 + 15x^5$

80. $6x^3 - 21x^2 - 22x + 77$

81. $12x^3 - 142x^2 + 60x + 14$

82. $30x^4 - 40x^3 + 21x^2 - 8x + 3$

83. $-84x^3 - 164x^2 - 79x - 9$

84. $x^4 - 4x^3 - 30x^2 + 116x - 99$

85. $-2x^5 + 12x^4 - 5x^3 + 58x^2 - 129x - 11$

86. $6x^4 + 65x^3 + 53x^2 - 156x + 60$

87. $x^3 - 28x - 48$

88. $-3x^3 + 10x^2 - 9x + 2$

89. $-120x^3 + 50x^2 + 535x - 90$

90. $x = -2, x = 0, x = 6$

91. $y = -3, x = -\frac{3}{2}, x = \frac{3}{2}, x = 0$

92. $w = -1, w = 1, w = -4, w = 4$

93. $a = -3, a = 3, a = 5$

94. $p = -2, p = 2, p = -\frac{1}{2}$

Answers

95. $r = -\frac{5}{2}, r = -2, r = 2, r = 0$

96. $f(x) = x^3 + 2x^2 - 15x$

97. $f(x) = x^3 - 2x^2 - 31x - 28$

98. $f(x) = x^3 - 2x^2 + 49x - 98$

99. $f(x) = x^4 - 4x^3 + 14x^2 - 36x + 45$

100. $f(x) = x^5 - 7x^4 + 6x^3 + 32x^2 - 72x + 60$

101. $f(x) = x^4 + 9x^4 - 10x^3 - 214x^2 - 203x + 833$

Chapter 4

4.1 Start Thinking

Example	Expanded Form	Simplest Form
$x^2 + x^2$	$x \bullet x + x \bullet x$	$2x^2$
$x^4 \bullet x^4$	$(x \bullet x \bullet x \bullet x) \bullet (x \bullet x \bullet x \bullet x)$	x^8
$\frac{x^8}{x^5}$	$\frac{x \bullet x \bullet x \bullet x \bullet x \bullet x \bullet x \bullet x}{x \bullet x \bullet x \bullet x \bullet x}$	x^3

$2^{-2} = \frac{1}{4}$; Because $2^2 = 4$, they are reciprocals.

4.1 Warm Up

1. k^5 2. $24u^{10}v^3$ 3. $25a^6b^{20}c^2$

4. $729x^8y^{10}z^5$ 5. $8g^9h^5j^{13}$ 6. $2xy^9$

4.1 Cumulative Review Warm Up

1. $y = 0.06(x - 4)^2 + 1$

2. $y = -6(x + 4)^2 + 8$

4.1 Practice A

1. 5 2. ±7 3. ±3

4. 3 5. 2 6. 8

7. 125 8. 100 9. 2

10. 7 11. 5 12. 0.44

13. 2.47 14. 59,049 15. 0.03

16. 6.05 in. 17. 5.6 m 18. $x = \pm 4$

19. $x = 5$ 20. $x = 12.32, x = -0.32$

21. $x = 3$ 22. $x = -2.57$

23. $x = \pm 1.57$ 24. 2.57%

4.1 Practice B

1. 7 2. no real roots 3. −3

4. 216 5. 8 6. 4 7. −3125

8. $\frac{1}{1024}$ 9. $\frac{1}{81}$ 10. 0.51 11. 7.42

12. 8869.01 13. 0.07 14. 25.64

15. −12.41 16. 7.6 in. 17. 7 m

18. $x = \pm 1.78$ 19. $x = -2.97$ 20. $x = \pm 3$

21. $x = 3.42$ 22. $x = \pm 5$ 23. $x = -7$

24. a. about 0.72 au
    b. 12 years

4.1 Enrichment and Extension

1. 3.17 2. 6.17 3. 10.20

4. 11.23 5. 9.11 6. 7.21

7. 11.88 8. 12.81 9. 3.63

10. 14.67 11. 15.78 12. 19.88

4.1 Puzzle Time

I WISH YOU WOULD NOT ASK ME THAT I DO NOT KNOW ONE BUG FROM ANOTHER

4.2 Start Thinking

$x^2 \bullet x^2 = x^4$

When multiplying numbers with exponents, the exponents are added. Most students will probably expect $\sqrt{x} \bullet \sqrt{x}$ to yield an answer less than the value of x.

$$\sqrt{x} \bullet \sqrt{x} = x^{1/2} \bullet x^{1/2} = x^{1/2+1/2} = x^1 = x$$

The answer is not greater than or less than x; It is equal to x. This answer may surprise some students whose intuition is to think of multiplying fractions.

Answers

4.2 Warm Up

1. $\frac{1}{x}$
2. $\frac{3}{x^6}$
3. $4b^7$
4. ab^4
5. y^2
6. p^3

4.2 Cumulative Review Warm Up

1. $b^2 - 4ac = 309$; two real solutions
2. $b^2 - 4ac = -491$; two imaginary solutions
3. $b^2 - 4ac = 144$; two real solutions
4. $b^2 - 4ac = 1409$; two real solutions
5. $b^2 - 4ac = 89$; two real solutions
6. $b^2 - 4ac = -23$; two imaginary solutions

4.2 Practice A

1. $7^{1/2}$
2. $14^{3/2}$
3. $\frac{1}{5^{4/5}}$
4. $10^{3/4}$
5. $\frac{3}{2}$
6. $7^{1/2}$
7. 15
8. 9
9. $2\sqrt[4]{6}$
10. 3
11. 2
12. $\frac{1}{4}$
13. $2\sqrt[4]{13}$
14. $\frac{\sqrt[3]{18}}{2}$
15. $\frac{\sqrt{15}}{9}$
16. $2 - \sqrt{3}$
17. $\frac{24 + 6\sqrt{5}}{11}$
18. $\frac{-8\sqrt{2} + 8\sqrt{5}}{3}$
19. $13\sqrt[4]{2}$
20. $-10\sqrt[5]{13}$
21. $11\left(9^{1/4}\right)$
22. $-3\sqrt{2}$
23. $44\sqrt{7}$
24. $5\sqrt[4]{5}$
25. a. $s = V^{1/3}$

    b. $S = 6V^{2/3}$

    c. $24\left(10^{2/3}\right)$ cm^2

4.2 Practice B

1. $\frac{1}{2^{3/5}}$
2. 4
3. $11^{1/3}$
4. $9^{2/5}$
5. $9^{3/4}$ or $3^{3/2}$
6. $5^{4/3}$
7. 25
8. $3\sqrt[5]{2}$
9. 2
10. $\frac{1}{10}$
11. 5
12. 2
13. $\frac{\sqrt[3]{12}}{3}$
14. $\frac{\sqrt[3]{20}}{5}$
15. $\frac{7\sqrt[4]{4}}{2}$
16. $\frac{35 + 7\sqrt{3}}{22}$
17. $\frac{-6\sqrt{2} + 6\sqrt{7}}{5}$
18. $\frac{\sqrt{30} + \sqrt{6}}{12}$
19. $4\left(25^{2/3}\right)$
20. $-5\sqrt{6}$
21. $8\sqrt[3]{3}$
22. $13\sqrt[5]{2}$
23. $5\left(3^{1/4}\right)$
24. $13\left(7^{1/3}\right)$
25. a. $r = \frac{\sqrt{\pi V}}{3\pi}$

    b. $S = 6\sqrt{\pi V} + \frac{V}{9}$

    c. $\left(36\sqrt{3\pi} + 12\right) \approx 122.52$ m^2

4.2 Enrichment and Extension

1. $x^{1/2}(x - 8)$
2. $\left(3x^{2/3} - 2\right)\left(4x^{2/3} + 1\right)$
3. $5x^{1/3}\left(2x^{1/24} - 3 + 4x^{1/2}\right)$
4. $4x^{5/3}\left(2x^{2/3} - x^2 + 3\right)$
5. $\left(x^{1/5} + 5\right)\left(x^{2/5} - 3\right)$
6. $\left(x^{1/6} - 3\right)\left(x^{1/3} - 3\right)$
7. $(x + 2)^{3/2}(x + 1)(x + 3)$
8. $x(x + 1)^{1/2}(x - 3)$
9. $3x^{-1/2}(x - 2)(x - 1)$
10. $x^{-3/2}(2x + 1)(x + 2)$

4.2 Puzzle Time

HE WAS NOT SO BRIGHT

Answers

4.3 Start Thinking

Sample answer:

$f(x) = |x|$ is a function with a domain of all real numbers, but because the values of $f(x)$ will be non-negative, the range is non-negative. When constructing this function, it is necessary to consider operations that make numbers positive and can include zero.

The function $f(x) = \sqrt{x}$ has a domain of all real non-negative numbers because you cannot get a real number value of $f(x)$ if $x < 0$. Its range is also all real non-negative numbers.

4.3 Warm Up

1. translation 9 units down
2. vertical shrink by a factor of $\frac{1}{4}$ and a translation 5 units up
3. vertical shrink by a factor of $\frac{1}{2}$
4. translation 0.5 unit down
5. translation 3 units up
6. reflection in the x-axis, a vertical stretch by a factor of 2, and a translation 1 unit up

4.3 Cumulative Review Warm Up

1. no **2.** no **3.** yes **4.** yes

4.3 Practice A

1.

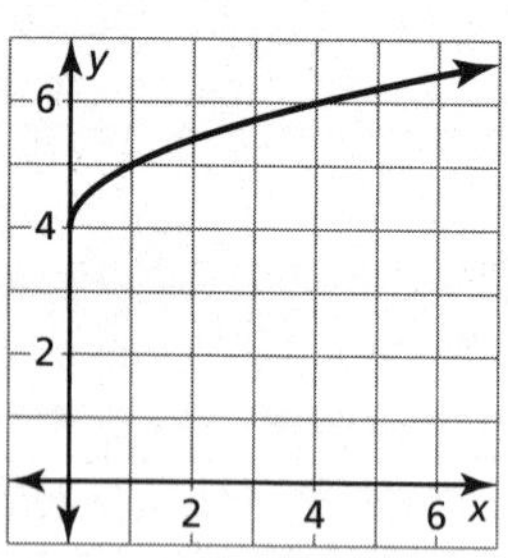

$x \geq 0; y \geq 4$

2.

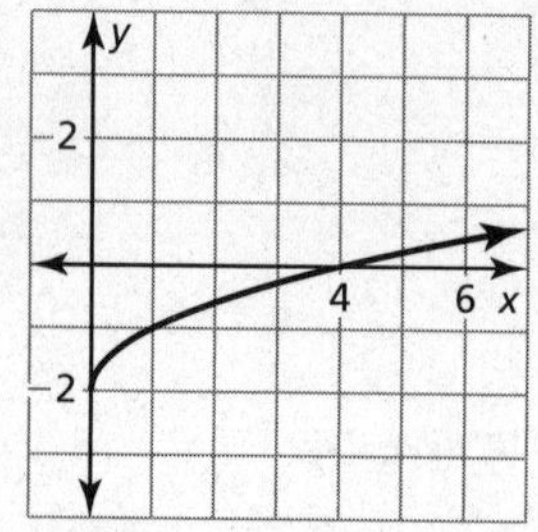

$x \geq 0; y \geq -2$

3.

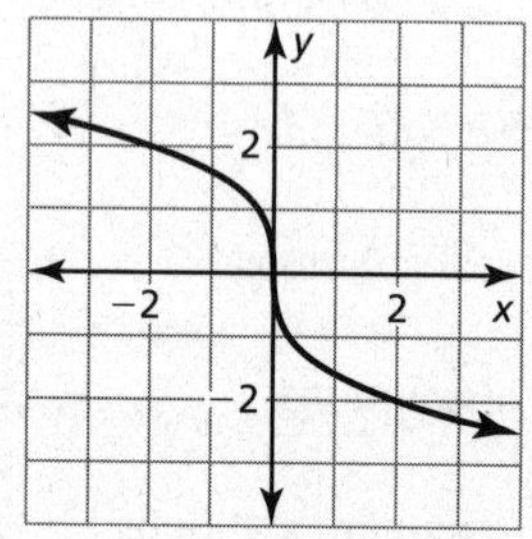

x: all real numbers; y: all real numbers

4.

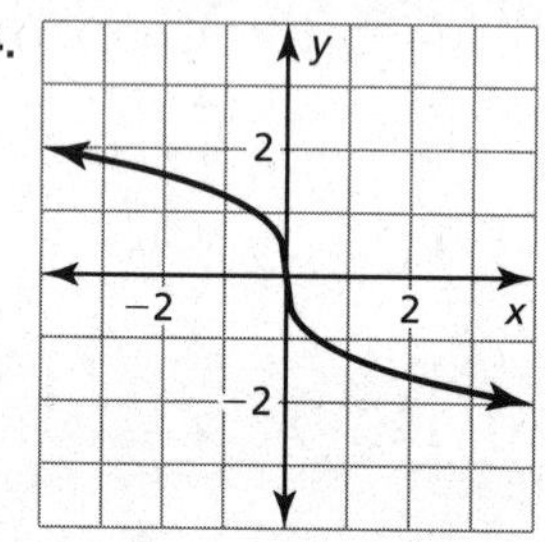

x: all real numbers; y: all real numbers

5.

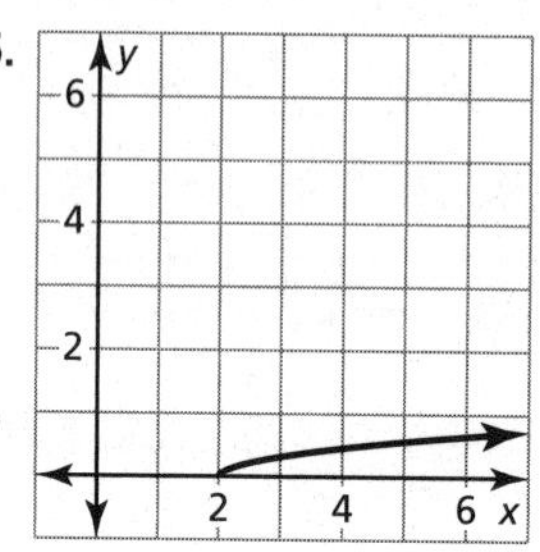

$x \geq 2; y \geq 0$

6.

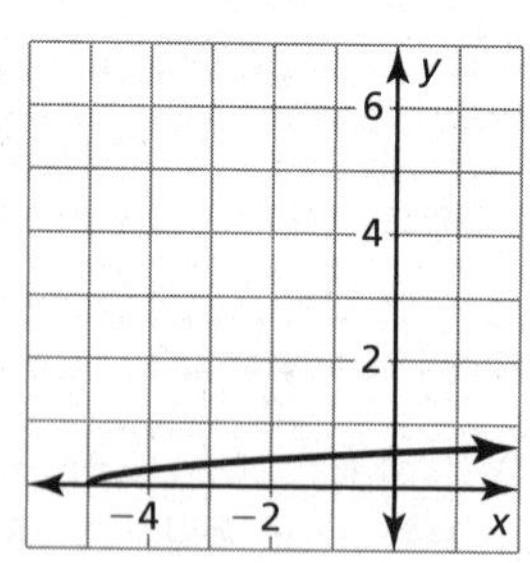

$x \geq -5; y \geq 0$

Answers

7. translation 1 unit right and 4 units up

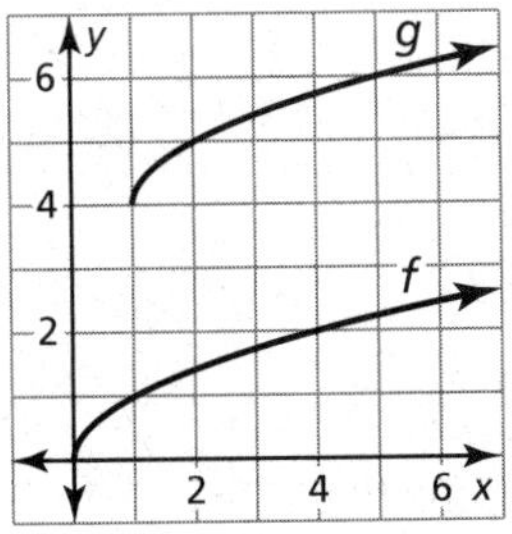

8. translation 2 units left and a vertical stretch by a factor of 3

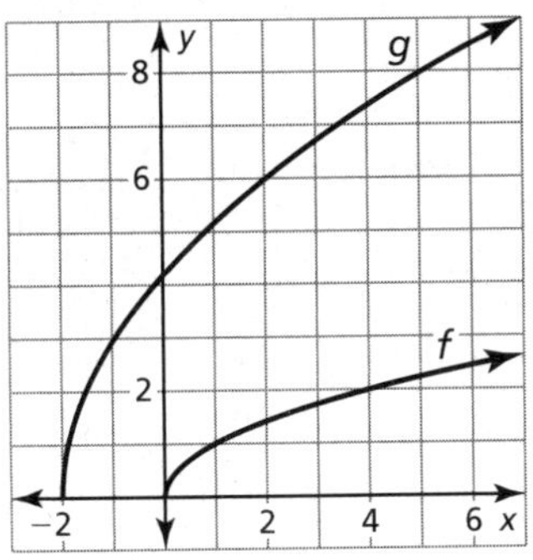

9. vertical stretch by a factor of 2 and a reflection in the x-axis

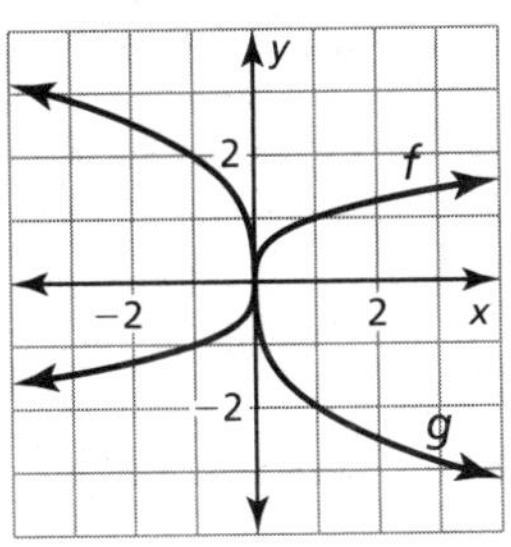

10. translation 1 unit right and 3 units up

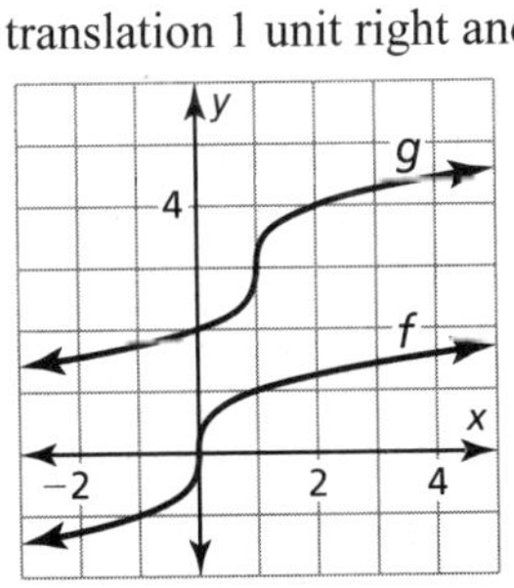

11. reflection in the y-axis and a vertical stretch by a factor of 3

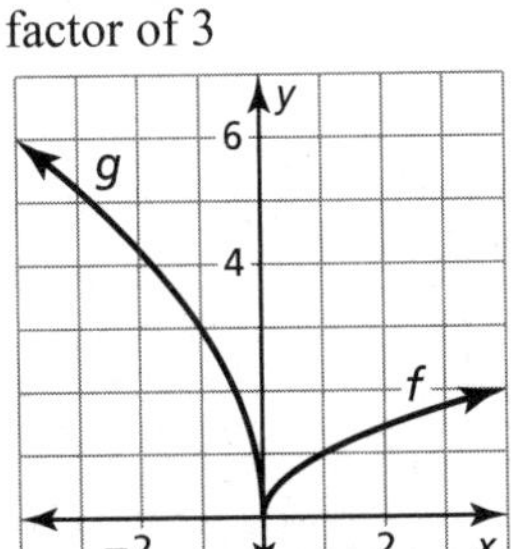

12. reflection in the x-axis and a vertical shrink by a factor of $\frac{1}{3}$

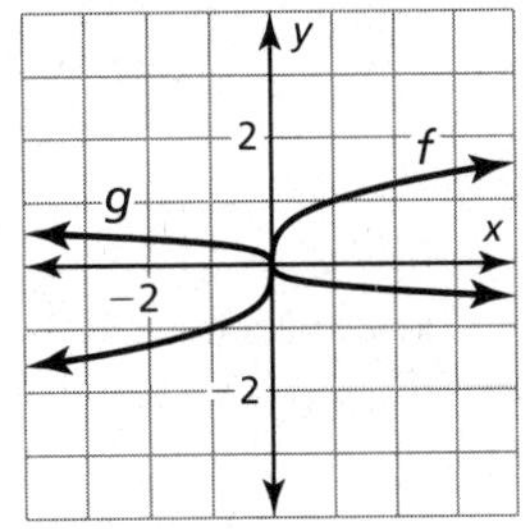

13.

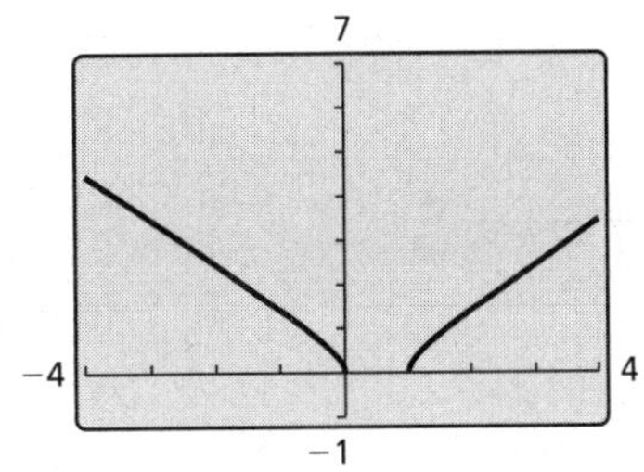

$x \leq 0, x \geq 1; y \geq 0$

14.

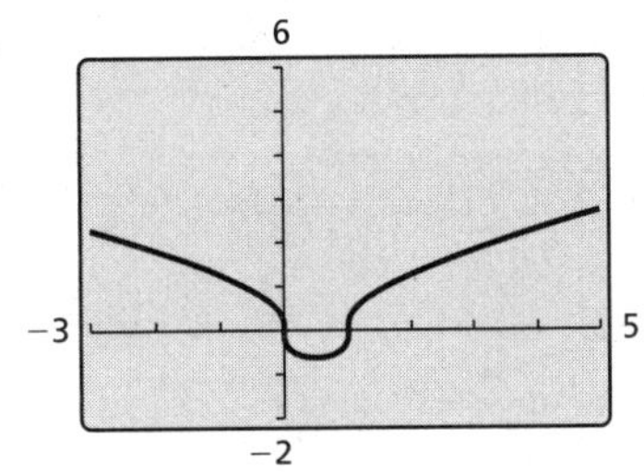

x: all real numbers; $y \geq -0.63$

15.

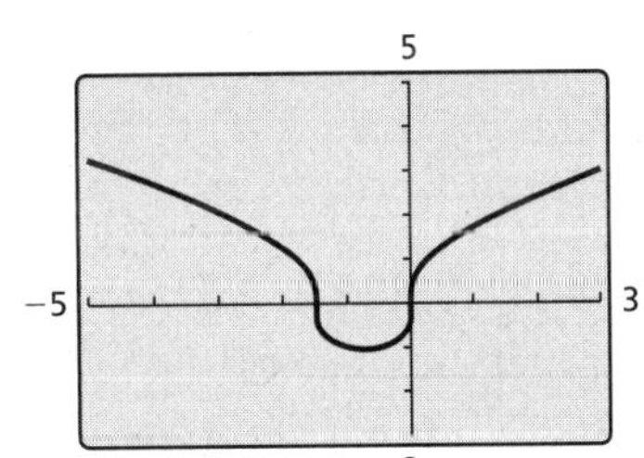

x: all real numbers; $y \geq -1.04$

16. $g(x) = \frac{1}{3}\sqrt{x+2}$

17. $g(x) = -5\sqrt{x} - 5$

18.

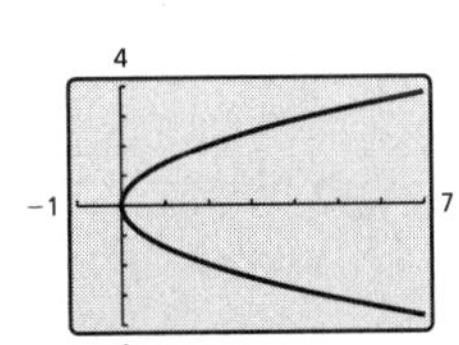

vertex: $(0, 0)$; opens right

19.

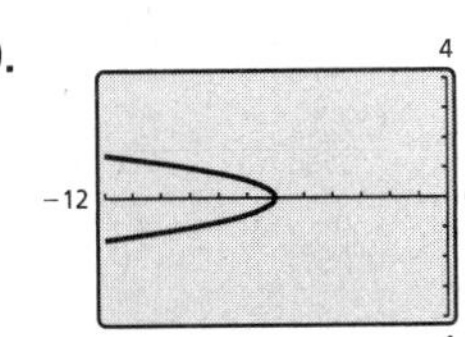

vertex: $(-6, 0)$; opens left

Answers

20.

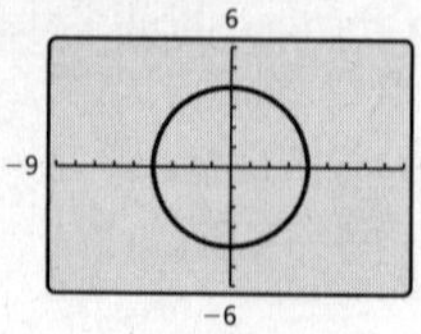

center: $(0, 0)$; $r = 4$;

x-intercepts: ± 4;

y-intercepts: ± 4

21.

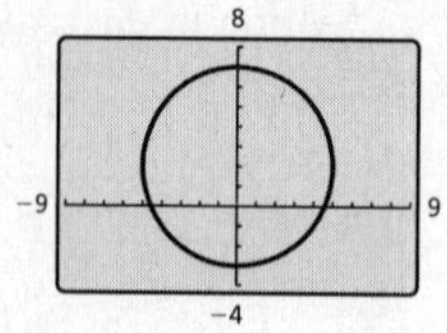

center: $(0, 2)$; $r = 5$;

x-intercepts: $\pm\sqrt{21}$;

y-intercepts: -3, 7

4.3 Practice B

1.

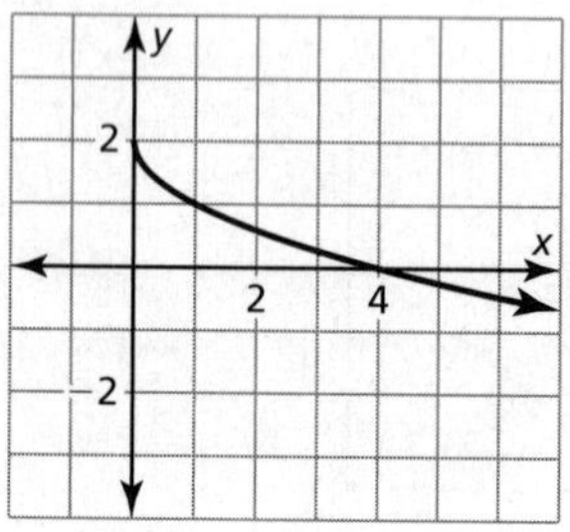

$x \geq 0$; $y \leq 2$

2.

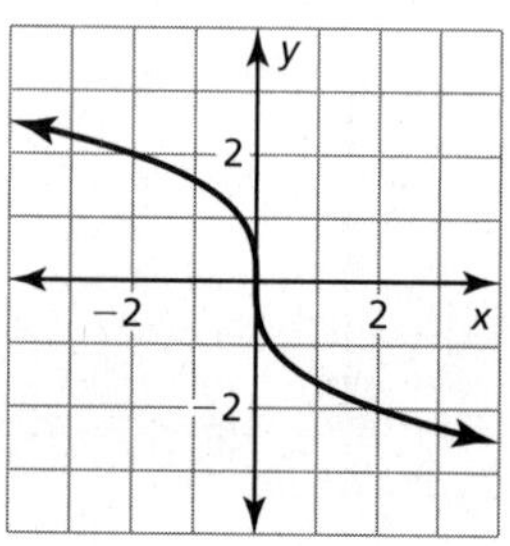

x: all real numbers; y: all real numbers

3.

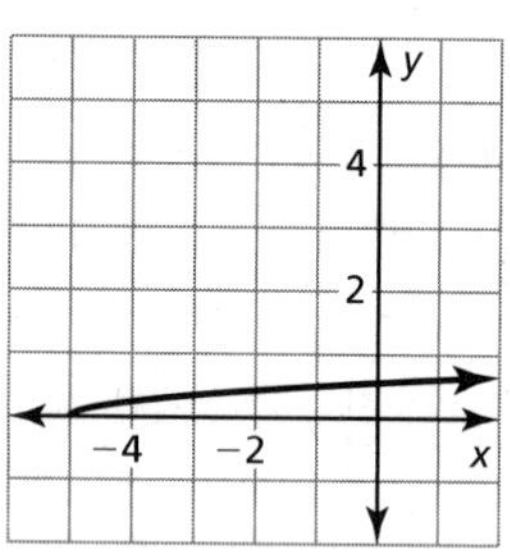

$x \geq -5$; $y \geq 0$

4.

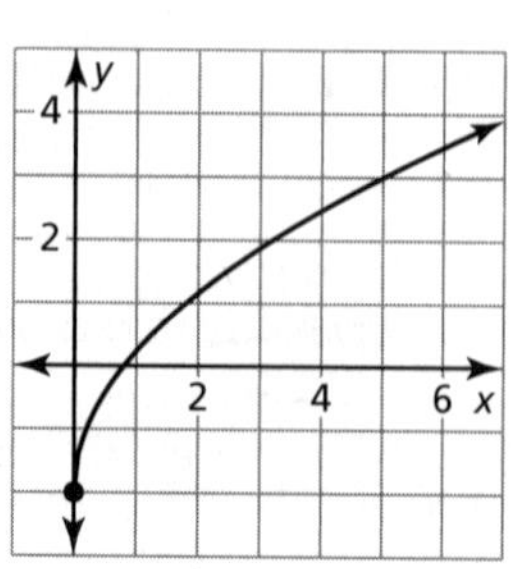

$x \geq 0$; $y \geq -2$

5.

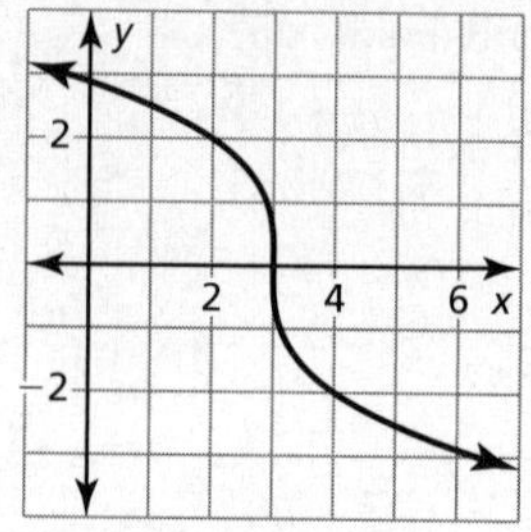

x: all real numbers; y: all real numbers

6.

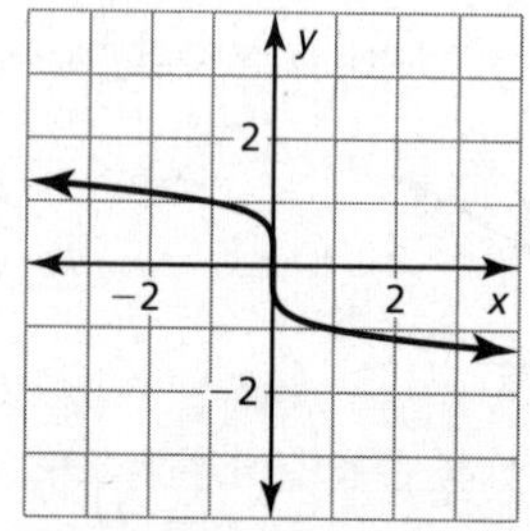

x: all real numbers; y: all real numbers

7. translation 2 units right and a vertical stretch by a factor of 4

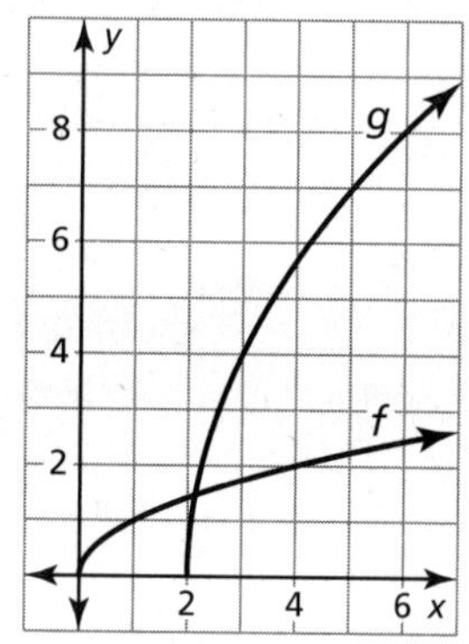

8. translation 5 units right and 1 unit down

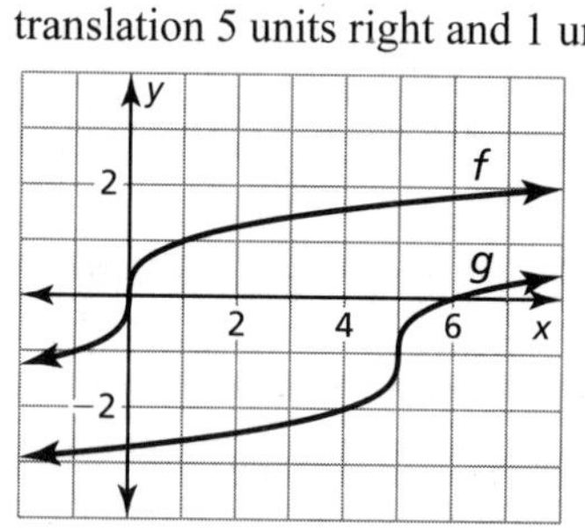

9. vertical shrink by a factor of $\frac{1}{3}$ and a reflection in the y-axis

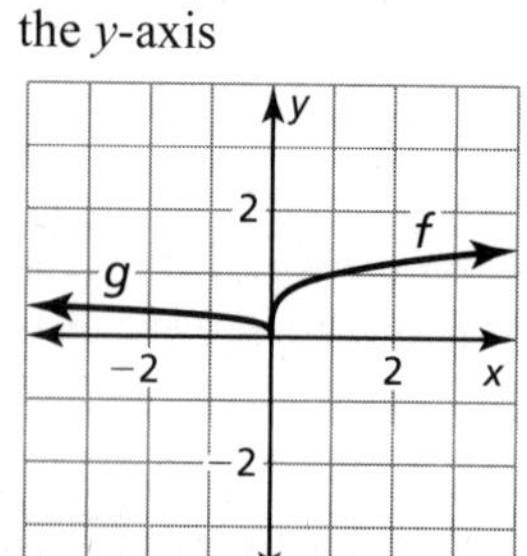

Answers

10. vertical shrink by a factor of $\frac{1}{2}$ and a translation 3 units down

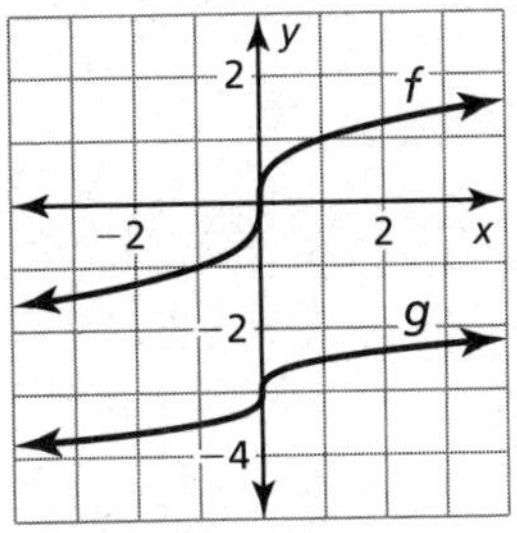

11. reflection in the x-axis and a translation 1 unit right and 3 units up

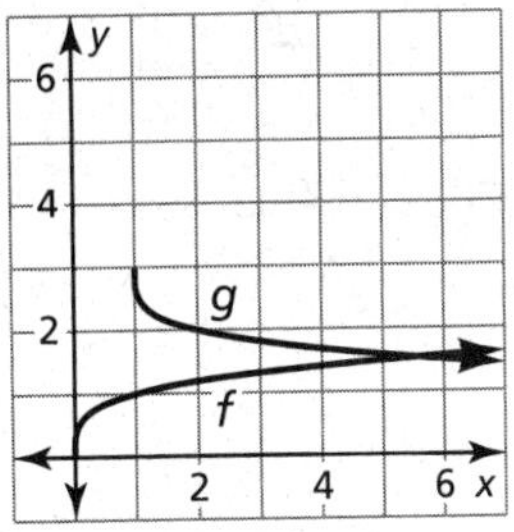

12. reflection in the y-axis, a horizontal shrink by a factor of $\frac{1}{243}$, and a translation 2 units down

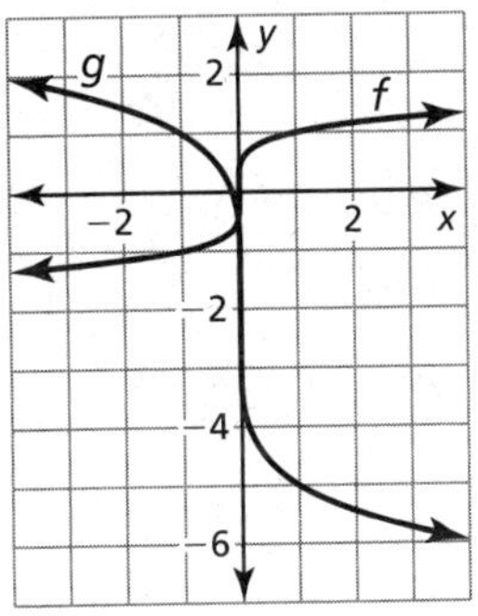

13.

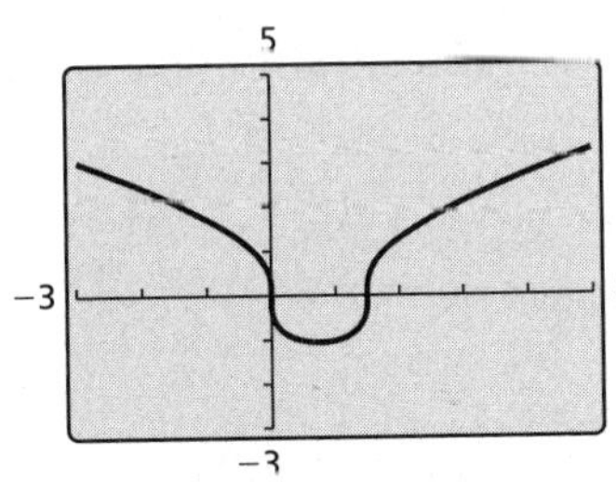

x: all real numbers; $y \geq -1.04$

14.

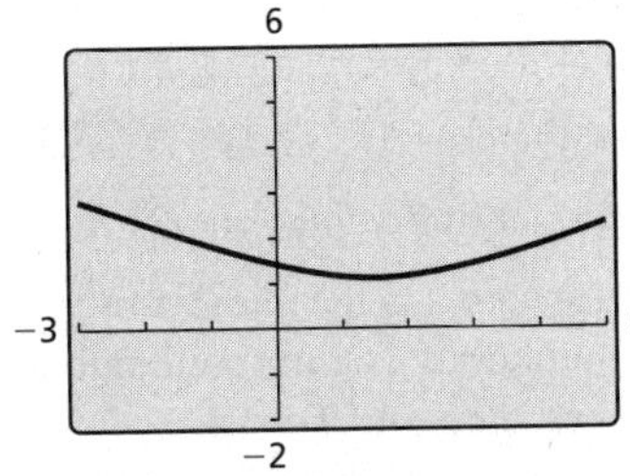

x: all real numbers; $y \geq 1.12$

15.

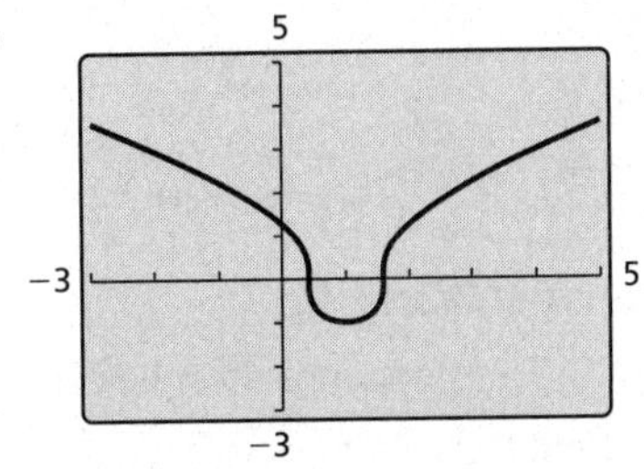

x: all real numbers; $y \geq -1$

16. $g(x) = \sqrt{\frac{3}{2}x} + 2$

17. $g(x) = \sqrt{x - 4} + \frac{1}{2}$

18.

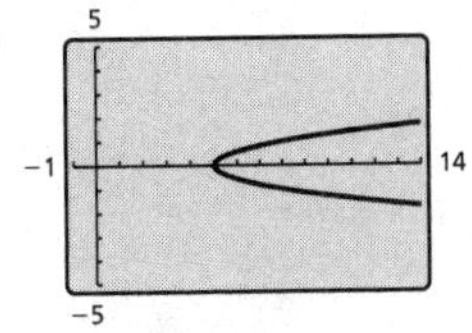

vertex: $(5, 0)$;
opens right

19.

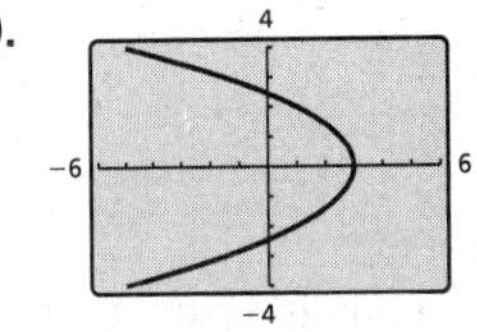

vertex: $(3, 0)$;
opens left

20.

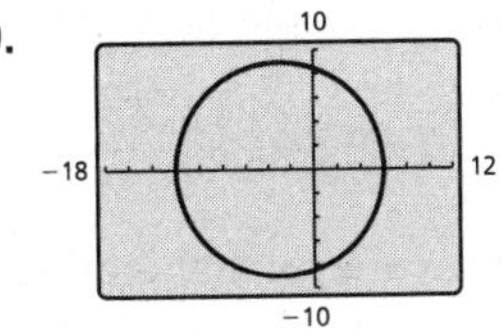

center: $(-3, 0)$; $r = 9$;
x-intercepts: $-12, 6$;
y-intercepts: $\pm 6\sqrt{2}$

21.

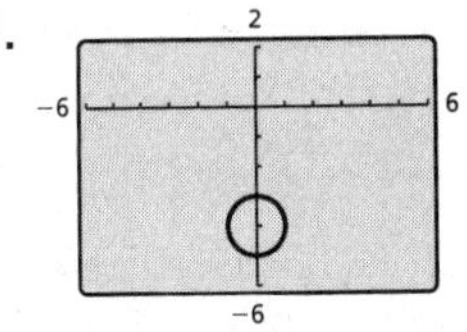

center: $(0, -4)$; $r = 1$;
There are no x-intercepts;
y-intercepts: $-5, -3$

4.3 Enrichment and Extension

1. $(10, 53.13°)$
2. $(1.41, 45°)$
3. $(8.94, 26.57°)$
4. $(10.20, 168.69°)$
5. $(5, 233.13°)$
6. $(8.60, 234.46°)$

4.3 Puzzle Time

A POTATO

4.4 Start Thinking

You can manipulate the Pythagorean Theorem to show the following.

$$a^2 + b^2 = c^2$$
$$a^2 + b^2 = 13^2$$
$$\sqrt{a^2 + b^2} = \sqrt{13^2}$$
$$\sqrt{a^2 + b^2} = 13$$

Answers

4.4 Warm Up

1. yes **2.** no **3.** yes

4. no **5.** yes **6.** no

4.4 Cumulative Review Warm Up

1. $y = 150x + 2000$; The slope indicates the car costs \$150 per month to lease.

2. $y = 3x$; The slope indicates the helicopter is traveling at a rate of speed of 3 miles per minute.

4.4 Practice A

1. $x = 9$ **2.** $x = \frac{40}{3}$ **3.** $x = 54$

4. $x = 216$ **5.** $x = 4$ **6.** $x = 5$

7. about 46 years **8.** $x = 16$ **9.** $x = 7, 9$

10. $x = 3$ **11.** $x = 0, -\frac{3}{2}$ **12.** $x = \frac{1}{3}$

13. $x = 7$ **14.** $x = 27$ **15.** no solution

16. $x = 4$

17. The cube root, instead of the cube, of the right side of the equation was found.

$$\sqrt[3]{2x + 1} = 8$$
$$\left(\sqrt[3]{2x + 1}\right)^3 = (8)^3$$
$$2x + 1 = 512$$
$$2x = 511$$
$$x = \frac{511}{2}$$

18. $x \geq 9$ **19.** $3 \leq x \leq 52$

20. $x > 5$

21. a. about 0.11 in.

b. about 0.18 in.

c. no; It is about 1.59 times as long.

4.4 Practice B

1. $x = 6$ **2.** $x = \frac{25}{16}$ **3.** $x = -256$

4. $x = \frac{3}{16}$ **5.** $x = 32$ **6.** $x = 32$

7. no solution **8.** $x = 0, x = \frac{4}{3}$

9. $x = \frac{1}{5}$ **10.** $x = 4$ **11.** $x = -3$

12. no solution **13.** $x = 64$ **14.** $x = \frac{4}{3}$

15. $x = 2$ **16.** $0 \leq x \leq 25$

17. $x \geq 26$ **18.** $x > -2$ **19.** 25 ft

20. $(-2, 0), (-1, 1)$

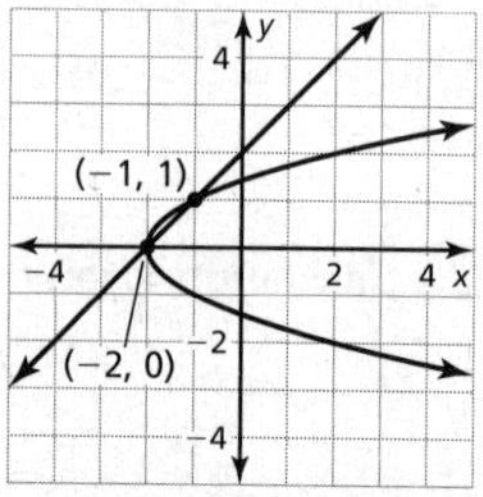

21. $(3, 2), (-2, -3)$

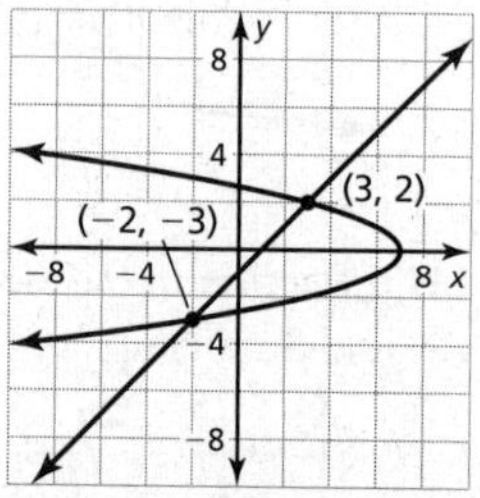

22. $(0, -3), (3, 0)$

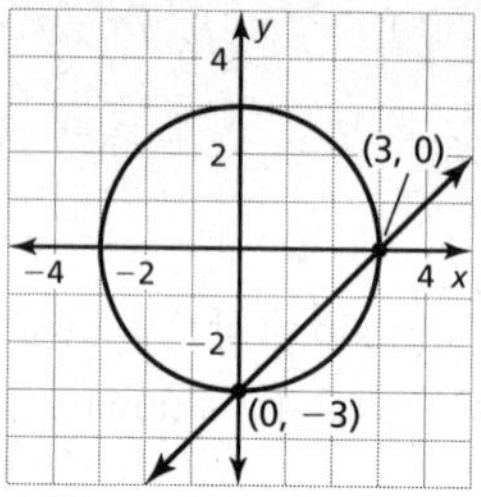

23. $(-4, 0), (0, 4)$

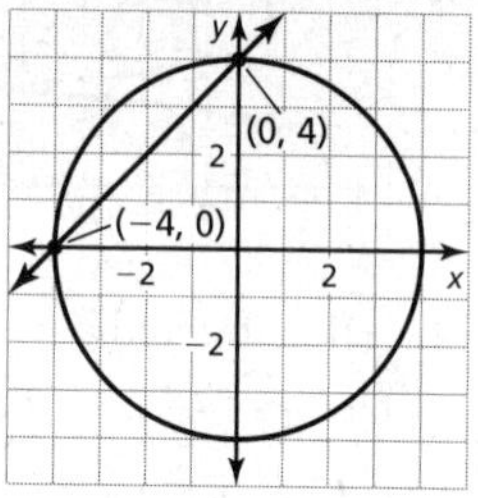

24. about 44.4 ft

4.4 Enrichment and Extension

1. $x = -21$ **2.** $x = 3$

3. $x \approx 23.39$ **4.** $x \approx -2.87$

5. $x \geq 19$ **6.** $0 \leq x \leq 1.33$

4.4 Puzzle Time

WELL THIS IS A FINE PICKLE YOU'VE GOTTEN US INTO

4.5 Start Thinking

The equations as a system are

$$f(x) = 3x - 750$$
$$g(x) = 15x - 1300$$

The equation is only logical because the company sold the same number of bracelets and T-shirts. If the numbers were different, the new equation would not work. The combined equation $h(x) = 18x - 2050$ represents the total amount of earnings, $h(x)$, on bracelets and T-shirts, where x is the number of bracelets sold or the number of T-shirts sold, but not the combined number of bracelets and T-shirts.

Answers

4.5 Warm Up

1. $3x^4 + 6x^2$
2. $-\frac{7}{x^6}$
3. $\frac{x^2}{x - 1}$
4. a^4b^4
5. $a^2 - 5ab - 6b^2$
6. $-6x^2 + 6x$

4.5 Cumulative Review Warm Up

1. $y = \frac{3}{88}(x + 7)(x - 11)$
2. $y = \frac{19}{9}(x - 10)(x)$
3. $y = \frac{71}{28}(x + 15)(x + 3)$
4. $y = -\frac{1}{5}(x + 6)(x + 4)$

4.5 Practice A

1. $(f + g)(x) = 12\sqrt[4]{x};\ x \geq 0;$
$(f - g)(x) = -18\sqrt[4]{x};\ x \geq 0;$
$(f + g)(81) = 36;\ (f - g)(81) = -54$

2. $(f + g)(x) = 3x^2 + 6x + 7;$ x: all real numbers;
$(f - g)(x) = x^2 + 12x - 7;$ x: all real numbers;
$(f + g)(1) = 16;\ (f - g)(1) = 6$

3. $(fg)(x) = 2x^2\sqrt{x};\ x \geq 0;\ \left(\frac{f}{g}\right)(x) = \frac{x^2}{2\sqrt{x}};$
$x > 0;\ (fg)(9) = 486;\ \left(\frac{f}{g}\right)(9) = \frac{27}{2}$

4. $(fg)(x) = 40x^{14/3};$ x: all real numbers;
$\left(\frac{f}{g}\right)(x) = \frac{5}{2}x^{4/3};$ x: all real numbers, $x \neq 0;$
$(fg)(8) = 655{,}360;\ \left(\frac{f}{g}\right)(8) = 40$

5. $(fg)(x) = 8x;$ x: all real numbers; $\left(\frac{f}{g}\right)(x) = 2x^{1/3};$
x: all real numbers, $x \neq 0;\ (fg)(-27) = -216;$
$\left(\frac{f}{g}\right)(-27) = -6$

6. $(f + g)(5) = 654.91;\ (f - g)(5) = 595.09;$
$(fg)(5) = 18{,}691.86;\ \left(\frac{f}{g}\right)(5) = 20.90$

7. $(f + g)(5) = 148.49;\ (f - g)(5) = -125.10;$
$(fg)(5) = 1600.00;\ \left(\frac{f}{g}\right)(5) = 0.09$

8. Raising a number to the power $\frac{1}{2}$ is the same as taking the square root of a number, and you cannot take the square root of a negative number.

$f(x) = 4x^{1/2} + 2$ and $g(x) = -4x^{1/2}$

The domain of $(f + g)(x)$ is $x \geq 0$.

9. **a.** $(A - B)(t) = \frac{1}{2}t^{2/3}$

b. the difference in the amount of mold in the two specimens

4.5 Practice B

1. $(f + g)(x) = -8\sqrt[3]{4x};$ x: all real numbers;
$(f - g)(x) = 10\sqrt[3]{4x};$ x: all real numbers;
$(f + g)(-2) = 16;\ (f - g)(-2) = -20$

2. $(f + g)(x) = -x^3 + x^2 - x;$ x: all real numbers;
$(f - g)(x) = -x^3 - 11x^2 + 7x;$ x: all real numbers; $(f + g)(-1) = 3;\ (f - g)(-1) = -17$

3. $(fg)(x) = 3x^3\sqrt[3]{x^2};$ x: all real numbers;
$\left(\frac{f}{g}\right)(x) = 3x^2\sqrt[3]{x};$ x: all real numbers, $x \neq 0;$
$(fg)(-8) = -6144;\ \left(\frac{f}{g}\right)(-8) = -384$

4. $(fg)(x) = 15x^{9/4};\ x \geq 0;\ \left(\frac{f}{g}\right)(x) = \frac{3}{5}x^{7/4};$
$x > 0;\ (fg)(16) = 7680;\ \left(\frac{f}{g}\right)(16) = \frac{384}{5}$

5. $(fg)(x) = 20x^{7/6};\ x \geq 0;\ \left(\frac{f}{g}\right)(x) = 5x^{1/2};$
$x > 0;\ (fg)(64) = 2560;\ \left(\frac{f}{g}\right)(64) = 40$

6. $(f + g)(5) = 3.81;\ (f - g)(5) = -14.07;$
$(fg)(5) = -45.88;\ \left(\frac{f}{g}\right)(5) = -0.57$

Answers

7. $(f+g)(5) = 26.77$; $(f-g)(5) = 13.35$;

$(fg)(5) = 134.58$; $\left(\frac{f}{g}\right)(5) = 2.99$

8. $g(0)$ is equal to 0, and 0 cannot be in the denominator of a fraction.

$f(x) = 4x^{7/3}$ and $g(x) = 2x^{2/3}$

The domain of $\left(\frac{f}{g}\right)(x)$ is all real numbers except $x = 0$.

9. $(f+g)(5) = -5$, $(f-g)(0) = -46$,

$(fg)(3) = 24$, and $\left(\frac{f}{g}\right)(2) = \frac{1}{2}$

4.5 Enrichment and Extension

1. true; The remainder is 4 when both 18 and 60 are divided by 7.
2. false; The remainder is 3 when 21 is divided by 6, and the remainder is 2 when 44 is divided by 6.
3. false; The remainder is 7 when 15 is divided by 8, and the remainder is 1 when 25 is divided by 8.
4. true; The remainder is 2 when both 10 and 14 are divided by 4.

5. 3 **6.** 1 **7.** 0 **8.** 4

9. Sunday **10.** Wednesday

11. *Sample answer:* January 25, 1990: Thursday

4.5 Puzzle Time

DINNER IS ON ME

4.6 Start Thinking

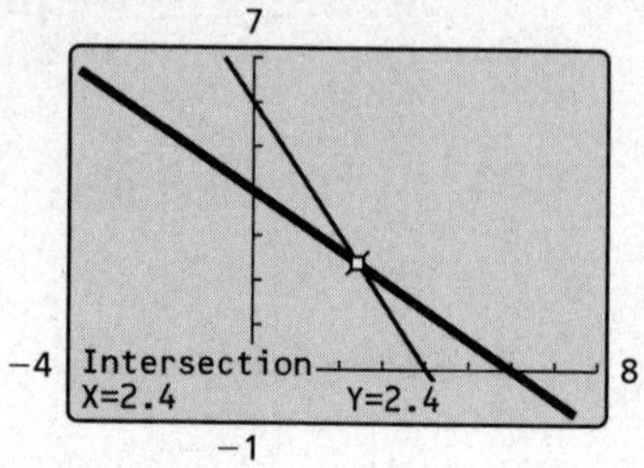

When solved for x, the equation is $x = -\frac{3}{2}y + 6$. Switching x and y, it is $y = -\frac{3}{2}x + 6$, and the system is $y = -\frac{2}{3}x + 4$

$y = -\frac{3}{2}x + 6$.

The point of intersection is $(2.4, 2.4)$. The point of intersection represents the only x-value that, when substituted into the original equation, yields the same y-value.

4.6 Warm Up

1. $y = 3x - 4$ **2.** $y = \frac{3}{2}x - 5$

3. $y = -\frac{5}{6}x + \frac{3}{2}$ **4.** $y = -4x + 7$

5. $y = \frac{1}{5}x - \frac{11}{5}$ **6.** $y = \frac{1}{6}x - \frac{3}{2}$

4.6 Cumulative Review Warm Up

1. $5i$ **2.** $2i\sqrt{5}$ **3.** $9i\sqrt{2}$

4. $-18i$ **5.** $2i$ **6.** $2i\sqrt{11}$

4.6 Practice A

1. $x = \frac{y-3}{2}$; $x = -3$
2. $x = 3y + 6$; $x = -3$
3. $x = \frac{1}{2}\sqrt[3]{y}$; $x = -\frac{1}{2}\sqrt[3]{3}$
4. $g(x) = \frac{x}{4}$

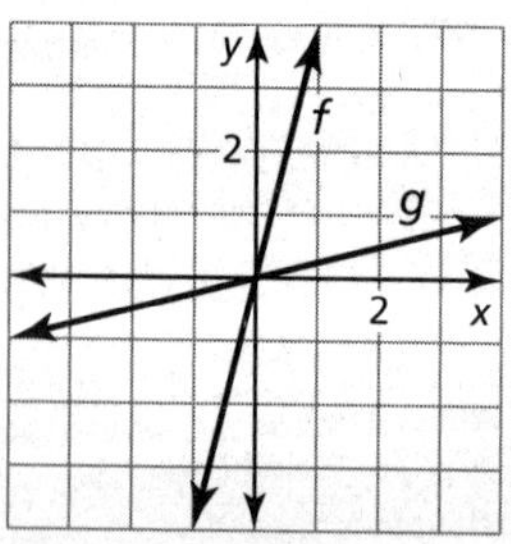

Answers

5. $g(x) = \frac{1}{4}x + \frac{1}{4}$

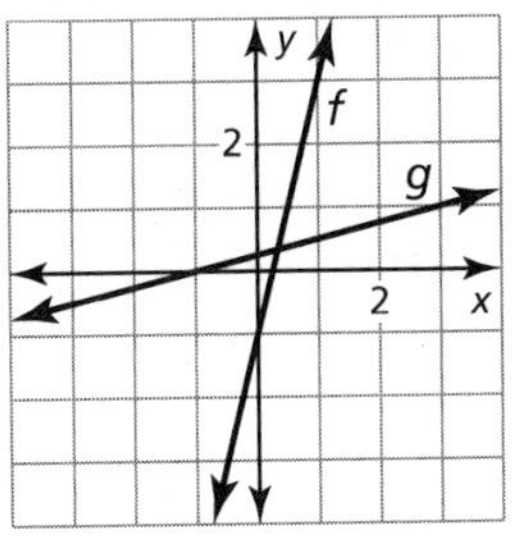

6. $g(x) = 2x + 10$

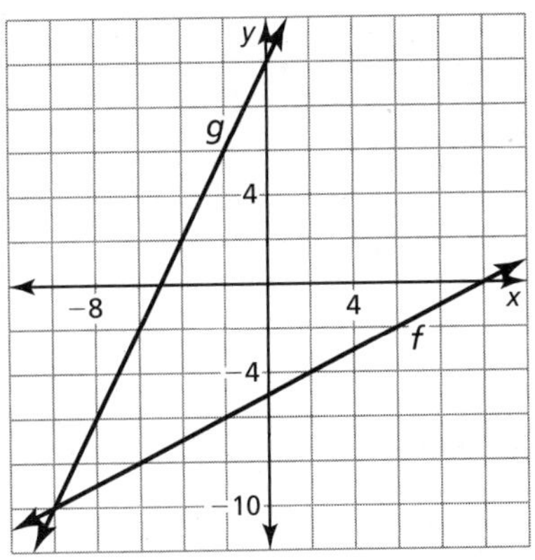

7. $g(x) = 5x + 10$; *Sample answer:* switching the positions of x and y and solving for y; less mental math

8. a. no; The coordinates of $(0, 9)$ cannot be switched to obtain $(0, 0)$.

   b. yes; The coordinates of the corresponding points are switched.

9. $g(x) = \frac{\sqrt{x}}{3}$

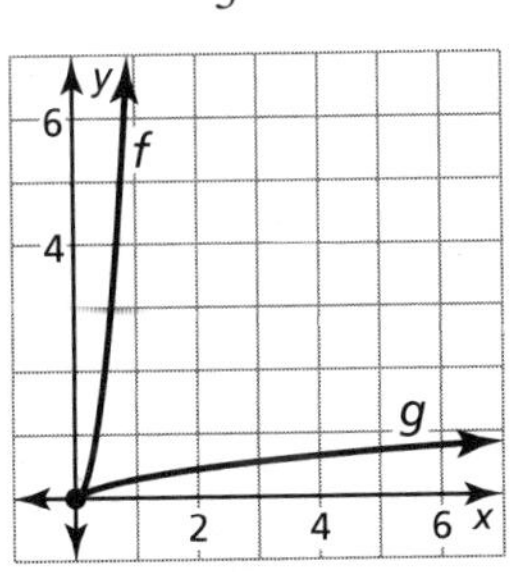

10. $g(x) = -\frac{\sqrt{x}}{4}$

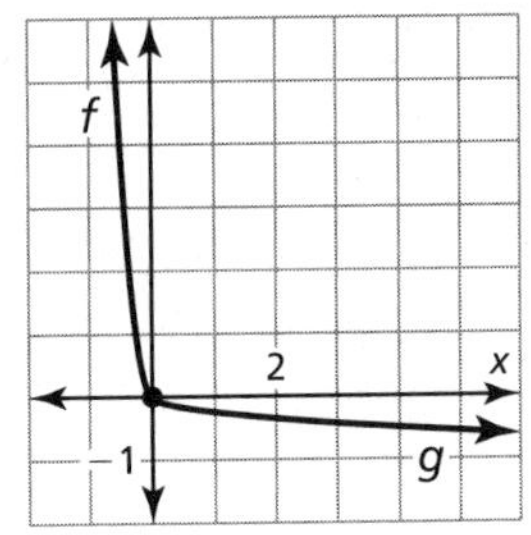

11. $g(x) = \sqrt[3]{x} - 2$

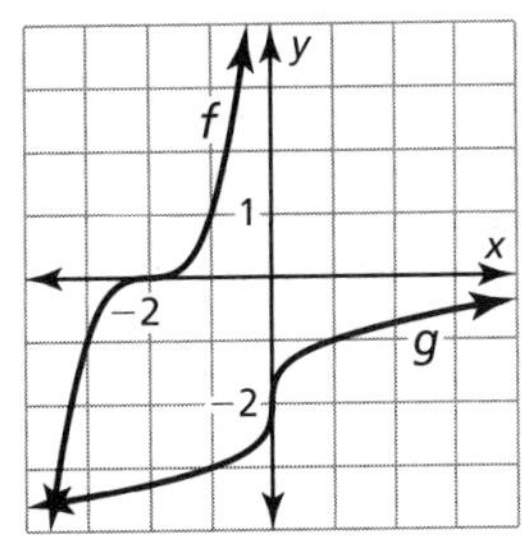

12. yes; passes the horizontal line test

13. no; does not pass the horizontal line test

4.6 Practice B

1. $x = -\frac{3}{4}y + \frac{3}{2}; x = \frac{15}{4}$

2. $x = \frac{\sqrt[4]{25y}}{5}$; The output cannot be -3.

3. $x = \pm\sqrt{y + 4} + 3; x = 2, x = 4$

4. $g(x) = \frac{-x + 4}{3}$

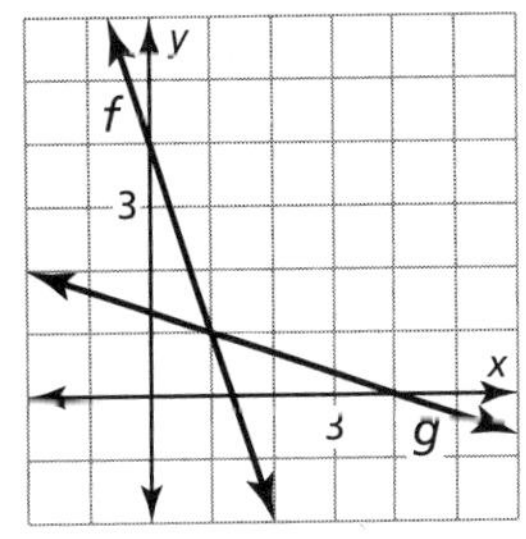

5. $g(x) = -3x + 3$

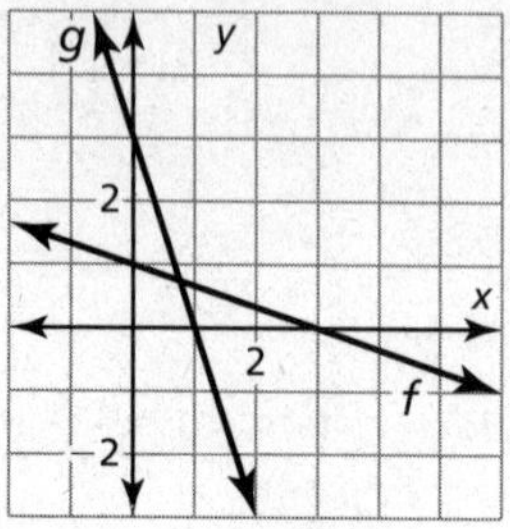

6. $g(x) = \frac{5}{2}x + \frac{1}{2}$

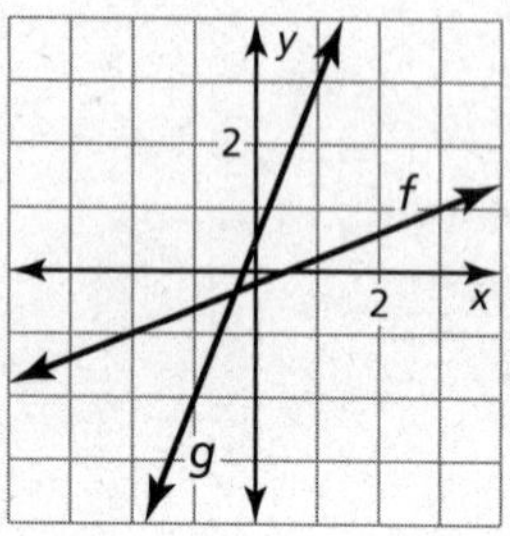

7. After switching the positions of x and y, y was not solved for.

$f(x) = 3x - 8$

$y = 3x - 8$

$x = 3y - 8$

$x + 8 = 3y$

$y = \dfrac{x + 8}{3}$

$g(x) = \dfrac{x + 8}{3}$

8. $g(x) = -\dfrac{\sqrt{-x}}{3}$

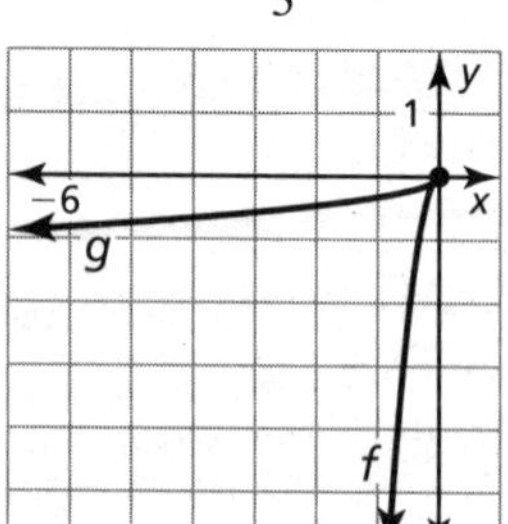

9. $g(x) = \sqrt[3]{x} + 1$

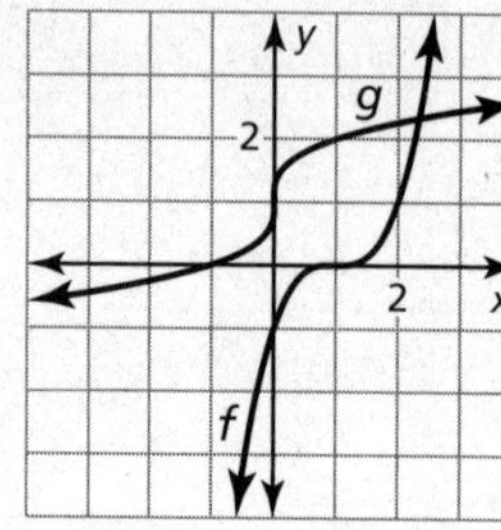

10. $g(x) = -\sqrt[6]{x}$

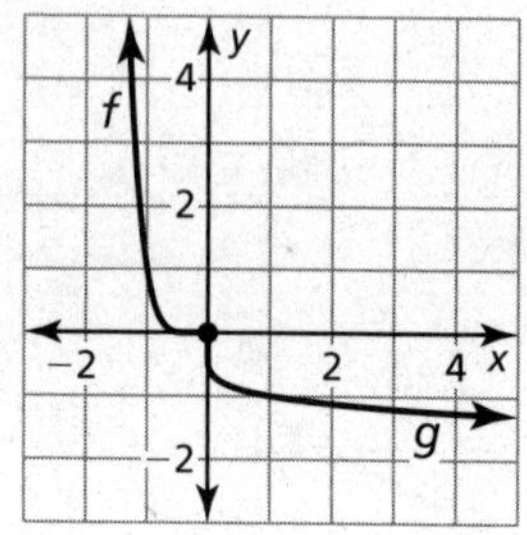

11. $g(x) = \dfrac{\sqrt[3]{x}}{2}$; *Sample answer:* switching the positions of x and y and solving for y; less mental math

12. no 13. yes 14. yes 15. no

16. a. $r = \sqrt[3]{\dfrac{3V}{4\pi}}$; The new equation represents the radius in terms of the volume.

b. $r \approx 3.27\,\text{m}$

4.6 Enrichment and Extension

1. $g(x) = 1 + \sqrt{x + 1}$

2. $g(x) = -4 + \sqrt{x + 16}$

3. $g(x) = \dfrac{3}{2} + \sqrt{x + \dfrac{9}{4}}$

4. $g(x) = 4 + \sqrt{x + 4}$

5. $g(x) = 6 + \sqrt{x + 41}$

6. $g(x) = -5 + \sqrt{x + 10}$

7. $g(x) = -3 + \sqrt{x + 8}$

8. $g(x) = 2 + \sqrt{x - 8}$

9. $g(x) = \frac{1}{2} + \sqrt{\frac{x+1}{4}}$

10. $g(x) = -1 + \sqrt{\frac{x+6}{6}}$

11. $g(x) = 2 + \sqrt{\frac{x+31}{4}}$

12. $g(x) = 1 + \sqrt{\frac{x+4}{9}}$

4.6 Puzzle Time

A BOMB SQUID

Cumulative Review

1. w^{11} **2.** b **3.** $\frac{1}{g^2}$

4. c^5 **5.** m^7 **6.** v^6

7. 1 **8.** $\frac{1}{p^5}$ **9.** x^2

10. $5a^8$ **11.** $-4y^5$ **12.** $3n^5$

13. $\frac{32m^{15}}{n^5}$ **14.** $\frac{343y^3}{q^{12}}$ **15.** $\frac{256x^{20}}{j^{12}}$

16. $\frac{x^{10}}{b^{10}}$ **17.** $81r^4c^8$ **18.** $\frac{d^{15}}{w^6}$

19. $y = -3x + 7$ **20.** $y = 9$

21. $y = 2x + 4$ **22.** $y = 3x + 4$

23. $y = \frac{1}{5}x - 4$ **24.** $y = 2x - 12$

25. $y = 2x + 6$ **26.** $y = 2x - 8$

27. $y = -4x + 6$ **28.** $y = 16x + 8$

29. $y = \frac{4}{x+3}$ **30.** $y = \frac{8}{2x-1}$

31. a. 84 in. **b.** 432 in.^{2}

32. a. $100 - 18m$

b. \$28; You have \$28 left over after purchasing 4 movies.

c. \$10; You have \$10 left over after purchasing 5 movies.

33. 64 **34.** 3125 **35.** 216

36. 9 **37.** $\frac{1}{3}$ **38.** $\frac{1}{49}$

39. 256 **40.** 81 **41.** $\frac{1}{m^{12}}$

42. $\frac{1}{p^{10}}$ **43.** k^{28} **44.** j^{24}

45. $\frac{16y^2}{25}$ **46.** $\frac{8}{27z^3}$ **47.** $\frac{36}{25w^2}$

48. $\frac{64x^3}{27}$ **49.** $x = \frac{9}{5}$ and $x = 3$

50. $x = -10$ and $x = 6$ **51.** $x = -4$ and $x = \frac{52}{7}$

52. $x = -\frac{51}{5}$ and $x = 5$ **53.** no solution

54. no solution **55.** $x = \frac{12}{5}$ and $x = 4$

56. $x = -8$ and $x = -4$

57. $x = 4$ and $x = 12$ **58.** $x = -\frac{5}{3}$ and $x = 13$

59. $x = -29$ and $x = -\frac{1}{3}$

60. $x = \frac{9}{5}$ and $x = 3$ **61.** function

62. not a function **63.** not a function

64. not a function **65.** function

66. not a function **67.** function

68. function

69. a. $3x + 225$ **b.** \$243

70. a. $5.85m + 28.84t$ **b.** \$115.77

71. $g(x) = x + 2$ **72.** $g(x) = x + 2$

73. $g(x) = |2x + 7| + 3$

74. $g(x) = |3x - 14| - 1$

Answers

75. $g(x) = 3(x + 1) + 4$

76. $g(x) = 2x - 5$

77. $g(x) = -\frac{1}{4}(x + 3) + 3$

78. $(3, -1, 5)$ **79.** $(2, 4, -3)$

80. There are an infinite number of solutions.

81. $(-3, 3, 4)$ **82.** $(-2, 4, -3)$ **83.** $(2, 7, -6)$

84. There are an infinite number of solutions.

85. $(-3.8, -2.6, 2.6)$ **86.** $(1, -2, 3)$

87. \$142,500 **88.** 8% **89.** \$125

90. \$68 **91.** 6%

92. $x = -6$ and $x = 6$

93. $x = -8$ and $x = 8$

94. $x = -7$ and $x = 7$

95. $x = -4$ and $x = 4$

96. $x = -6$ and $x = 6$

97. $x = -8$ and $x = 8$

98. $x = -2$ and $x = 5$

99. $x = -1$ and $x = 3$

100. $x = 1$ and $x = 6$

101. $x = 4$ and $x = 7$

102. $x = 1$ and $x = 8$

103. $x = -7$ and $x = -3$

104. $x = -4$ and $x = \frac{1}{2}$

105. $x = -4$ and $x = \frac{2}{3}$

106. $x = -1$ and $x = -\frac{3}{5}$

107. $25 - 9i$ **108.** $-21 + 3i$

109. $9 + 19i$ **110.** $-15 - 2i$

111. $-6 + 10i$ **112.** $-8 - 16i$

113. $19 + 59i$ **114.** $44 - 18i$

115. $32 + 24i$ **116.** $-7 - 24i$

117. **a.** 1.25 h **b.** 1 h

118. **a.** 0.5 h **b.** 1.6 h

119. yes; $g(x) = \sqrt[3]{x + 2}$

120. yes; $g(x) = x^2 - 3, x \geq 0$

121. yes; $g(x) = \frac{x^3}{27} + 2$

122. yes; $g(x) = \frac{3x^3 - 2}{4}$

Chapter 5

5.1 Start Thinking

semi-annual balance $=$ \$7828.59, quarterly balance $=$ \$7831.04, monthly balance $=$ \$7832.68, daily balance $=$ \$7833.48

Sample answer: The balance after 5 years will increase by a small amount as the number of compoundings increases. However, it seems there is a limit as to how high the balance will increase. The balance only increased by \$0.80 when the number of compoundings jumped from 12 to 365.

5.1 Warm Up

1. growth; initial value 1; percent increase 20%

2. decay; initial value 1; percent decrease 22%

3. decay; initial value 1; percent decrease 37.5%

4. growth; initial value 28; percent increase 3%

5. decay; initial value 25,000; percent decrease 5%

6. growth; initial value 1; percent increase 100%

Answers

5.1 Cumulative Review Warm Up

1. $y = 2x + 10$

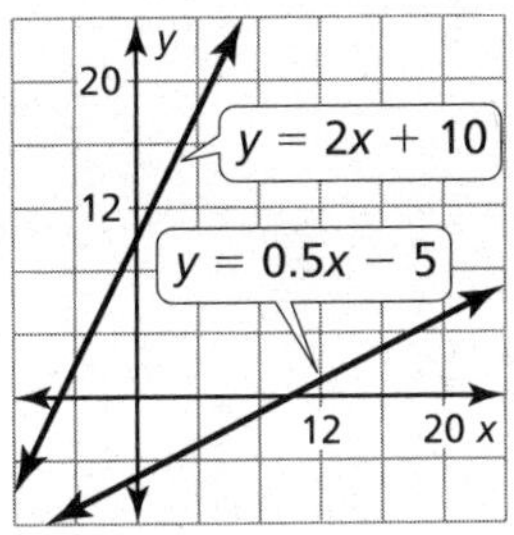

2. $y = -\frac{1}{3}x + \frac{7}{3}$

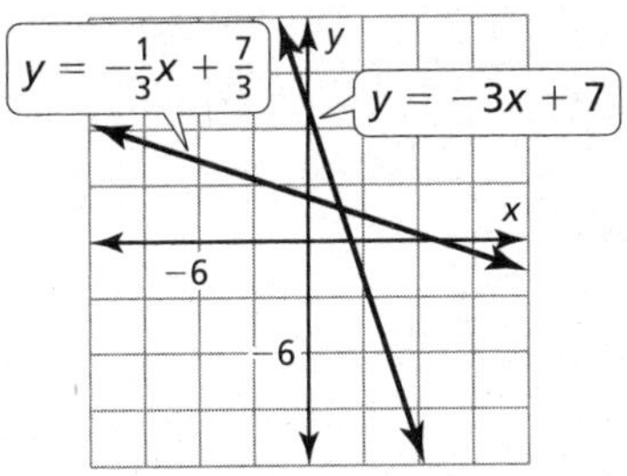

3. $y = 5x + 9$

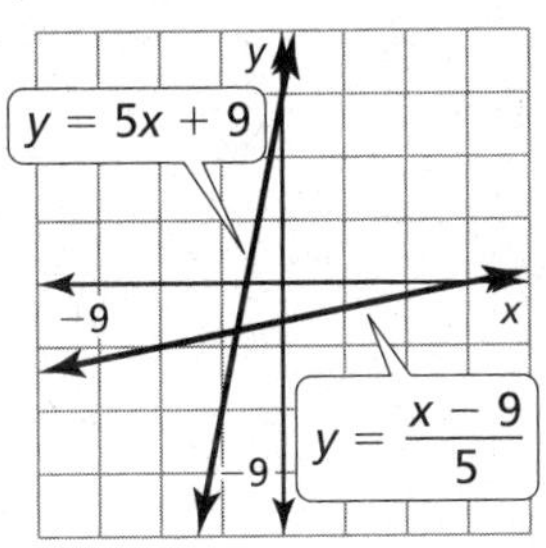

4. $y = \sqrt[3]{x} - 2$

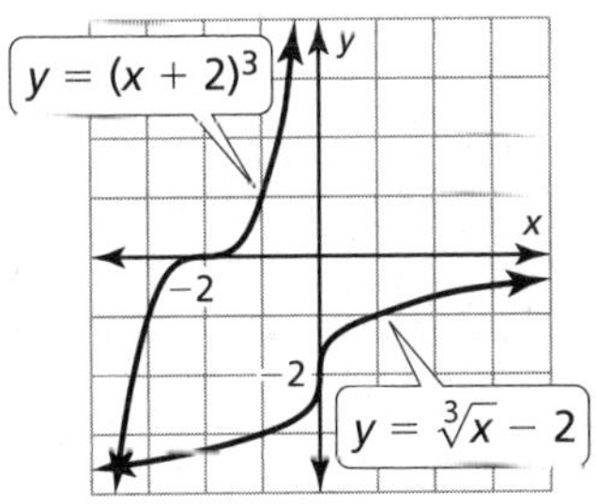

5. $y = \sqrt[3]{2x + 6}$

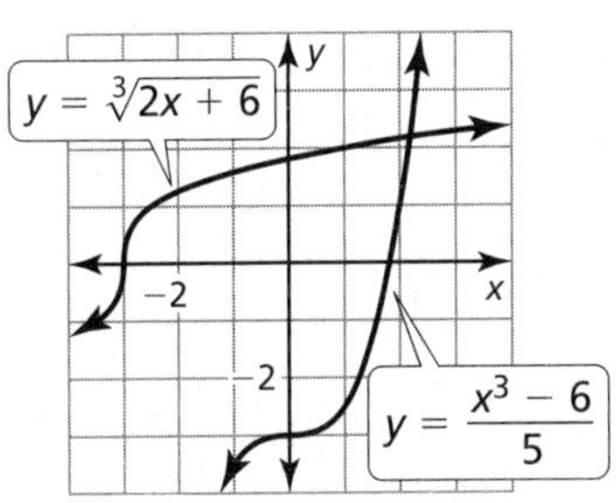

6. $y = \sqrt{-\frac{1}{2}x},\ x \leq 0$

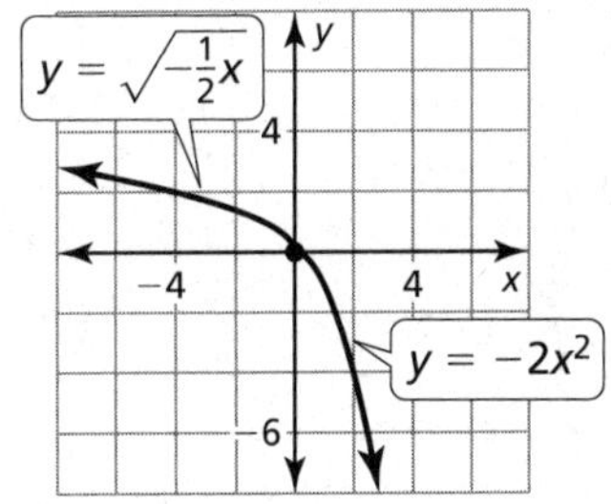

5.1 Practice A

1. e^7
2. e^5
3. $\frac{e^3}{3}$
4. $\frac{5}{e^5}$
5. $9e^{6x}$
6. $4e^{5x}$
7. incorrectly squared the exponent $3x$, instead of multiplying by 2

$$\left(2x^{3x}\right)^2 = (2)^2\left(e^{3x}\right)^2$$
$$= 4e^{6x}$$

8. exponential growth

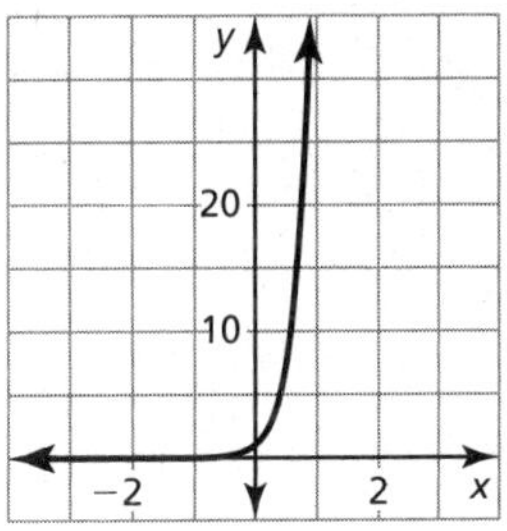

9. exponential decay

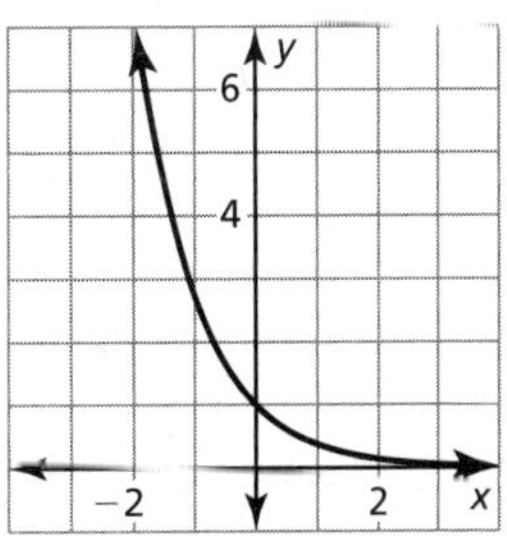

10. exponential decay

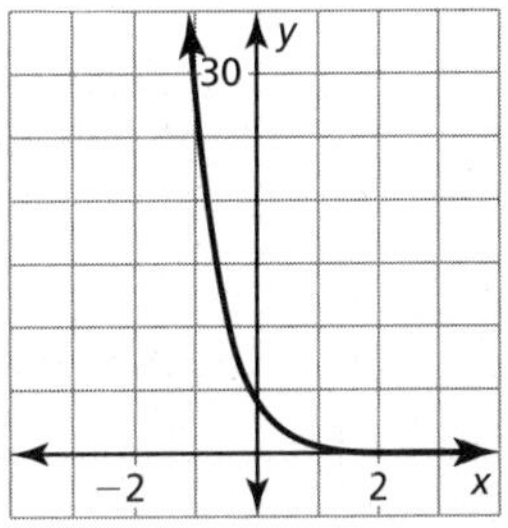

Answers

11. $y = (1 - 0.393)^x$; 39.3%

12. $y = 2(1 - 0.181)^x$; 18.1%

13. $y = 5(1 + 0.822)^x$; 82.2%

14.

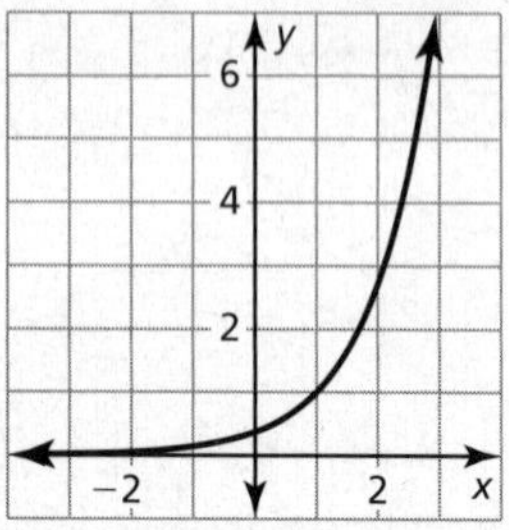

domain: $-\infty < x < \infty$; range: $y > 0$

15.

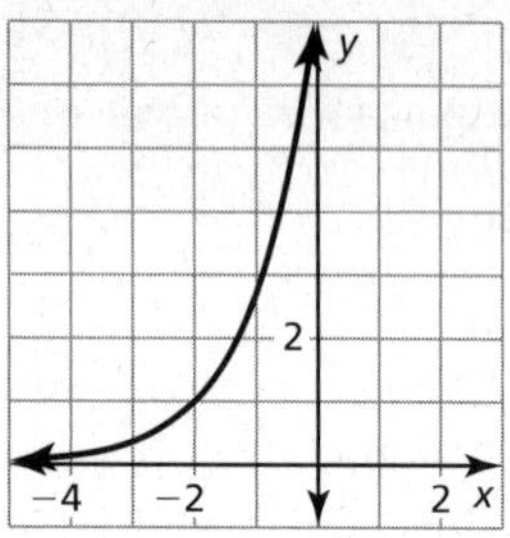

domain: $-\infty < x < \infty$; range: $y > 0$

16.

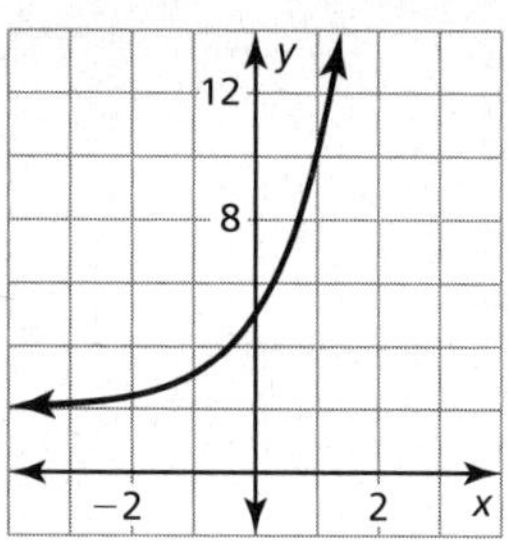

domain: $-\infty < x < \infty$; range: $y > 2$

17. **a.** \$4415.25 **b.** \$4376.70 **c.** 4.6 years

5.1 Practice B

1. $\frac{1}{e^2}$ **2.** $\frac{3}{2e^3}$ **3.** $\frac{125}{e^{12x}}$

4. $2\sqrt{5}e^{4x}$ **5.** $4e^{3x}$ **6.** e^{3x+3}

7. incorrectly placed 2^4 in the denominator, although the exponent is not negative

$$\left(2e^{-3x}\right)^4 = \frac{16}{e^{12x}}$$

8. exponential growth

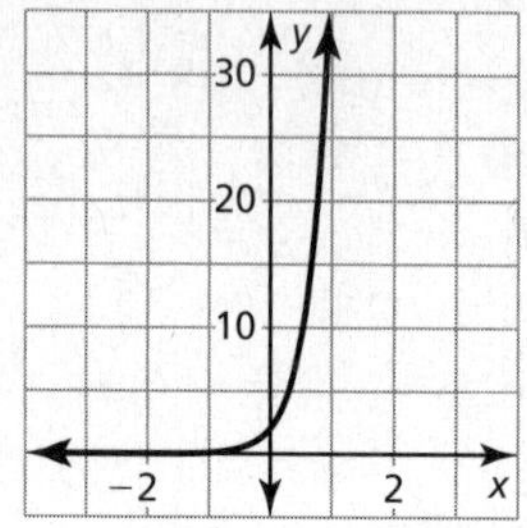

9. exponential decay

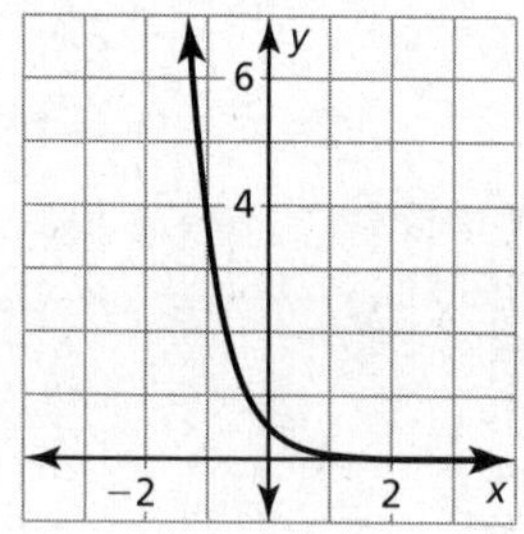

10. exponential growth

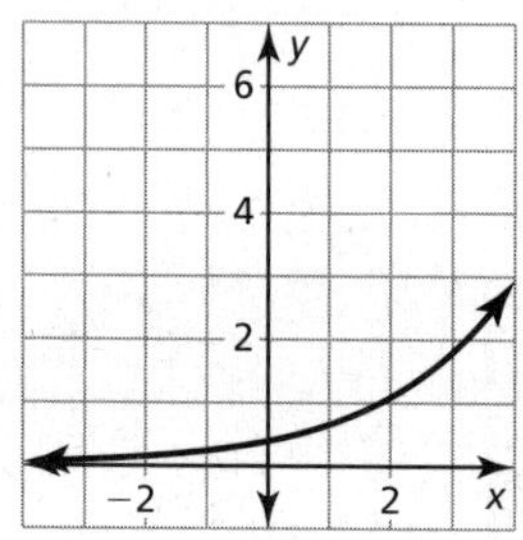

11. $y = (1 + 0.284)^x$; 28.4%

12. $y = 3(1 - 0.478)^x$; 47.8%

13. $y = 0.25(1 + 1.4596)^x$; 145.96%

14.

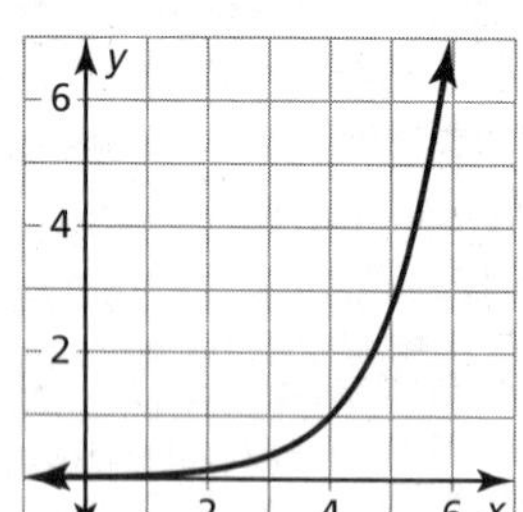

domain: $-\infty < x < \infty$; range: $y > 0$

Answers

15. 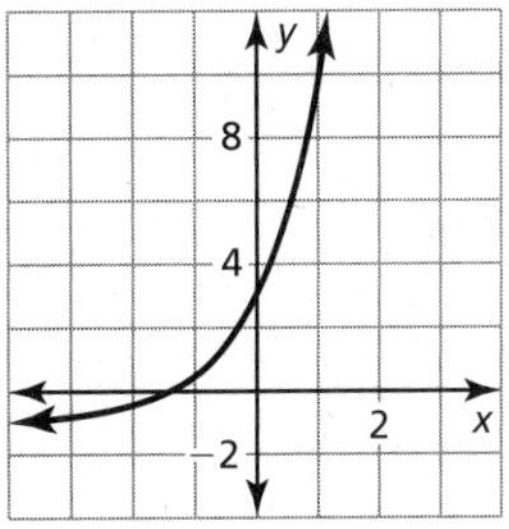

domain: $-\infty < x < \infty$; range: $y > -1$

16. 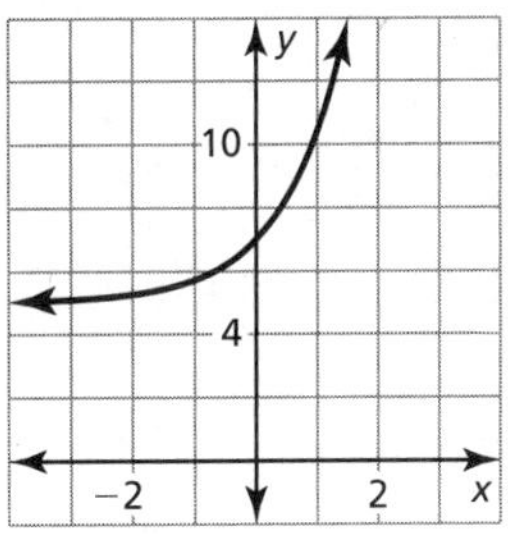

domain: $-\infty < x < \infty$; range: $y > 5$

17. a. \$5415.71 **b.** \$5416.44

c. \$4.96 **d.** \$49.60; 10 times more

5.1 Enrichment and Extension

Sum	Approximation
$\sqrt{12}\left(1 - \frac{1}{3 \bullet 3} + \frac{1}{5 \bullet 3^2} - \frac{1}{7 \bullet 3^3} + \ldots\right)$	≈ 3.137852892
$\sqrt{12}\left(1 - \frac{1}{3 \bullet 3} + \frac{1}{5 \bullet 3^2} - \frac{1}{7 \bullet 3^3} + \frac{1}{9 \bullet 3^4} - \ldots\right)$	≈ 3.142604746
$\sqrt{12}\left(1 - \frac{1}{3 \bullet 3} + \frac{1}{5 \bullet 3^2} - \frac{1}{7 \bullet 3^3} + \frac{1}{9 \bullet 3^4} - \frac{1}{11 \bullet 3^5} + \ldots\right)$	≈ 3.141308785
$\sqrt{12}\left(1 - \frac{1}{3 \bullet 3} + \frac{1}{5 \bullet 3^2} - \frac{1}{7 \bullet 3^3} + \frac{1}{9 \bullet 3^4} - \frac{1}{11 \bullet 3^5} + \frac{1}{13 \bullet 3^6} \ldots\right)$	≈ 3.141674313

1. The number π is a mathematical constant that is the ratio of a circle's circumference to its diameter.

2. $\pi = 3.14159265359$

3. The fraction $\frac{22}{7} \approx 3.142857$ is only an approximation of π, just like 3.14.

4. In mathematics, a transcendental number is a number that is *not* the root of any nonzero polynomial equation; yes

5. yes; no; All transcendental numbers are irrational, but not all irrational numbers are transcendental. For example, the irrational number $\sqrt{2}$ is a root of the polynomial equation $x^2 - 2 = 0$, so it is not transcendental.

6. Euler's Identity

5.1 Puzzle Time

VACUUM CLEANER

5.2 Start Thinking

The graph of $f^{-1}(x)$ is a reflection of $f(x)$ in the line $y = x$. The domain of $f(x)$ is the range of $f^{-1}(x)$ and the range of $f(x)$ is the domain of $f^{-1}(x)$.

5.2 Warm Up

1. 2 **2.** 0 **3.** 3

4. -1 **5.** $\frac{1}{2}$ **6.** $-\frac{1}{2}$

5.2 Cumulative Review Warm Up

1. a. 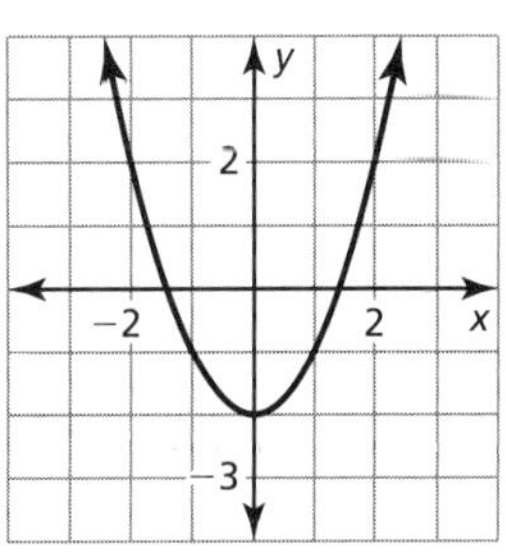

b. 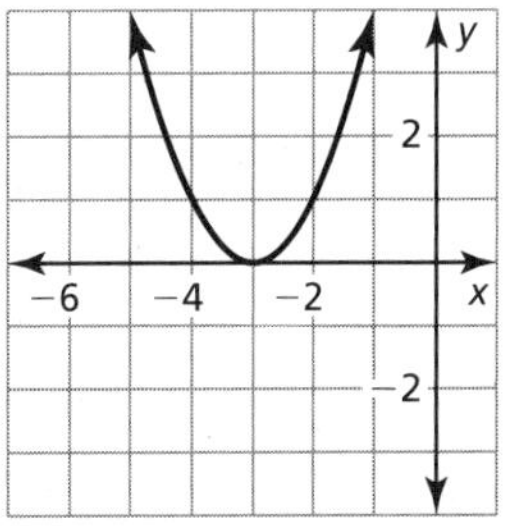

Answers

c. 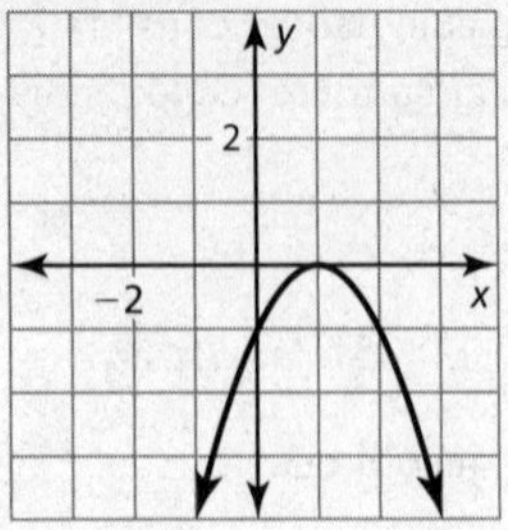

2. a. 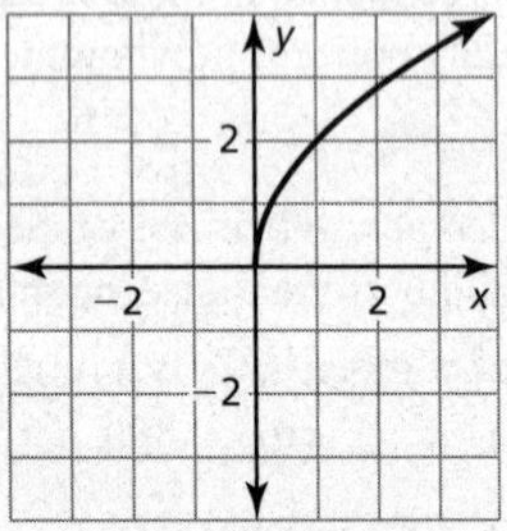

b. 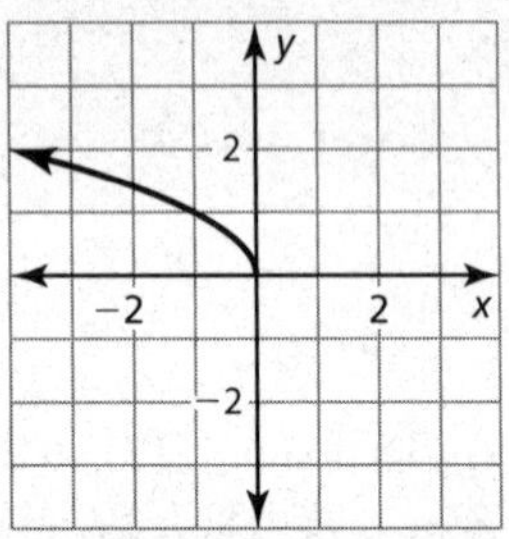

c. 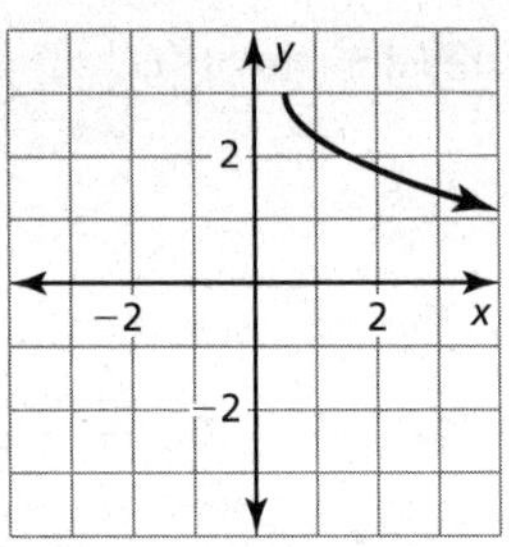

5.2 Practice A

1. $2^3 = 8$ 2. $7^1 = 7$ 3. $5^2 = 25$

4. $\log_4 16 = 2$ 5. $\log_5 1 = 0$ 6. $\log_6 \frac{1}{6} = -1$

7. 4 8. 3 9. 1

10. −1 11. 0 12. −3

13. 0.699 14. 2.639 15. −0.602

16. 30 decibels 17. x 18. $2x$

19. $3x$ 20. $y = \log_{1.1} x$ 21. $y = \log_3 x$

22. $y = 3^x$ 23. $y = 3(10^x)$

24. $y = \frac{1}{3}e^x$ 25. $y = \frac{1}{5}\ln x$

26. a. 208.6 mi/h b. 8.2 mi

5.2 Practice B

1. $9^0 = 1$ 2. $6^3 = 216$

3. $2^{-2} = \frac{1}{4}$ 4. $\log_{13} \frac{1}{169} = -2$

5. $\log_4 8 = \frac{3}{2}$ 6. $\log_{81} 9 = \frac{1}{2}$

7. 2 8. 5 9. 0

10. −4 11. −3 12. −2

13. −0.699 14. 0.673 15. −2.916

16. yes; The sound is 130 decibels.

17. $7x$ 18. 18 19. $3x$

20. $y = \log_{0.75} x$ 21. $y = \left(\frac{3}{4}\right)^x$

22. $y = 2(10^x)$ 23. $y = e^x - 2$

24. $y = \ln x + 3$ 25. $y = \log_6 (x - 2)$

26. a. 116.83 in.; 9.74 ft
b. 1651.8 lb

5.2 Enrichment and Extension

1.

d	1	2	3	4	5
P	0.3010	0.1761	0.1249	0.0969	0.0792

d	6	7	8	9
P	0.0669	0.0580	0.0512	0.0458

2. 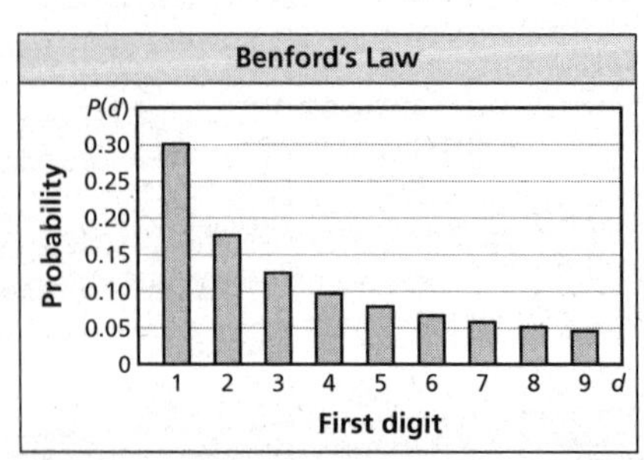

3. about 0.0458 4. about 0.699

Answers

5. about 0.2273

6. no; According to Benford's Law, you would expect about $\frac{0.3010}{0.0458} \approx 7$ times as many purchases to begin with the number 1 than purchases that begin with the number 9.

7. no; There is not a wide range of costs available at a car wash.

8. Answers will vary.

5.2 Puzzle Time

AT A SNOOPER MARKET

5.3 Start Thinking

1.

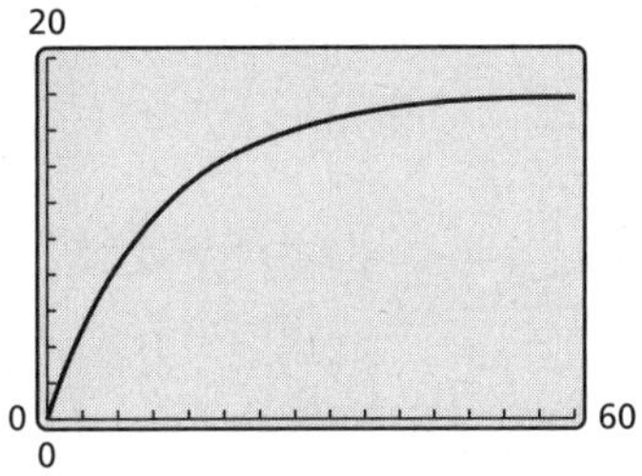

Sample answer: $x\text{min} = 0$, $x\text{max} = 60$, $x\text{scl} = 4$, $y\text{min} = 0$, $y\text{max} = 20$, $y\text{scl} = 2$

2. approximately 6 windows after 5 days; approximately 10 windows after 10 days

3. approximately 17 windows in approximately 36 days; Note that the graph asymptotically approaches 18 from below, but never reaches it.

5.3 Warm Up

1. reflection in x axis, followed by a translation 3 units up

2. vertical shrink by a factor of $\frac{1}{2}$, followed by a translation 2 units left and 5 units down

3. reflection in y-axis, and a horizontal shrink by a factor of $\frac{1}{3}$

4. translation 2 units right, and a vertical stretch by a factor of 4

5.3 Cumulative Review Warm Up

1. $y = 3x + 5$

2. $y = -5x - 2$

3. $y = -2x + 9$

4. $y = \frac{11}{5}x + \frac{62}{5}$

5.3 Practice A

1. translation 3 units up

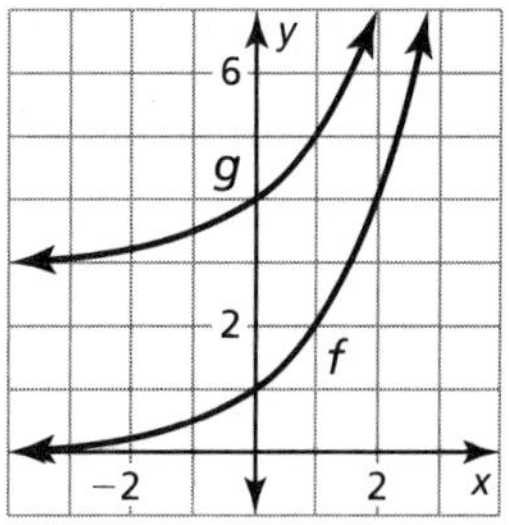

2. translation 2 units down

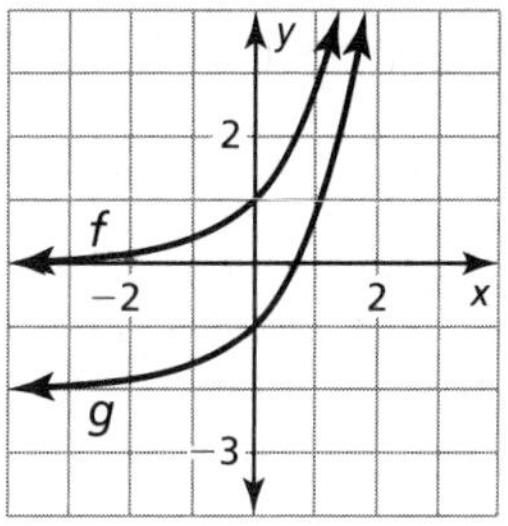

3. translation 1 unit right

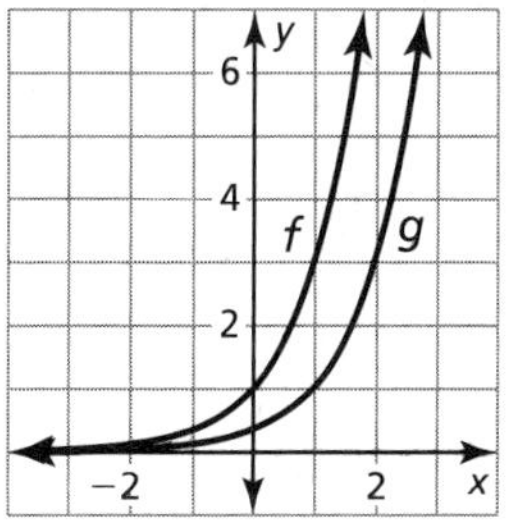

4. translation 4 units up

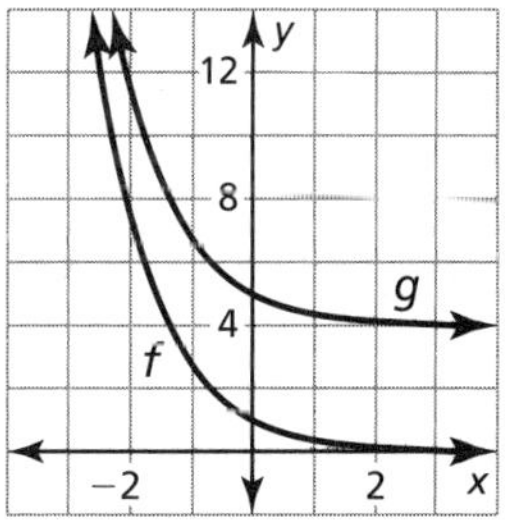

5. horizontal shrink by a factor of $\frac{1}{3}$

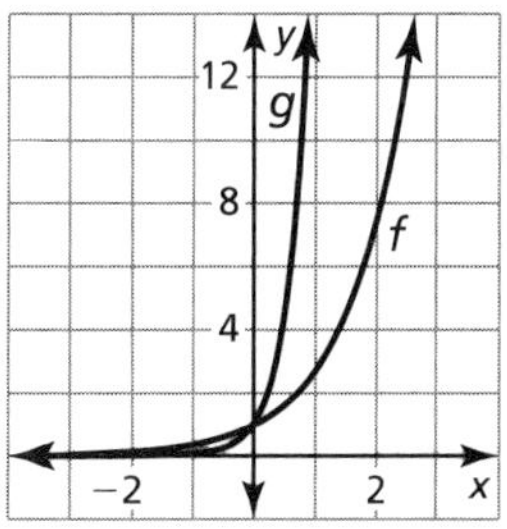

Answers

6. vertical stretch by a factor of $\frac{3}{2}$

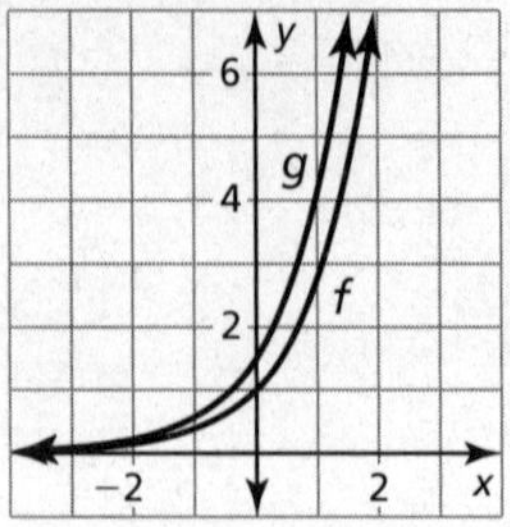

7. reflection in the x-axis, followed by a horizontal translation 2 units left

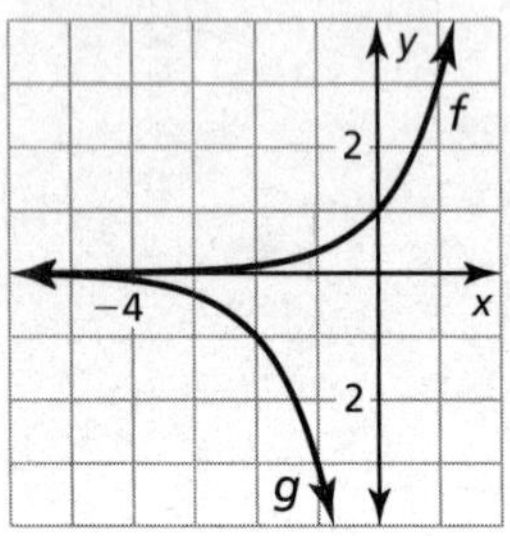

8. horizontal shrink by a factor of $\frac{1}{5}$ and a vertical stretch by a factor of 2

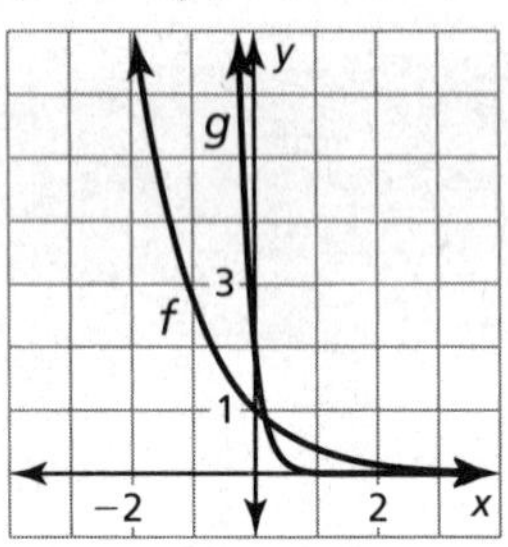

9. incorrectly evaluates as $(2e)^x$, but in $f(x) = 2e^x$, the 2 is the leading coefficient and not part of the base

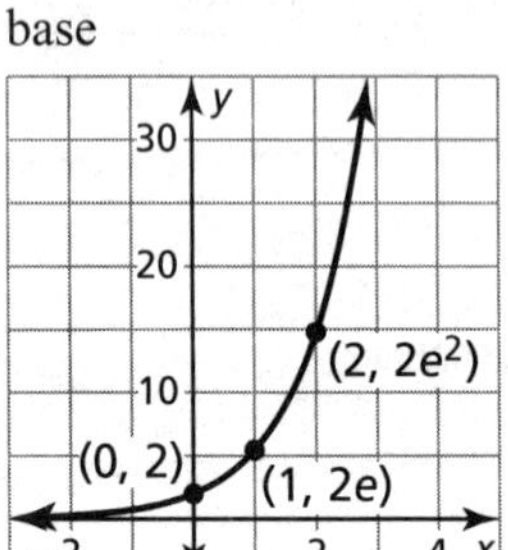

10. vertical stretch by a factor of 4, followed by a translation 1 unit down

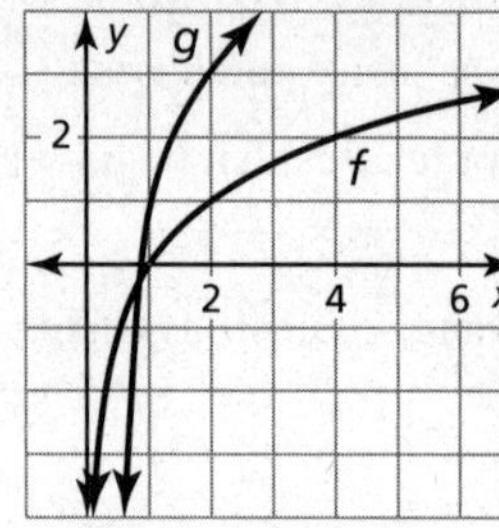

11. reflection in the x-axis, followed by a translation 3 units up

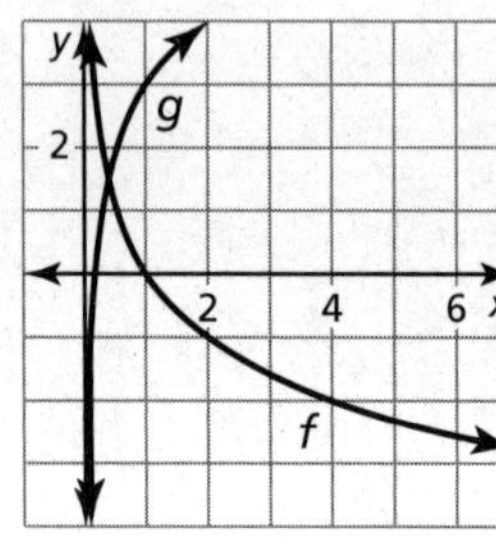

12. $g(x) = -3^{x+3} - 1$ **13.** $g(x) = \frac{1}{4}e^x + 5$

14. $g(x) = \log_8(-(x + 4))$

15. $g(x) = 9\log_{1/6}(x - 2) - 3$

5.3 Practice B

1. translation 4 units down

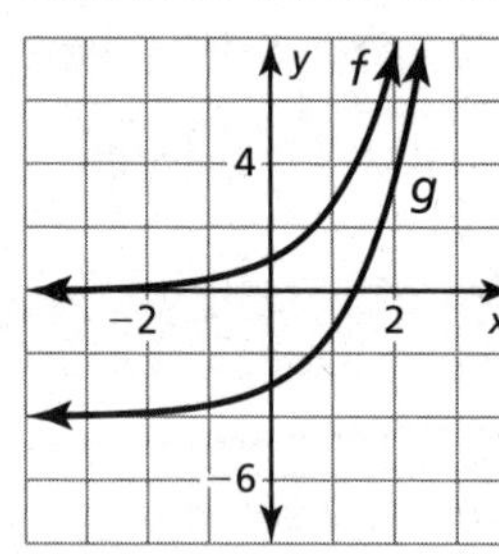

2. translation 2 units left

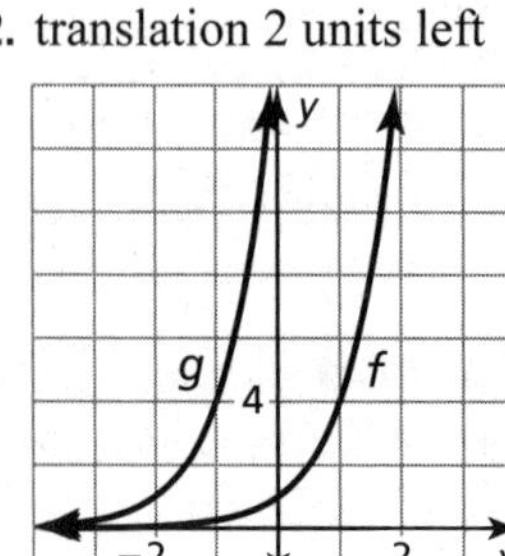

Answers

3. translation 5 units down

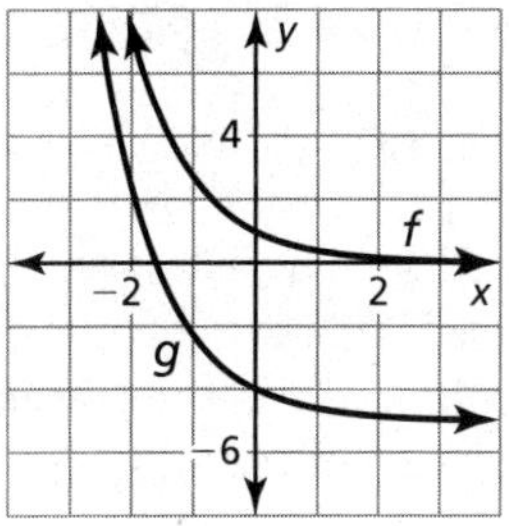

4. translation 2 units up

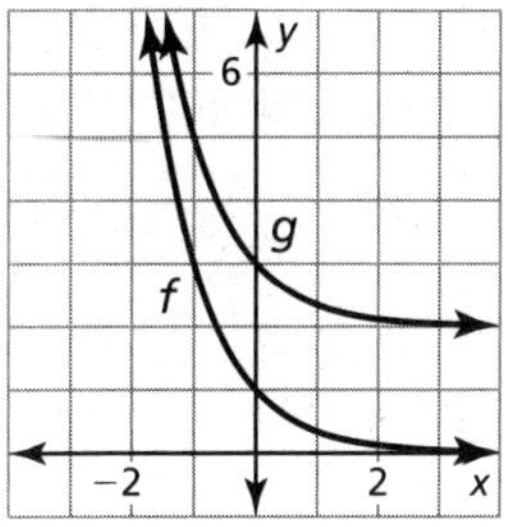

5. horizontal shrink by a factor of $\frac{1}{2}$, followed by a translation 1 unit down

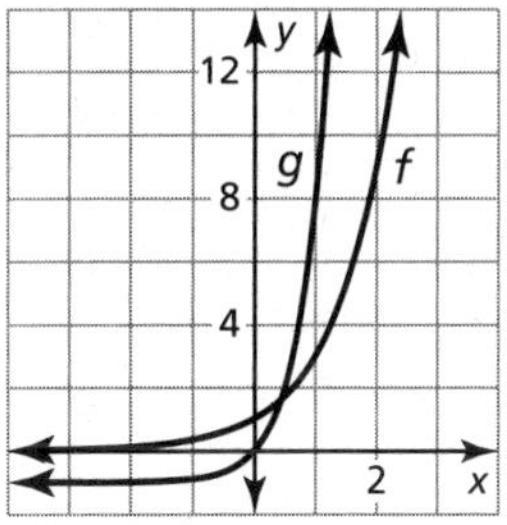

6. reflection in the x-axis, followed by a translation 2 units left

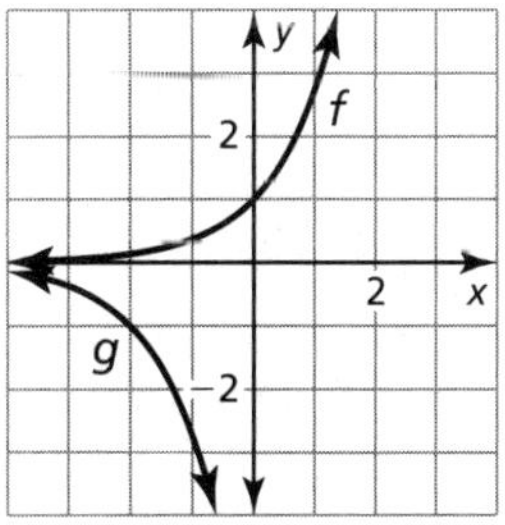

7. translation 1 unit right, followed by a horizontal shrink by a factor of $\frac{1}{4}$

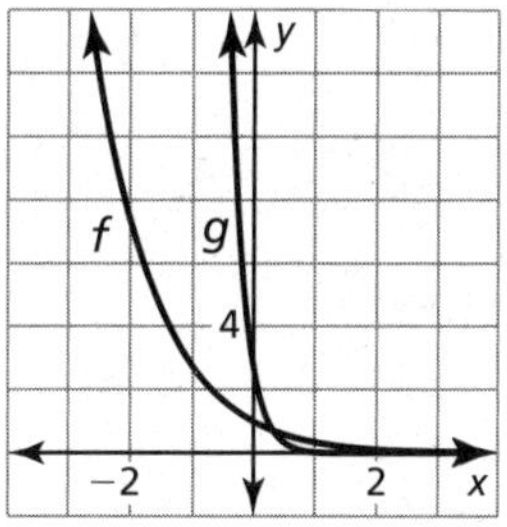

8. translation 2 units right and 3 units up

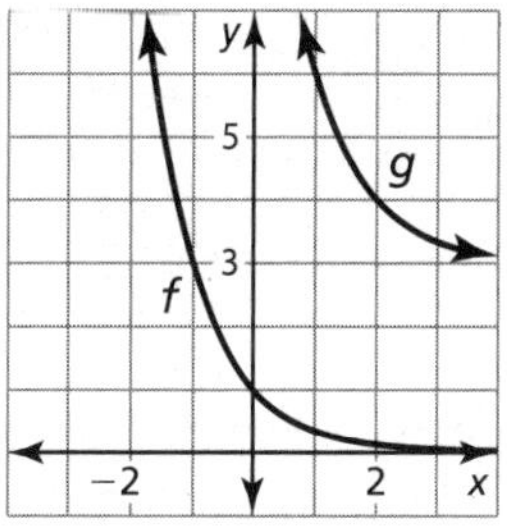

9. incorrectly evaluated as $2^x + 3$, but in $f(x) = 2^{x+3}$, the 3 is in the exponent and is added to x first

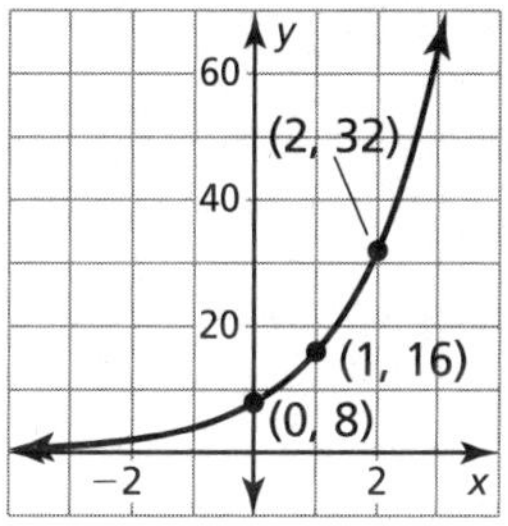

10. translation 2 units right and 4 units up

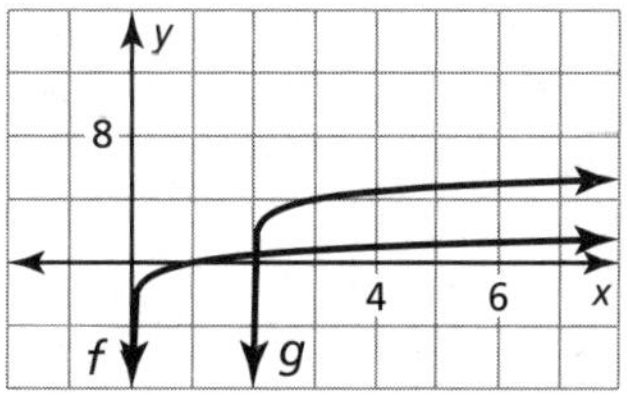

11. reflection in the y-axis, followed by a reflection in the x-axis

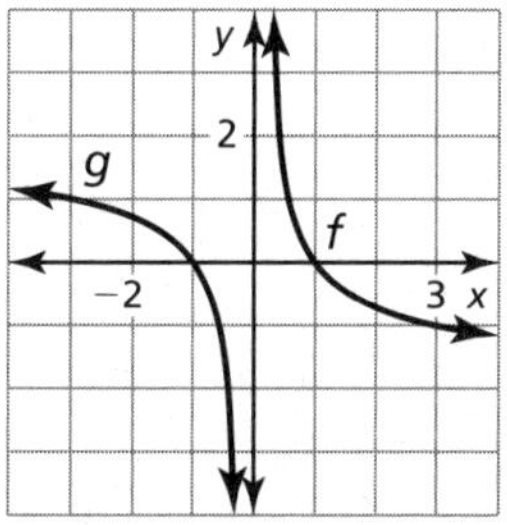

Answers

12. $g(x) = \left(\frac{2}{5}\right)^{-x/2} - 4$

13. $g(x) = 2\left(e^{-(x+2)} + 3\right) = 2e^{-x-2} + 6$

14. $g(x) = -\log_{12}(x - 5) + 2$

5.3 Enrichment and Extension

1. K	2. D	3. E	4. M
5. L	6. G	7. I	8. J
9. F	10. A	11. N	12. C
13. B	14. H		

5.3 Puzzle Time

LUNCH

5.4 Start Thinking

Sample answer: In each group, the right side of the third equation is the sum of the previous two, whereas the left side of the third equation involves a product inside the log.

$\log_2(3 \bullet 4) = \log_2(3) + \log_2(4)$

$\log_3\left(15 \bullet \frac{1}{3}\right) = \log_3(15) + \log_3\left(\frac{1}{3}\right)$

$\log_4(9 \bullet 2) = \log_4(9) + \log_4(2)$

Following this pattern:

$$\begin{aligned}\log_5(25 \bullet x) &= \log_5 25 + \log_5 x\\ &= 2 + \log_5 x\\ &\approx 2 + 1.4\\ &= 3.4\end{aligned}$$

5.4 Warm Up

1. $n = 7$ 2. $n = 6$ 3. $n = \frac{1}{2}$

4. $n = \frac{1}{4}$ 5. $n = 2$ 6. $n = -3$

5.4 Cumulative Review Warm Up

1. 5, 1 2. $-\frac{4}{3}, \frac{5}{2}$ 3. $2 \pm 2\sqrt{3}$

4. $1 \pm 2i$ 5. 3 6. $\frac{-1 \pm \sqrt{10}}{3}$

5.4 Practice A

1. 0.43 2. 1.796 3. 1.366

4. $\log_2 5 + \log_2 x$ 5. $\log 7 + 4 \log x$

6. $\log_6 2 + \log_6 x - \log_6 y$

7. confuses logarithm rules for products and powers; The log of a product is the sum of the logs, not another product.

   $\log_4 3x = \log_4 3 + \log_4 x$

8. $\log_7 \frac{3}{5}$ 9. $\log 2$ 10. $\ln\left(x^3 y^9\right)$

11. $\log_2\left(9y^{1/2}\right)$ 12. 0.683 13. 3.459

14. 1.285 15. yes; $\ln x = \frac{\log x}{\log e}$

16. about 7 decibel levels

5.4 Practice B

1. 2.732 2. −1.113 3. −0.43

4. $\log_3 12 + 7 \log_3 x$

5. $\log_6 5 + 2 \log_6 x - 3 \log_6 y$

6. $\log_8 6 + \frac{1}{2}\log_8 x + \frac{1}{2}\log_8 y$

7. does not distribute the $\frac{1}{3}$ power to y as well as x

   $\ln \sqrt[3]{xy} = \frac{1}{3}\ln x + \frac{1}{3}\ln y$

8. $\log_9\left(\frac{x^5}{4}\right)$ 9. $\log_8\left(5\sqrt[4]{x}\right)$

10. $\ln\left(16x^5 y^3\right)$ 11. $\log_6\left(\frac{1}{x^3}\right)$

12. 1.302 13. 3.096 14. −0.544

15. yes; $\frac{\ln y}{\ln 3} = \log_3 y$

16. about 4.77 decibel levels

Answers

5.4 Enrichment and Extension

1. 9
2. 6
3. 3
4. −1.9041
5. −9.4717
6. 1
7. 1.4307
8. 400
9. 50
10. −4
11. 2.0402
12. 2.9307
13. 68.6309
14. 0.5
15. 161
16. 9983.6667

5.4 Puzzle Time

A SILLYMANDER

5.5 Start Thinking

$10{,}000 = 5000e^{0.0525t}$; $t \approx 13.2$ years

5.5 Warm Up

1. $\log_2 5 = x - 1$; $x = 1 + \log_2 5$
2. $\ln 11 = -2x$; $x = -\frac{1}{2}\ln 11$
3. $\log_3 4 = 3 - 2x$; $x = \frac{3}{2} - \frac{1}{2}\log_3 4$
4. $2^{-1} = x + 5$; $x = -\frac{9}{2}$
5. $10^2 = 3 - x$; $x = -97$
6. $e^{1/3} = 2x$; $x = \frac{1}{2}\sqrt[3]{e}$

5.5 Cumulative Review Warm Up

1. $f(x) \to -\infty$ as $x \to -\infty$ and $f(x) \to -\infty$ as $x \to +\infty$
2. $g(x) \to +\infty$ as $x \to -\infty$ and $g(x) \to -\infty$ as $x \to +\infty$
3. $h(x) \to -\infty$ as $x \to -\infty$ and $h(x) \to +\infty$ as $x \to +\infty$
4. $p(x) \to +\infty$ as $x \to -\infty$ and $p(x) \to +\infty$ as $x \to +\infty$

5.5 Practice A

1. $x = -10$
2. $x = 2$
3. $x = 7$
4. $x = \log_2 5 \approx 2.322$
5. $x = \log_8 35 \approx 1.71$
6. $x = -\frac{1}{5}$
7. **a.** 24 years **b.** 246.1 cm
8. $x = 7$
9. $x = 1$
10. $x = 6$
11. $x = \frac{32}{5}$
12. $x = 100.7$
13. $x = 1$
14. $x = 4$; extraneous solution: $x = -1$
15. $x = 1$; extraneous solution: $x = -\frac{3}{2}$
16. $x = 2.91$; extraneous solution: $x = -6.91$
17. $x = \pm\sqrt{6}$
18. **a.** 14.2 years
    b. 13.95 years
    c. 13.864 years
    d. 13.863 years
19. $x < 1.92$
20. $x \geq 2.89$
21. $x > 9$
22. $x \approx 4.32$
23. $x \approx 1.03$

5.5 Practice B

1. $x = -3$
2. $x = \log_7 32 \approx 1.781$
3. $x = -4$
4. $x = \frac{1}{4}\log_6 13 \approx 0.358$
5. $x = \frac{1}{3}\ln 2 \approx 0.231$
6. $x = \frac{1}{2}\ln 2 \approx 0.347$
7. **a.** 2130.9 years
   b. more years; 942.95 more years
8. $x = 2$
9. $x = \frac{1}{3}$
10. $x = \frac{4}{3}$
11. $x = 1$
12. $x = 5, -3$
13. $x = -5, 4$
14. $x = 13.22$; extraneous solution: $x = -11.22$

Answers

15. $x = \pm\frac{5}{4}$

16. $x = -1.35, -6.65$

17. $x = 0.772$; extraneous solution: $x = -7.772$

18. $x < 2 + 3\ln 3 \approx 4.08$

19. $x > e^5 \approx 148.4$

20. $x \geq \frac{1}{81} \approx 0.012$

21. 8.5 years

5.5 Enrichment and Extension

1. $x \approx 1.442$
2. $x \approx 0.828$
3. $x \approx 7.389$
4. $x = -3$
5. $x = 6.25$
6. $x = -5$
7. $x = 5$
8. $x \approx 2.727$

5.5 Puzzle Time

SO THEY HAVE SOMEWHERE TO PUT THE GROCERIES WHEN THEY GO SHOPPING

5.6 Start Thinking

1. exponential; *Sample answer*: An account balance increases over time.
2. linear; *Sample answer*: A scientist introduces a specific amount of chemical to a solution each hour.
3. quadratic; *Sample answer*: A model rocket is shot from a platform.

5.6 Warm Up

1. $y = 1.8x - 3.333$
2. $y = -1.626x + 3390.974$

5.6 Cumulative Review Warm Up

1. $x = \frac{17}{2}$
2. $x = 8$
3. $x = \frac{4}{7}$
4. $x = \frac{13}{4}$

5.6 Practice A

1. exponential; common ratio $= \frac{1}{3}$
2. quadratic; common second difference
3. $y = 3(2)^x$
4. $y = 5(4)^x$
5. $y = 2(3)^x$
6. $y = 2(5)^x$
7. $y = 7(2)^x$
8. $y = 5(3)^x$
9. The x-values must have a common difference before a common ratio in the y-values can be used.

x	0	1	3	5	7
y	2	4	8	16	32

×2 ×2 ×2 ×2

The outputs have a common ratio of 2, but the inputs do not have a common difference, so the data do not represent a recognizable function.

10. exponential relationship; $y = 96(2)^x$
11. linear relationship; For every increase of x by 1, y increases by 5.
12. $y = 5(1.2)^x$

5.6 Practice B

1. exponential; common ratio $= 4$
2. exponential; common ratio $= \frac{3}{2}$
3. $y = 5(2)^x$
4. $y = 6(3)^x$
5. $y = 9(2)^x$
6. $y = 3(5)^x$
7. $y = \frac{1}{10}(6)^x$
8. $y = \frac{1}{2}(4)^x$
9. assumed that the output of any exponential function is always positive, because b^x is always positive, but forgot that the coefficient a can be negative

x	0	2	4	6	8
y	−2	−4	−8	−16	−32

×2 ×2 ×2 ×2

The outputs have a common ratio of 2, so the data represent an exponential function.

10. exponential relationship; $y = 64\left(\frac{\sqrt{2}}{2}\right)^{x-1}$
11. linear relationship; For every increase of x by 10, y increases by 15.
12. $y = 3.4(1.5)^x$

Answers

5.6 Enrichment and Extension

1. 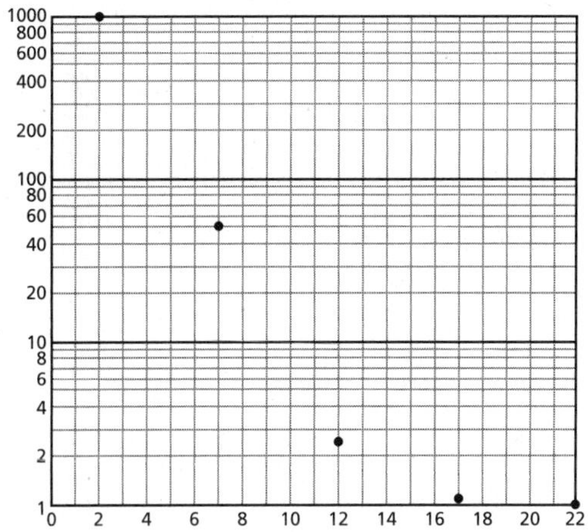

2. 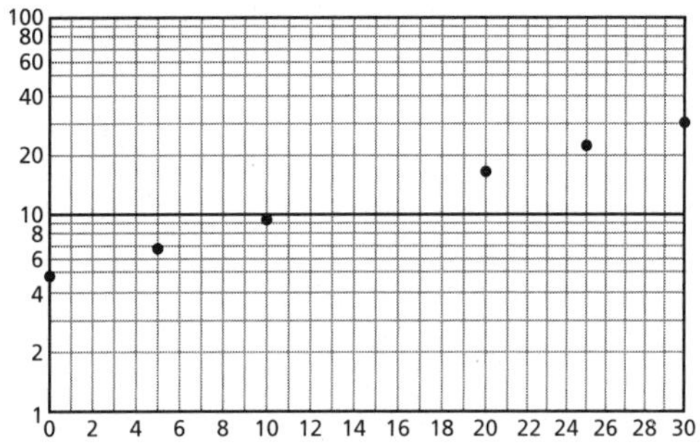

5.6 Puzzle Time

SHEEP SHAPE

Cumulative Review

1. 245 **2.** 48 **3.** 162

4. −32 **5.** −224 **6.** −320

7. −8 **8.** 9 **9.** 16

10. $\frac{8}{27}$ **11.** $\frac{16}{25}$ **12.** $\frac{1}{81}$

13. $-\frac{8}{125}$ **14.** $\frac{1}{256}$ **15.** $-\frac{27}{64}$

16. $-\frac{1}{32}$ **17.** $-\frac{16}{81}$ **18.** $-\frac{27}{125}$

19. 5×10^{-4} **20.** 3.82×10^{11}

21. -1.32×10^{-2} **22.** 5.7×10^{-6}

23. 2.3×10^{6} **24.** -9.2×10^{-10}

25. -3.59×10^{-4} **26.** 5.13×10^{4}

27. 4.1×10^{-5} **28.** 4.5×10^{3}

29. 1.2×10^{7} **30.** 7.31×10^{5}

31. -8.17×10^{-7} **32.** 4.17×10^{11}

33. -6.2×10^{4} **34.** 8.3×10^{12}

35. 8.9×10^{-8} **36.** 2×10^{-3}

37. a. Bowling Alley A: $19, Bowling Alley B: $25
b. Bowling Alley A: $22.50, Bowling Alley B: $25
c. 6 games or more

38. a. $22.50 **b.** $25 **c.** $100

39. 0.00185 **40.** −89,000

41. 0.0000000384 **42.** 0.000000257

43. 460,000 **44.** −0.00334

45. 0.0000000305 **46.** 44,900

47. 590,000 **48.** −2,250,000,000

49. 0.308 **50.** −0.0000000106

51. 55,100,000 **52.** 618,000,000

53. 0.000000184 **54.** 890,000

55. 0.000000092 **56.** −0.0000732

57. $y = -\frac{2}{5}x - 12$ **58.** $y = \frac{5}{11}x - 4$

59. $y = \frac{3}{5}x - 9$ **60.** $y = \frac{1}{4}x - 4$

61. $y = -\frac{7}{8}x + 3$ **62.** $y = \frac{3}{5}x - 12$

63. $y = -2x + 7$ **64.** $y = 3x - 20$

65. $y = -\frac{1}{2}x - 10$ **66.** $y = \frac{5}{6}x + \frac{11}{2}$

67. $y = -3x + 13$ **68.** $y = 4x + 24$

69. $x = 0$ and $x = 7$ **70.** $x = -3$ and $x = 0$

71. $x = -4$ and $x = 0$

Answers

72. $x = 0$ and $x = 8$

73. $x = 1$ and $x = 2$

74. $x = -9$ and $x = 3$

75. $x = -6$ and $x = -5$

76. $x = -2$ and $x = 8$

77. $x = -\frac{7}{2}$ and $x = 5$

78. $x = -5$ and $x = \frac{1}{3}$

79. $x = -4$ and $x = \frac{7}{2}$

80. $x = -\frac{1}{3}$ and $x = 7$

81. $x = -\frac{1}{5}$ and $x = 3$

82. $x = \frac{7}{3}$ and $x = 9$

83. $4 + 4i$

84. $4 - 3i$

85. $11 - 12i$

86. $-12 - i$

87. $10 + 6i$

88. $8 - 20i$

89. $1 - 3i$

90. $-2 + 14i$

91. $-14 - 3i$

92. $21 - 5i$

93. $5 + 8i$

94. $-10 + 6i$

95. $18 + 21i$

96. $-17 - 5i$

97. $-6 + 12i$

98. $-10 - 5i$

99. $-28 - 21i$

100. $-6 - 7i$

101. $-81 + 69i$

102. $-42 + 66i$

103. $-36 + 32i$

104. $-50 - 30i$

105. $-41 + 126i$

106. $-31 + 53i$

107. $-79 + 14i$

108. $27 + 3i$

109. $x = \frac{-3 \pm i}{2}$

110. $x = \frac{1 \pm i\sqrt{35}}{6}$

111. $x = \frac{-3 \pm i\sqrt{87}}{8}$

112. $x = \frac{5 \pm i\sqrt{167}}{12}$

113. $x = 2 \pm -i\sqrt{3}$

114. $x = \frac{3 \pm i\sqrt{6}}{5}$

115. $x = \frac{9 \pm i\sqrt{15}}{8}$

116. $x = \frac{5 \pm i\sqrt{7}}{4}$

117. **a.** $P = 10x$

b. $A = 4x^2 + 30x - 100$

c. $x = 12$

d. 836 ft^2

118. $x = \frac{-5 \pm \sqrt{41}}{2}$

119. $x = \frac{-2 \pm \sqrt{2}}{2}$

120. $x = 4 \pm \sqrt{21}$

121. $x = 1$ and $x = 2$

122. $x = 1$ and $x = \frac{3}{2}$

123. $x = \frac{5 \pm \sqrt{17}}{2}$

124. $x = \frac{3 \pm 2\sqrt{3}}{3}$

125. $x = \frac{-3 \pm \sqrt{89}}{10}$

126. $x = \frac{1}{3}$ and $x = \frac{1}{2}$

127. $x = \frac{7 \pm \sqrt{73}}{4}$

128. $x = \frac{-7 \pm \sqrt{37}}{2}$

129. $x = \frac{1 \pm \sqrt{129}}{8}$

130. $x = \frac{5 \pm \sqrt{13}}{6}$

131. $x = \frac{9 \pm \sqrt{129}}{4}$

132. **a.** $P = 6x + 8$

b. $A = 2x^2 + 8x$

c. $x = 8$

d. 192 ft^2

133. **a.** $A = 40.5x$

b. $x = 6$

c. 243 ft^2

d. $605.07

Answers

134. a. $A = 32x$

b. $x = 11$

c. 352 ft^2

d. \$1615.68

Chapter 6

6.1 Start Thinking

1. The quotient of x and y is the same for each point, $\frac{x}{y} = \frac{1}{2}$. The graph is the line $y = 2x$.

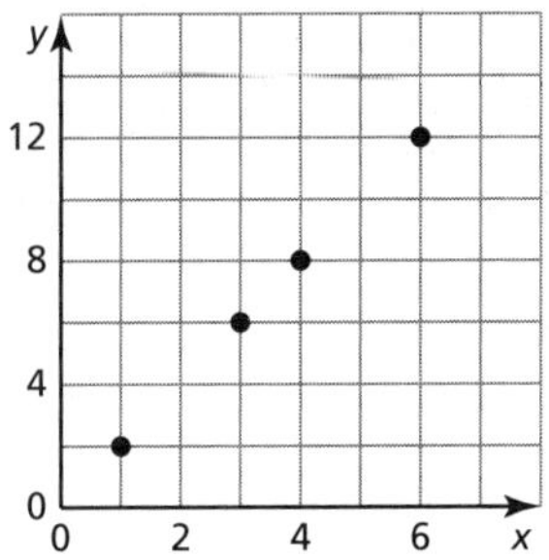

2. The product of x and y is the same for each point, $xy = 2$. The graph is not a line but curved.

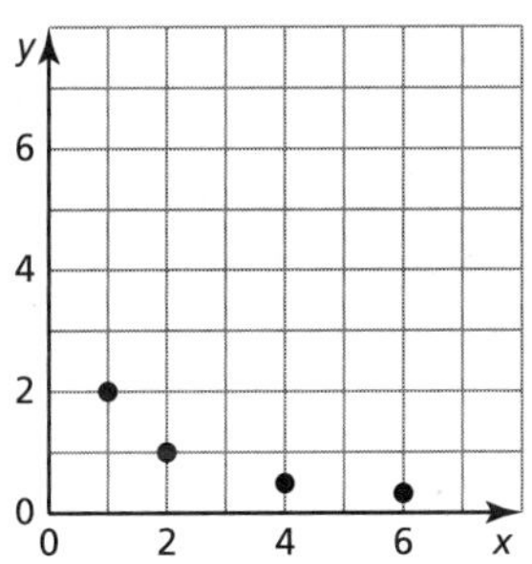

3. The product of x and y is the same for each point, $xy = 3$. The graph is not a line but curved.

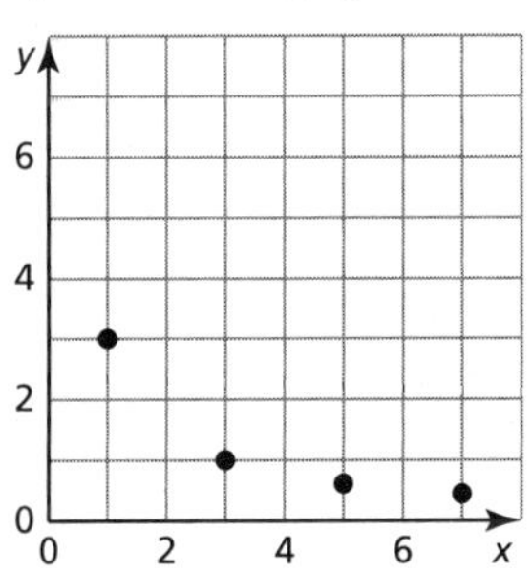

6.1 Warm Up

1. $-\frac{3}{4}$ **2.** 10 **3.** 0.8

4. -26 **5.** -26 **6.** $-\frac{3}{2}$

6.1 Cumulative Review Warm Up

1. $\left(-\frac{1}{2}, 3\right)$ **2.** $\left(\frac{21}{2}, -3\right)$

3. $\left(\sqrt{3}, 3\right), \left(-\sqrt{3}, 3\right)$

6.1 Practice A

1. inverse variation **2.** inverse variation

3. direct variation **4.** direct variation

5. neither **6.** direct variation

7. inverse variation **8.** neither

9. direct variation **10.** inverse variation

11. $y = -\frac{30}{x}; y = -10$ **12.** $y = \frac{7}{x}; y = \frac{7}{3}$

13. $y = \frac{2}{x}; y = \frac{2}{3}$

14. incorrectly used the equation $\frac{y}{x} = a$, which implied a direct variation between x and y, not inverse variation;

$xy = a$

$(5)(6) = a$

$a = 30$

So, $y = \frac{30}{x}$.

15. 3750 songs

6.1 Practice B

1. inverse variation **2.** inverse variation

3. direct variation **4.** neither

5. direct variation **6.** inverse variation

7. direct variation **8.** inverse variation

9. neither **10.** inverse variation

11. $y = -\frac{12}{x}; y = -4$ **12.** $y = -\frac{10}{3x}; y = -\frac{10}{9}$

13. $y = \frac{2}{x}; y = \frac{2}{3}$

Answers

14. The final equation given is a direct variation, not an inverse variation;

$xy = a$

$\frac{1}{3} \bullet 2 = a$

$a = \frac{2}{3}$

So, $y = \frac{2}{3x}$.

15. 6.4 amperes

6.1 Enrichment and Extension

1. $y = -\frac{5}{3}$

2. $w = \frac{15}{4}$

3. $h \approx 24.7$ m

4. As pressure increases, volume decreases; $n = \frac{PV}{aT}$

6.1 Puzzle Time

THE SPACE BAR

6.2 Start Thinking

1.

x	-1	$-\frac{1}{2}$	$-\frac{1}{10}$	$-\frac{1}{100}$	$\frac{1}{100}$	$\frac{1}{10}$	$\frac{1}{2}$	1
$g(x)$	-1	-2	-10	-100	100	10	2	1

$g(x)$ approaches $-\infty$ as x approaches 0 from the left and $g(x)$ approaches $+\infty$ as x approaches 0 from the right.

2.

x	-1000	-100	-10	10	100	1000
$g(x)$	$-\frac{1}{1000}$	$-\frac{1}{100}$	$-\frac{1}{10}$	$\frac{1}{10}$	$\frac{1}{100}$	$\frac{1}{1000}$

$g(x)$ approaches 0 as x approaches both $+\infty$ and $-\infty$.

6.2 Warm Up

1.

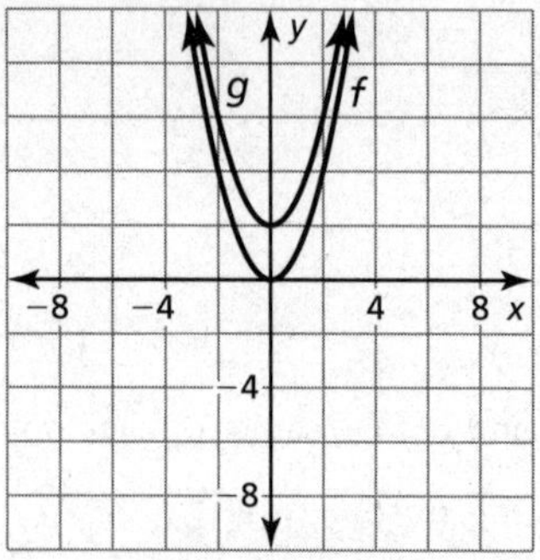

2.

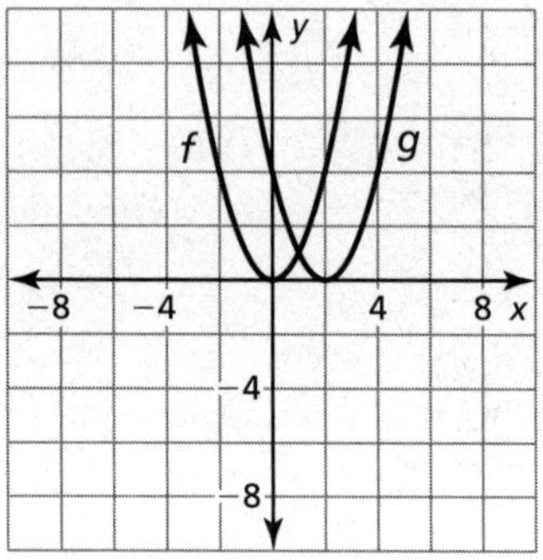

3.

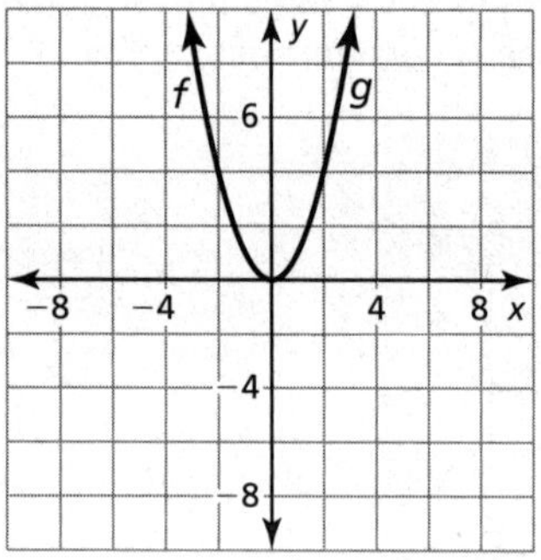

4.

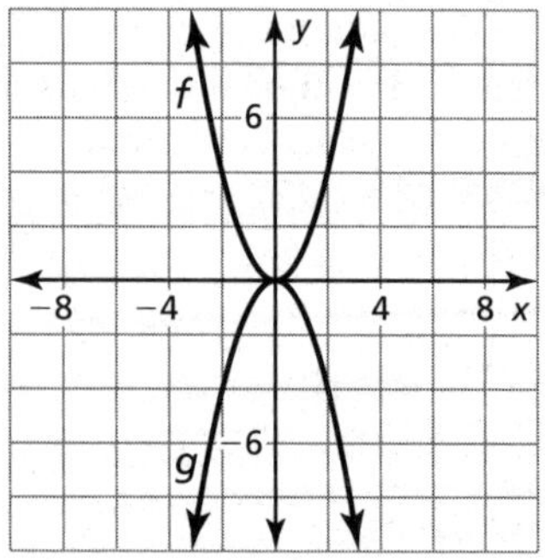

Answers

5. 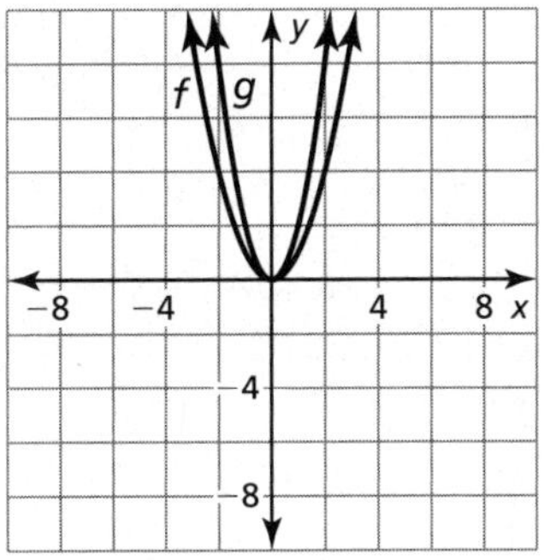

6. 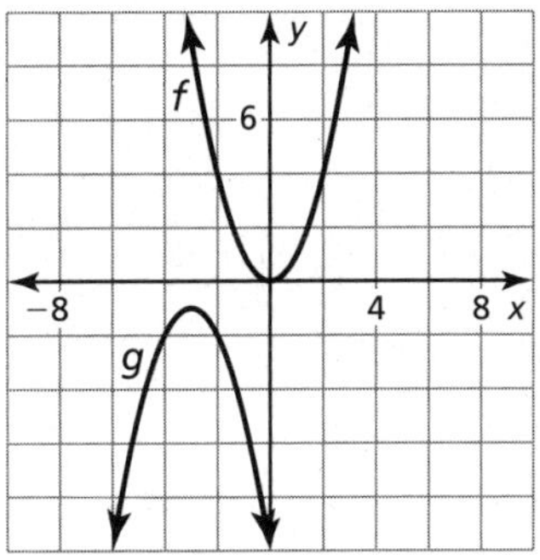

6.2 Cumulative Review Warm Up

1. maximum: $\frac{25}{8}$

2. minimum: $\frac{47}{12}$

3. maximum: $\frac{41}{8}$

6.2 Practice A

1. 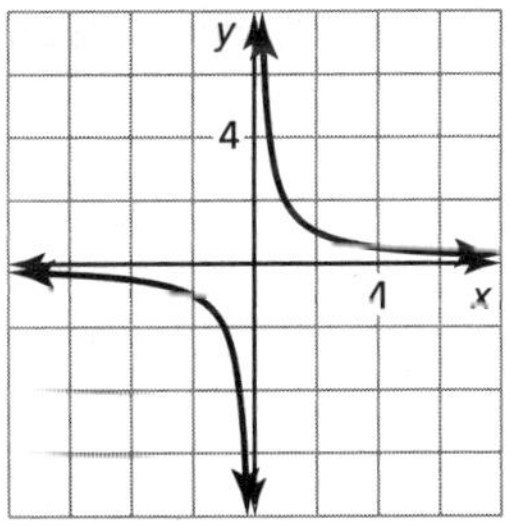

The graph of h lies farther from the axes than the graph of f. Both graphs lie in the first and third quadrants and have the same asymptotes, domain, and range.

2. 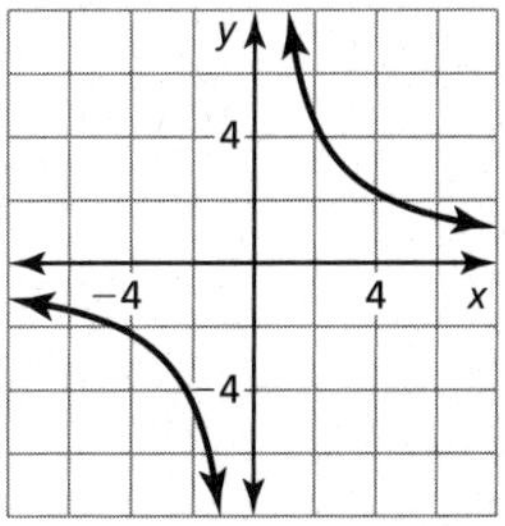

The graph of g lies farther from the axes than the graph of f. Both graphs lie in the first and third quadrants and have the same asymptotes, domain, and range.

3. 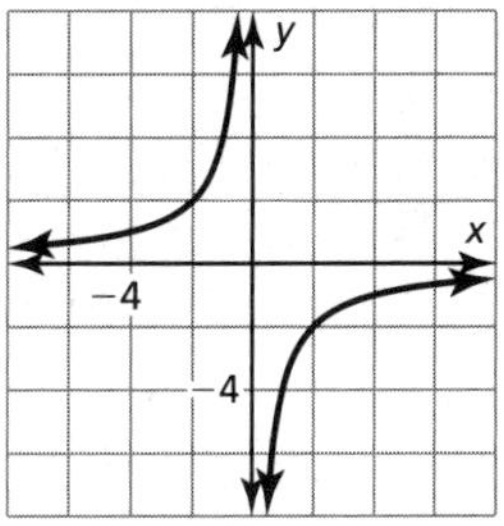

The graph of h is a reflection in the x-axis of the graph of f, and lies farther from the axes than the graph of f. The graphs have the same asymptotes, domain, and range.

4. 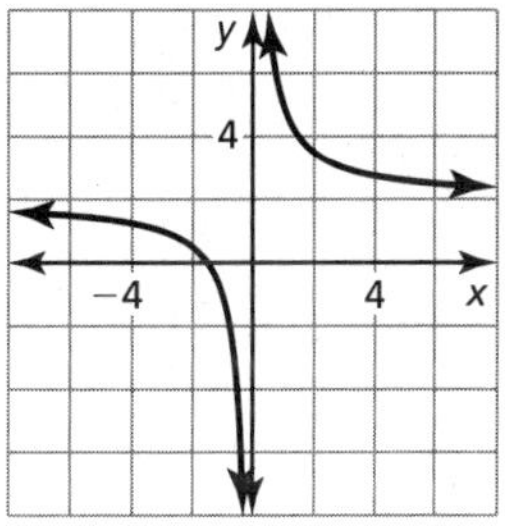

domain: all real numbers except 0, range: all real numbers except 2

5. 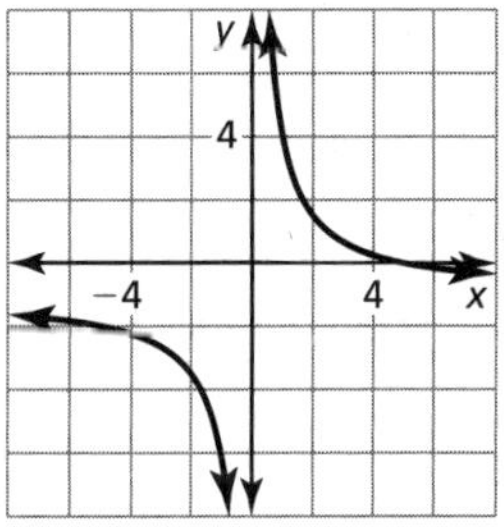

domain: all real numbers except 0, range: all real numbers except −1

Answers

6. 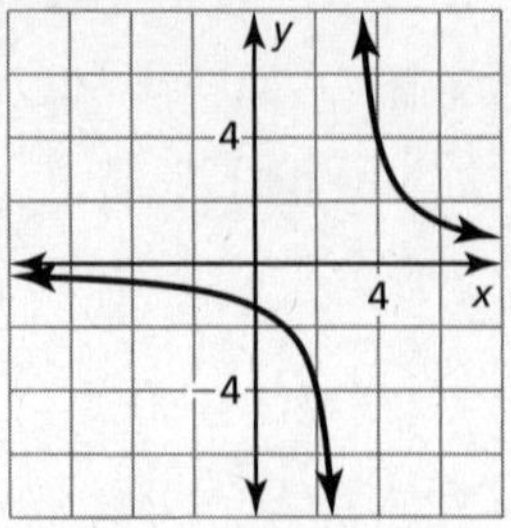

domain: all real numbers except 3, range: all real numbers except 0

7. 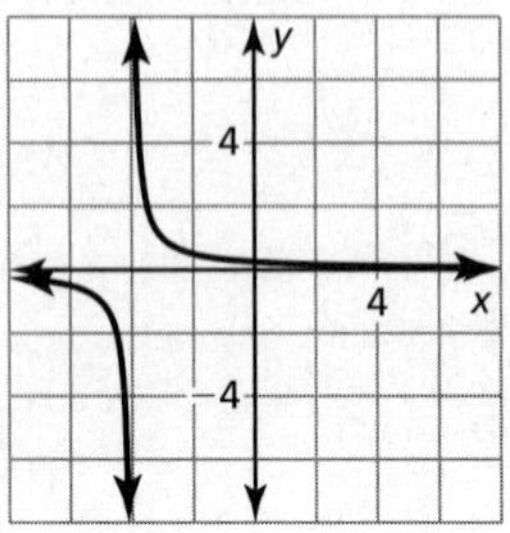

domain: all real numbers except -4, range: all real numbers except 0

8. 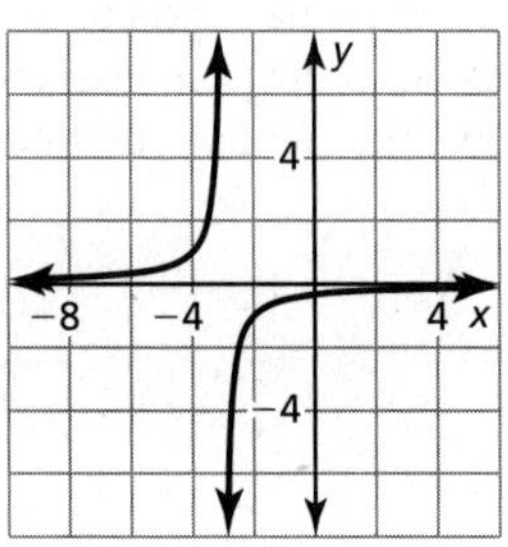

domain: all real numbers except -3, range: all real numbers except 0

9. 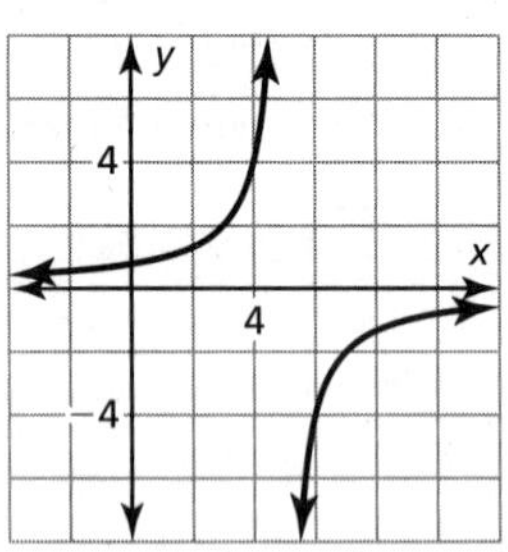

domain: all real numbers except 5, range: all real numbers except 0

10. 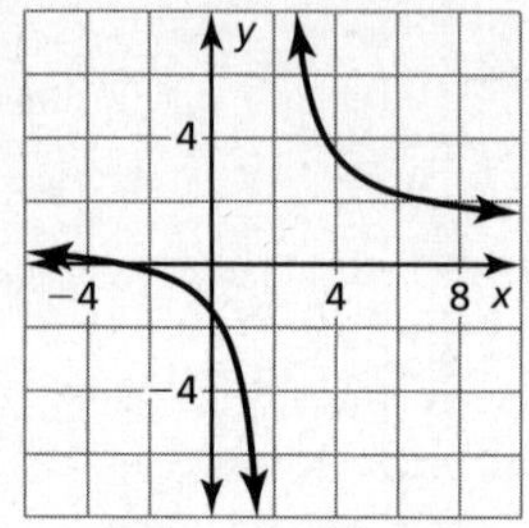

domain: all real numbers except 2, range: all real numbers except 1

11. 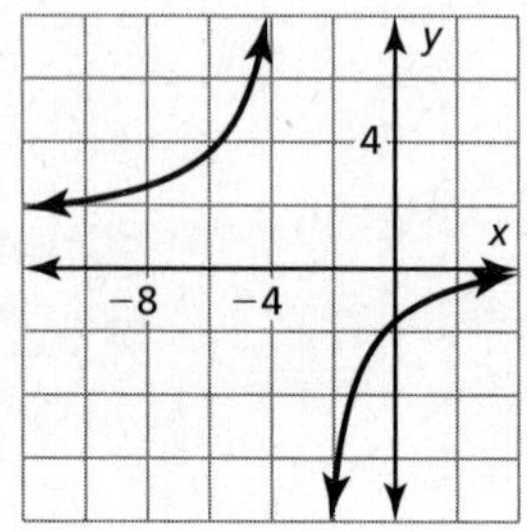

domain: all real numbers except -3, range: all real numbers except 1

12. 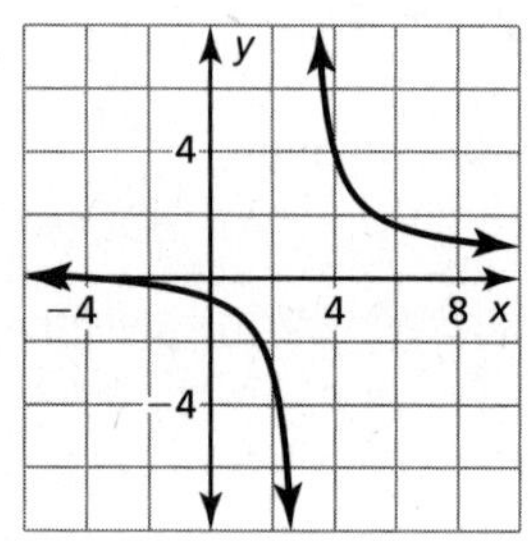

domain: all real numbers except 3, range: all real numbers except $\frac{1}{2}$

13. 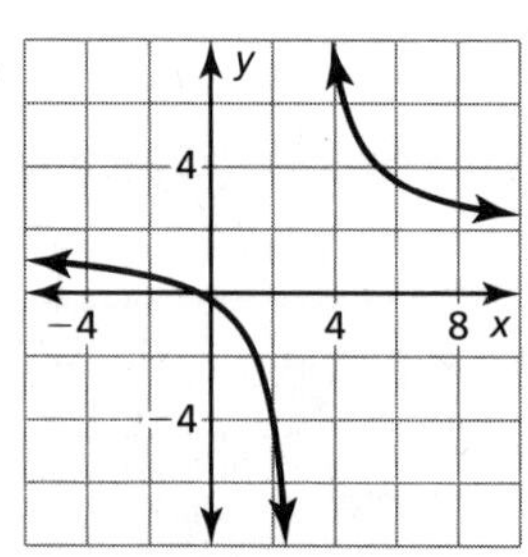

domain: all real numbers except 3, range: all real numbers except $\frac{5}{3}$

Answers

14. 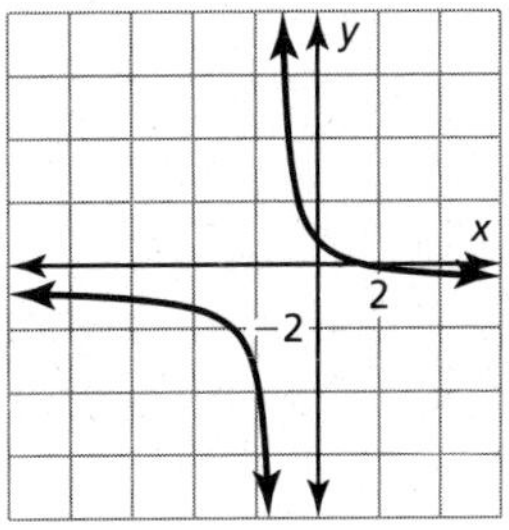

domain: all real numbers except $-\frac{4}{3}$, range: all real numbers except $-\frac{2}{3}$

15. 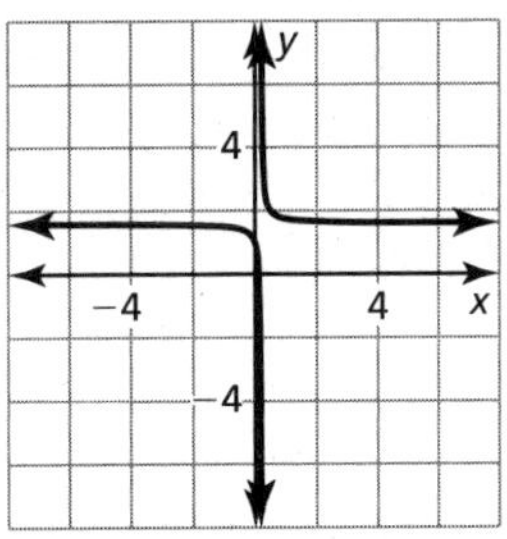

domain: all real numbers except $\frac{1}{5}$, range: all real numbers except $\frac{8}{5}$

16. $g(x) = \dfrac{1}{x+1} + 4$

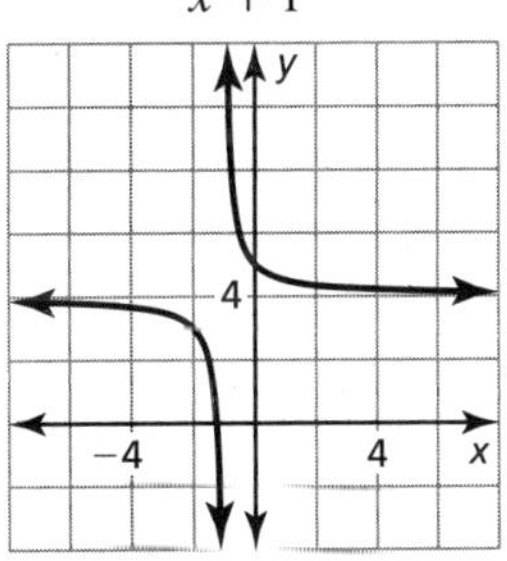

translation 1 unit left and 4 units up

17. $g(x) = \dfrac{17}{x-2} + 6$

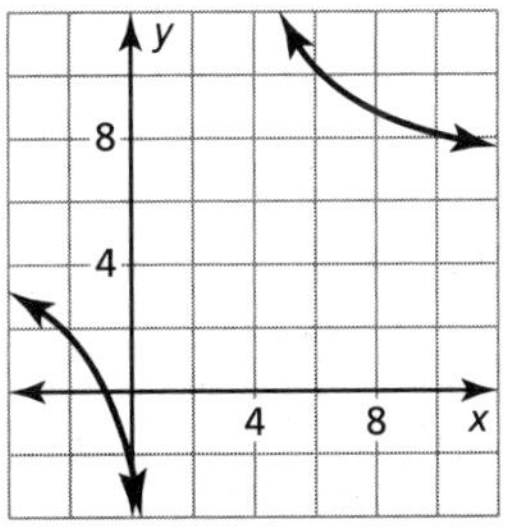

translation 2 units right and 6 units up

18. $g(x) = \dfrac{6}{x-4} + 3$

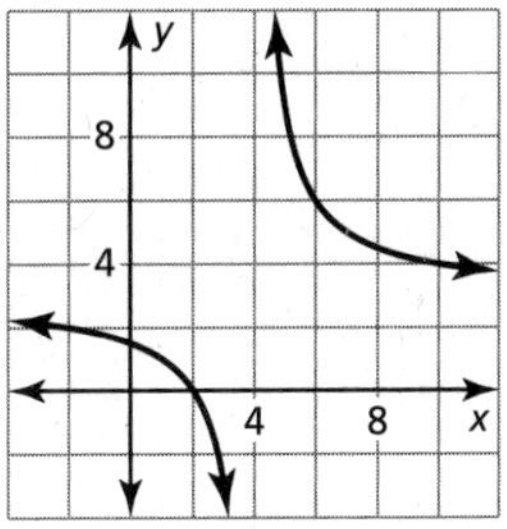

translation 4 units right and 3 units up

19. $g(x) = \dfrac{-22}{x+2} + 5$

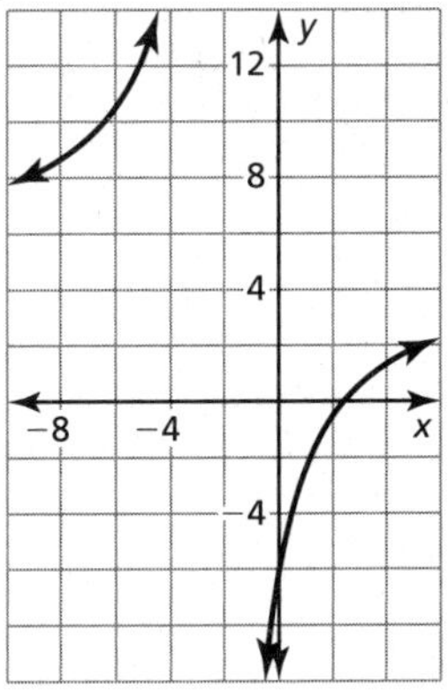

translation 2 units left and 5 units up

20. $g(x) = \dfrac{20}{x-5} + 1$

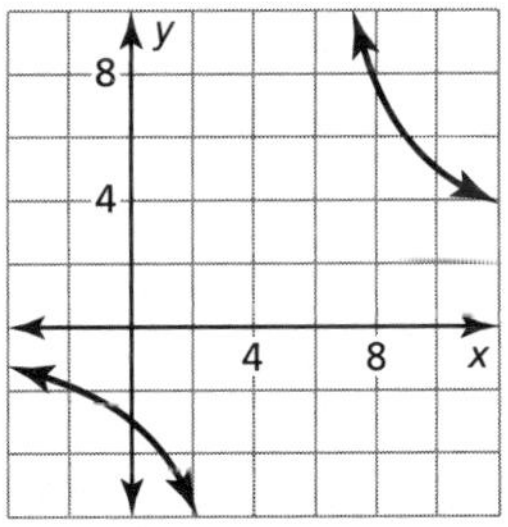

translation 5 units right and 1 unit up

21. $g(x) = \dfrac{12}{x-9} + 1$

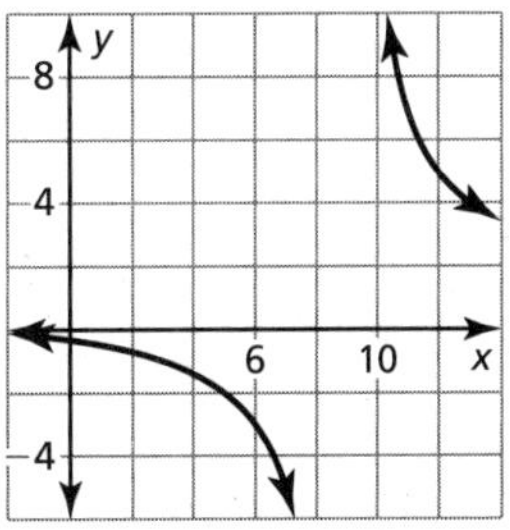

translation 9 units right and 1 unit up

Answers

22. a. 20 students

b. The average cost approaches \$150.

23.

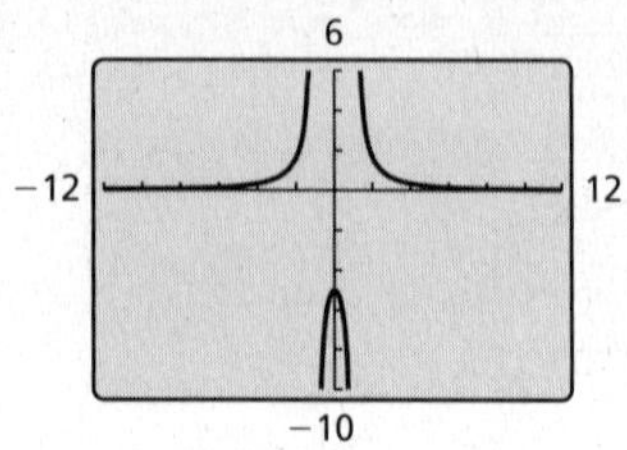

even

24.

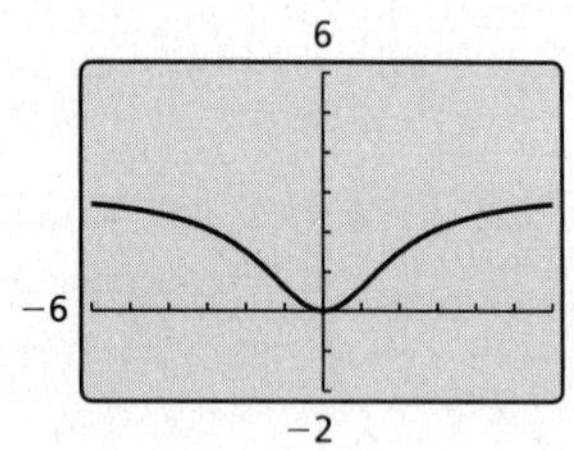

even

25.

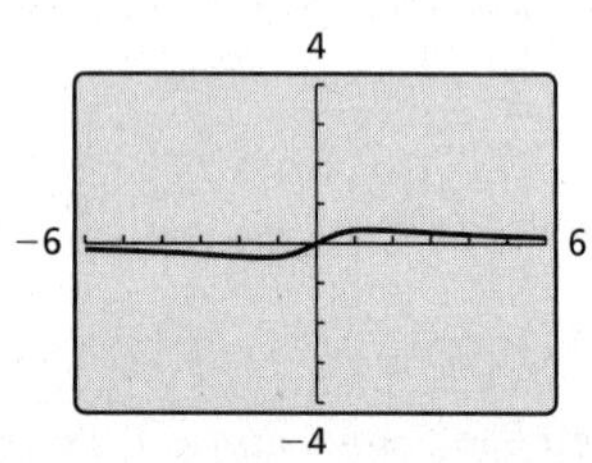

odd

6.2 Practice B

1.

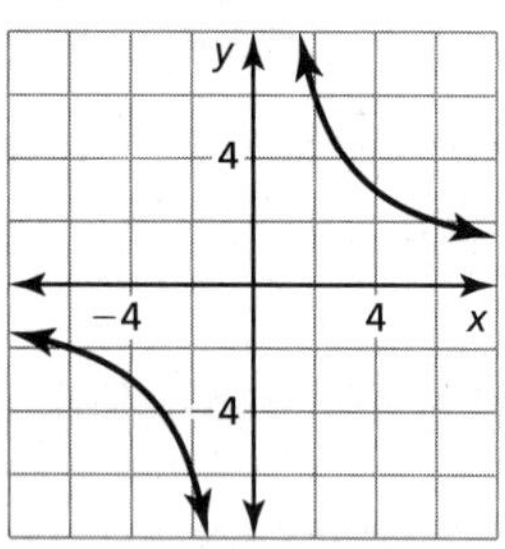

The graph of g lies farther from the axes than the graph of f. Both graphs lie in the first and third quadrants and have the same asymptotes, domain, and range.

2.

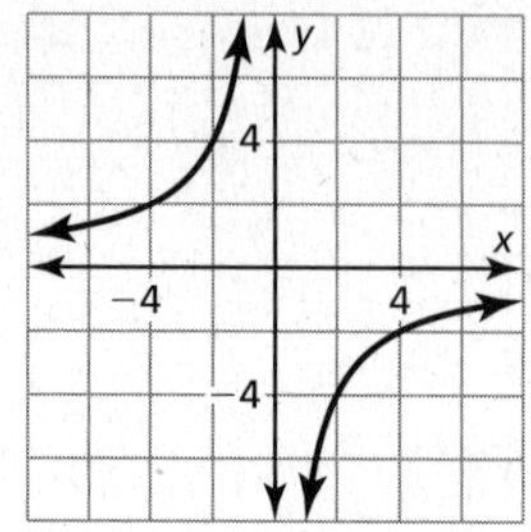

The graph of h is a reflection in the x-axis of the graph of f, and lies farther from the axes than the graph of f. The graphs have the same asymptotes, domain, and range.

3.

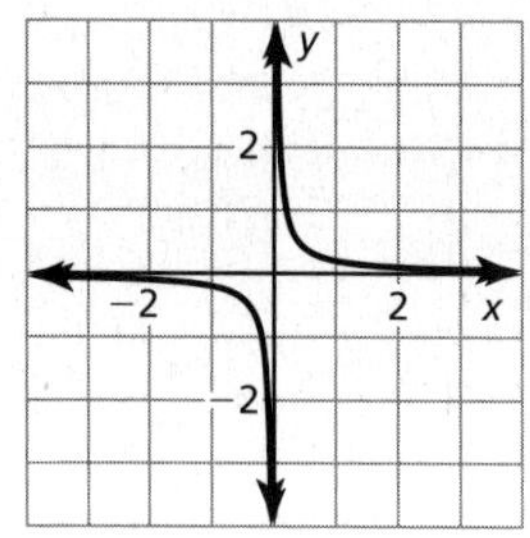

The graph of g lies closer to the axes than the graph of f. Both graphs lie in the first and third quadrants and have the same asymptotes, domain, and range.

4.

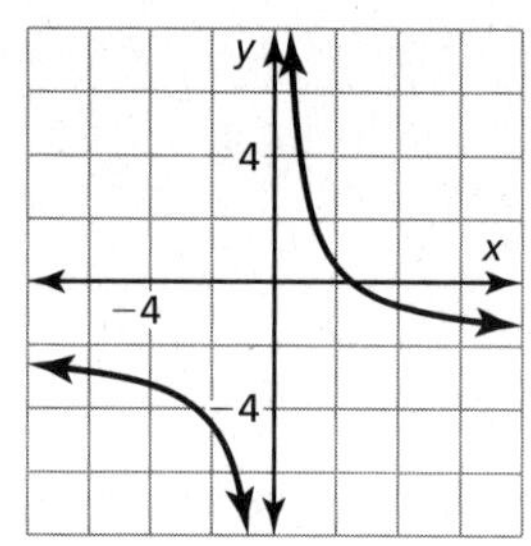

domain: all real numbers except 0, range: all real numbers except -2

5.

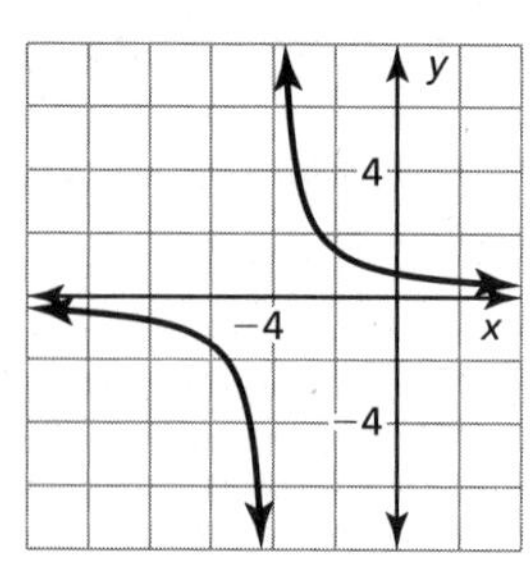

domain: all real numbers except -4, range: all real numbers except 0

Answers

6. 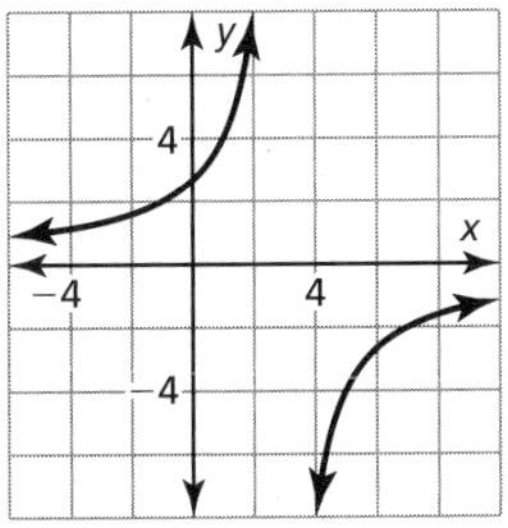

domain: all real numbers except 3, range: all real numbers except 0

7. 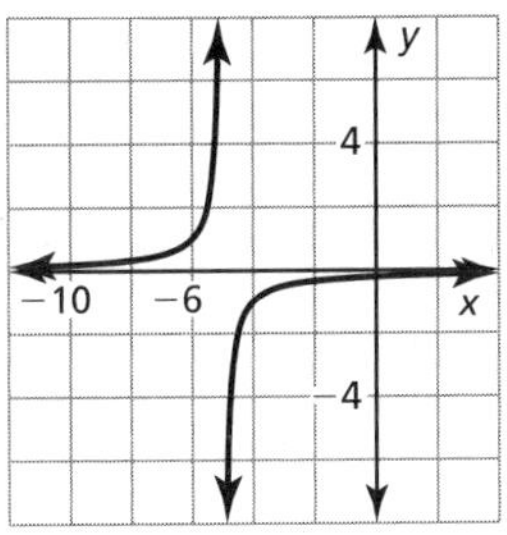

domain: all real numbers except -5, range: all real numbers except 0

8. 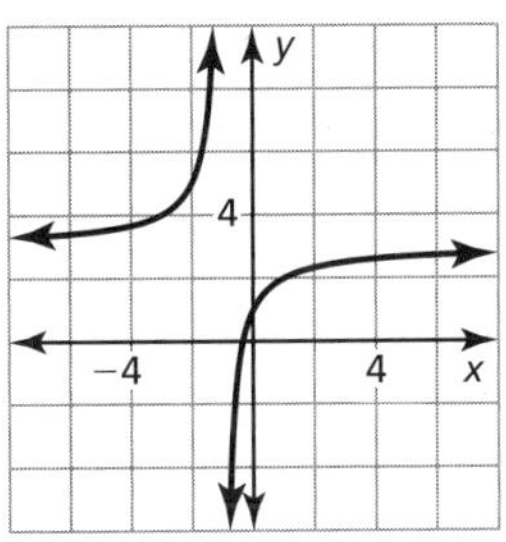

domain: all real numbers except -1, range: all real numbers except 3

9. 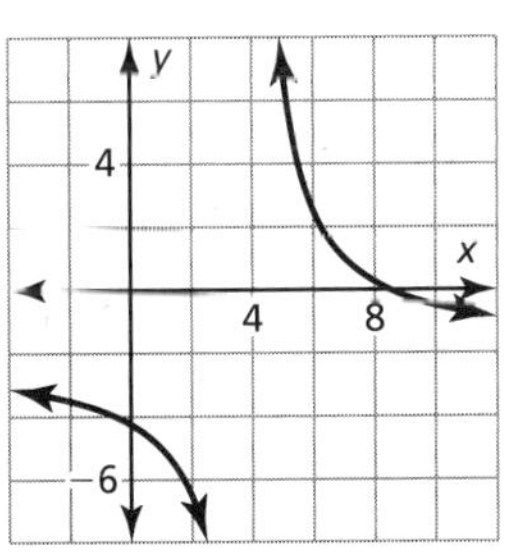

domain: all real numbers except 4, range: all real numbers except -2

10. 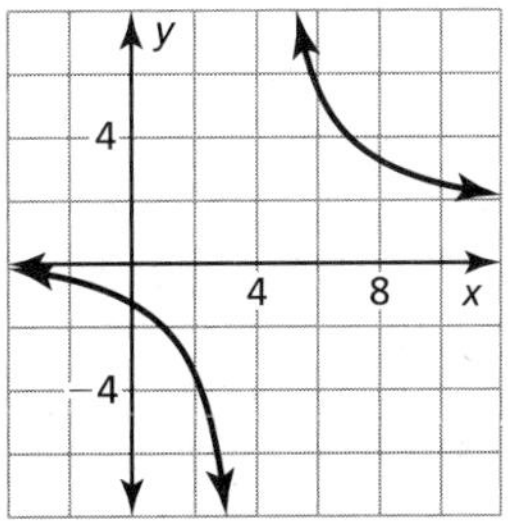

domain: all real numbers except 4, range: all real numbers except 1

11. 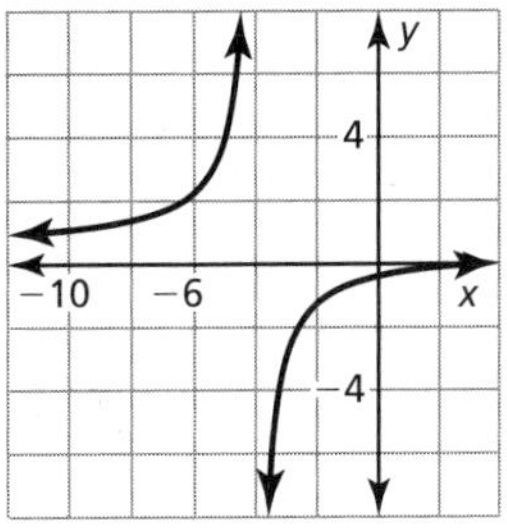

domain: all real numbers except -4, range: all real numbers except $\frac{1}{2}$

12. 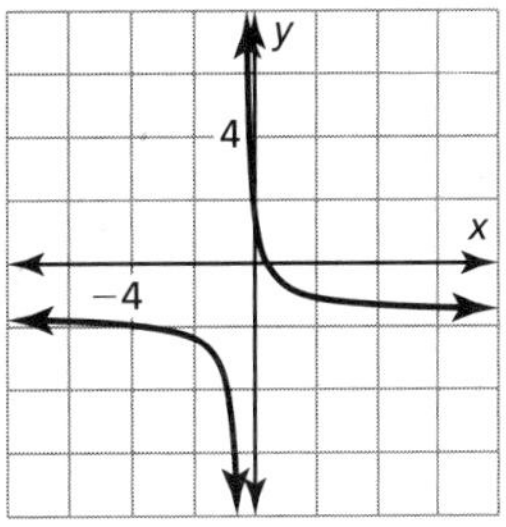

domain: all real numbers except $-\frac{2}{5}$, range: all real numbers except $-\frac{8}{5}$

13. 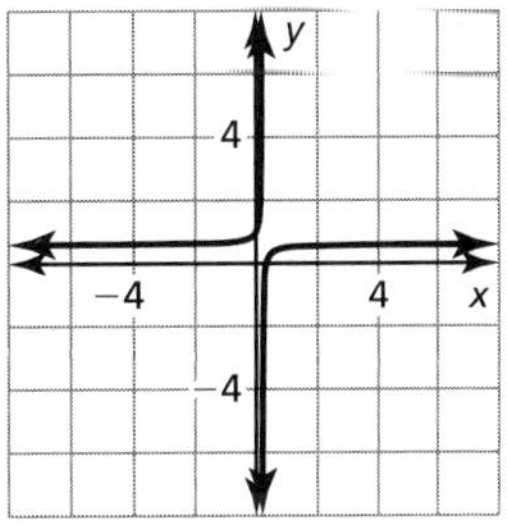

domain: all real numbers except $\frac{1}{5}$, range: all real numbers except $\frac{3}{5}$

Answers

14. 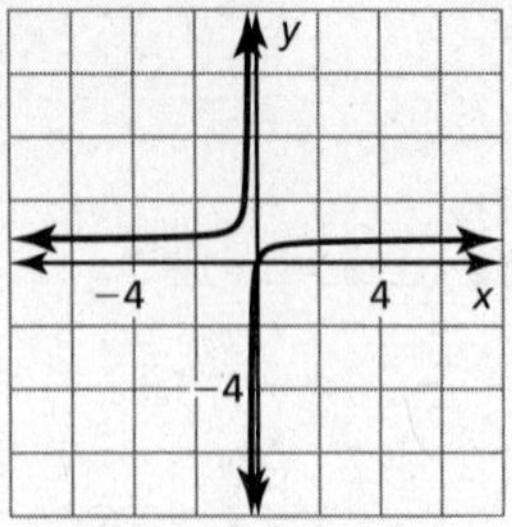

domain: all real numbers except $-\frac{1}{4}$, range: all real numbers except $\frac{3}{4}$

15. 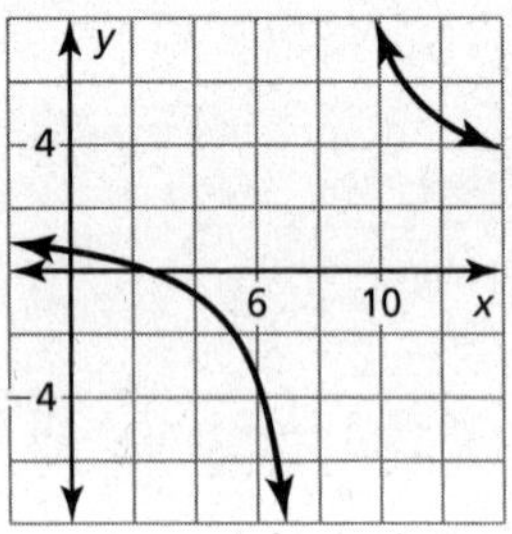

domain: all real numbers except 8, range: all real numbers except 2

16. $g(x) = \dfrac{1}{x + 2} + 3$

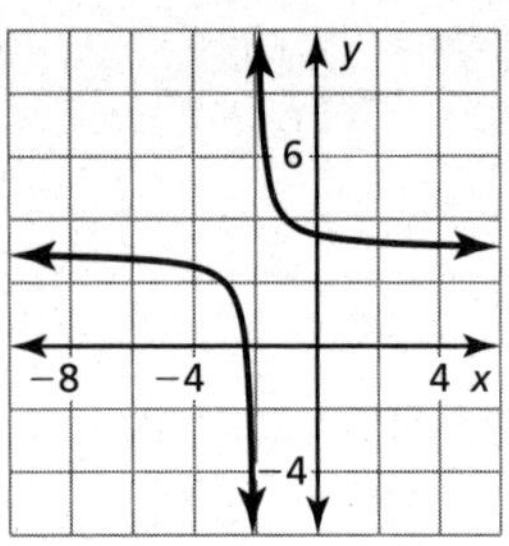

translation 2 units left and 3 units up

17. $g(x) = \dfrac{10}{x - 3} + 4$

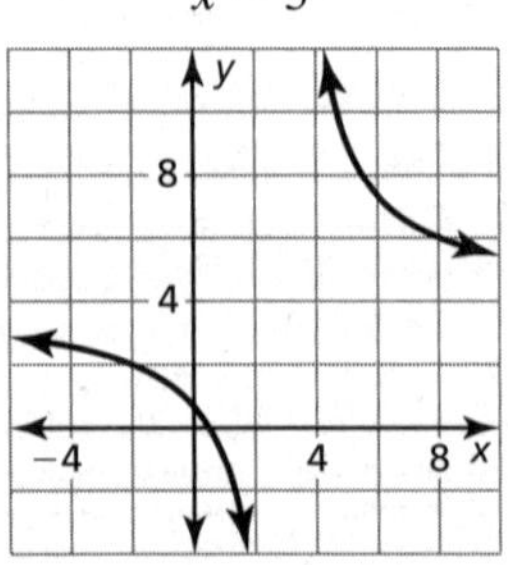

translation 3 units right and 4 units up

18. $g(x) = \dfrac{-30}{x + 5} + 4$

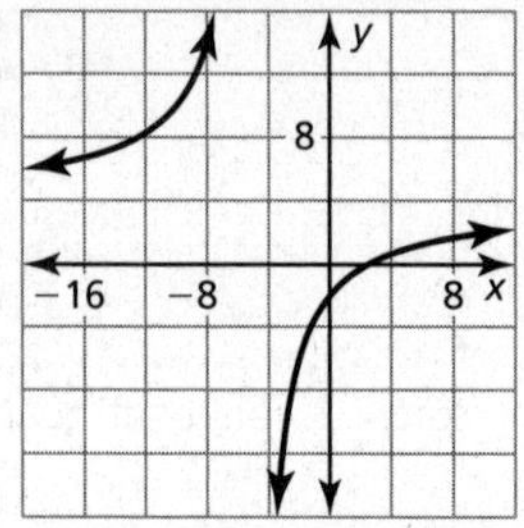

translation 5 units left and 4 units up

19. $g(x) = \dfrac{15}{x - 3} + 1$

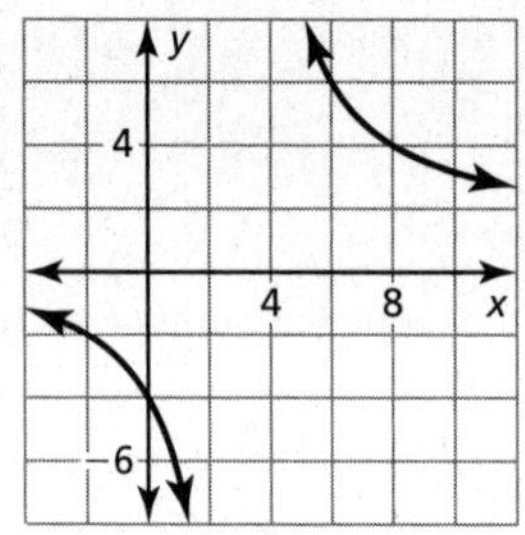

translation 3 units right and 1 unit up

20. $g(x) = \dfrac{-50}{x + 4} + 5$

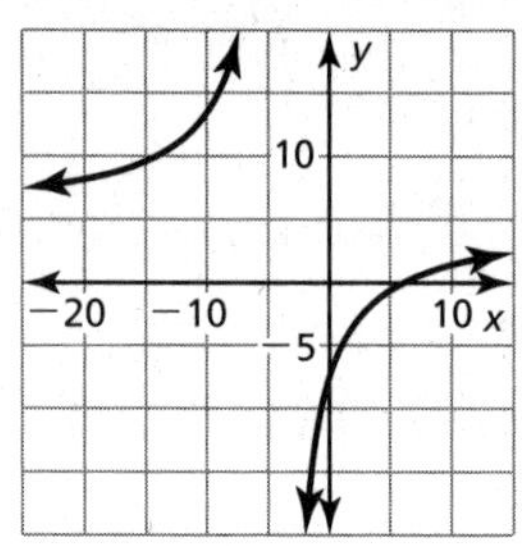

translation 4 units left and 5 units up

21. $g(x) = \dfrac{-44}{x + 6} + 7$

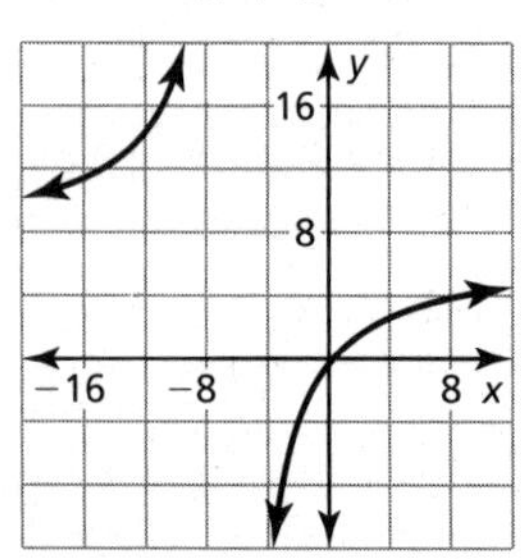

translation 6 units left and 7 units up

22. **a.** 38 statues

b. The average cost decreases, approaching $22.

Answers

23. a. 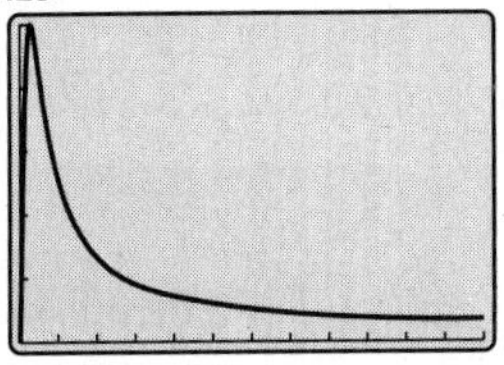

Sample answer: $0 < x \leq 24;\ 0 < y \leq 0.25$

b. $t = \frac{1}{2}\text{h}$

6.2 Enrichment and Extension

1. a. $-\infty$

b. ∞

c. $-\infty$

d. ∞

e. 0

f. 0

2. a. ∞

b. $-\infty$

c. $-\infty$

d. ∞

e. 3

f. 3

3. a. $-\infty$

b. ∞

c. $\frac{3}{2}$

d. $\frac{3}{2}$

e. 1

f. 1

4. a. ∞

b. ∞

c. -2

d. -2

6.2 Puzzle Time

A COIN

6.3 Start Thinking

1. -8 **2.** $0, 5$ **3.** $-\frac{3}{2}, \frac{5}{3}$

6.3 Warm Up

1. $\frac{10}{21}$ **2.** $\frac{36}{25}$ **3.** $\frac{6}{11}$

4. $\frac{2x - 2}{25}$ **5.** $\frac{x^2}{x^2 - 1}, x \neq \pm 1$

6. $\frac{-3x - 3}{2x^2 + x}, x \neq -1, -\frac{1}{2}, 0$

6.3 Cumulative Review Warm Up

1. \$7269.23 **2.** \$7715.86

3. \$7613.38 **4.** \$7500.05

5. \$7902.47 **6.** \$7401.70

6.3 Pratice A

1. $\frac{3x}{5x + 2}, x \neq 0$ **2.** $\frac{6x - 1}{2x}$

3. $\frac{x + 1}{x - 2}, x \neq 5$ **4.** cannot be simplified

5. $\frac{x + 1}{x^2 + 2x + 4}, x \neq 2$ **6.** $\frac{x - 4}{x^2 - x + 1}, x \neq -1$

7. $\frac{6x^2}{y^3}, x \neq 0$ **8.** $\frac{(x + 2)(x - 3)}{x}, x \neq 1$

9. $\frac{(x - 5)(x - 1)}{4}, x \neq 0, x \neq -7$

10. $(x - 5)(x + 1), x \neq 0, x \neq -3$

11. $\frac{(x + 3)(x - 3)}{4}, x \neq 0, x \neq 2$

12. $\frac{2x(x - 5)}{(x + 3)(x + 2)}, x \neq -1$

13. The only difference between the two functions is the domain. The domain of f is all real numbers except $-\frac{1}{4}$. The domain of g is all real numbers.

14. $\frac{56x^9}{y^{15}}, x \neq 0$

15. $2, x \neq 0, x \neq -2, x \neq 3$

16. $5, x \neq 0, x \neq 1, x \neq -3$

17. $\dfrac{x + 7}{(x + 3)(x - 2)}$

18. a. $\dfrac{2h + 2r}{rh}$

b. old: $\dfrac{2h + 2r}{rh}$; new: $\dfrac{h + 2r}{rh}$

c. yes; $h + 2r < 2h + 2r$

6.3 Practice B

1. $\dfrac{4x^2}{3x^2 + 7}, x \neq 0$

2. $\dfrac{x + 2}{x - 1}, x \neq -3$

3. cannot be simplified

4. $\dfrac{x - 5}{x^2 - 4x + 16}, x \neq -4$

5. $\dfrac{x^2 - 4}{5x - 3}$

6. $\dfrac{x - 1}{12x^2 - 10}$

7. $\dfrac{(x - 4)(x - 2)}{x}, x \neq -3$

8. $(x + 6)(x + 2), x \neq 0, x \neq 4$

9. $\dfrac{(x - 2)(x + 1)}{3}, x \neq 0, x \neq -5$

10. $\dfrac{3x(x + 2)}{(x + 1)(x + 4)}, x \neq 3$

11. $\dfrac{(x + 7)(x - 4)(x - 3)}{x + 5}, x \neq 5$

12. $(x + 5)(x - 4), x \neq 3, x \neq -3$

13. $1, x \neq 0, x \neq -5, x \neq 4$

14. $\dfrac{x - 3}{(x + 2)(x - 7)}$

15. $\dfrac{(x - 3)(x + 2)}{(x - 2)(x + 3)}, x \neq -1, x \neq -4, x \neq -2$

16. $1, x \neq 3, x \neq -3, x \neq -4, x \neq 2$

17. $\dfrac{4}{x + 3}$

18. $x + 3$

6.3 Enrichment and Extension

1. $f(x) = \dfrac{x^2 + x - 6}{x^2 - 2x - 35}; y = 1$

2. $h(x) = \dfrac{x^3 + 2x^2 - 16x - 32}{x^2 + 6x + 9}$; There is no horizontal asymptote.

3. $k(x) = \dfrac{-4x^3 + 5x^2 + 2}{7x^3 + 2x^2 + 5x - 9}; y = -\dfrac{4}{7}$

4. $m(x) = \dfrac{2x - 3}{3x^2 + 5x - 2}; y = 0$

5. $t(x) = \dfrac{3x^2 - 22x + 7}{10x^2 + 7x - 12}; y = \dfrac{3}{10}$

6.3 Puzzle Time

A NEWSPAPER

6.4 Start Thinking

$3x(x - 2)(x + 3); x = 0, x = 2, x = -3$

6.4 Warm Up

1. $\frac{23}{21}$

2. $\frac{23}{10}$

3. $-\frac{1}{7}$

4. $\frac{5}{8}$

5. $-\frac{11}{16}$

6. $\frac{33}{100}$

6.4 Cumulative Review Warm Up

1. $-x^2 - x + 4$

2. $3x^3 - 10x^2 + 8x$

3. $\dfrac{4 - 3x}{2x - x^2}$

4. $-9x^2 + 18x - 8$

5. 2

6. 13

6.4 Practice A

1. $\dfrac{3}{x}$

2. $\dfrac{x - 3}{9x^2}$

3. $\dfrac{7 - 3x}{x - 2}$

4. $x^2(6x - 18)$

5. $5x(x + 2)$

6. $x^2 - 9$

7. $x^2 - 3x - 10$

Answers

8. The calculation shown simply subtracts the numerators of the two rational expressions, without first rewriting each rational expression with a common denominator.

$$\frac{x}{x+3} - \frac{2}{x-1} = \frac{x(x-1)}{(x+3)(x-1)} - \frac{2(x+3)}{(x+3)(x-1)}$$

$$= \frac{(x^2 - x) - (2x + 6)}{(x+3)(x-1)}$$

$$= \frac{x^2 - 3x - 6}{(x+3)(x-1)}$$

9. $\frac{21 - 8x}{6x^2}$ **10.** $\frac{6x}{(x-1)(x+2)}$

11. $\frac{-5x^2 - 14x - 18}{(x+4)(x-3)}$ **12.** $\frac{6x+2}{(x+9)(x-2)}$

13. never; The LCD is a multiple of each denominator, not a sum.

14. sometimes; This happens if one denominator divides into the other; For example, the LCD of $\frac{1}{x(x-1)}$ and $\frac{1}{x-1}$ is $x(x-1)$.

15. $g(x) = \frac{3}{x-2} + 4$

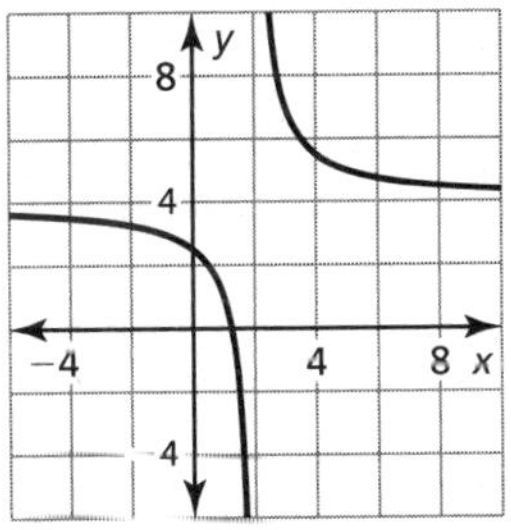

translation 2 units right and 4 units up

16. $g(x) = \frac{-17}{x+4} + 5$

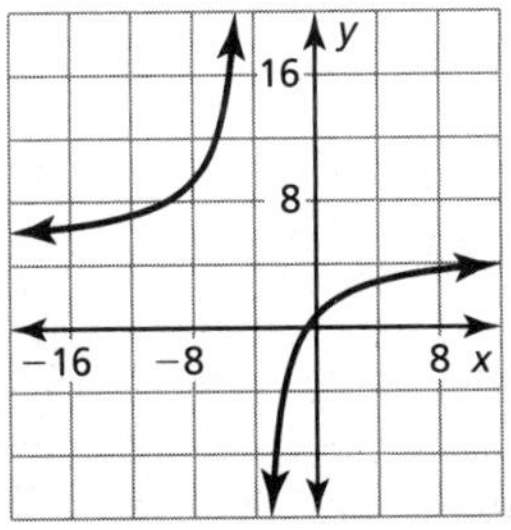

translation 4 units left and 5 units up

17. $g(x) = \frac{30}{x-3} + 10$

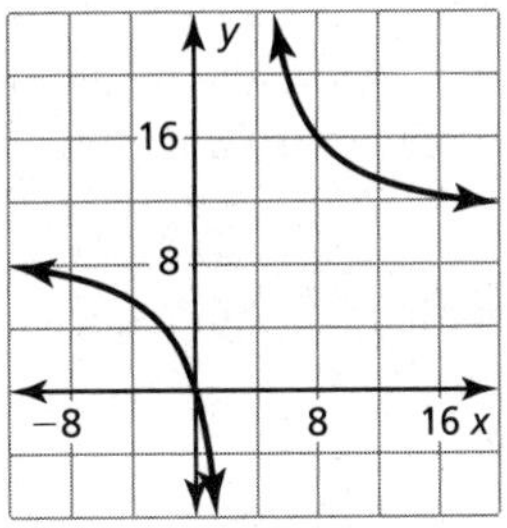

translation 3 units right and 10 units up

18. $g(x) = \frac{4}{x} + 3$

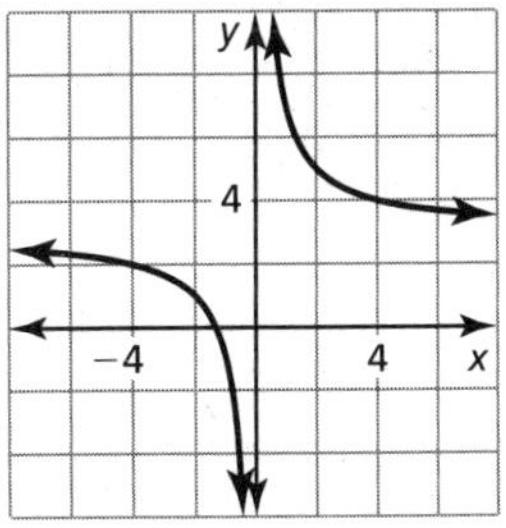

translation 3 units up

6.4 Practice B

1. $\frac{x-5}{25x^2}$ **2.** $\frac{2x^2 + 8x}{x+6}$

3. 3 **4.** $36x^2(x-2)$

5. $x^2 - 100$

6. $(25x^2 - 4)(3x^2 - 10x - 8)$

7. $x^2 + 7x - 18$

8. By not using the least common denominator, the resulting rational expression is not in simplest form. The numerator and denominator still share a common factor of x.

$$\frac{4}{7x} + \frac{5}{x^3} = \frac{4(x^2)}{7x(x^2)} + \frac{5(7)}{x^3(7)} = \frac{4x^2 + 35}{7x^3}$$

9. $\frac{4x^2 - 13x + 7}{(x-5)(x+1)}$ **10.** $\frac{3x+16}{(x-8)(x+3)}$

11. $\frac{3x+37}{(x+2)(x-8)}$

12. $\dfrac{8x^3 - 15x^2 - 17x - 9}{x(x - 3)(2x + 1)}$

13. always; If one denominator divides the other, then any multiple of the larger one is also a multiple of the smaller one.

14. sometimes; This happens if one denominator divides the other.

15. $g(x) = \dfrac{-17}{x + 4} + 5$

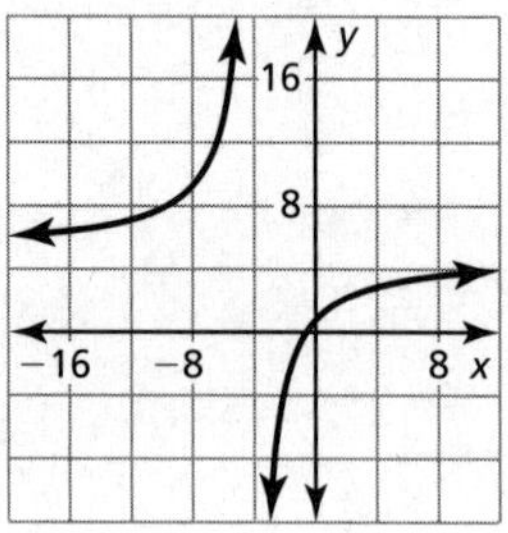

translation 4 units left and 5 units up

16. $g(x) = \dfrac{-108}{x + 12} + 9$

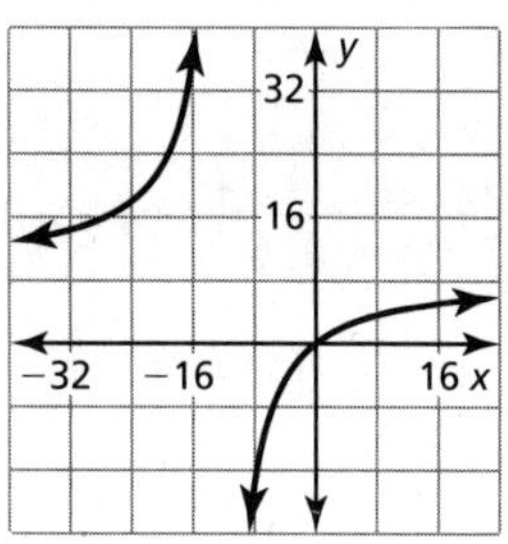

translation 12 units left and 9 units up

17. $g(x) = \dfrac{-4}{x} + 5$

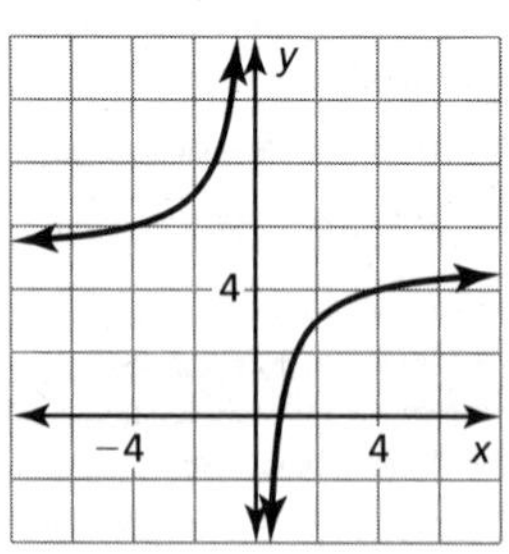

translation 5 units up

18. $g(x) = \dfrac{61}{x - 6} + 8$

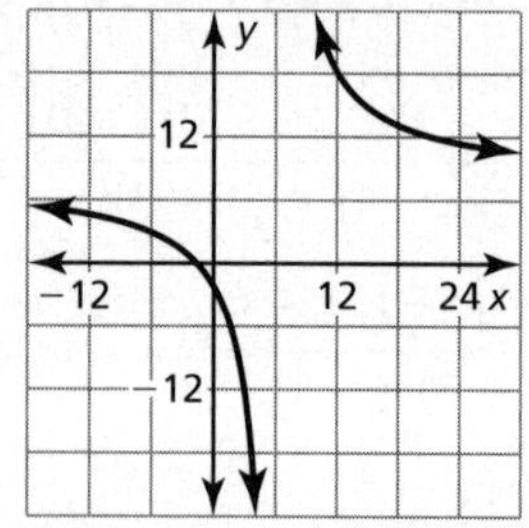

translation 6 units right and 8 units up

6.4 Enrichment and Extension

1. $\dfrac{5}{x + 1} - \dfrac{2}{x + 4}$

2. $\dfrac{1}{x} + \dfrac{2}{x + 2} + \dfrac{3}{x - 2}$

3. $\dfrac{\frac{8}{3}}{x - 3} + \dfrac{\frac{10}{3}}{x + 3}$

4. $\dfrac{\frac{26}{11}}{x - 7} - \dfrac{\frac{4}{11}}{x + 4}$

5. $\dfrac{\frac{4}{9}}{x} + \dfrac{\frac{35}{18}}{x - 3} - \dfrac{\frac{7}{18}}{x + 3}$

6. $\dfrac{5}{x + 1} + \dfrac{3}{x - 2} - \dfrac{1}{x + 3}$

6.4 Puzzle Time

GLASS CLIPPERS

6.5 Start Thinking

\$200 membership: $y = \dfrac{200}{x}$, \$120 membership: $y = \dfrac{120 + 10x}{x}$

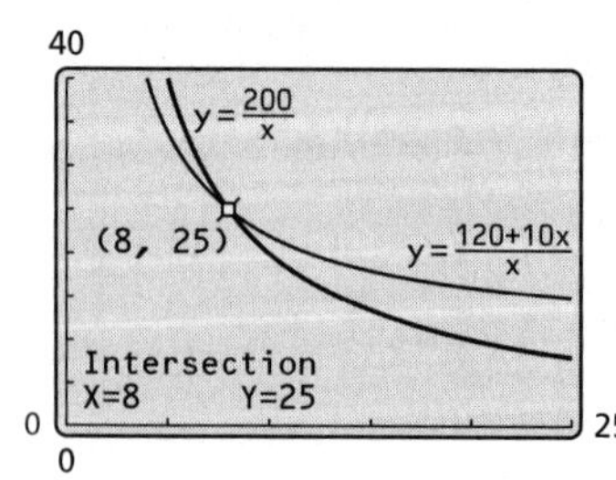

The point of intersection (8, 25) indicates that with 8 court rentals, both memberships will have the same cost per rental, \$25.

If more than 8 rentals per year are expected, the \$200 membership is a better deal. You can see the graph that represents the cost per rental for the \$200 membership lies below the \$120 membership graph for values of x greater than 8.

Answers

6.5 Warm Up

1. $-\frac{1}{7}$
2. 1, 2
3. −12
4. 2, 5
5. $-\frac{2}{3}, 2$
6. $0, \frac{7}{2}$

6.5 Cumulative Review Warm Up

1. $-2 + i$
2. $14 - 22i$
3. $16 + 30i$
4. $-74 + 21i$
5. $15 + 12i$
6. $-4 + 3i$

6.5 Practice A

1. $x = -6$
2. $x = -10$
3. $x = -4$
4. $x = 9$
5. $x(x - 2)$
6. $x(x + 5)$
7. $x = \frac{3}{4}$
8. $x = 8$
9. $x = \frac{5}{2}$
10. $x = 1$
11. $x = 8, x = -2$
12. $x = 3$
13. did not multiply the right side of the equation by the LCD, $2x$

$$\frac{4}{x} + \frac{1}{2} = 1$$

$$2x \bullet \frac{4}{x} + 2x \bullet \frac{1}{2} = 2x \bullet 1$$

14. $\frac{3}{5} + \frac{3}{t} = 1; t = 7.5$ h
15. yes; $g(x) = \frac{5}{x} + 2$
16. yes; $g(x) = \frac{4}{x + 1}$

6.5 Practice B

1. $x = -8$
2. $x = 2, x = 7$
3. $x = 0, x = -\frac{1}{3}$
4. $x = 8$
5. $5(x + 3)(x + 2)$
6. $4(x - 8)(3x - 2)$
7. $x = 8$
8. $x = 1$
9. $x = \frac{9}{2}$
10. $x = \frac{8}{3}$
11. $x = 2, x = -1$
12. $x = 2, x = -4$
13. incorrectly multiplied by $(x + 2)$, which is not the common denominator $x(x + 2)$

$$\frac{3}{x + 2} + 5 = \frac{1}{x}$$

$$x(x + 2) \bullet \frac{3}{x + 2} + x(x + 2) \bullet 5 = x(x + 2) \bullet \frac{1}{x}$$

14. $\frac{1.4}{3} + \frac{1.4}{t} = 1; t = 2.625$ h
15. yes; $g(x) = \frac{3}{2x} + \frac{7}{2}$
16. yes; $g(x) = \frac{1}{\sqrt[3]{x - 9}}$

6.5 Enrichment and Extension

1. *Sample answer:* $\left(\frac{11}{3}, 1\right), (7, -1), \left(\frac{9}{2}, 0\right)$. If point (x, y) lies on the graph of f, then point (y, x) lies on the graph of f^{-1}.
2. no; The point $\left(2, \frac{11}{3}\right)$ does not lie on the graph of f.
3. $f^{-1}(x) = \frac{9 - 2x}{x - 2}$
4. $f\left(f^{-1}(x)\right) = \frac{2\left(\frac{9 - 2x}{x - 2}\right) + 9}{\left(\frac{9 - 2x}{x - 2}\right) + 2} = \frac{\frac{5x}{x - 2}}{\frac{5}{x - 2}} = x$ and

$$f^{-1}\left(f(x)\right) = \frac{9 - 2\left(\frac{2x + 9}{x + 2}\right)}{\left(\frac{2x + 9}{x + 2}\right) - 2} = \frac{\frac{5x}{x + 2}}{\frac{5}{x + 2}} = x$$

5. 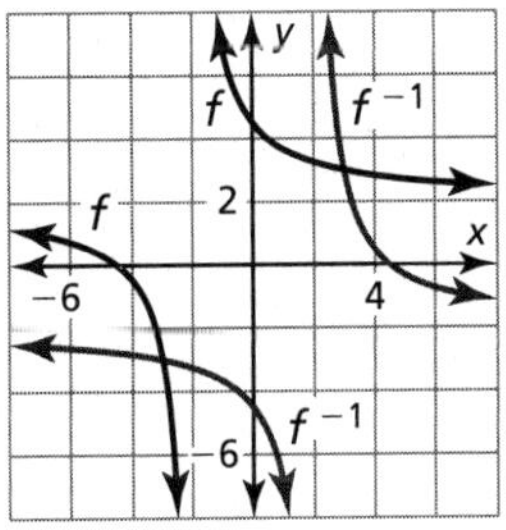

6. domain: all real numbers, $x \neq -2$; range: all real numbers, $x \neq 2$

Answers

7. domain: all real numbers, $x \neq 2$; range: all real numbers, $x \neq -2$; The domain of f is the range of f^{-1} and the range of f is the domain of f^{-1}.

8. The vertical asymptote of the graph of f is the horizontal asymptote of the graph of f^{-1}. The horizontal asymptote of the graph of f is the vertical asymptote of the graph of f^{-1}.

9. The graphs of f and f^{-1} intersect at points that lie on the line $y = x$.

10. no; The zero of f^{-1} is the x-value where $f^{-1}(x) = 0$, or equivalently $f(0) = x$. This is the y-intercept of the graph of f, not the asymptotes.

11. yes; $g^{-1}(x) = \dfrac{-x-7}{2x-1}$

12. no; Rational functions of the form $f(x) = \dfrac{ax+b}{cx+d}$ will have an inverse, but not all rational functions will have an inverse. To have an inverse, the function must pass the horizontal line test, and not all rational functions pass the horizontal line test.

6.5 Puzzle Time

ELVIS PARSLEY

Cumulative Review

1. $(x+3)(x-7)$
2. $(x-4)(x+12)$
3. $(x+1)(x+8)$
4. $(x-4)(x-5)$
5. $(x+3)(x-9)$
6. $(x-4)(x-8)$
7. $(2x-1)(x+3)$
8. $2(x-4)(x+3)$
9. $(3x-4)(x+9)$
10. $(5x+3)(x+4)$
11. $(3x-7)(x-6)$
12. $(4x-3)(x-1)$
13. $2(2x+7)(x+4)$
14. $(3x-5)(2x+1)$
15. $(2x+1)(4x-3)$
16. $(5x-7)(x+1)$
17. $2(4x-1)(x+3)$
18. $6(x-2)(2x+1)$
19. $2^2 \bullet 3$
20. $3 \bullet 67$
21. $2^3 \bullet 11$
22. $2^3 \bullet 7$
23. $5^2 \bullet 3$
24. $5^2 \bullet 2^2 \bullet 3$
25. prime
26. $3^2 \bullet 11$
27. $2^2 \bullet 47$
28. prime
29. $5 \bullet 17$
30. prime
31. prime
32. 13^2
33. prime
34. $13 \bullet 5$
35. $x = 3$
36. $x = -\dfrac{7}{2}$
37. $x = \dfrac{154}{25}$
38. $x = -80$
39. $x = 6$
40. $x = -\dfrac{2}{5}$
41. $x = \dfrac{75}{4}$
42. $x = -\dfrac{168}{19}$
43. $x = \dfrac{45}{62}$
44. $x = \dfrac{400}{231}$
45. $x = \dfrac{54}{115}$
46. $x = \dfrac{325}{42}$
47. $x + 3$ and $x - 8$
48. $2x - 5$ and $x + 4$
49. $3x + 5$ and $2x - 3$
50. $x = -7$ and $x = 5$
51. $x = -6$ and $x = 3$
52. $x = -11$ and $x = 9$
53. $x = -6$ and $x = 8$
54. $x = 3$ and $x = 5$
55. $x = -5$ and $x = 9$
56. $x = -4$ and $x = -1$
57. $x = -5$ and $x = 5$
58. $x = -4$ and $x = 8$
59. $x = -3$ and $x = 0$
60. $x = -8$ and $x = -\dfrac{1}{2}$
61. $x = \dfrac{5}{3}$ and $x = 7$
62. $x = -5$ and $x = \dfrac{3}{2}$

Answers

63. $x = 0$ and $x = \frac{7}{2}$

64. 64; two real solutions

65. −15; two imaginary solutions

66. 0; one real solution

67. 177; two real solutions

68. 164; two real solutions

69. −479; two imaginary solutions

70. −159; two imaginary solutions

71. 0; one real solution

72. 40; two real solutions

73. 132; two real solutions

74. −92; two imaginary solutions

75. 0; one real solution

76. $x = 1$

77. $x = \frac{5 \pm 1}{6} = \frac{2}{3}, 1$

78. $x = \frac{-2 \pm 2\sqrt{21}}{5}$

79. $x = \frac{1 \pm i\sqrt{53}}{3}$

80. $x = \frac{-9 \pm \sqrt{141}}{2}$

81. $x = \frac{-5 \pm \sqrt{57}}{8}$

82. $x = \frac{4 \pm 5i\sqrt{2}}{6}$

83. $x = \frac{7 \pm i\sqrt{239}}{-6}$

84. $x = \frac{-1 \pm i\sqrt{179}}{5}$

85. $x = \frac{3 \pm \sqrt{73}}{2}$

86. $x = \frac{-7 \pm i\sqrt{111}}{16}$

87. $x = \frac{1 \pm i\sqrt{15}}{2}$

88. $x + 8$

89. $2x + 5$

90. $4x - 1$

91. $(21, 12)$

92. $(-3, 11)$

93. $(-4, -21)$

94. $(-7, -6)$

95. $(15, 13)$

96. $(-9, -11)$

97. $(6, 36)$

98. $(1, 9)$

99. $(1, -5)$

100. $(-5, -6)$ and $(1, 0)$

101. $(-3, 0)$ and $(6, 9)$

102. $(-7, -5)$ and $(-2, 0)$

103. $(-2, -12)$

104. $(-12, 7)$

105. $\left(\frac{7}{3}, -2\right)$

106. $(8, 5)$

107. $\left(-\frac{17}{2}, -1\right)$

108. $\left(6, \frac{11}{4}\right)$

109. $\left(\frac{1 + \sqrt{17}}{2}, -21 + \sqrt{17}\right)$ and $\left(\frac{1 - \sqrt{17}}{2}, -21 - \sqrt{17}\right)$

110. $\left(-1 + \sqrt{11}, \frac{47 - 13\sqrt{11}}{2}\right)$ and $\left(-1 - \sqrt{11}, \frac{47 + 13\sqrt{11}}{2}\right)$

111. $\left(-3 + \sqrt{6}, \frac{16 - 11\sqrt{6}}{3}\right)$ and $\left(-3 - \sqrt{6}, \frac{16 + 11\sqrt{6}}{3}\right)$

112. $\left(\frac{-3 + \sqrt{417}}{6}, \frac{108 - \sqrt{417}}{6}\right)$ and $\left(\frac{-3 - \sqrt{417}}{6}, \frac{108 + \sqrt{417}}{6}\right)$

113. 2448 ft^3

114. **a.** 54 in.^2 **b.** 27 in.^3

115. **a.** $196\pi \text{ cm}^2$ **b.** $\frac{1372}{3}\pi \text{ cm}^3$

116. 158

117. −2034

118. $\frac{1777}{32}$

119. $10x^4 - 12x^2 - 2x - 1$

120. $9x^4 - 2x^3 - x^2 - 5x + 4$

121. $8x^5 - 4x^4 + 5x^3 - 13x^2 - 8x + 10$

122. $-6x^3 + 18x - 10$

123. $-24x^4 + 4x^3 + 16x - 11$

124. $-17x^5 + 2x^4 - 7x^3 - 10x + 5$

125. direct variation **126.** direct variation

127. neither **128.** inverse variation

129. $\dfrac{3(x - 7)}{x - 4}, x \neq -5, x \neq -3, x \neq 4$

130. $\dfrac{-5(x + 6)}{4(x - 11)}, x \neq -3, x \neq -7, x \neq 11$

131. $\dfrac{2x^2 + 13x + 3}{3(x + 2)(x + 5)}$

132. $\dfrac{-9x^2 - 65x - 6}{(x + 6)(x - 6)(x + 7)}$

133. $x = 4$ **134.** $x = -8$

Chapter 7

7.1 Start Thinking

1. Start with 5, then add 2 to the prior term to find each consecutive term.
2. Start with $\frac{2}{3}$, then take the prior term and multiply the numerator by 2 and add 4 to the denominator to find each consecutive term.
3. Start with 8, then subtract $\frac{3}{2}$ from the prior term to find each consecutive term.
4. Start with 3, then multiply the prior term by -2 to find each consecutive term.

7.1 Warm Up

1. $f(1) = 0, f(2) = 2, f(3) = 4, f(4) = 6, f(5) = 8$
2. $f(1) = 2, f(2) = 4, f(3) = 8, f(4) = 16, f(5) = 32$
3. $f(1) = 1, f(2) = 3, f(3) = 5, f(4) = 7, f(5) = 9$
4. $f(1) = -5, f(2) = 5, f(3) = -5, f(4) = 5, f(5) = -5$
5. $f(1) = \frac{1}{3}, f(2) = \frac{2}{3}, f(3) = \frac{4}{3}, f(4) = \frac{8}{3}, f(5) = \frac{16}{3}$
6. $f(1) = 1, f(2) = -4, f(3) = 9, f(4) = -16, f(5) = 25$

7.1 Cumulative Review Warm Up

1. $x = 4$
2. $x = 4$
3. $x = \frac{1}{2} \ln 4 = \ln 2$
4. $x = 1$
5. $x = 2 - \frac{1}{e^2} \approx 1.865$
6. $x = 3$

7.1 Practice A

1. $-2, -1, 0, 1, 2, 3$
2. $3, 2, 1, 0, -1, -2$
3. $1, 8, 27, 64, 125, 216$
4. $-4, -1, 4, 11, 20, 31$
5. $4, 16, 64, 256, 1024, 4096$
6. $0, -3, -8, -15, -24, -35$
7. Start with 1, then each consecutive term is 3 more than the previous term: 13, $a_n = 3n - 2$.
8. Start with 1, then each consecutive term is 3 raised to a power 1 more than the power 3 is raised to in the previous term: 81, $a_n = 3^{n-1}$.
9. Start with 1.5, then each consecutive term is 1.5 more than the previous term: 7.5, $a_n = 1.5n$.
10. Start with 4.2, then each consecutive term is 1.6 more than the previous term: 10.6, $a_n = 2.6 + 1.6n$.
11. Start with 4.7, then each consecutive term is 1.3 less than the previous term: -0.5, $a_n = 6 - 1.3n$.

Answers

12. Start with -5, then the absolute value of each consecutive term is 5 more than the absolute value of the previous term, and the signs alternate: -25, $a_n = (-1)^n 5n$.

13. Start with $\frac{1}{6}$, then the numerator of each consecutive term is 1 more than the numerator of the previous term: $\frac{5}{6}$, $a_n = \frac{n}{6}$.

14. Start with $\frac{3}{2}$, then the denominator of each consecutive term is 2 more than the denominator of the previous term: $\frac{3}{10}$, $a_n = \frac{3}{2n}$.

15. $a_n = 10(2^{n-1})$

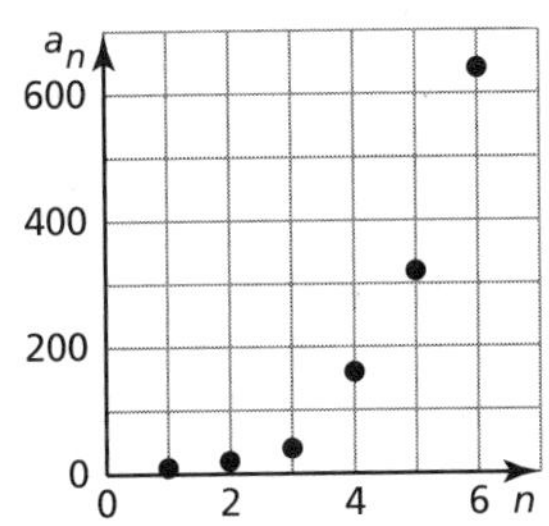

16. $\sum_{n=1}^{5} 4n$ **17.** $\sum_{n=1}^{5}(6n - 3)$ **18.** $\sum_{n=1}^{\infty}(6n - 5)$

19. $\sum_{n=1}^{\infty}(2n - 3)$ **20.** $\sum_{n=1}^{\infty}\frac{1}{5^n}$ **21.** $\sum_{n=1}^{\infty}\frac{n}{n + 6}$

22. 30 **23.** 90 **24.** 55

25. 93 **26.** 73 **27.** $\frac{153}{28}$

28. 105; $\sum_{n=1}^{6}(28 - 3n)$

7.1 Practice B

1. 5, 8, 13, 20, 29, 40 **2.** $\frac{1}{3}$, 1, 3, 9, 27, 81

3. 1, 0, 1, 4, 9, 16

4. 2, −1, −6, −13, −22, −33

5. $-\frac{1}{6}, -\frac{2}{7}, -\frac{3}{8}, -\frac{4}{9}, -\frac{1}{2}, -\frac{6}{11}$

6. $\frac{1}{2}, \frac{2}{5}, \frac{3}{8}, \frac{4}{11}, \frac{5}{14}, \frac{6}{17}$

7. Start with 2, then each consecutive term is 5 more than the previous term: 22, $a_n = 5n - 3$.

8. Start with 2.7, then each consecutive term is 3.5 more than the previous term: 16.7, $a_n = 3.5n - 0.8$.

9. Start with 6.4, then each consecutive term is 2.9 less than the previous term: -5.2, $a_n = 9.3 - 2.9n$.

10. Start with $-\frac{1}{7}$, then each consecutive term is $\frac{1}{7}$ less than the previous term: $-\frac{5}{7}$, $a_n = -\frac{n}{7}$.

11. Start with $\frac{5}{2}$, then the denominator of each consecutive term is twice the denominator of the previous term: $\frac{5}{32}$, $a_n = \frac{5}{2^n}$.

12. Start with $\frac{3}{1}$, then the numerator of each consecutive term is 3 times the numerator of the previous term, and the denominator of each term is 1 more than the denominator of the previous term: $\frac{243}{5}$, $a_n = \frac{3^n}{n}$.

13. Start with $\frac{2}{3}$, then each consecutive term is $\frac{2}{3}$ times the previous term: $\frac{32}{243}$, $a_n = \left(\frac{2}{3}\right)^n$.

14. Start with 2, then each consecutive term is 10 times the previous term: 20,000, $a_n = 2(10^{n-1})$.

15. $a_n = 100 + 7n$

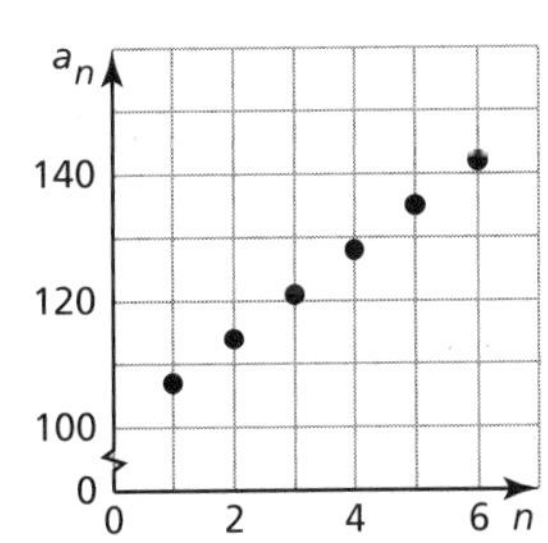

16. $\sum_{n=1}^{5}(7n - 1)$ **17.** $\sum_{n=1}^{\infty}(3n + 7)$

18. $\sum_{n=1}^{\infty}(n^2 - 4)$

19. $\sum_{n=1}^{\infty}\frac{n}{4^n}$

20. $\sum_{n=1}^{\infty}\frac{2}{n + 4}$

21. $\sum_{n=1}^{5}(-1)^n(2n - 1)$

22. 189 **23.** 450 **24.** 103

25. $\frac{107}{210}$ **26.** 45 **27.** 300

28. 294; $\sum_{i=1}^{7}(34 + 2i)$

7.1 Enrichment and Extension

1. a. 2 **b.** $2n - 1$ **c.** $S_n = n^2$

2. a. 3 **b.** $3n$ **c.** $S_n = \frac{3}{2}n(n + 1)$

3. a. k **b.** nk **c.** $S_n = \frac{k}{2}n(n + 1)$

4. The series is equivalent to $\log 10 + \log 10^2 \log 10^3 + \ldots + \log 10^n$ which by the Power Property of Logarithms, is equivalent to $\log 10 + 2 \log 10 + 3 \log 10 + \ldots + n \log 10$. The common difference is $\log 10$, so the series is arithmetic. By Exercise 3, a general rule for the sum is $S_n = \frac{\log 10}{2}n(n + 1)$. Because $\log 10 = 1$, this simplifies to $S_n = \frac{n}{2}(n + 1)$.

5. $S_n = \frac{\ln e}{2}n(n + 1) = \frac{n}{2}(n + 1)$

6. $S_n = \frac{\log_b a}{2}n(n + 1)$

7.1 Puzzle Time

RASH DECISIONS

7.2 Start Thinking

1. 2, 4, 6, 8, 10; $a_n = 2n$

2. 9.5, 9, 8.5, 8, 7.5; $a_n = 10 - \frac{1}{2}n$

3. 2.5, 4, 5.5, 7, 8.5; $a_n = 1 + \frac{3}{2}n$

Sample answer: All the graphs appear to be linear and the rules are linear equations.

7.2 Warm Up

1. $(-2, -5)$ **2.** $(3, 2)$ **3.** $(-4, 8)$

4. $(-2, 2)$ **5.** $\left(\frac{3}{2}, -\frac{5}{2}\right)$ **6.** $\left(0, \frac{1}{6}\right)$

7.2 Cumulative Review Warm Up

1. $f(x) = x^3 + 3x^2 - 2x - 6$

2. $f(x) = x^3 + 3x^2 - 3x - 1$

3. $f(x) = 2x^4 - x^3 - 12x^2 + 6x$

4. $f(x) = 2x^4 + 3x^3 - 32x^2 - 48x$

5. $f(x) = x^4 + 4x^3 - 6x^2 - 20x - 7$

6. $f(x) = 10x^4 - 69x^3 + 80x^2 + 75x$

7.2 Practice A

1. yes; $d = -3$

2. no; no common difference

3. yes; $d = \frac{1}{3}$

4. no; no common difference

5. a. $a_n = 4n - 9$ **b.** $a_n = 12 - 3n$

6. $a_n = 7n + 8; a_{20} = 148$

7. $a_n = 71 - 9n; a_{20} = -109$

8. $a_n = 15n - 40; a_{20} = 260$

9. $a_n = \frac{3}{2}n - \frac{9}{2}; a_{20} = \frac{51}{2}$

10. The number 27 was used as the first term in the formula instead of -27.

Use $a_1 = -27$ and $d = 15$.

$a_n = -27 + (n - 1)15$

$a_n = -42 + 15n$

Answers

11. $a_n = 4n - 1$

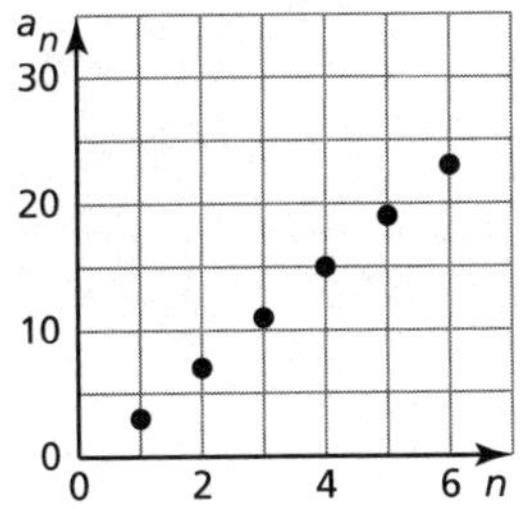

12. $a_n = 28 - 4n$

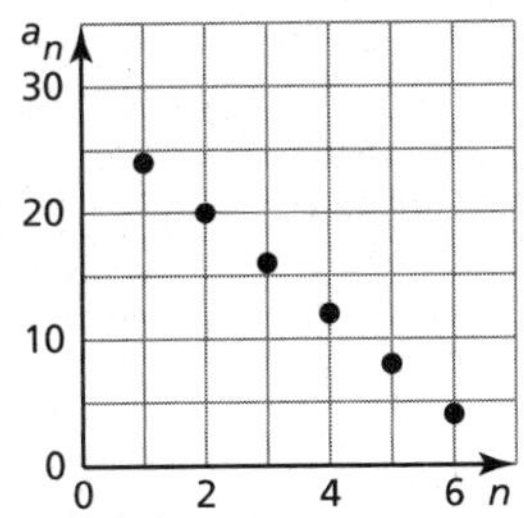

13. $a_n = 4n + 13$

14. $a_n = 6n + 18$

15. $a_n = 72 - 10n$

16. $a_n = 15 - 7n$

17. 15,500; $a_n = 2n$; $a_{124} = 248$; $\sum_{n=1}^{124} 2n = 15{,}500$

7.2 Practice B

1. no; no common difference

2. yes; $d = -4$ 3. yes; $d = \frac{1}{6}$ 4. yes; $d = \frac{3}{10}$

5. a. $a_n = 19 - 7n$ b. $a_n = 10n - 18$

6. $a_n = 45 - 8n$; $a_{20} = -115$

7. $a_n = \frac{4}{3}n - \frac{16}{3}$; $a_{20} = \frac{64}{3}$

8. $a_n = 2.1n - 1.9$; $a_{20} = 40.1$

9. $a_n = 2.9 - 0.7n$; $a_{20} = -11.1$

10. The formula $a_n = n_1 - (n - 1)d$ was used instead of the formula $a_n = n_1 + (n - 1)d$.

Use $a_1 = -27$ and $d = 15$.

$a_n = -27 + (n - 1)15$

$a_n = -42 + 15n$

11. $a_n = 4n + 15$

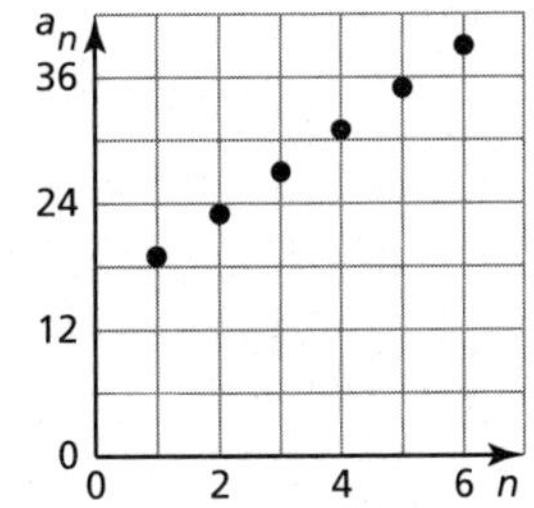

12. $a_n = \frac{1}{2}n + \frac{11}{2}$

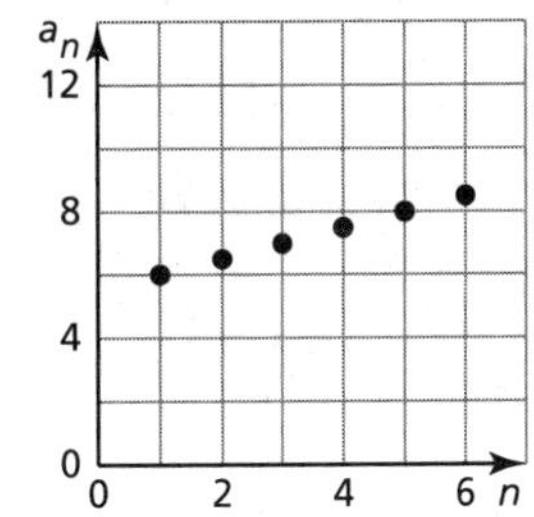

13. $a_n = 5n + 24$

14. $a_n = 80 - 17n$

15. $a_n = 6n - 27$

16. $a_n = \frac{2}{3}n + 18$

17. 62,500; $a_n = 2n - 1$; $a_{250} = 499$;

$\sum_{n=1}^{250}(2n - 1) = 62{,}500$

7.2 Enrichment and Extension

1. 54 2. 23 3. 85

4. 38 5. 42 6. 91

7. 75 8. 8181 9. 7876

10. $t_n = (a - 3b) + n(a + b)$

11. $x = 3, y = 2$

7.2 Puzzle Time

POULTRYGEIST

Answers

7.3 Start Thinking

1. 2, 4, 8, 16; $a_n = 2^n$

2. $\frac{1}{2}, \frac{1}{4}, \frac{1}{8}, \frac{1}{16}; a_n = \left(\frac{1}{2}\right)^n$

3. $-2, 4, -8, 16; a_n = (-1)^n(2)^n = (-2)^n$

Sample answer: The first two graphs appear to be exponential growth and decay. The third graph oscillates between negative and positive values. All the equations involve exponential expressions.

7.3 Warm Up

1. 30 **2.** −27 **3.** 1270

4. −6 **5.** 15 **6.** $-\frac{11}{16}$

7.3 Cumulative Review Warm Up

1.

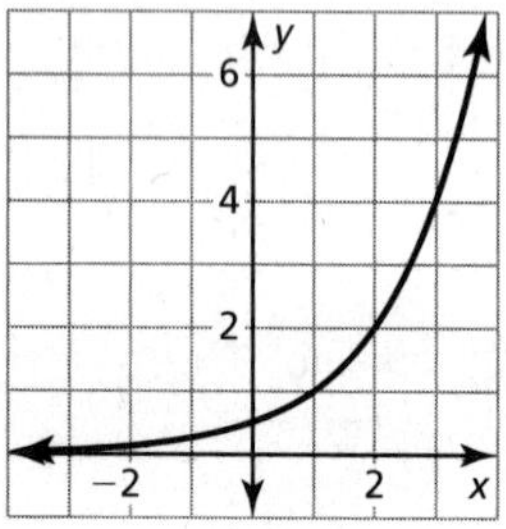

2.

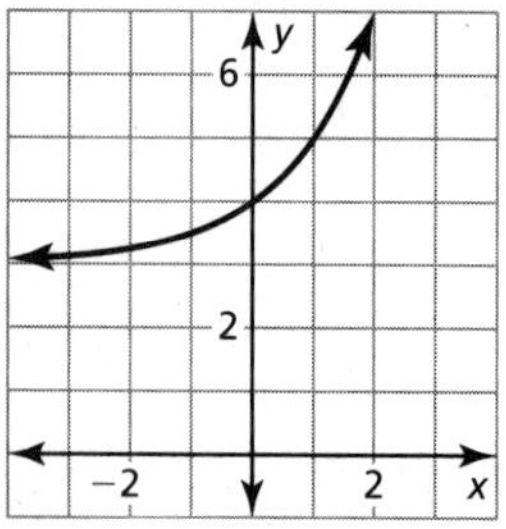

3.

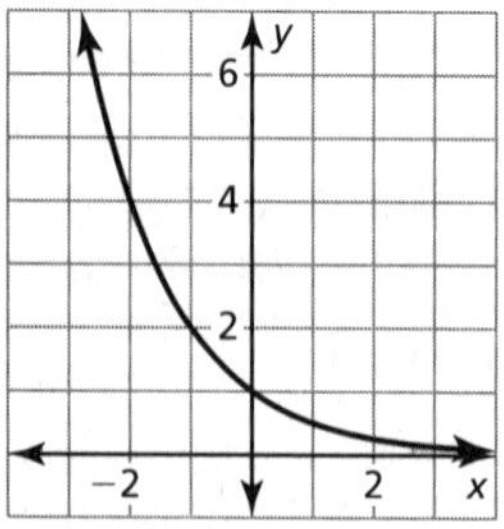

4.

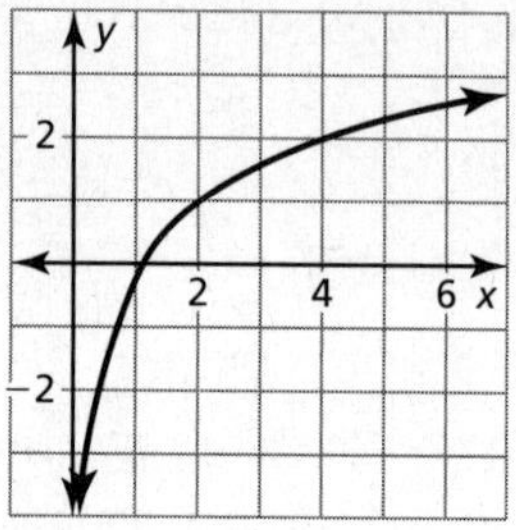

5.

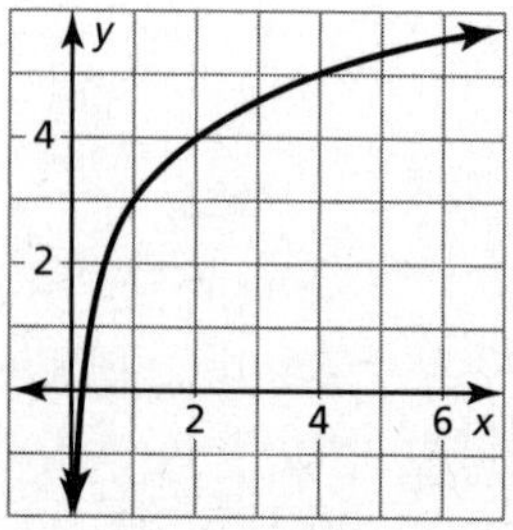

6.

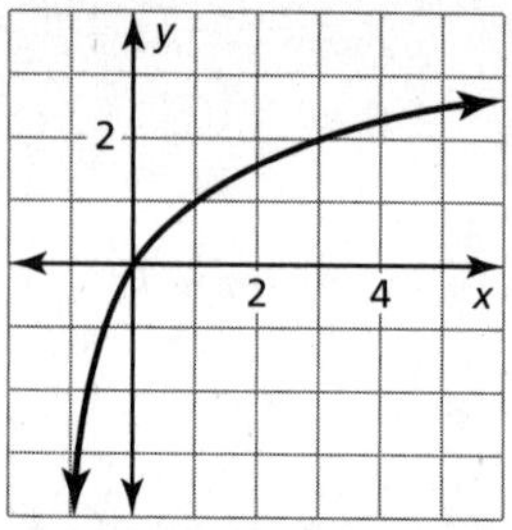

7.3 Practice A

1. yes; $r = \frac{1}{2}$ **2.** no; no common ratio

3. no; no common ratio **4.** yes; $r = 6$

5. a. $a_n = -5(3)^{n-1}$ **b.** $a_n = 54\left(\frac{1}{6}\right)^{n-1}$

6. $a_n = 3(2)^{n-1}; a_7 = 192$

7. $a_n = 7(3)^{n-1}; a_7 = 5103$

8. $a_n = 192\left(\frac{1}{2}\right)^{n-1}; a_7 = 3$

9. $a_n = 36\left(\frac{2}{3}\right)^{n-1}; a_7 = \frac{256}{81}$

Answers

10. $a_n = 3^{n-1}$

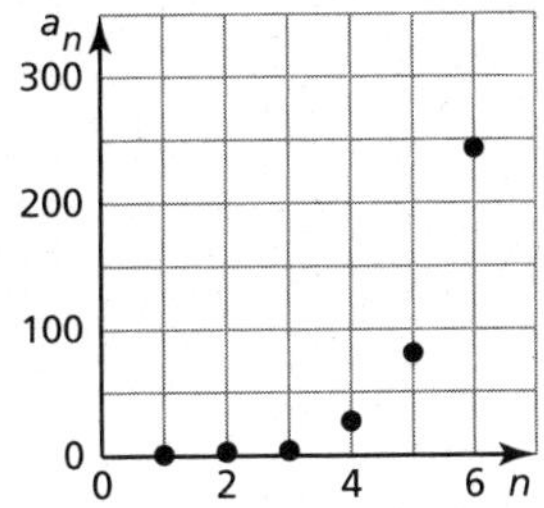

11. $a_n = 3(4)^{n-1}$

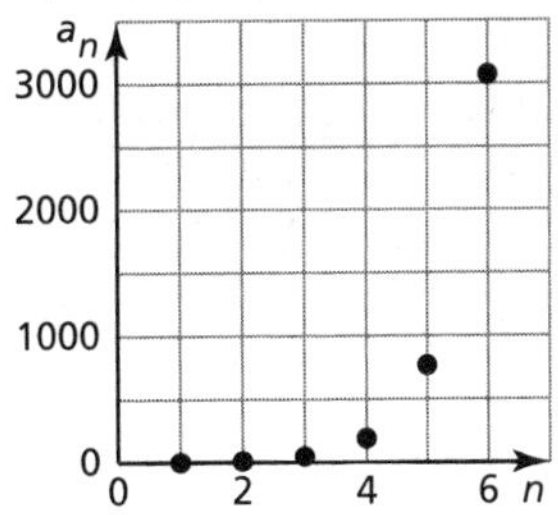

12. $a_n = 40\left(\frac{1}{2}\right)^{n-1}$

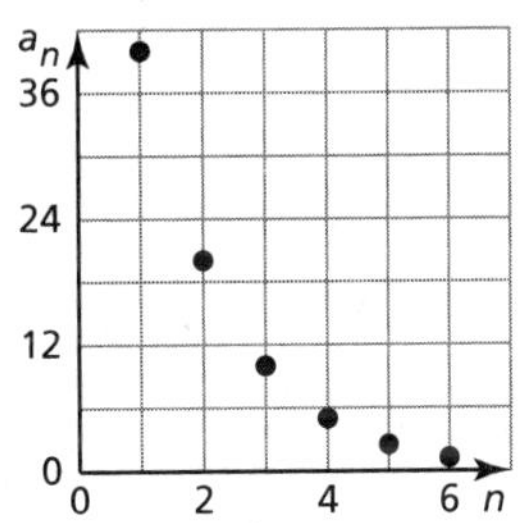

13. $a_n = -13(2)^{n-1}$

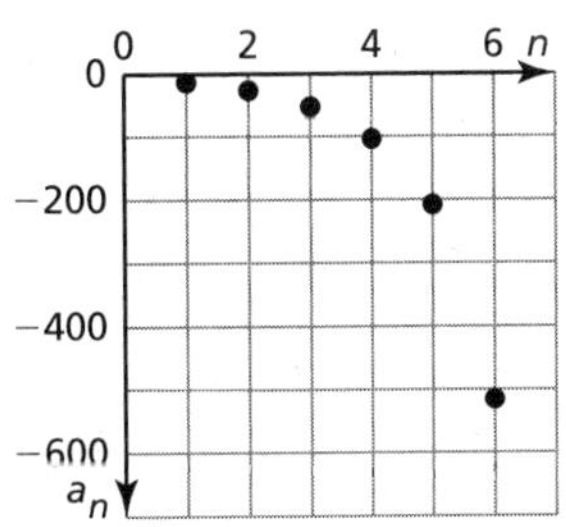

14. r, not a_1, should be raised to the $(n-1)$ power.

$a_n = a_1 r^{n-1}$

$147 = a_1 7^2$

$3 = a_1$

$a_n = 3(7)^{n-1}$

15. \$304.22

7.3 Practice B

1. no; no common ratio

2. yes; $r = \frac{1}{3}$ **3.** yes; $r = 5$ **4.** yes; $r = 2$

5. a. $a_n = -12(7)^{n-1}$ **b.** $a_n = 62\left(\frac{1}{2}\right)^{n-1}$

6. $a_n = 9(2)^{n-1}$; $a_7 = 576$

7. $a_n = 80\left(\frac{1}{4}\right)^{n-1}$; $a_7 = \frac{5}{256}$

8. $a_n = 3\left(\frac{2}{5}\right)^{n-1}$; $a_7 = \frac{192}{15,625}$

9. $a_n = 1.2(-2)^{n-1}$; $a_7 = 76.8$

10. $a_n = 2(5)^{n-1}$

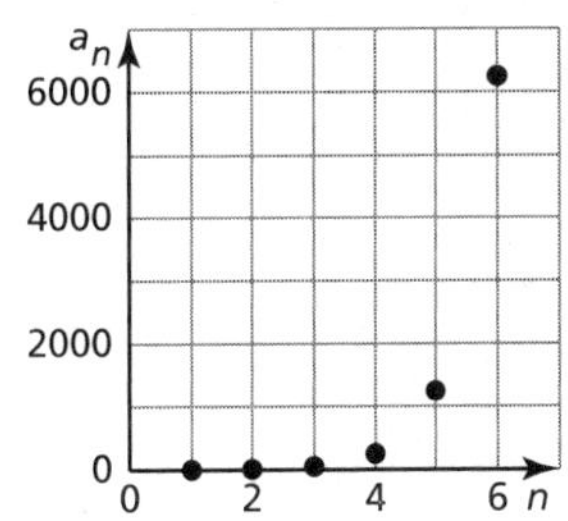

11. $a_n = 54\left(\frac{1}{3}\right)^{n-1}$

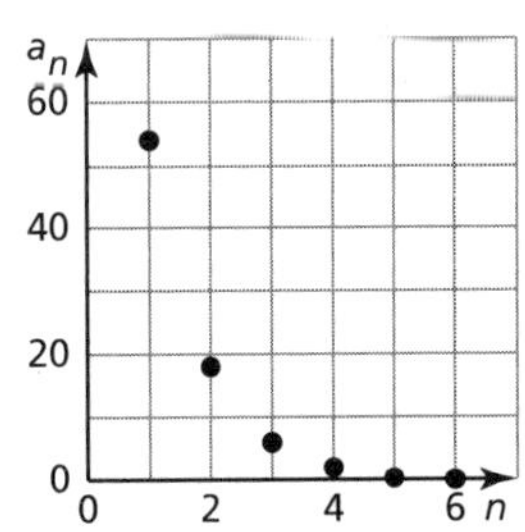

12. $a_n = -14(3)^{n-1}$

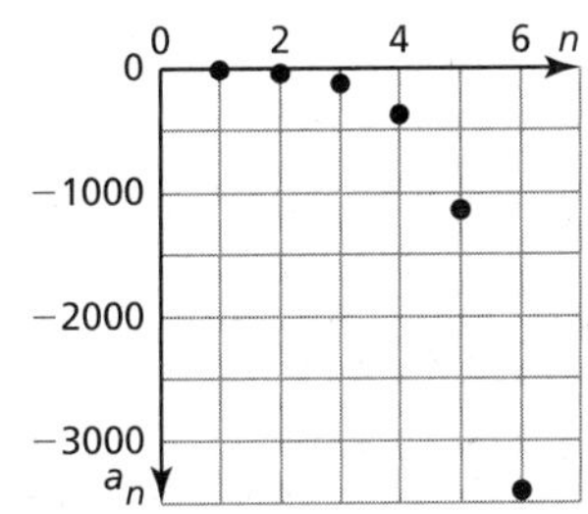

Answers

13. $a_n = 256\left(-\frac{1}{4}\right)^{n-1}$

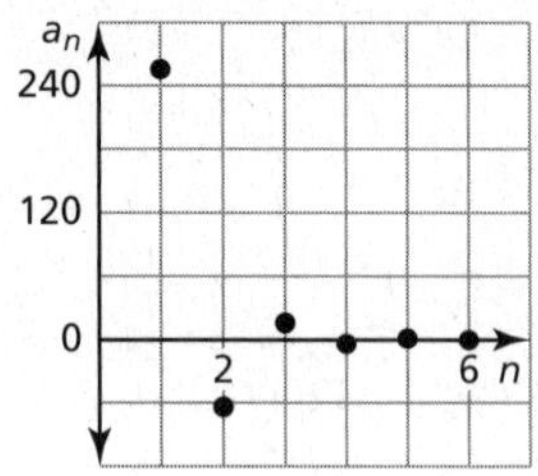

14. The number 147 is the third term of the sequence, not the first term.

$a_n = a_1 r^{n-1}$

$147 = a_1 7^2$

$3 = a_1$

$a_n = 3(7)^{n-1}$

15. \$368.33

7.3 Enrichment and Extension

1. 160, 80, 40

2. 7, 10, 13

3. 2, 4, 6, 9 or $2, \frac{1}{4}, -\frac{3}{2}, 9$

4. When $d = 0, a_2 = 2$; When $d = 1, a_2 = 3$.

5. $x = \dfrac{ac - b^2}{2b - c - a}$

6. *Sample answer*: Arithmetic mean is $\dfrac{a^2 + b^2}{2}$.

Geometric mean is ab or $-ab$.

Show: $\dfrac{a^2 + b^2}{2} \geq ab$ and $\dfrac{a^2 + b^2}{2} \geq -ab$

Because all squares are nonnegative, $(a - b)^2 \geq 0$; So $a^2 - 2ab + b^2 \geq 0$ and $\dfrac{a^2 + b^2}{2} \geq -ab$. Because all squares are nonnegative, $(a + b)^2 \geq 0$; So $a^2 + 2ab + b^2 \geq 0$ and $\dfrac{a^2 + b^2}{2} \geq -ab$.

7.3 Puzzle Time

SEE A MOOOVIE

7.4 Start Thinking

0.1, 0.01, 0.001, 0.0001, 0.00001, 0.000001, 0.0000001, 0.00000001, 0.000000001, 0.0000000001;

0.1111111111; $0.\bar{1} = \frac{1}{9}$

7.4 Warm Up

1. 31

2. 25

3. $\dfrac{25{,}220}{2187} \approx 11.53$

4. −531,440

5. −9050

6. 3,587,226.5

7.4 Cumulative Review Warm Up

1.

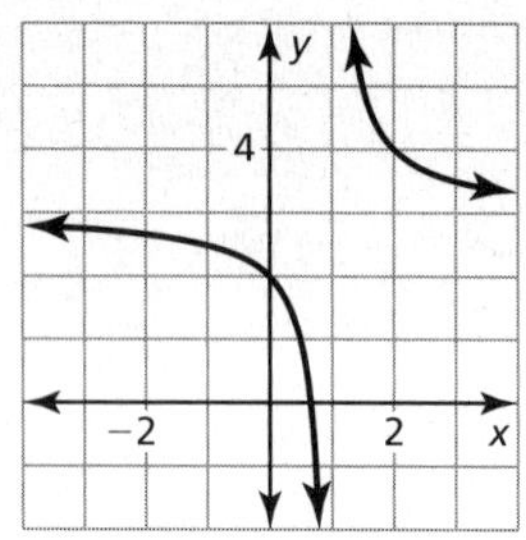

domain: all real numbers except 1, range: all real numbers except 3

2.

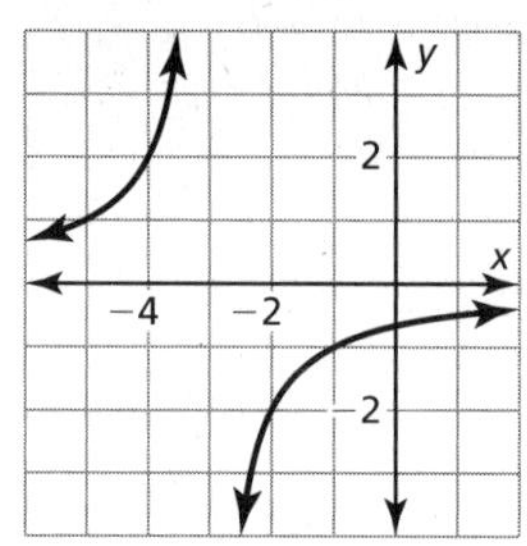

domain: all real numbers except −3, range: all real numbers except 0

3.

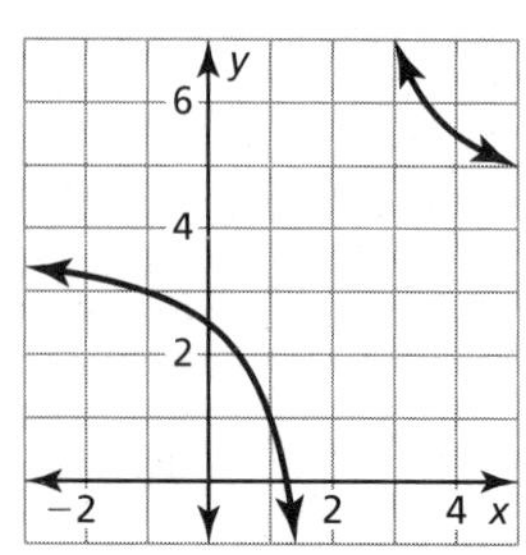

domain: all real numbers except 2, range: all real numbers except 4

Answers

4.

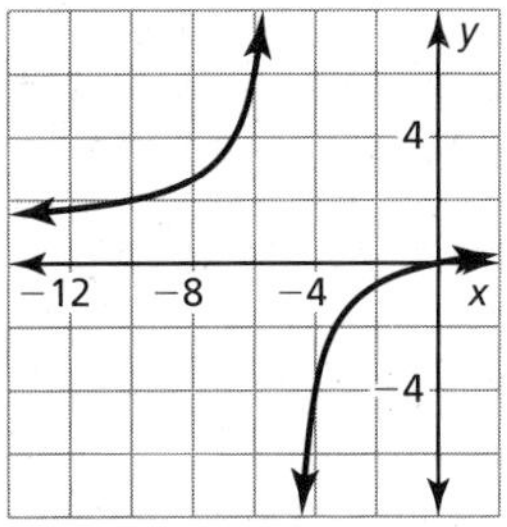

domain: all real numbers except −5, range: all real numbers except 1

5.

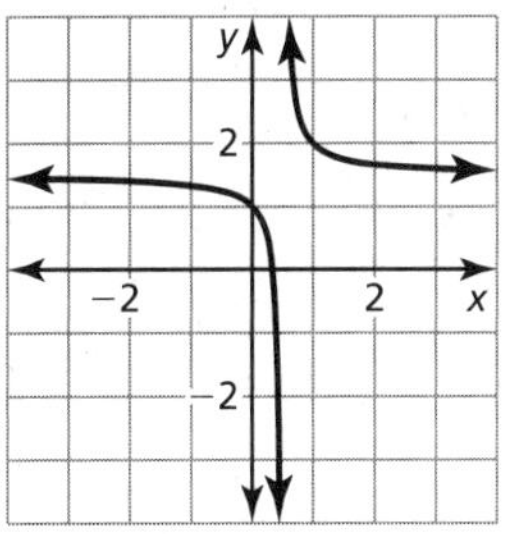

domain: all real numbers except $\frac{1}{2}$, range: all real numbers except $\frac{3}{2}$

6.

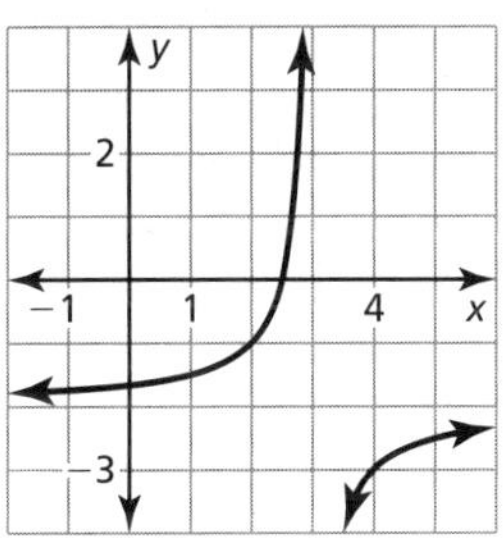

domain: all real numbers except 3, range: all real numbers except −2

7.4 Practice A

1. $S_1 = \frac{1}{3}, S_2 = \frac{1}{2}, S_3 = \frac{7}{12}, S_4 = \frac{5}{8}, S_5 = \frac{31}{48}$

As n increases, $S \approx \frac{2}{3}$.

2. $S_1 = 5, S_2 = \frac{25}{3}, S_3 = \frac{95}{9}, S_4 = \frac{325}{27},$

$S_5 = \frac{1055}{81}$

As n increases, $S \approx 15$.

3. $S = \frac{28}{3}$

4. does not exist

5. $S = \frac{15}{2}$

6. $S = -18$

7. a can be any value, but the absolute value of r must be less than 1.

For this series, $a_1 = \frac{5}{2}$ and $r = \frac{1}{3}$.

$$S = \frac{a_1}{1-r} = \frac{\frac{5}{2}}{1-\frac{1}{3}} = \frac{\frac{5}{2}}{\frac{2}{3}} = \frac{5}{2} \bullet \frac{3}{2} = \frac{15}{4}$$

8. 40 ft **9.** $\frac{2}{11}$ **10.** $\frac{5}{9}$ **11.** $\frac{5}{3}$

12. about \$8,333,333

7.4 Practice B

1. $S_1 = \frac{3}{4}, S_2 = 1, S_3 = \frac{13}{12}, S_4 = \frac{10}{9}, S_5 = \frac{121}{108}$

As n increases, $S \approx \frac{9}{8}$.

2. $S_1 = 6, S_2 = 10, S_3 = \frac{38}{3}, S_4 = \frac{130}{9}, S_5 = \frac{422}{27}$

As n increases, $S \approx 18$.

3. $S = \frac{20}{3}$

4. does not exist

5. does not exist

6. $S = \frac{3}{5}$

7. $1 - r$ is equal to $\frac{2}{3}$, not $\frac{1}{3}$.

For this series, $a_1 = \frac{5}{2}$ and $r = \frac{1}{3}$.

$$S = \frac{a_1}{1-r} = \frac{\frac{5}{2}}{1-\frac{1}{3}} = \frac{\frac{5}{2}}{\frac{2}{3}} = \frac{5}{2} \bullet \frac{3}{2} = \frac{15}{4}$$

8. yes; $a_1 = 2, r = \frac{1}{2}, S = \frac{2}{\frac{1}{2}} = 4$

9. $\frac{5}{11}$ **10.** $\frac{5}{99}$ **11.** $\frac{13}{9}$ **12.** \$6000

Answers

7.4 Enrichment and Extension

1. $-1 < x < 1; \dfrac{1}{1 - x^2}$ **2.** $-\dfrac{1}{3} < x < \dfrac{1}{3}; \dfrac{1}{1 - 3x}$

3. $2 < x < 4; \dfrac{1}{4 - x}$ **4.** $0 < x < 2; \dfrac{1}{x}$

5. $x < -2$ or $x > 2; \dfrac{x}{x + 2}$

6. $-\sqrt{2} < x < \sqrt{2}; \dfrac{2x^2}{6 + 3x^2}$

7. The series is an infinite geometric series with common ratio $r = \sin^2 x$. $|\sin^2 x| < 1$ is true when $x \neq \dfrac{\pi}{2} + n\pi$, so

$$S = \frac{\sin^2 x}{1 - \sin^2 x} = \frac{\sin^2 x}{\cos^2 x} = \tan^2 x.$$

8. The series is an infinite geometric series with common ratio $r = -\tan^2 x$. $|-\tan^2 x| < 1$ simplifies to $\tan^2 x < 1$, which is true when $-\dfrac{\pi}{4} < x < \dfrac{\pi}{4}$. So,

$$S = \frac{\tan^2 x}{1 - (-\tan^2 x)} = \frac{\tan^2 x}{\sec^2 x} = \frac{\sin^2 x \cos^2 x}{\cos^2 x}$$

$$= \sin^2 x;$$

$\tan^2 x < 1$ is also true when $\dfrac{3\pi}{4} < x < \dfrac{5\pi}{4}$.

9. When the values are substituted into the sum formula, r will be greater than 1.

7.4 Puzzle Time

BLOODHOUND

7.5 Start Thinking

1. To determine each term in the sequence, you add the two terms that come before it. For example, $a_3 = a_1 + a_2$, or $3 = 1 + 2$. Then to find a_7, you find the sum of a_5 and a_6, or $8 + 13 = 21$.

2. To determine each term in the sequence, you multiply the two terms that come before it. For example, $a_3 = a_1 \cdot a_2$, or $2 = 1 \cdot 2$. Then to find a_7, you find the product of a_5 and a_6, or $8 \bullet 32 = 256$.

3. To determine each term in the sequence, you find the difference of the two terms that come before it. For example, $a_3 = a_1 - a_2$, or $5 = 10 - 5$. Then to find a_7, you find the difference a_5 and a_6, or $5 - (-5) = 10$.

4. To determine each term in the sequence, you multiply the two terms that come before it. For example, $a_3 = a_1 \bullet a_2$ or $2 = (-2)(-1)$. Then to find a_7, you find the product of a_5 and a_6, or $(-4)(8) = -32$.

7.5 Warm Up

1. $\dfrac{7}{2}, 3, \dfrac{5}{2}, 2, \dfrac{3}{2}, 1$ **2.** $4, \dfrac{7}{2}, \dfrac{10}{3}, \dfrac{13}{4}, \dfrac{16}{5}, \dfrac{19}{6}$

3. 2, 6, 12, 20, 30, 42 **4.** $2, 3, \dfrac{9}{2}, \dfrac{27}{4}, \dfrac{81}{8}, \dfrac{243}{16}$

5. $-9, -2, 17, 54, 115, 206$

6. 1, 4, 7, 10, 13, 16

7.5 Cumulative Review Warm Up

1. $x = \dfrac{1 \pm \sqrt{7}}{3}$ **2.** $x = 4 \pm \sqrt{14}$

3. $x = 2, 1$ **4.** $x = \pm\sqrt{2}$

5. $x = 5 \pm \sqrt{19}$ **6.** $x = \dfrac{8}{3}, -2$

7.5 Practice A

1. 1, 6, 11, 16, 21, 26

2. $1, -3, -7, -11, -15, -19$

3. 3, 12, 48, 192, 768, 3072

4. $12, 4, \dfrac{4}{3}, \dfrac{4}{9}, \dfrac{4}{27}, \dfrac{4}{81}$

5. 1; 3; 11; 123; 15,131; 228,947,163

6. 2, 2, 2, 2, 2, 2

7. $a_1 = 32, a_n = a_{n-1} - 8$

8. $a_1 = -47, a_n = a_{n-1} + 12$

9. $a_1 = 2, a_n = 3a_{n-1}$

10. $a_1 = 5, a_n = -2a_{n-1}$

Answers

11. $a_1 = 21, a_n = \frac{1}{3}a_{n-1}$

12. $a_1 = 1, a_n = a_{n-1} + 6$

13. $a_1 = 2, a_2 = 3, a_n = a_{n-2} + a_{n-1}$

14. $a_1 = 2, a_2 = 3, a_n = (a_{n-2})(a_{n-1})$

15. $a_1 = 7, a_n = a_{n-1} + 2$

16. $a_1 = -7, a_n = a_{n-1} - 3$

17. $a_1 = 2, a_n = a_{n-1} - 13$

18. $a_1 = 8, a_n = 10a_{n-1}$

19. $a_1 = -2, a_n = 7a_{n-1}$

20. $a_1 = 1, a_n = a_{n-1} - 0.8$

21. $a_1 = 95, a_n = a_{n-1} + 20$

22. $a_n = 8 - 3n$ 23. $a_n = 9 + 5n$

24. $a_n = -3(2)^{n-1}$ 25. $a_n = 20\left(\frac{1}{2}\right)^{n-1}$

7.5 Practice B

1. 1, 10, 19, 28, 37, 46 2. $32, 8, 2, \frac{1}{2}, \frac{1}{8}, \frac{1}{32}$

3. 24, 36, 54, 81, 121.5, 182.25

4. 1, 0, −1, 0, −1, 0 5. 1, 4, −3, 7, −10, 17

6. $256, 2, 128, \frac{1}{64}, 8192, \frac{1}{524{,}288}$

7. $a_1 = 30, a_n = a_{n-1} - 9$

8. $a_1 = 3, a_n = -5a_{n-1}$ 9. $a_1 = 28, a_n = \frac{1}{7}a_{n-1}$

10. $a_1 = 1, a_n = a_{n-1} + 11$

11. $a_1 = 2, a_2 = 6, a_n = (a_{n-2})(a_{n-1})$

12. $a_1 = 1, a_2 = 7, a_n = a_{n-2} + a_{n-1}$

13. $a_1 = 61, a_2 = 39, a_n = a_{n-2} - a_{n-1}$

14. $a_1 = -5, a_n = a_{n-1} + n$

15. $a_1 = -4, a_n = a_{n-1} + 3$

16. $a_1 = 6, a_n = 15a_{n-1}$

17. $a_1 = -16, a_n = 9a_{n-1}$

18. $a_1 = -2.1, a_n = a_{n-1} + 0.3$

19. $a_1 = -\frac{1}{3}, a_n = \frac{1}{5}a_{n-1}$

20. $a_1 = \frac{1}{2}, a_n = 7a_{n-1}$

21. $a_1 = 26, a_n = 1.002a_{n-1}$

22. $a_n = -26.2 + 7.2n$ 23. $a_n = -7(0.45)^{n-1}$

24. $a_n = \frac{23}{6} + \frac{1}{6}n$ 25. $a_n = -9\left(\frac{1}{3}\right)^{n-1}$

7.5 Enrichment and Extension

1. 3, 5, 2, −3, −5, −2, 3, 5, …; −3

2. $4, 8, 2, \frac{1}{4}, \frac{1}{8}, \frac{1}{2}, 4, 8, \ldots; \frac{1}{4}$

3. a. 5

   b. $n - 2$ additional diagonals; $d_4 = 2$, $d_n = d_{n-1} + n - 2$

4. $S_{n+1} = S_n + 2n + 1$

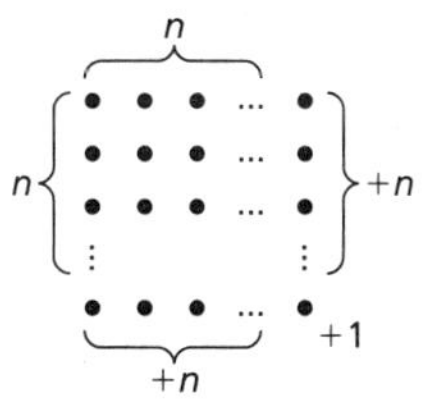

7.5 Puzzle Time

KING CONGA

Cumulative Review

1.

x	y
4	−18
8	−62
12	−106
16	−150

2.

x	y
2	2
4	38
6	74
8	110

Answers

3.

x	y
6	49
12	85
18	121
24	157

4.

x	y
2	12
3	37
4	72
5	117

5.

x	y
2	−8
3	−23
4	−44
5	−71

6.

x	y
2	−3
3	17
4	45
5	81

7.

x	y
0	−2
1	−1
2	1
3	5

8.

x	y
0	4
1	3
2	1
3	−3

9.

x	y
0	−11
1	−9
2	−3
3	15

10. $x = 7$ **11.** $x = 3$ **12.** $x = -11$

13. $x = 60$ **14.** $x = 48$ **15.** $x = 39$

16. $x = -9$ **17.** $x = -4$ **18.** $x = 2$

19. $x = 6$ **20.** $x = 4$ **21.** $x = 3$

22. $T = 51 + 3h$ **23.** $T = 44 - 2h$

24. 22,302 **25.** −4091 **26.** 46,348

27. −29,222 **28.** −62,598 **29.** −706

30. $-2x^2 + 4x + 24$ **31.** $3x^3 + 7x^2 + 3x + 5$

32. $-11x^5 + 10x^4 + 4x^3 + x^2 + 6x - 4$

33. $7x^5 + 7x^4 + 2x^3 + 7x^2 - 14x - 6$

34. $2x^3 - 5x^2 + 15x - 9$

35. $6x^4 + 15x^3 - 8x^2 + 11x - 10$

36. $-6x^4 - 13x^3 + 6x^2 + 3x - 8$

37. $-5x^5 + 22x^4 - 20x^3 - 3x^2 + 27x + 9$

38. $21x^4 - 35x^3 + 21x^2$

39. $56x^6 - 120x^5 + 88x^4 + 24x^3 - 56x^2 - 72x$

40. $-30x^3 + 31x^2 + 51x + 14$

41. $44x^4 + 147x^3 - 64x^2 - 7x$

42. $56x^4 - 22x^3 + 93x^2 - 34x + 3$

43. $2x^4 - 6x^3 - 31x^2 + 97x - 12$

44. $x^3 + 3x^2 - 18x - 40$

45. $x^3 + 2x^2 - 13x + 10$

46. $x^3 + 4x^2 - 16x - 64$

47. $x^3 - 11x^2 + 38x - 40$

48. $20x^3 + 27x^2 + x - 6$

49. $24x^3 + 58x^2 - 7x - 5$

50. $x^2 + 12x + 36$ **51.** $x^2 - 49$

52. $x^2 - 16x + 64$ **53.** $4x^2 + 44x + 121$

54. $36x^2 - 132x + 121$ **55.** $81x^2 + 180x + 100$

56. $2x - 11 + \dfrac{51}{x + 4}$ **57.** $5x - 29 + \dfrac{196}{x + 7}$

58. $9x + 17 + \dfrac{32}{x - 2}$

59. $4x^2 + 33x + 225 + \dfrac{1581}{x - 7}$

60. $5x^3 - 23x^2 + 87x - 346 + \dfrac{1376}{x + 4}$

61. $3x^3 - 20x^2 + 100x - 496 + \dfrac{2475}{x + 5}$

62. a. $(2x + 4)$ ft **b.** 40 ft^2

Answers

63. a. $(4x + 5)$ ft **b.** 153 ft^2 **c.** 9 ft **d.** 17 ft

64. $x(x + 8)(x - 3)$ **65.** $c(c + 9)(c - 1)$

66. $m(m - 5)(m - 8)$ **67.** $a(a - 1)(a + 3)$

68. $(d + 8)(d^2 - 8d + 64)$

69. $(g - 3)(g^2 + 3g + 9)$

70. $(y - 5)(y^2 + 5y + 25)$

71. $(n + 7)(n^2 - 7n + 49)$

72. $(2b + 5)(4b^2 - 10b + 25)$

73. $(3w - 7)(9w^2 + 21w + 49)$

74. $(f^2 + 3)(f - 7)$ **75.** $(r^2 - 2)(3r + 5)$

76. $(h + 1)(h - 1)(5h - 8)$

77. $(s^2 + 4)(3s - 7)$ **78.** $(v^3 + 7)(4v - 1)$

79. $(p^3 - 9)(7p + 2)$

80. $z = -6, z = 0$, and $z = 2$

81. $x = -3, x = 0$, and $x = 11$

82. $m = -4, m = 0$, and $m = 11$

83. $h = -10, h = -3$, and $h = 0$

84. $v = -9, v = -5$, and $v = 0$

85. $f = 0, f = 4$, and $f = 8$

86. $x = -4, x = 0$, and $x = 7$

87. $x = -8, x = 0$, and $x = 3$

88. $x = 0, x = 10$, and $x = 12$

89. $x = 0, x = 2$, and $x = 5$

90. $x = 0, x = 2$, and $x = 10$

91. $x = -3, x = 0$, and $x = 11$

92. a. $(3x + 5)$ in. **b.** 64 in.^2

93. a. $(6x + 11)$ in. **b.** 400 in.^2 **c.** 20 in.

Chapter 8

8.1 Start Thinking

Sample answer:

$x_1 \approx 2\frac{5}{8}$ in., $y_1 \approx 1\frac{1}{2}$ in., $x_2 \approx 3\frac{7}{16}$ in., $y_2 \approx 2$ in.

1. $\frac{y_1}{3} = \frac{1}{2} = 0.5$ **2.** $\frac{x_1}{3} = \frac{7}{8} \approx 0.88$

3. $\frac{y_1}{x_1} = \frac{4}{7} \approx 0.57$ **4.** $\frac{y_2}{4} = \frac{1}{2} = 0.5$

5. $\frac{x_2}{4} = \frac{55}{64} \approx 0.86$ **6.** $\frac{y_2}{x_2} = \frac{32}{55} \approx 0.58$

It appears that regardless of the size of the 30°-60°-90° triangle, the ratios of corresponding sides are equal or approximately equal.

8.1 Warm Up

1. $3\sqrt{5}$ **2.** $\sqrt{39}$ **3.** $4\sqrt{6}$

8.1 Cumulative Review Warm Up

1. $a_n = -6n + 17$ **2.** $a_n = 4(-2)^{n-1}$

3. $a_n = 8n - 8$ **4.** $a_n = (-1)^n(2n - 1)$

5. $a_n = \frac{n + 1}{n^2}$ **6.** $a_n = \frac{3^{n-1}}{3n - 1}$

8.1 Practice A

1. $\sin\theta = \frac{12}{13}$, $\cos\theta = \frac{5}{13}$, $\tan\theta = \frac{12}{5}$,

$\csc\theta = \frac{13}{12}$, $\sec\theta = \frac{13}{5}$, $\cot\theta = \frac{5}{12}$

2. $\sin\theta = \frac{1}{2}$, $\cos\theta = \frac{\sqrt{3}}{2}$, $\tan\theta = \frac{\sqrt{3}}{3}$,

$\csc\theta = 2$, $\sec\theta = \frac{2\sqrt{3}}{3}$, $\cot\theta = \sqrt{3}$

Answers

3. $\cos\theta = \frac{2\sqrt{10}}{7}$, $\tan\theta = \frac{3\sqrt{10}}{20}$, $\csc\theta = \frac{7}{3}$,

   $\sec\theta = \frac{7\sqrt{10}}{20}$

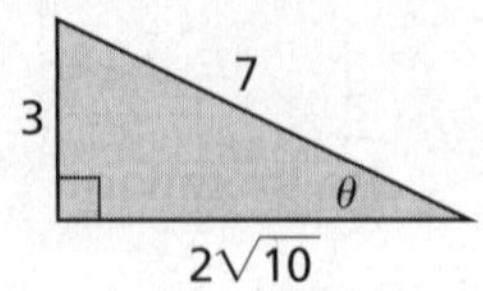

4. $\cos\theta = \frac{\sqrt{105}}{11}$, $\tan\theta = \frac{4\sqrt{105}}{105}$, $\csc\theta = \frac{11}{4}$,

   $\sec\theta = \frac{11\sqrt{105}}{105}$, $\cot\theta = \frac{\sqrt{105}}{4}$

5. $\sin\theta = \frac{\sqrt{11}}{6}$, $\tan\theta = \frac{\sqrt{11}}{5}$, $\csc\theta = \frac{6\sqrt{11}}{11}$,

   $\sec\theta = \frac{6}{5}$, $\cot\theta = \frac{5\sqrt{11}}{11}$

6. $\sin\theta = \frac{3}{5}$, $\cos\theta = \frac{4}{5}$, $\csc\theta = \frac{5}{3}$,

   $\sec\theta = \frac{5}{4}$, $\cot\theta = \frac{4}{3}$

7. tangent is opposite over adjacent, not adjacent over opposite: $\tan\theta = \frac{\text{opp.}}{\text{adj.}} = \frac{12}{9} = \frac{4}{3}$

8. $x = 5$ **9.** $x = 5.5$ **10.** $h \approx 47.0$ ft

8.1 Practice B

1. $\sin\theta = \frac{5}{13}$, $\cos\theta = \frac{12}{13}$, $\tan\theta = \frac{5}{12}$,

   $\csc\theta = \frac{13}{5}$, $\sec\theta = \frac{13}{12}$, $\cot\theta = \frac{12}{5}$

2. $\sin\theta = \frac{\sqrt{65}}{9}$, $\cos\theta = \frac{4}{9}$, $\tan\theta = \frac{\sqrt{65}}{4}$,

   $\csc\theta = \frac{9\sqrt{65}}{65}$, $\sec\theta = \frac{9}{4}$, $\cot\theta = \frac{4\sqrt{65}}{65}$

3. $\sin\theta = \frac{12}{13}$, $\cos\theta = \frac{5}{13}$, $\tan\theta = \frac{12}{5}$,

   $\csc\theta = \frac{13}{12}$, $\sec\theta = \frac{13}{5}$, $\cot\theta = \frac{5}{12}$;

   Sine and cosine switch values; Tangent and cotangent switch values; Secant and cosecant switch values.

4. $\sin\theta = \frac{4\sqrt{6}}{11}$, $\tan\theta = \frac{4\sqrt{6}}{5}$, $\csc\theta = \frac{11\sqrt{6}}{24}$,

   $\sec\theta = \frac{11}{5}$, $\cot\theta = \frac{5\sqrt{6}}{24}$

5. $\sin\theta = \frac{8\sqrt{113}}{113}$, $\cos\theta = \frac{7\sqrt{113}}{113}$, $\tan\theta = \frac{8}{7}$,

   $\csc\theta = \frac{\sqrt{113}}{8}$, $\sec\theta = \frac{\sqrt{113}}{7}$

6. $\sin\theta = \frac{2\sqrt{10}}{11}$, $\cos\theta = \frac{9}{11}$, $\tan\theta = \frac{2\sqrt{10}}{9}$,

   $\csc\theta = \frac{11\sqrt{10}}{20}$, $\cot\theta = \frac{9\sqrt{10}}{20}$

7. secant is hypotenuse over adjacent, not adjacent over hypotenuse; $\sec\theta = \frac{\text{hyp.}}{\text{adj.}} = \frac{15}{9} = \frac{5}{3}$

8. $x = 8$ **9.** $x = 10$ **10.** $h \approx 14.1$ ft

8.1 Enrichment and Extension

1. **a.** 220 m **b.** 454 m **c.** 234 m

2. **a.** 70 ft **b.** 54 ft

8.1 Puzzle Time

DODGEBALL

8.2 Start Thinking

Sample answer:

In each case, the central angle is approximately 57°. It seems that regardless of the size of the radius, this angle is the same.

8.2 Warm Up

1.

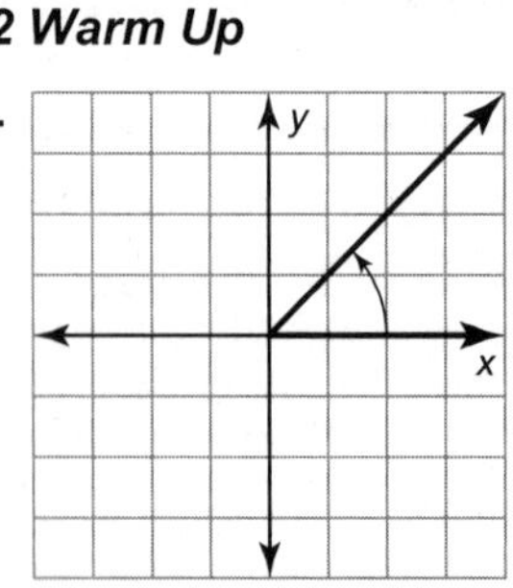

Answers

2.

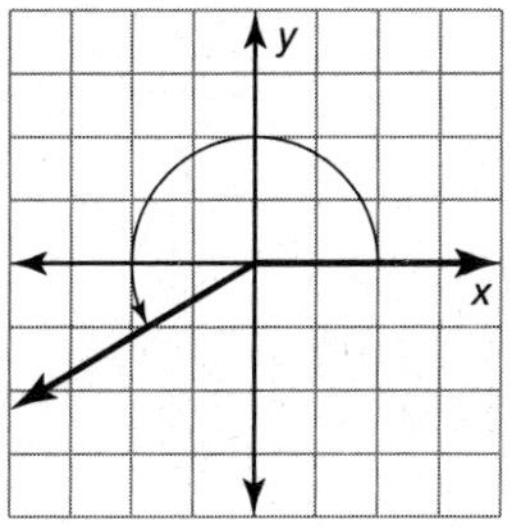

3.

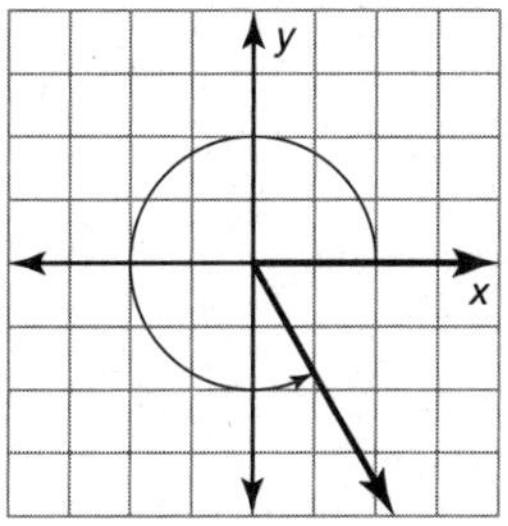

8.2 Cumulative Review Warm Up

1. $(-4,-3)$ **2.** $(3,-1)$

3. infinitely many solutions

4. $(-3,0)$ and $\left(-\frac{9}{5},\frac{12}{5}\right)$ **5.** $(-1,2)$ and $(3,6)$

6. $(1,2)$ and $(-2,11)$

8.2 Practice A

1.

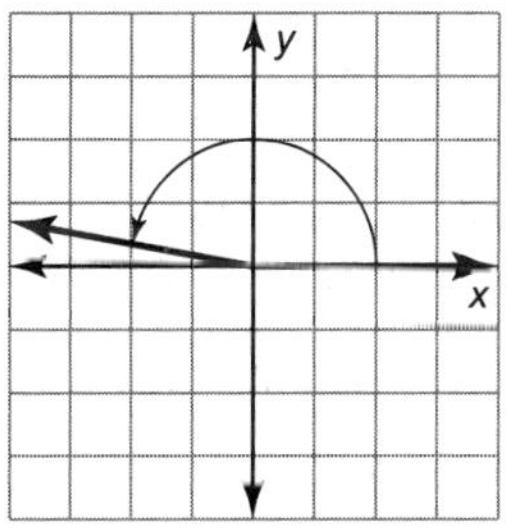

2.

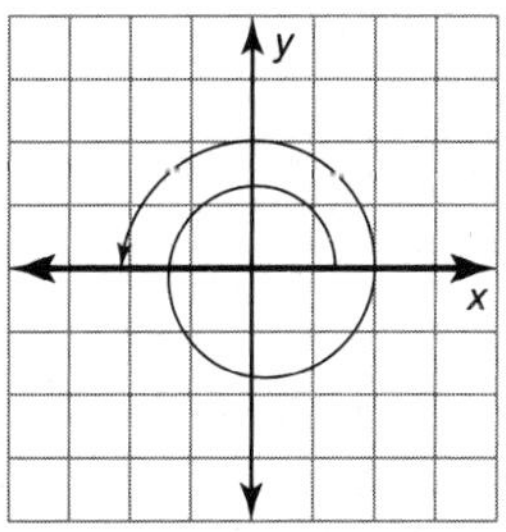

3.

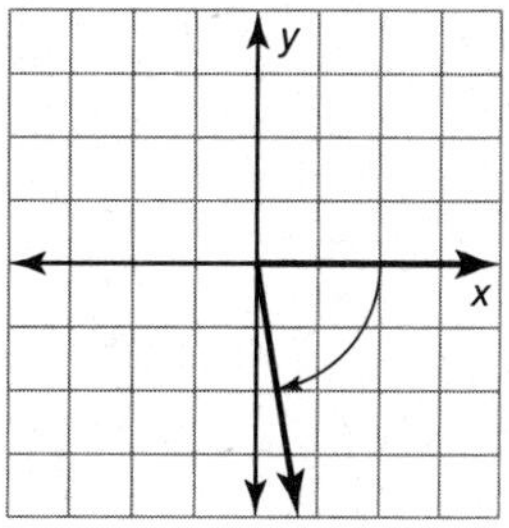

4.

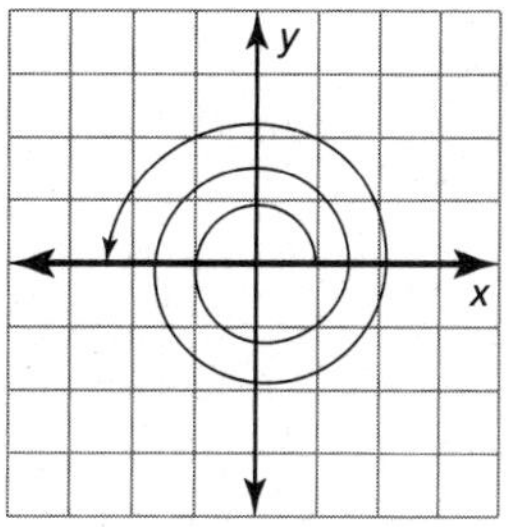

5.

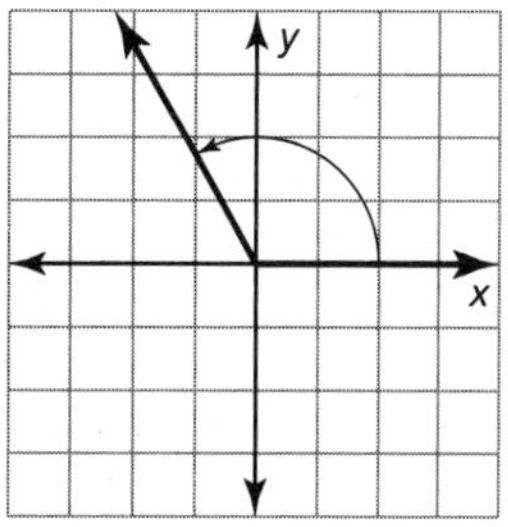

6.

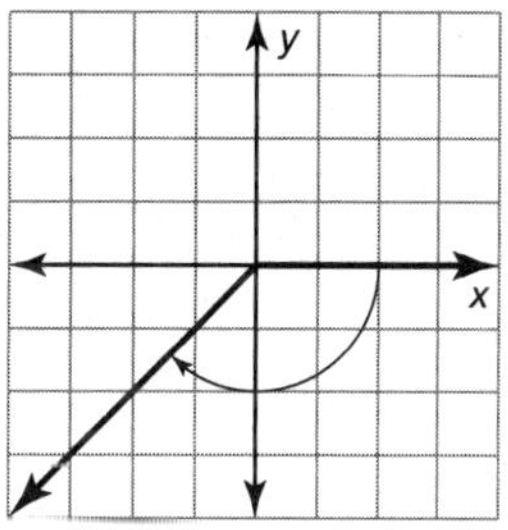

7. B **8.** A **9.** C

10. $435°, -285°$ **11.** $320°, -400°$

12. $\frac{13\pi}{3}, -\frac{5\pi}{3}$ **13.** $\frac{5\pi}{4}$

14. $\frac{5\pi}{18}$ **15.** $-\frac{8\pi}{9}$

16. $15°$ **17.** $100°$ **18.** $-22.5°$

Answers

19. hour hand: $105° = \frac{7\pi}{12}$ radians

minute hand: $1260° = 7\pi$ radians

The minute degrees and radians are 12 times the hour degrees and radians.

8.2 Practice B

1. 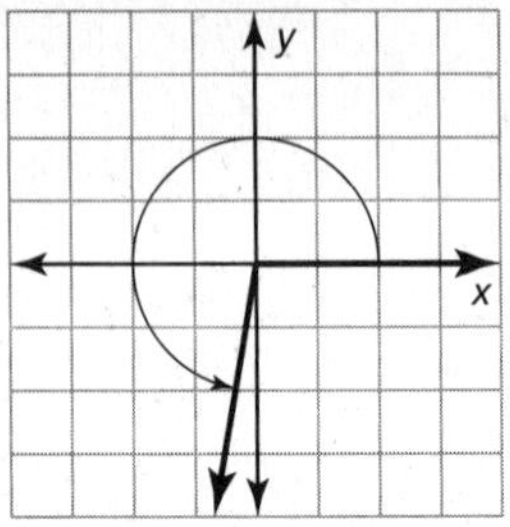

2. 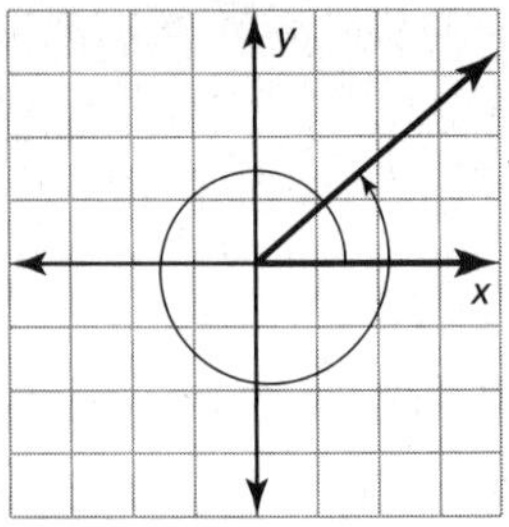

3. 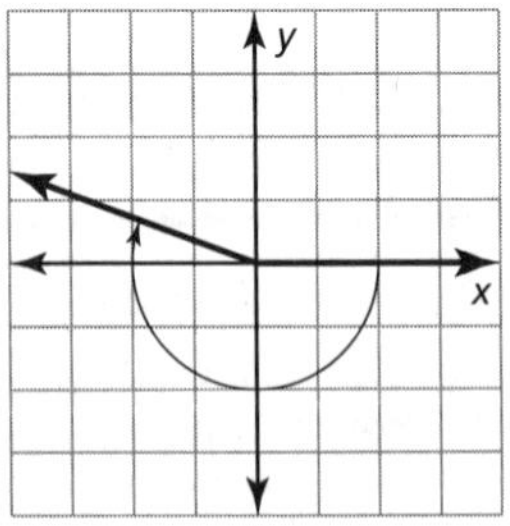

4. 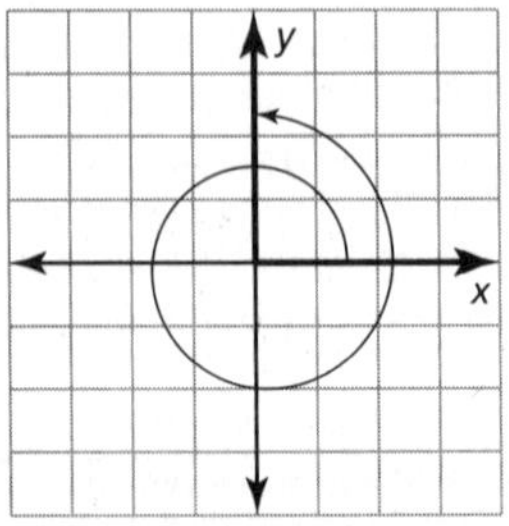

5. 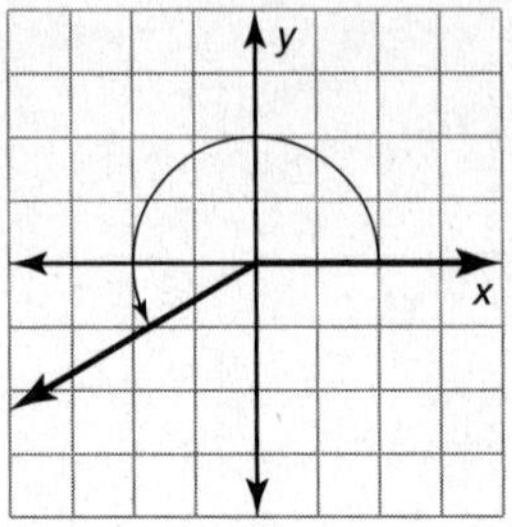

6. 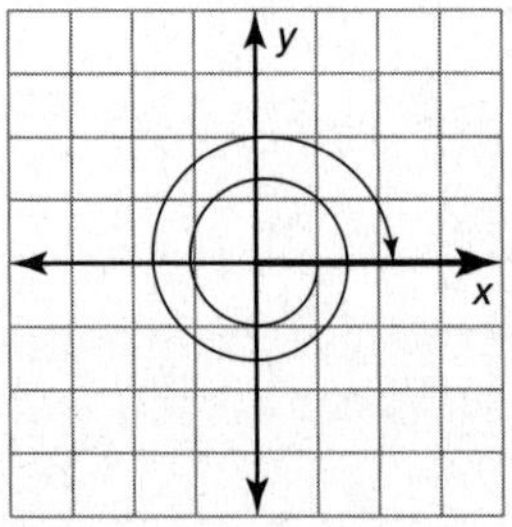

7. A 8. C 9. B

10. $\frac{13\pi}{4}, -\frac{3\pi}{4}$ 11. $300°, -60°$

12. $\frac{\pi}{2}, -\frac{3\pi}{2}$ 13. $\frac{10\pi}{9}$

14. $\frac{\pi}{180}$ 15. $-\frac{95\pi}{36}$

16. $54°$ 17. $-75°$

18. $\left(\frac{1080}{\pi}\right)^{\circ} \approx 343.8°$

19. a. $160°\,26'\,24''$

b. $98.26°$

8.2 Enrichment and Extension

1. > 2. < 3. = 4. =

5. = 6. = 7. > 8. <

9. = 10. > 11. < 12. <

13. sin 4, sin 3, sin 1, sin 2

14. cos 3, cos 4, cos 2, cos 1

15. 1.5 cm or 2 cm 16. 2 cm or 4 cm

Answers

17. a. $20 = 2r + s = 2r + r\theta;\ \theta = \dfrac{20 - 2r}{r}$

$A = \dfrac{1}{2}r^2\theta = \dfrac{1}{2}r^2\left(\dfrac{20}{r} - 2\right) = 10r - r^2$

b. 5 cm

c. 2 radians

8.2 Puzzle Time

HE RAN OUT OF BOUNDS

8.3 Start Thinking

Sample answer:

$(x, y) = \left(\dfrac{\sqrt{2}}{2}, \dfrac{\sqrt{2}}{2}\right)$, $\sin 45° = \dfrac{y}{1} = y$,

$\cos 45° = \dfrac{x}{1} = x$, $\tan 45° = \dfrac{y}{x}$

$\sin\theta = 0.95$, $\cos\theta = -0.31$, $\tan\theta = \dfrac{0.95}{-0.31} \approx -3.06$

For angles in the third quadrant, because both x-values and y-values are negative, the sine and cosine would be negative and the tangent would be positive. For angles in the fourth quadrant, because x-values are positive and y-values are negative, both the sine and the tangent will be negative and the cosine will be positive.

8.3 Warm Up

1. $\sqrt{74}$ **2.** 10 **3.** $\dfrac{\sqrt{89}}{6}$ **4.** $\sqrt{6.28}$

8.3 Cumulative Review Warm Up

1. $-2x^2 + 2x + 8$

2. $2x^3 - 13x^2 - 12x + 35$

3. $4x^2 - 4x + 1$ **4.** $9 - 25x^2$

5. $9 - 6x^2 + x^4$

6. $27x^3 - 54x^2 + 36x - 8$

8.3 Practice A

1. $\sin\theta = \dfrac{3\sqrt{13}}{13}$, $\cos\theta = -\dfrac{2\sqrt{13}}{13}$, $\tan\theta = -\dfrac{3}{2}$,
$\csc\theta = \dfrac{\sqrt{13}}{3}$, $\sec\theta = -\dfrac{\sqrt{13}}{2}$, $\cot\theta = -\dfrac{2}{3}$

2. $\sin\theta = \dfrac{\sqrt{17}}{17}$, $\cos\theta = -\dfrac{4\sqrt{17}}{17}$, $\tan\theta = -\dfrac{1}{4}$,
$\csc\theta = \sqrt{17}$, $\sec\theta = -\dfrac{\sqrt{17}}{4}$, $\cot\theta = -4$

3. $\sin\theta = -\dfrac{2\sqrt{29}}{29}$, $\cos\theta = \dfrac{5\sqrt{29}}{29}$, $\tan\theta = -\dfrac{2}{5}$,
$\csc\theta = -\dfrac{\sqrt{29}}{2}$, $\sec\theta = \dfrac{\sqrt{29}}{5}$, $\cot\theta = -\dfrac{5}{2}$

4. $\sin\theta = -\dfrac{3\sqrt{10}}{10}$, $\cos\theta = -\dfrac{\sqrt{10}}{10}$, $\tan\theta = 3$,
$\csc\theta = -\dfrac{\sqrt{10}}{3}$, $\sec\theta = -\sqrt{10}$, $\cot\theta = \dfrac{1}{3}$

5. $\sin\theta = 0$, $\cos\theta = -1$, $\tan\theta = 0$,
$\csc\theta =$ undefined, $\sec\theta = -1$,
$\cot\theta =$ undefined

6. $\sin\theta = 1$, $\cos\theta = 0$, $\tan\theta =$ undefined,
$\csc\theta = 1$, $\sec\theta =$ undefined, $\cot\theta = 0$

7. $\sin\theta = -1$, $\cos\theta = 0$, $\tan\theta =$ undefined,
$\csc\theta = -1$, $\sec\theta =$ undefined, $\cot\theta = 0$

8. 10° **9.** 50° **10.** 40°

11. $\dfrac{\pi}{6}$ **12.** $\dfrac{\pi}{4}$ **13.** $\dfrac{\pi}{3}$

14. 2 **15.** 1 **16.** $-\dfrac{\sqrt{3}}{2}$

17. 173.2 ft; 57.7 yd

8.3 Practice B

1. $\sin\theta = -\dfrac{3}{5}$, $\cos\theta = -\dfrac{4}{5}$, $\tan\theta = \dfrac{3}{4}$,
$\csc\theta = -\dfrac{5}{3}$, $\sec\theta = -\dfrac{5}{4}$, $\cot\theta = \dfrac{4}{3}$

2. $\sin\theta = \dfrac{3\sqrt{13}}{13}$, $\cos\theta = \dfrac{2\sqrt{13}}{13}$, $\tan\theta = \dfrac{3}{2}$,
$\csc\theta = \dfrac{\sqrt{13}}{3}$, $\sec\theta = \dfrac{\sqrt{13}}{2}$, $\cot\theta = \dfrac{2}{3}$

3. $\sin\theta = -\frac{2\sqrt{5}}{5}$, $\cos\theta = \frac{\sqrt{5}}{5}$, $\tan\theta = -2$,

   $\csc\theta = -\frac{\sqrt{5}}{2}$, $\sec\theta = \sqrt{5}$, $\cot\theta = -\frac{1}{2}$

4. $\sin\theta = \frac{3\sqrt{10}}{10}$, $\cos\theta = -\frac{\sqrt{10}}{10}$, $\tan\theta = -3$,

   $\csc\theta = \frac{\sqrt{10}}{3}$, $\sec\theta = -\sqrt{10}$, $\cot\theta = -\frac{1}{3}$

5. $\sin\theta = 0$, $\cos\theta = -1$, $\tan\theta = 0$,
   $\csc\theta =$ undefined, $\sec\theta = -1$,
   $\cot\theta =$ undefined

6. $\sin\theta = 0$, $\cos\theta = 1$, $\tan\theta = 0$,
   $\csc\theta =$ undefined, $\sec\theta = 1$,
   $\cot\theta =$ undefined

7. $\sin\theta = -1$, $\cos\theta = 0$, $\tan\theta =$ undefined,
   $\csc\theta = -1$, $\sec\theta =$ undefined, $\cot\theta = 0$

8. 70° 9. 70° 10. 50°

11. $\frac{\pi}{4}$ 12. $\frac{\pi}{6}$ 13. $\frac{\pi}{3}$

14. $\frac{\sqrt{3}}{3}$ 15. $-\frac{\sqrt{2}}{2}$ 16. $-\frac{2\sqrt{3}}{3}$

17. yes; $d = \frac{22^2}{32}\sin(2 \bullet 45^\circ) = 15.125$ ft

8.3 Enrichment and Extension

1. about 1.52 2. 1.0

3. 1.97×10^8 m/sec 4. 2.42

8.3 Puzzle Time

A TENNIS MATCH

8.4 Start Thinking

θ	0	$\frac{\pi}{2}$	π	$\frac{3\pi}{2}$	2π
$f(\theta) = \sin\theta$	0	1	0	−1	0

θ	$\frac{5\pi}{2}$	3π	$\frac{7\pi}{2}$	4π
$f(\theta) = \sin\theta$	1	0	−1	0

The highest value of $f(\theta)$ is 1. The lowest value of $f(\theta)$ is −1. The high and low value repeat every 2π units.

Sample answer:

The graph of $f(\theta) = \sin\theta$ on the interval $0 \le \theta \le 4\pi$ starts at $(0, 0)$ then moves up to the point $\left(\frac{\pi}{2}, 1\right)$, then down to $(\pi, 0)$, then down further to $\left(\frac{3\pi}{2}, -1\right)$, and then up to $(2\pi, 0)$. This same cycle repeats on the interval $2\pi \le \theta \le 4\pi$.

8.4 Warm Up

1. vertical stretch by a factor of 4
2. vertical shrink by a factor of $\frac{1}{2}$, reflection in x-axis
3. horizontal shrink by a factor of $\frac{1}{3}$
4. vertical stretch by a factor of 7, translation 2 units right
5. vertical stretch by a factor of 3
6. translation $\frac{\pi}{2}$ units left

8.4 Cumulative Review Warm Up

1. $a_n = \frac{1}{2}n - \frac{1}{2}$ 2. $a_n = -9n + 16$

3. $a_n = \frac{2n + 3}{2^n}$ 4. $a_n = -3\left(-\frac{2}{3}\right)^{n-1}$

Answers

8.4 Practice A

1. amplitude $= 1$, period $= 4\pi$

2. amplitude $= 2$, period $= 4$

3. amplitude $= 2$, period $= 2\pi$

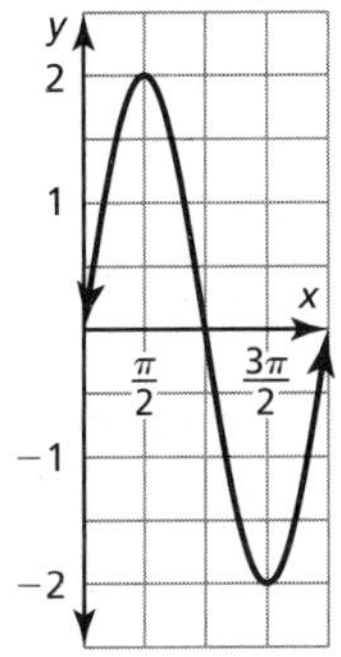

vertical stretch by a factor of 2

4. amplitude $= 4$, period $= 2\pi$

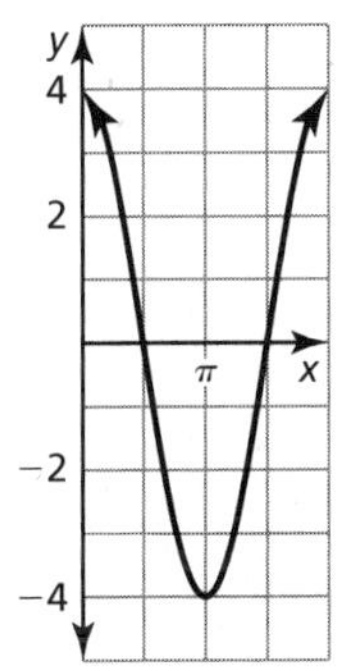

vertical stretch by a factor of 4

5. amplitude $= 1$, period $= \pi$

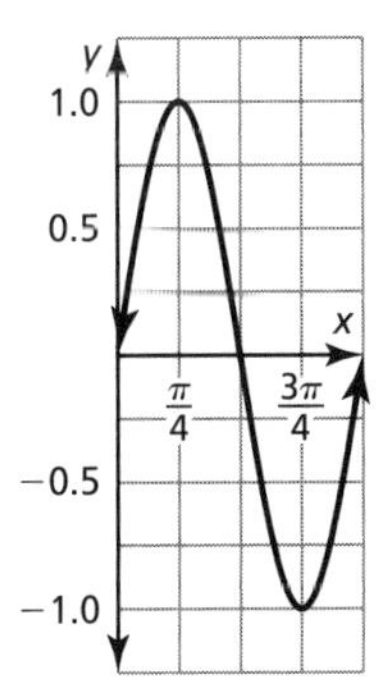

horizontal shrink by a factor of $\frac{1}{2}$

6. amplitude $= 3$, period $= 2$

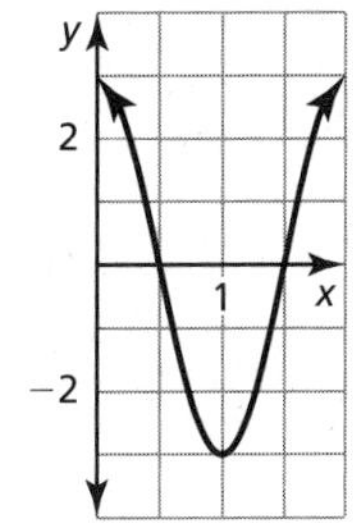

vertical stretch by a factor of 3 and a horizontal shrink by a factor of $\frac{1}{\pi}$

7. A, D

8. The amplitude is 3, so the pendulum swings 3 inches to the left and 3 inches to the right. The period is $\frac{1}{3}$, so it takes one-third of a second to go left, right, and back to a position of rest.

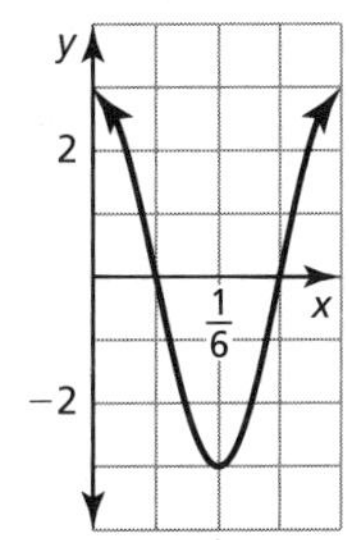

9.
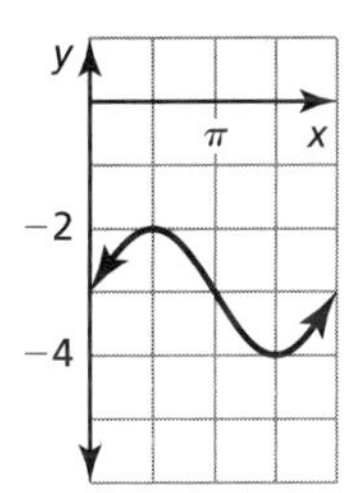

10.
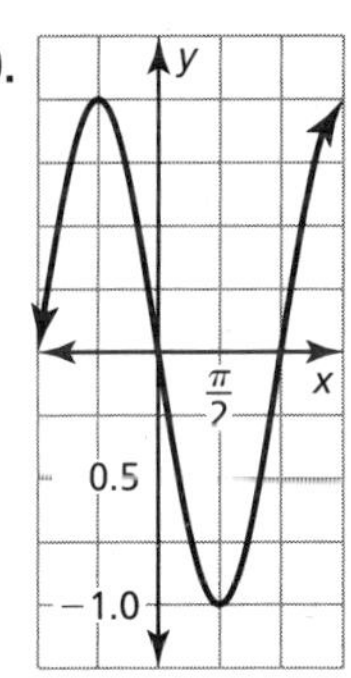

11.
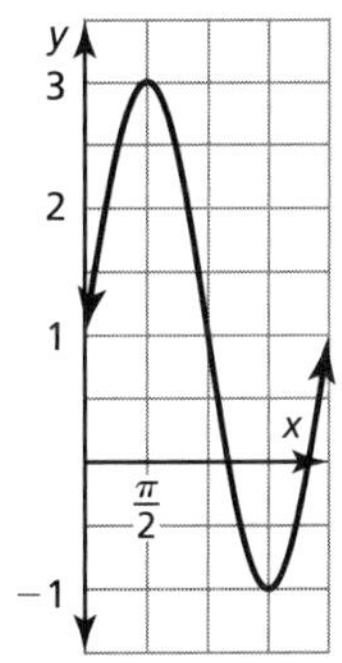

12.
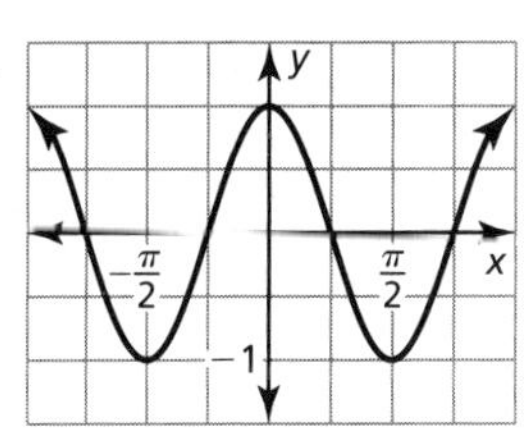

13. $f(x) = 2\cos(x + \pi) - 4$

Answers

14. $f(x) = \sin(4x - 8) + 1$

8.4 Practice B

1. amplitude $= 3$, period $= 12$

2. amplitude $= 0.5$, period $= \dfrac{\pi}{2}$

3. amplitude $= 4$, period $= 2\pi$

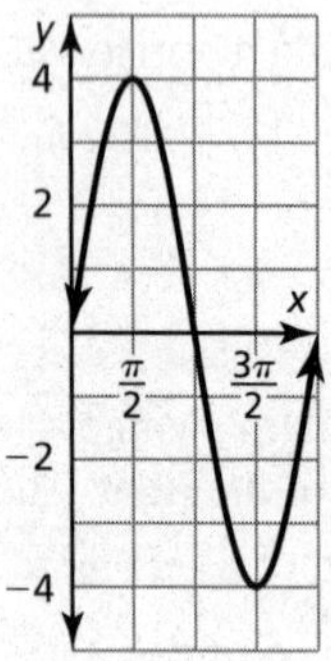

vertical stretch by a factor of 4

4. amplitude $= 1$, period $= 2$

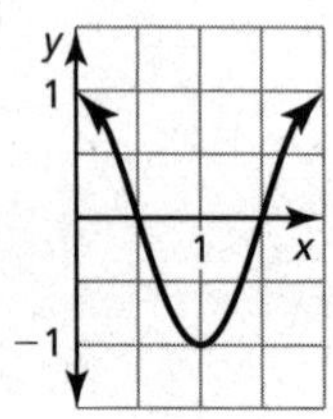

horizontal shrink by a factor of $\dfrac{1}{\pi}$

5. amplitude $= 5$, period $= \dfrac{\pi}{2}$

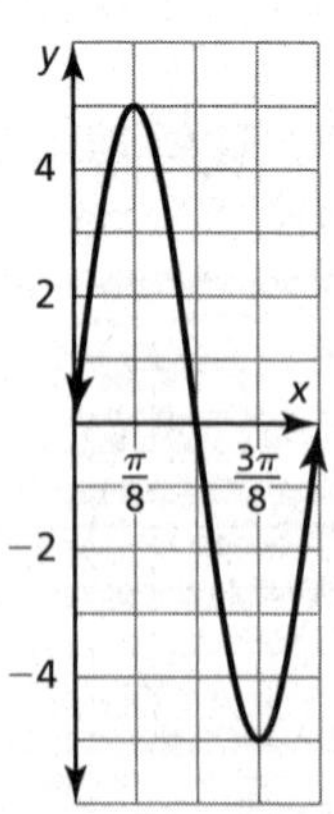

vertical stretch by a factor of 5 and a horizontal shrink by a factor of $\dfrac{1}{4}$

6. amplitude $= \dfrac{1}{4}$, period $= \pi$

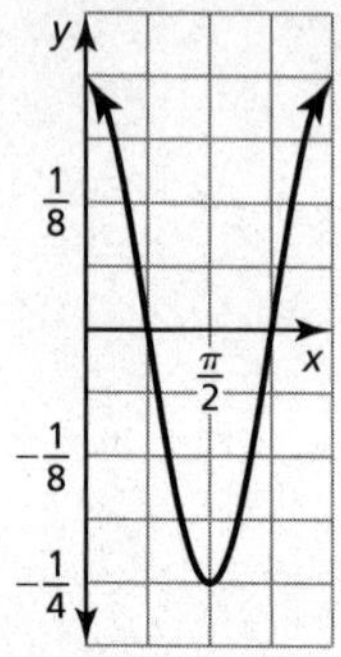

vertical shrink by a factor of $\dfrac{1}{4}$ and a horizontal shrink by a factor of $\dfrac{1}{2}$

7. a. $y = \cos\dfrac{2\pi}{3}x$

b. $y = 3\cos\dfrac{\pi}{2}x$

c. $y = 12\cos x$

d. $y = \dfrac{1}{3}\cos 2x$

8.

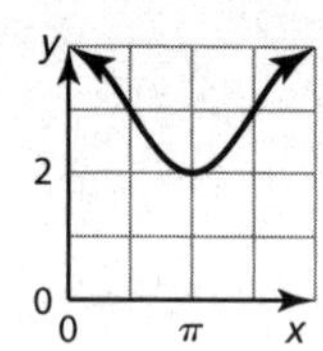

9.

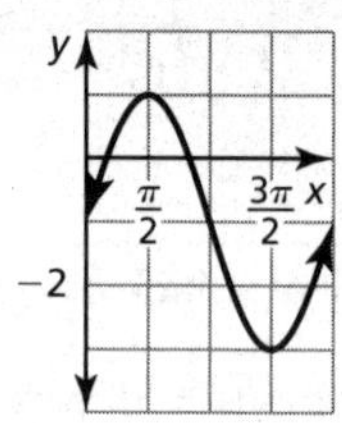

10.

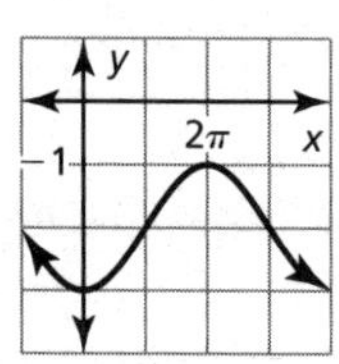

11.

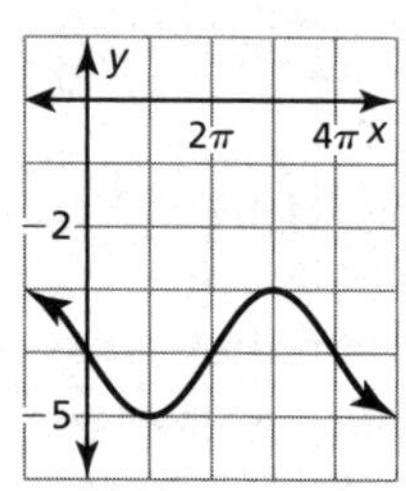

12. $f(x) = -\dfrac{1}{2}\cos 3x + 2$

13. $f(x) = -\dfrac{1}{3}\sin \pi x - 3$

Answers

8.4 Enrichment and Extension

1. 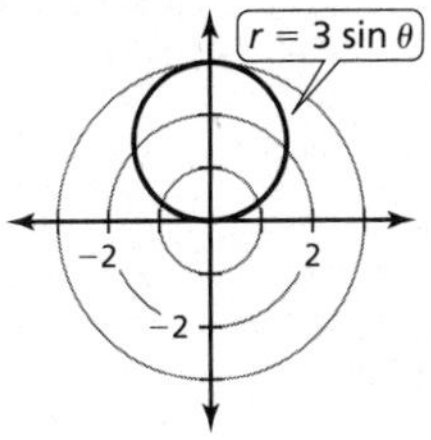

2. 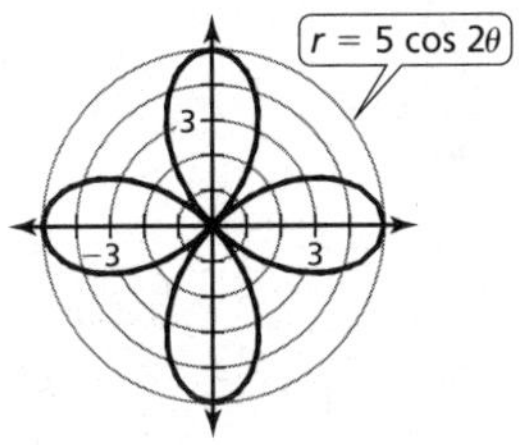

3. 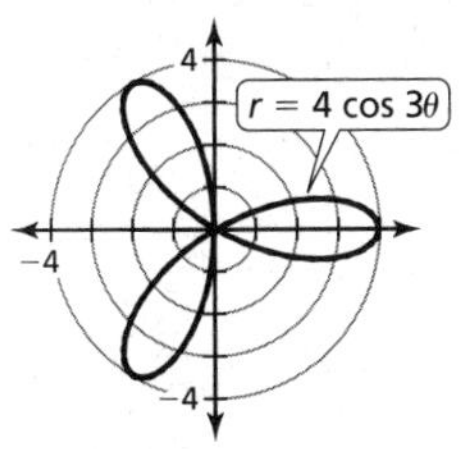

4. 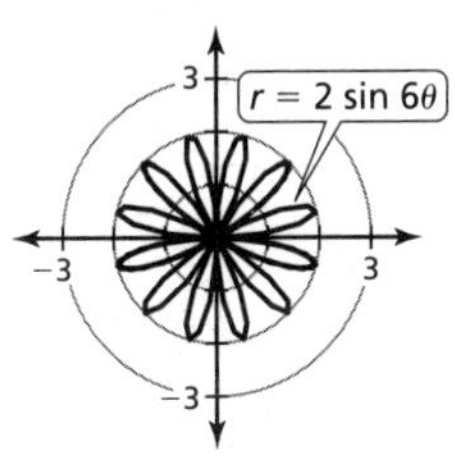

5. When n is odd, the rose curve has n petals. When n is even, the rose curve has $2n$ petals.

8.4 Puzzle Time

TRAFFIC JAM

8.5 Start Thinking

$0 \le \theta \le \frac{\pi}{2}$ or $0° \le \theta \le 90°$; $-\frac{\pi}{2} \le \theta \le 0$ or $-90° \le \theta \le 0$; As the angles move from 0 to $\frac{\pi}{2}$, the x-values decrease from 1 to 0. The x-values do the same thing as the angles move from 0 to $-\frac{\pi}{2}$. As the angles move from 0 to $\frac{\pi}{2}$, the y-values increase from 0 to 1. The y-values decrease from 0 to -1 as the angles move from 0 to $-\frac{\pi}{2}$;

$\tan \theta = \frac{y}{x}$;

Sample answer: As the angles move from 0 to $\frac{\pi}{2}$, the value of $\tan \theta = \frac{y}{x}$ is going to increase because the denominator (x-value) is approaching zero and y is positive. As the angles move from 0 to $-\frac{\pi}{2}$, the value of $\tan \theta = \frac{y}{x}$ is going to decrease because the denominator (x-value) is approaching zero and y is negative; The graph of $f(\theta) = \tan \theta$ will cross the x-axis (have an x-intercept) every time $y = 0$, which occurs when the angle is 0 and when the angle is any multiple of π added to 0. That is $0 + n\pi = n\pi$, where n is any integer. The graph of $f(\theta) = \tan \theta$ will have a vertical asymptote every time $x = 0$, which occurs when the angle is $\frac{\pi}{2}$ and when the angle is any multiple of π added to $\frac{\pi}{2}$. That is $\frac{\pi}{2} + n\pi$, where n is any integer.

8.5 Warm Up

1. 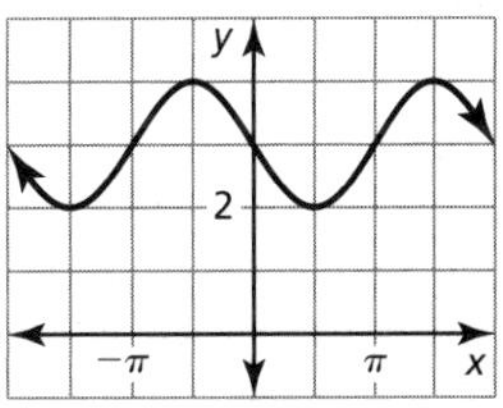

2. 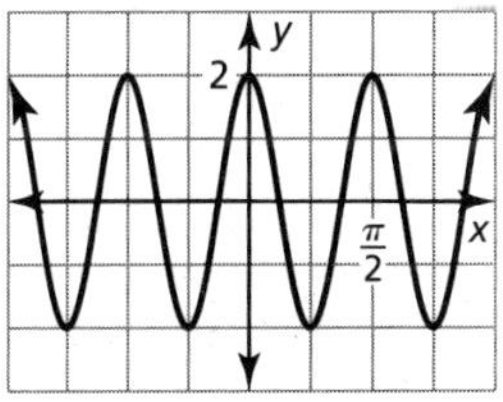

Answers

3. 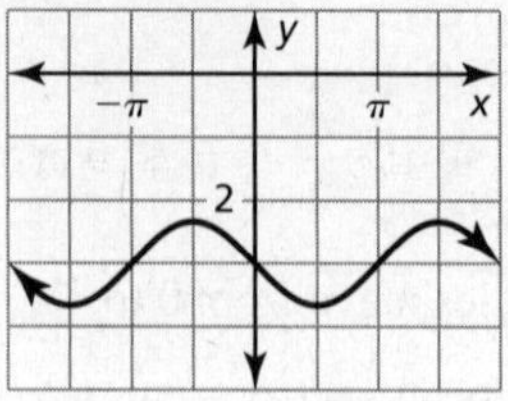

4. 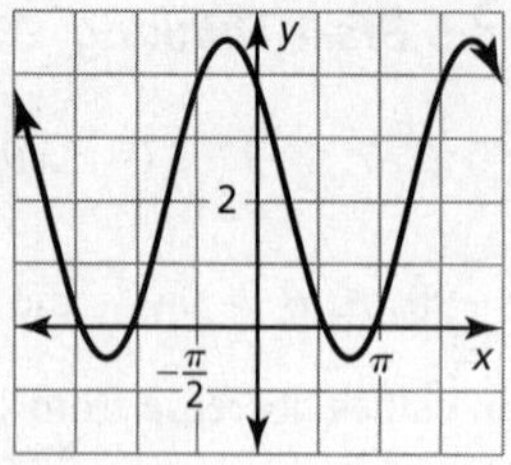

8.5 Cumulative Review Warm Up

1. $\log_2 3 + 2 \log_2 x$

2. $\log(2 - x) + 3 \log x$

3. $\frac{1}{2} \ln 2 + \frac{3}{2} \ln x - \ln 5$

4. $\ln \frac{8}{x^2}$

5. $\log 7\sqrt{x - 1}$

6. $\log_3 \frac{(x - 2)^3}{x^2}$

8.5 Practice A

1. 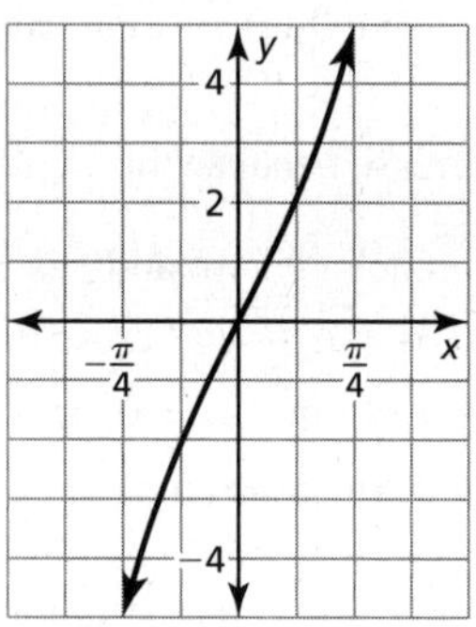

vertical stretch by a factor of 4

2. 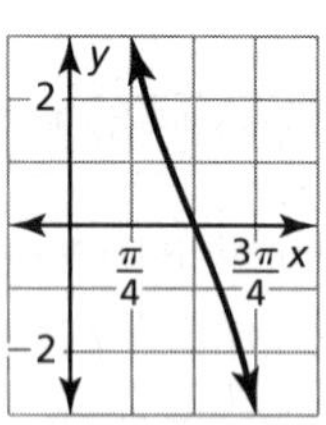

vertical stretch by a factor of 3

3. 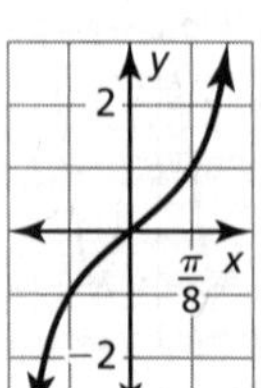

horizontal shrink by a factor of $\frac{1}{2}$

4. 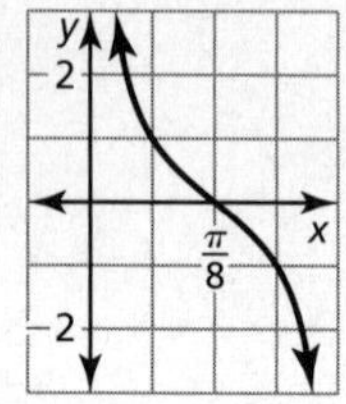

horizontal shrink by a factor of $\frac{1}{4}$

5. The value of b is 2π and not 2;

Period: $\frac{\pi}{|b|} = \frac{\pi}{|2\pi|} = \frac{\pi}{2\pi} = \frac{1}{2}$

6. a. 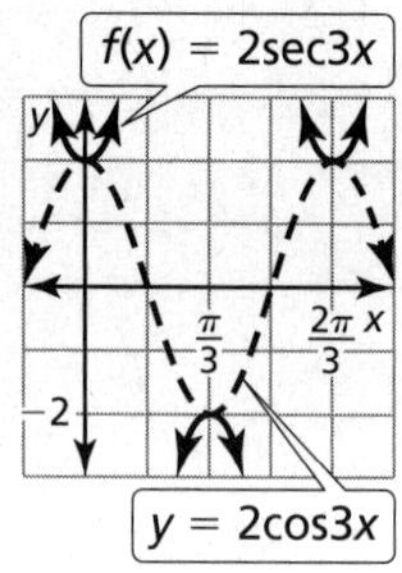

b. 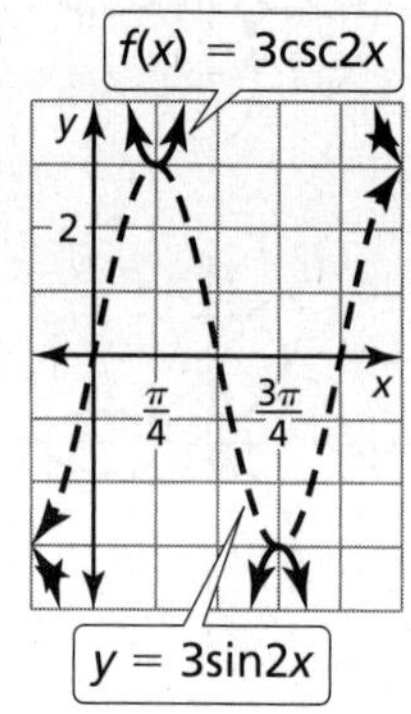

7. 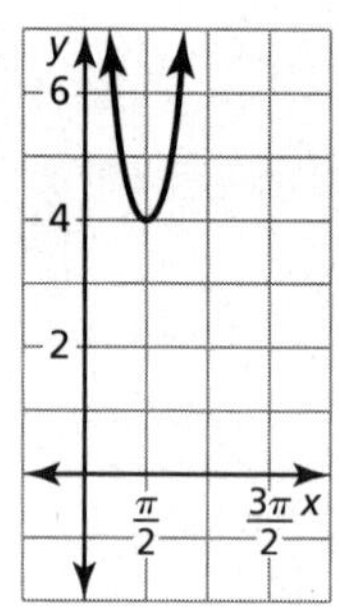

vertical stretch by a factor of 4

8. 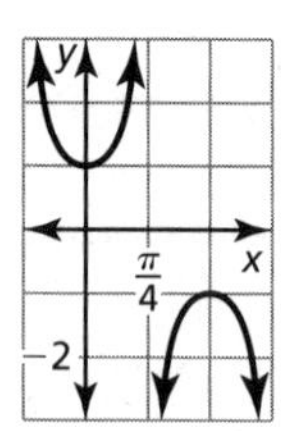

horizontal shrink by a factor of $\frac{1}{2}$

Answers

9. 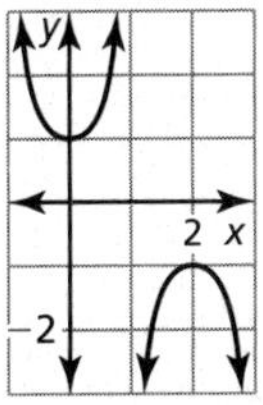

vertical shrink by a factor of $\frac{1}{3}$ and a horizontal shrink by a factor of $\frac{1}{2\pi}$

10. 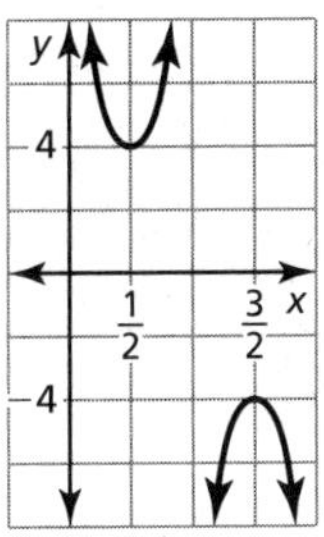

vertical stretch by a factor of 4 and a horizontal shrink by a factor of $\frac{1}{\pi}$

8.5 Practice B

1. 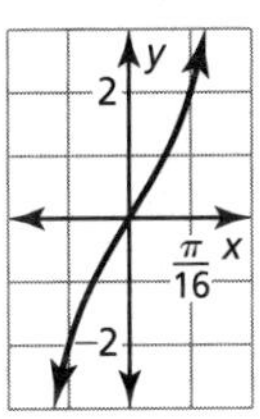

vertical stretch by a factor of 2 and a horizontal shrink by a factor of $\frac{1}{4}$

2. 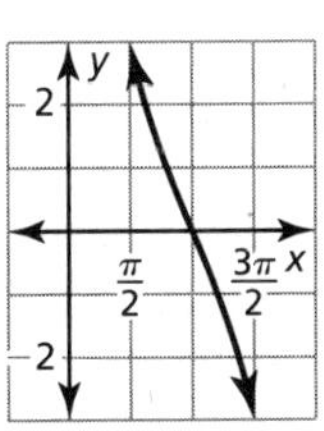

vertical stretch by a factor of 3 and a horizontal stretch by a factor of 2

3. 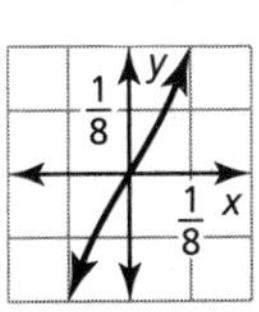

vertical shrink by a factor of $\frac{1}{4}$ and a horizontal shrink by a factor of $\frac{1}{2\pi}$

4. 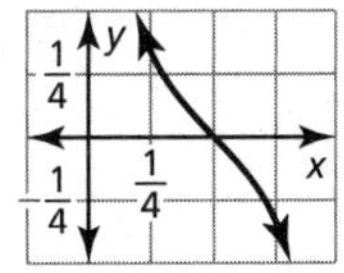

vertical shrink by a factor of $\frac{1}{3}$ and a horizontal shrink by a factor of $\frac{1}{\pi}$

5. There is a horizontal stretch instead of a horizontal shrink; a vertical stretch by a factor of 4 and a horizontal stretch by a factor of 2

6. a. 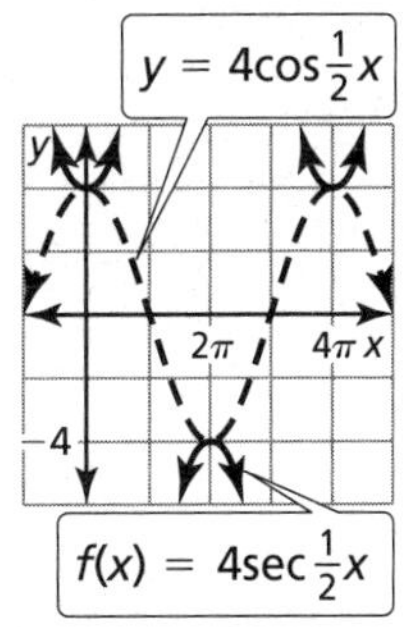

b. 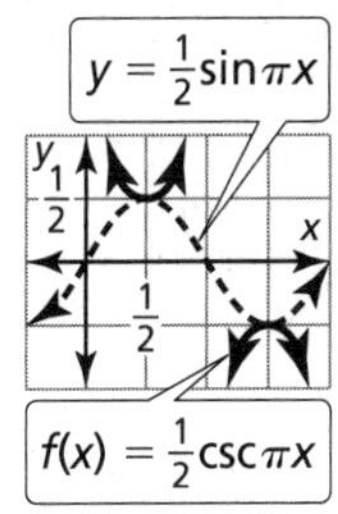

7. 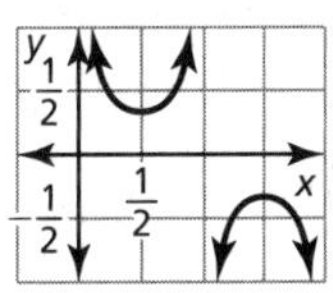

vertical shrink by a factor of $\frac{1}{3}$ and a horizontal shrink by a factor of $\frac{1}{\pi}$

8. 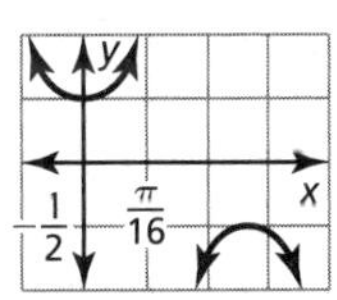

vertical shrink by a factor of $\frac{1}{2}$ and a horizontal shrink by a factor of $\frac{1}{6}$

9. 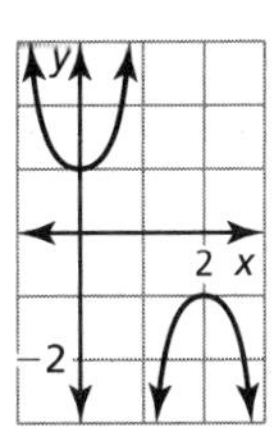

horizontal shrink by a factor of $\frac{2}{\pi}$

Answers

10. 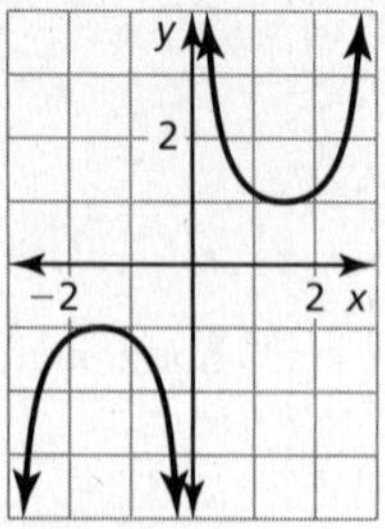

horizontal shrink by a factor of $\frac{3}{\pi}$

8.5 Enrichment and Extension

1. $y = \cot 0.25(x + 3) + 2$
2. $y = 3 \sec 2(x - 1) - 3$
3. $y = \tan 3(x - 2) - 1$
4. $y = \csc 0.5(x + 1) + 2$
5. $y = \cot 4(x - 1) + 3$
6. $y = \sec 0.5x - 1$

8.5 Puzzle Time

MEMORY

8.6 Start Thinking

The x-intercepts occur at $(0, 0)$, $(40, 0)$, and $(80, 0)$. They represent the time you load the Ferris wheel car and then each time when the car is at ground level. The Ferris wheel is approximately 80 feet tall. The ride takes 160 seconds or 2 minutes and 40 seconds. The graph of $h = \cos t$ has been reflected in the t-axis, shifted 40 units up, vertically stretched by a factor of 40, and horizontally stretched so that the period is now 40.

8.6 Warm Up

1. amplitude: 2, period: 2π
2. amplitude: $\frac{2}{3}$, period: 1
3. amplitude: 1, period: $\frac{2\pi}{3}$
4. amplitude: $\frac{\sqrt{3}}{2}$, period: π
5. amplitude: 1.4, period: 2π
6. amplitude: 2.4, period: $\frac{2\pi}{1.8} = \frac{10\pi}{9}$

8.6 Cumulative Review Warm Up

1. domain: all real numbers, range: $f(x) \geq -2$
2. domain: $x > 2$, range: all real numbers
3. domain: $x \geq -8$, range: $h(x) \geq -4$
4. domain: all real numbers, range: all real numbers
5. domain: all real numbers, range: $-6 \leq g(x) \leq -2$
6. domain: all real numbers, range: $-3 \leq h(x) \leq 3$

8.6 Practice A

1. $f = \frac{1}{\pi}$
2. $f = \frac{3}{2\pi}$
3. $f = \frac{2}{\pi}$
4. $f = \frac{5}{2}$
5. $P = 4 \sin 524\pi t$

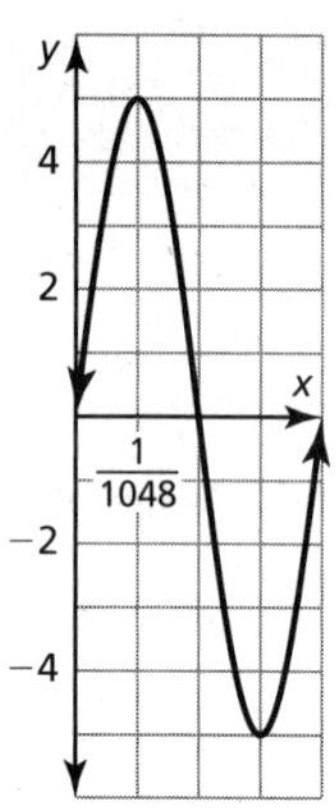

6. $y = 3 \sin 2x + 1$
7. $y = 2 \sin\left(x - \frac{\pi}{4}\right) - 1$
8. a. $y = 1.3 \sin(0.5x - 0.437) + 3.1$

   b. 4.37 ft

8.6 Practice B

1. $f = \frac{3}{2\pi}$
2. $f = \frac{2}{\pi}$
3. $f = \frac{1}{4}$
4. $f = \frac{0.2}{\pi}$

Answers

5. $P = 4 \sin 55\pi t$

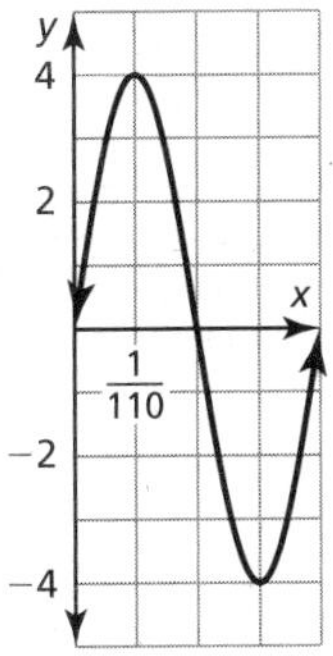

6. $y = 3 \sin \pi\left(x + \frac{1}{2}\right)$

7. $y = \sin \frac{\pi}{4}(x - 1) + 4$

8. a. $y = 7 \sin(0.628x - 1.57) + 7.998$

   b. 5.8 m

8.6 Enrichment and Extension

1. *Sample answer*: $y = 2.5 \cos \frac{\pi}{6}(x - 4) + 319.5$

2. *Sample answer*: $y = 3 \cos \frac{\pi}{6}(x - 5) + 345$

3. *Sample answer*: $y = 3 \cos \frac{\pi}{6}(x - 4) + 377$

4. Seasonal changes in temperature result in cyclical patterns in plant photosynthesis and carbon dioxide levels from summer to winter; The midline of the sinusoid increases over time. That is, there is more carbon dioxide in the atmosphere; People could be harming the environment.

8.6 Puzzle Time

A MEWTATION

Cumulative Review

1. $-|4|, |7 - 8|, |-1 + 6|, |-6|$

2. $-|4 + 1|, |-4 + 3|, |-8 + 3|, |9|$

3. $\frac{|-6|}{|8|}, |2|, |-2 + 5|, |-6|$

4. $|-5 + 5|, \frac{|5|}{|-7|}, |4 - 9|, |-8|$

5. $|9 - 9|, |-6| - |4|, |8| + |-4|, |-7 \bullet -6|$

6. $|-4 + 6|, |-8| + |4|, |9 + 6|, |4 \bullet -4|$

7. $|7 - 5|, |3^2|, |9| + |-4|, |-2 \bullet -9|$

8. $|-8| + |-7|, |-4^2|, |7 \bullet 3|, |-6 \bullet -6|$

9. $(10, -3)$ 10. $(-9, -13)$ 11. $(0, 6)$

12. $(10, -6)$ 13. $(-3, -2)$ 14. $(0, 0)$

15. $(7, -16)$ 16. $(-14, 6)$ 17. $(0, 5)$

18. $(-2, -1)$ 19. $2\sqrt{149}$ 20. $3\sqrt{85}$

21. $3\sqrt{53}$ 22. $5\sqrt{26}$ 23. $\sqrt{10}$

24. $3\sqrt{10}$ 25. $3\sqrt{85}$ 26. $\sqrt{409}$

27. $5\sqrt{5}$ 28. $2\sqrt{181}$

29. a. 5.7 ft b. 12 ft^3 c. 8 ft^3 d. \$9.52

30. $c = 20$ ft 31. $b = 40$ km 32. $a = 10$ in.

33. $b = 24$ mi 34. $a = 13.5$ cm

35. $c = 6.5$ m 36. $c = 24.4$ $in.^2$

37. a. 4.2 ft

   b. 4.5 ft^3

   c. 3.4 ft^3

   d. \$8.47, or \$8.40, using exact value from part (c)

38. 44π in. 39. 34π km 40. 2π mi

41. 20π cm 42. 6π ft 43. 14π mm

44. 32π yd 45. 24π m 46. 24π cm

47. 36π mi^2 48. 9π $in.^2$ 49. 16π km^2

50. 484π yd^2 51. 225π ft^2 52. 49π cm^2

53. 4π m^2 54. 84π mi^2 55. 4π yd^2

56. a. $\frac{1}{27}$ b. 9 57. a. $\frac{1}{125}$ b. 25

Answers

58. **a.** $\frac{2}{27}$ **b.** 18

59. **a.** $\frac{1}{2}$ **b.** 16

60. **a.** $\frac{3}{125}$ **b.** 75

61. **a.** $\frac{193}{64}$ **b.** 19

62. **a.** $\frac{17}{8}$ **b.** 6

63. **a.** $-\frac{107}{27}$ **b.** 5

64. **a.** $-\frac{55}{8}$ **b.** -3

65. e^{-11} **66.** e^{14} **67.** e^3

68. $3e^5$ **69.** $-\frac{e^2}{2}$ **70.** $6e^7$

71. $8e^{-24x}$ **72.** $27e^{-27x}$ **73.** $9e^{12x}$

74. $4e^{6x}$ **75.** $6e^{2x}$ **76.** $3e^{5x}$

77. e^{7x+5} **78.** e^{-2x+9} **79.** e^{3x+4}

80. **a.** exponential decay
b. 20% decrease
c. a little more than 4 years

81. 5 **82.** 6 **83.** 0

84. -5 **85.** -3 **86.** -3

87. -2 **88.** -2 **89.** -2

90. $\log_2 5 + \log_2 x$ **91.** $\log_7 8 + \log_7 x$

92. $\log_9 5 + \log_9 x$ **93.** $\log_4 8 + 3 \log_4 x$

94. $\log_2 4 + 5 \log_2 x$ **95.** $\log_2 3 + 4 \log_2 x$

96. $\ln 2x - \ln y$

97. $\ln x - (\ln 5 + \ln y)$

98. $\ln 4 + \ln x - (\ln 3 + \ln y)$

99. $\frac{1}{2} + \frac{1}{2} \log_9 x$

100. $\frac{1}{3}(\log_8 x + 4 \log_8 y)$

101. $\frac{1}{2} + \frac{3}{2} \log_4 x$ **102.** $\log_7 66$

103. $\ln 2$ **104.** $\ln 42$

105. $\log_8\left(\frac{x^3}{y^2}\right)$ **106.** $\log_5\left(y^9 z^4\right)$

107. $\log_9\left(w^{10} x^8\right)$ **108.** $\ln\left(5^7 \bullet 4^6\right)$

109. $\log\left(\frac{4^7}{5^4}\right)$ **110.** $\ln\left(49x^7 y^9\right)$

111. $\log_4(5x)$ **112.** $\ln\left(\frac{z^3}{y^4}\right)$

113. $\log_3\left(25x^6 y^4\right)$ **114.** $\frac{4}{3x^2 - 1}, x \neq 0$

115. $\frac{x}{3x^3 + 4}, x \neq 0$ **116.** $\frac{x^3}{3x - 7}, x \neq \frac{7}{3}$

117. $\frac{x + 1}{x - 2}, x \neq -3, x \neq 2$

118. $\frac{(x + 3)^2}{x + 6}, x \neq -6, x \neq -3$

119. $\frac{(x - 2)^2}{x + 8}, x \neq -8, x \neq 2$

120. 0.9613 **121.** 0.7002 **122.** 1.0785

123. 0.6428 **124.** 19.0811 **125.** 1.0306

126. 0.0175 **127.** 0.0349 **128.** 4.8097

Chapter 9

9.1 Start Thinking

$\sin^2 \theta + \cos^2 \theta = 1$; 1; $\tan \theta = \frac{\sin \theta}{\cos \theta}$

9.1 Warm Up

1. $\sin \theta = \frac{6\sqrt{61}}{61}$, $\cos \theta = \frac{5\sqrt{61}}{61}$, $\tan \theta = \frac{6}{5}$,
$\csc \theta = \frac{\sqrt{61}}{6}$, $\sec \theta = \frac{\sqrt{61}}{5}$, $\cot \theta = \frac{5}{6}$

2. $\sin \theta = \frac{5\sqrt{3}}{14}$, $\cos \theta = \frac{11}{14}$, $\tan \theta = \frac{5\sqrt{3}}{11}$,
$\csc \theta = \frac{14\sqrt{3}}{15}$, $\sec \theta = \frac{14}{11}$, $\cot \theta = \frac{11\sqrt{3}}{15}$

Answers

3. $\sin\theta = \frac{2}{3}$, $\cos\theta = \frac{\sqrt{5}}{3}$, $\tan\theta = \frac{2\sqrt{5}}{5}$,

$\csc\theta = \frac{3}{2}$, $\sec\theta = \frac{3\sqrt{5}}{5}$, $\cot\theta = \frac{\sqrt{5}}{2}$

9.1 Cumulative Review Warm Up

1. $\frac{x+1}{x+4}$

2. $\frac{4(x+2)}{(x-4)(x+3)}$

3. $\frac{-4x(x-4)}{x(x+5)(x-5)}$

4. $\frac{-3x^3 - 4x^2 + 14x - 7}{2x^2(2x-1)}$

9.1 Practice A

1. $\cos\theta = \frac{\sqrt{5}}{3}$, $\tan\theta = \frac{2\sqrt{5}}{5}$, $\csc\theta = \frac{3}{2}$,

$\sec\theta = \frac{3\sqrt{5}}{5}$, $\cot\theta = \frac{\sqrt{5}}{2}$

2. $\sin\theta = \frac{4}{5}$, $\tan\theta = -\frac{4}{3}$, $\csc\theta = \frac{5}{4}$,

$\sec\theta = -\frac{5}{3}$, $\cot\theta = -\frac{3}{4}$

3. $\sin\theta = -\frac{4\sqrt{65}}{65}$, $\cos\theta = -\frac{7\sqrt{65}}{65}$,

$\csc\theta = -\frac{\sqrt{65}}{4}$, $\sec\theta = -\frac{\sqrt{65}}{7}$, $\cot\theta = \frac{7}{4}$

4. $\sin\theta = \frac{4\sqrt{41}}{41}$, $\cos\theta = -\frac{5\sqrt{41}}{41}$, $\tan\theta = -\frac{4}{5}$,

$\csc\theta = \frac{\sqrt{41}}{4}$, $\sec\theta = -\frac{\sqrt{41}}{5}$

5. $\sin\theta = -\frac{2\sqrt{10}}{7}$, $\cos\theta = \frac{3}{7}$, $\tan\theta = -\frac{2\sqrt{10}}{3}$,

$\csc\theta = -\frac{7\sqrt{10}}{20}$, $\cot\theta = -\frac{3\sqrt{10}}{20}$

6. $\sin\theta = -\frac{5}{6}$, $\cos\theta = -\frac{\sqrt{11}}{6}$, $\tan\theta = \frac{5\sqrt{11}}{11}$,

$\sec\theta = -\frac{6\sqrt{11}}{11}$, $\cot\theta = \frac{\sqrt{11}}{5}$

7. $\sin x$ **8.** $\cot x \cos x$ **9.** $\sec\theta$

10. $\cos^2\theta$ **11.** $\cos x$ **12.** $\cos^3 x$

13. $\sec^2 x$ is equal to $\tan^2 x + 1$ and not $\tan^2 x - 1$;

$$\begin{aligned}\tan^2 x + \sec^2 x &= \tan^2 x + (\tan^2 x + 1)\\ &= \tan^2 x + \tan^2 x + 1\\ &= 2\tan^2 x + 1\end{aligned}$$

14. $$\begin{aligned}\tan x \cot x &= \tan x \bullet \frac{1}{\tan x}\\ &= 1\end{aligned}$$

15. $$\begin{aligned}\sin\left(\frac{\pi}{2} - x\right)\sec x &= \cos x \bullet \frac{1}{\cos x}\\ &= 1\end{aligned}$$

16. $$\begin{aligned}\frac{\sin^2 x + (1 - \cos^2 x)}{\tan x \cos x} &= \frac{\sin^2 x + \sin^2 x}{\frac{\sin x}{\cos x} \bullet \cos x}\\ &= \frac{2\sin^2 x}{\sin x}\\ &= 2\sin x\end{aligned}$$

17. $$\begin{aligned}\frac{\cos^2(-x)\csc x}{\cot x} &= \frac{\cos^2 x \bullet \frac{1}{\sin x}}{\frac{\cos x}{\sin x}}\\ &= \frac{\frac{\cos^2 x}{\sin x}}{\frac{\cos x}{\sin x}}\\ &= \frac{\cos^2 x}{\sin x} \bullet \frac{\sin x}{\cos x}\\ &= \cos x\end{aligned}$$

18. increases; $\sin\theta$ and $\csc\theta$ are reciprocals of each other, so as one decreases, the other increases.

9.1 Practice B

1. $\cos\theta = -\frac{\sqrt{55}}{8}$, $\tan\theta = -\frac{3\sqrt{55}}{55}$, $\csc\theta = \frac{8}{3}$,

$\sec\theta = -\frac{8\sqrt{55}}{55}$, $\cot\theta = -\frac{\sqrt{55}}{3}$

2. $\sin\theta = -\frac{2\sqrt{2}}{3}$, $\tan\theta = 2\sqrt{2}$, $\csc\theta = -\frac{3\sqrt{2}}{4}$,

$\sec\theta = -3$, $\cot\theta = \frac{\sqrt{2}}{4}$

Answers

3. $\sin\theta = \frac{3\sqrt{10}}{10}$, $\cos\theta = \frac{\sqrt{10}}{10}$, $\csc\theta = \frac{\sqrt{10}}{3}$,

   $\sec\theta = \sqrt{10}$, $\cot\theta = \frac{1}{3}$

4. $\sin\theta = -\frac{3\sqrt{109}}{109}$, $\cos\theta = -\frac{10\sqrt{109}}{109}$,

   $\tan\theta = \frac{3}{10}$, $\csc\theta = -\frac{\sqrt{109}}{3}$, $\sec\theta = -\frac{\sqrt{109}}{10}$

5. $\sin\theta = \frac{2\sqrt{14}}{9}$, $\cos\theta = -\frac{5}{9}$, $\tan\theta = -\frac{2\sqrt{14}}{5}$,

   $\csc\theta = \frac{9\sqrt{14}}{28}$, $\cot\theta = -\frac{5\sqrt{14}}{28}$

6. $\sin\theta = -\frac{2}{7}$, $\cos\theta = \frac{3\sqrt{5}}{7}$, $\tan\theta = -\frac{2\sqrt{5}}{15}$,

   $\sec\theta = \frac{7\sqrt{5}}{15}$, $\cot\theta = -\frac{3\sqrt{5}}{2}$

7. $\csc x$

8. $-\csc\theta$

9. $\cos\theta$

10. $-\cot\theta$

11. 1

12. $\csc x + 1$

13. $\csc\theta$ is equal to $\frac{1}{\sin\theta}$ and not $\frac{1}{\cos\theta}$;

$$\begin{aligned}\csc\theta\cot\theta\cos\theta &= \frac{1}{\sin\theta}\bullet\frac{\cos\theta}{\sin\theta}\bullet\cos\theta\\ &= \frac{\cos^2\theta}{\sin^2\theta}\\ &= \cot^2\theta\end{aligned}$$

14.
$$\begin{aligned}\tan\left(\frac{\pi}{2}-x\right)\cos^2\left(\frac{\pi}{2}-x\right) &= \cot x\sin^2 x\\ &= \frac{\cos x}{\sin x}\bullet\sin^2 x\\ &= \cos x\sin x\end{aligned}$$

15.
$$\begin{aligned}\frac{1+\sin(-x)}{\cos x\tan x - 1} &= \frac{1-\sin x}{\cos x\bullet\frac{\sin x}{\cos x}-1}\\ &= \frac{1-\sin x}{\sin x - 1}\\ &= -1\end{aligned}$$

16.
$$\begin{aligned}\frac{1-\cos x}{\sin x} &= \frac{1}{\sin x}-\frac{\cos x}{\sin x}\\ &= \csc x - \cot x\end{aligned}$$

17.
$$\begin{aligned}&\frac{1+\sin\theta}{\cos\theta}+\frac{\cos\theta}{1-\sin\theta}\\ &= \frac{1+\sin\theta}{\cos\theta}\bullet\frac{1-\sin\theta}{1-\sin\theta}+\frac{\cos\theta}{1-\sin\theta}\bullet\frac{\cos\theta}{\cos\theta}\\ &= \frac{1-\sin^2\theta}{\cos\theta(1-\sin\theta)}+\frac{\cos^2\theta}{\cos\theta(1-\sin\theta)}\\ &= \frac{\cos^2\theta}{\cos\theta(1-\sin\theta)}+\frac{\cos^2\theta}{\cos\theta(1-\sin\theta)}\\ &= \frac{2\cos^2\theta}{\cos\theta(1-\sin\theta)}\\ &= \frac{2\cos\theta}{1-\sin\theta}\\ &= \frac{2\cos\theta}{1-\sin\theta}\bullet\frac{1+\sin\theta}{1+\sin\theta}\\ &= \frac{2\cos\theta(1+\sin\theta)}{1-\sin^2\theta}\\ &= \frac{2\cos\theta(1+\sin\theta)}{\cos^2\theta}\\ &= \frac{2(1+\sin\theta)}{\cos\theta}\\ &= \frac{2+2\sin\theta}{\cos\theta}\\ &= \frac{2}{\cos\theta}+\frac{2\sin\theta}{\cos\theta}\\ &= 2\sec\theta + 2\tan\theta\end{aligned}$$

18.
$$\begin{aligned}\tan x\cot x &= \frac{\sin x}{\cos x}\bullet\frac{\cos x}{\sin x}\\ &= \frac{\sin x\cos x}{\cos x\sin x}\\ &= 1\end{aligned}$$

9.1 Enrichment and Extension

1. $\frac{\pm\sqrt{1-\cos^2\theta}}{\cos\theta}$

2. $\frac{\cos\theta}{\pm\sqrt{1-\cos^2\theta}}$

3. $\frac{1}{\pm\sqrt{1-\cos^2\theta}}$

4. $\frac{1}{\tan\theta}$

5. $\frac{\tan\theta}{\pm\sqrt{1+\tan^2\theta}}$

6. $\frac{1}{\pm\sqrt{1+\tan^2\theta}}$

7. $\frac{\pm\sqrt{\csc^2\theta-1}}{\csc\theta}$

8. $\frac{1}{\pm\sqrt{\csc^2\theta-1}}$

9. $\dfrac{\csc \theta}{\pm\sqrt{\csc^2 \theta - 1}}$

10. $\pm\sqrt{\sec^2 \theta - 1}$

11. $\dfrac{\pm\sqrt{\sec^2 \theta - 1}}{\sec \theta}$

12. $\dfrac{1}{\pm\sqrt{\sec^2 \theta - 1}}$

9.1 Puzzle Time

PAPA SURF

9.2 Start Thinking

1. *Sample answer*: $75° = 30° + 45°$;

$$\begin{aligned}\sin 75° &= \sin(30° + 45°)\\ &= \sin 30° \cos 45° + \cos 30° \sin 45°\\ &= \frac{1}{2}\left(\frac{\sqrt{2}}{2}\right) + \frac{\sqrt{3}}{2}\left(\frac{\sqrt{2}}{2}\right)\\ &= \frac{\sqrt{2} + \sqrt{6}}{4}\end{aligned}$$

2. *Sample answer*: $\dfrac{\pi}{12} = \dfrac{\pi}{3} - \dfrac{\pi}{4}$;

$$\begin{aligned}\sin \frac{\pi}{12} &= \sin\left(\frac{\pi}{3} - \frac{\pi}{4}\right)\\ &= \sin \frac{\pi}{3} \cos \frac{\pi}{4} - \cos \frac{\pi}{3} \sin \frac{\pi}{4}\\ &= \frac{\sqrt{3}}{2}\left(\frac{\sqrt{2}}{2}\right) - \frac{1}{2}\left(\frac{\sqrt{2}}{2}\right)\\ &= \frac{\sqrt{6} - \sqrt{2}}{4}\end{aligned}$$

3. *Sample answer*: $\dfrac{19\pi}{12} = \dfrac{11\pi}{6} - \dfrac{\pi}{4}$;

$$\begin{aligned}\sin \frac{19\pi}{12} &= \sin\left(\frac{11\pi}{6} - \frac{\pi}{4}\right)\\ &= \sin \frac{11\pi}{6} \cos \frac{\pi}{4} - \cos \frac{11\pi}{6} \sin \frac{\pi}{4}\\ &= \frac{-1}{2}\left(\frac{\sqrt{2}}{2}\right) - \frac{\sqrt{3}}{2}\left(\frac{\sqrt{2}}{2}\right)\\ &= \frac{-\sqrt{2} - \sqrt{6}}{4}\end{aligned}$$

9.2 Warm Up

1. $\sin b = -\dfrac{\sqrt{39}}{8}$

2. $\cos a = -\dfrac{2\sqrt{10}}{7}$

3. $\tan b = -\dfrac{\sqrt{39}}{5}$

4. $\tan a = \dfrac{3\sqrt{10}}{20}$

5. $\csc b = -\dfrac{8\sqrt{39}}{39}$

6. $\sec a = -\dfrac{7\sqrt{10}}{20}$

9.2 Cumulative Review Warm Up

1. $A = \$5986.05$

2. $A = \$5470.18$

3. $A = \$5792.28$

4. $A = \$5943.14$

9.2 Practice A

1. $-2 - \sqrt{3}$

2. $-2 - \sqrt{3}$

3. $\dfrac{\sqrt{6} - \sqrt{2}}{4}$

4. $\dfrac{-\sqrt{2} - \sqrt{6}}{4}$

5. $\dfrac{-\sqrt{2} - \sqrt{6}}{4}$

6. $\dfrac{\sqrt{2} - \sqrt{6}}{4}$

7. $-\dfrac{63}{65}$

8. $\dfrac{33}{65}$

9. $-\dfrac{56}{65}$

10. $\dfrac{16}{65}$

11. $-\dfrac{63}{16}$

12. $-\dfrac{33}{56}$

13. $\cos x$

14. $\tan x$

15. $-\cos x$

16. The cosine of a sum is the difference of the products, not the sum;

$$\begin{aligned}\cos\left(x + \frac{3\pi}{2}\right) &= \cos x \cos \frac{3\pi}{2} - \sin x \sin \frac{3\pi}{2}\\ &= (0) \cos x - (-1) \sin x\\ &= \sin x\end{aligned}$$

17. $x = 0, \pi$

18. $x = \dfrac{\pi}{4}, \dfrac{5\pi}{4}$

19. $x = \dfrac{\pi}{4}, \dfrac{7\pi}{4}$

20. $x = \dfrac{\pi}{2}, \dfrac{3\pi}{2}$

21.
$$\begin{aligned}\tan(x - \pi) &= \frac{\tan x - \tan \pi}{1 + \tan x \tan \pi}\\ &= \frac{\tan x - 0}{1 + \tan x(0)}\\ &= \tan x\end{aligned}$$

9.2 Practice B

1. $-2 + \sqrt{3}$

2. $\dfrac{\sqrt{6} - \sqrt{2}}{4}$

3. $\dfrac{-\sqrt{2} - \sqrt{6}}{4}$

4. $\dfrac{\sqrt{6}-\sqrt{2}}{4}$ **5.** $\dfrac{\sqrt{6}-\sqrt{2}}{4}$ **6.** $2-\sqrt{3}$

7. $-\dfrac{171}{221}$ **8.** $-\dfrac{21}{221}$ **9.** $-\dfrac{220}{221}$

10. $\dfrac{140}{221}$ **11.** $-\dfrac{171}{140}$ **12.** $\dfrac{21}{220}$

13. $\tan x$ **14.** $\sin x$ **15.** $-\sin x$

16. The formula for $\sin(a+b)$ involves the product of sin with cosine, not sine with sine;

$$\begin{aligned}\sin\left(x+\frac{\pi}{2}\right) &= \sin x\cos\frac{\pi}{2}+\cos x\sin\frac{\pi}{2}\\ &= (0)\sin x+(1)\cos x\\ &= \cos x\end{aligned}$$

17. $x=\dfrac{7\pi}{6},\dfrac{11\pi}{6}$ **18.** $x=\dfrac{\pi}{2},\dfrac{3\pi}{2}$

19. $x=0,\dfrac{\pi}{2},\dfrac{3\pi}{2},\pi$ **20.** $x=\dfrac{\pi}{4},\dfrac{5\pi}{4}$

21. $$\begin{aligned}\tan(a+b) &= \frac{\tan a+\tan b}{1-\tan a\tan b} = \frac{\frac{\sin a}{\cos a}+\frac{\sin b}{\cos b}}{1-\frac{\sin a\sin b}{\cos a\cos b}}\\ &= \frac{\sin a\cos b+\sin b\cos a}{\cos a\cos b+\sin a\sin b}\\ &= \frac{\sin(a+b)}{\cos(a+b)}\end{aligned}$$

9.2 Enrichment and Extension

1. true; expand using identity for $\sin(a-b)$

2. true; expand using identities for $\cos(a+b)$, $\cos(n\pi)=(-1)^n$, and $\sin(n\pi)=0$

3. false; It should be $(-1)^n\sin\theta$.

4. false; It should be $(-1)^{n+1}\sin\theta$.

5. $\cos u\cos v\cos w-\sin u\sin v\cos w-\sin n\cos v\sin w-\cos u\sin v\sin w$

6. $\dfrac{\tan u+\tan v+\tan w-\tan u\tan v\tan w}{1-\tan u\tan v-\tan u\tan w-\tan v\tan w}$

7. Change the first term to $\sin(0-u)$. Then use the difference formula to expand and simplify.

8. Change the first term to $\cos(0-u)$. Then use the difference formula to expand and simplify.

9. Change the first term to $\tan(0-u)$. Then use the difference formula to expand and simplify.

10. Change the first term to $\sec(0-u)$. Use the fact that $\sec(0-u)=\dfrac{1}{\cos(0-u)}$. Then use the difference formula to expand and simplify.

11. The functions $y=\tan(x+\pi)$ and $y=\tan(x)$ are the same function. They coincide. The period of the tangent function is π.

9.2 Puzzle Time

STEREOTYPE

9.3 Start Thinking

1. 12 **2.** 12 **3.** 12

All three ratios are the same; *Sample answer:* yes

9.3 Warm Up

1. $a\approx 10.9$ **2.** $c\approx 23.3$ **3.** $b\approx 10.2$

4. $B\approx 57.8^\circ$ **5.** $A\approx 67.8^\circ$ **6.** $C\approx 49.5^\circ$

9.3 Cumulative Review Warm Up

1. $m\angle X=121^\circ, m\angle Z=121^\circ$

2. $m\angle X=112^\circ, m\angle Z=112^\circ$

3. $m\angle X=123^\circ, m\angle Z=143^\circ$

9.3 Practice A

1. −0.4226 **2.** −11.4301 **3.** 0.5299

4. 22.8 m^2 **5.** 143.9 ft^2

6. SSA; one triangle **7.** AAS; one triangle

8. SSA; no triangle **9.** SSA; two triangles

10. ASA; one triangle

11. $B=105^\circ, a\approx 5.9, c\approx 6.7$

12. $a\approx 90.5^\circ, A\approx 63.1^\circ, B\approx 36.9^\circ$

Answers

13. $B = 74°, b \approx 24.5, c \approx 24.0$

14. about 11.4 ft

9.3 Practice B

1. -1.0724 **2.** -0.6157 **3.** 0.9998

4. 60.4 cm^2 **5.** 12.7 in.^2

6. Draw a diagonal to form the SAS case of a triangle, use the area formula $A = \frac{1}{2}ab \sin C$, and then double this area; $A \approx 6578 \text{ m}^2$

7. $B = 33°, b \approx 8.3, c \approx 15.2$

8. $B \approx 64.5°, c \approx 2.4, C \approx 37.5°$

9. $a \approx 12.6, b \approx 5.2, C = 71°$

10. $x = 45°; y \approx 13.8$

9.3 Enrichment and Extension

1. one triangle **2.** no triangle **3.** one triangle

4. no triangle **5.** one triangle **6.** one triangle

7. one triangle **8.** two triangles **9.** two triangles

10. no triangle

9.3 Puzzle Time

BECAUSE THEN YOU'LL BE A MILE AWAY AND YOU'LL HAVE THEIR SHOES

9.4 Start Thinking

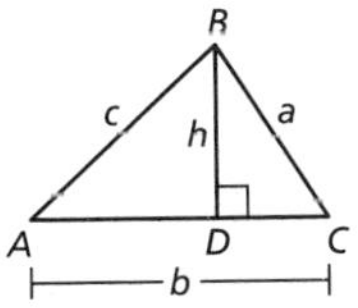

Sample answer: Let the length of $\overline{AD}$ be b_1. By the Pythagorean Theorem, $c^2 = b_1^2 + h^2$; $h = c\sin x°$

9.4 Warm Up

1. 8 **2.** 13.9

9.4 Cumulative Review Warm Up

1. $f(x) = 0.728x^3 + 0.82x^2 - 11.3x + 1$

2. $f(x) = 2x^3 - 9x^2 + 16x - 15$

3. $f(x) = -1.7860x^4 + 4.607x^3 - 0.12x^2 - 11.1x + 10$

9.4 Practice A

1. $A \approx 40.8°, B \approx 60.6°, C \approx 78.6°$

2. $A \approx 28.6°, b \approx 17.3, C \approx 20.4°$

3. $A \approx 48.2°, B \approx 106.6°, C \approx 25.2°$

4. 9.3 square units **5.** 2.7 square units

6.

Law of Sines	Law of Cosines	Neither
AAS, ASA, SSA	SSS, SAS	AAA

7. 127.2°

9.4 Practice B

1. $a \approx 12.0, B \approx 26.4°, C \approx 86.6°$

2. $A \approx 36.2°, B \approx 43.5°, C \approx 100.3°$

3. $A \approx 29.5°, B \approx 112.4°, C \approx 38.0°$

4. $A \approx 47.7°, B \approx 62.6°, C \approx 69.6°$

5. $A \approx 52.2°, C \approx 109.0°, b \approx 21.0$

6. 20.7 square units **7.** 181.0 square units

8. a. about 19.7 in.

b. about 83.8°

9.4 Enrichment and Extension

1. $A \approx 69.7°, B \approx 30.3°, c \approx 54.6$

2. $C = 78.9°, B = 39.1°, a = 12.6$

3. $a \approx 18.9, B \approx 38.9°, C \approx 75.1°$

9.4 Puzzle Time

IT WAS TWO TIRED

Cumulative Review

1. $x = -10$ **2.** no solution

3. $h \leq -6$ or $h \geq 2$ **4.** $x \geq -5$

5. $x < -4$ or $x > -2$ **6.** \$52.93

7. $3(x + 4) \geq 20$

Answers

8. 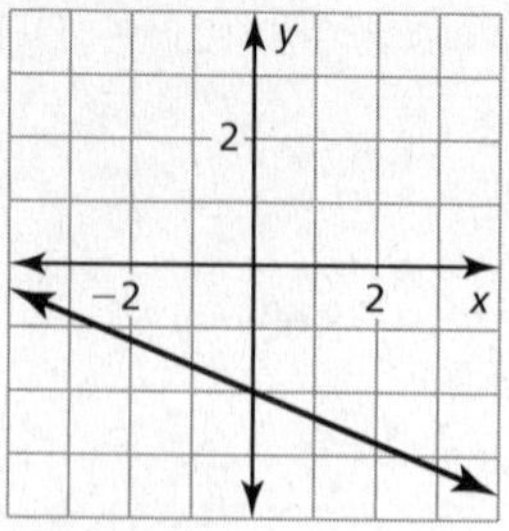

9. 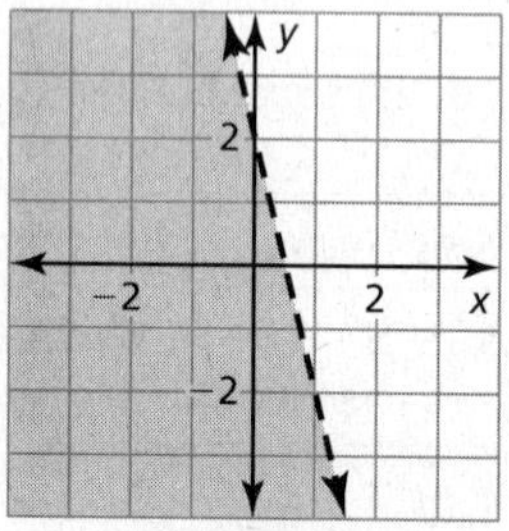

10. 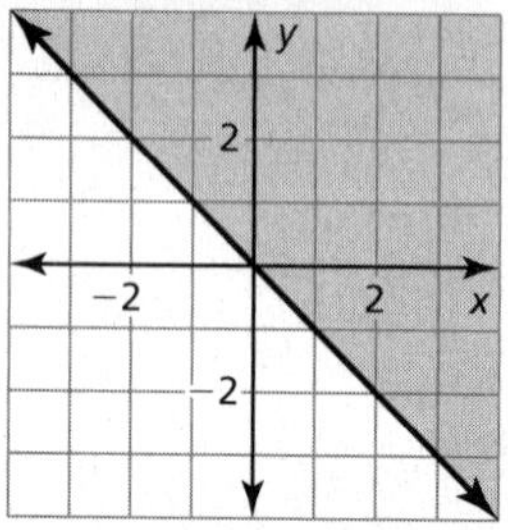

11. $y = 2x + 8$

12. $y = -\frac{1}{4}x + 2$

13. $y = 2x + 4$

14. $x \geq 0$

15. $x \geq 4$

16. $x \leq 2$

17. a. $y = 35 + 1.20x$

b. $35

c. $44.60

d. when you are downloading 46 songs or fewer or when you are downloading 72 songs or more

18. $\frac{1}{4x^9yz}$

19. $7m^3$

20. $\frac{yz}{x}$

21. 40

22. 285

23. 123

24. −78

25. $\frac{208}{105}$

26. 12

27. 1.24 sec

28. e^{10}

29. e^8

30. $6e^9$

31. $3e^{18}$

32. $64e^{36x}$

33. $729e^{-12x}$

34. a. $a_n = n + 2$

b. 33 dancers

c. 42 dancers

d. 9 dancers

35. a. 9 sec

b. If k is positive, then the time in part (a) will increase, and if k is negative, then the time in part (a) will decrease.

36. vertex: $(-7, 0)$, axis of symmetry: $x = -7$

37. vertex: $(9, 0)$, axis of symmetry: $x = 9$

38. vertex: $(1, -9)$, axis of symmetry: $x = 1$

39. 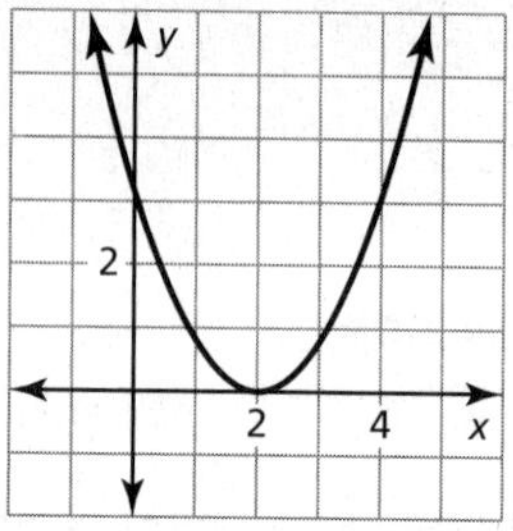

translation 2 units right and a vertical shrink by a factor of $\frac{7}{9}$

40. 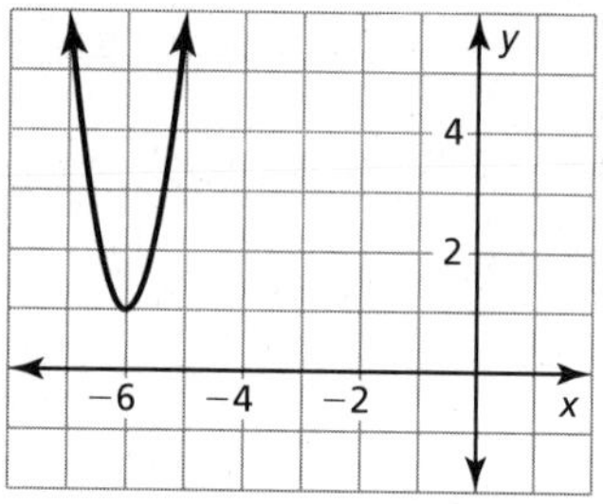

vertical stretch by a factor of 5 followed by a translation 6 units left and 1 unit up

41. $\frac{5x^2\sqrt[3]{4xy^2}}{6y^2}$

42. $\frac{20 - 4\sqrt{7}}{9}$

43. $-6\sqrt{3} - 2\sqrt{2}$

44. $x = -14, x = -3$

45. $x = -11, x = 18$

46. $x = \pm 5$

47. $x = \pm 1$

48. $x = \pm 7$

49. $x \approx -11.72, x \approx -1.28$

50. $y \approx -1.28, y \approx 3.28$

51. 17.5 ft by 17.5 ft by 35 ft

Answers

52. $x \approx -0.4, x \approx 1.2$

53. $y = -3, y = 1$

54. 8 in. by 5 in.

55. $x \geq 0$

56. $x \geq -9$

57. $x \leq -3$

58. $x \geq 0$

59. $x \geq 0$

60. $x \leq 5$

61.

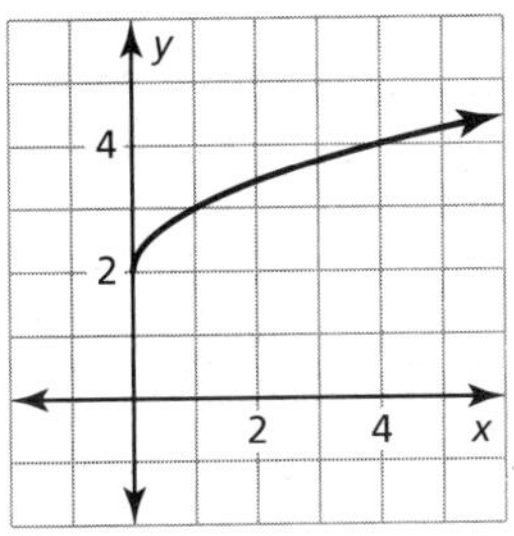

range: $y \geq 2$

62.

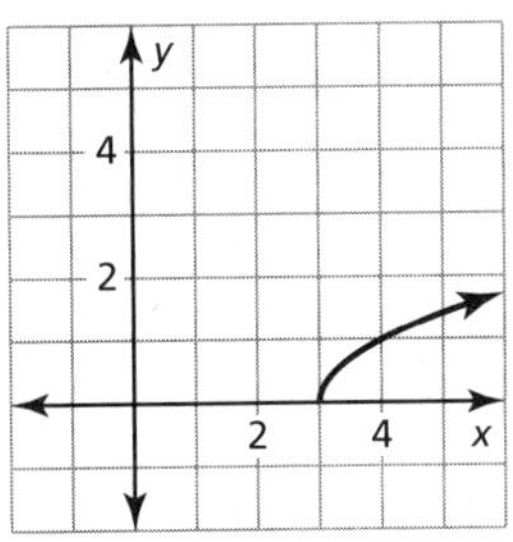

range: $y \geq 0$

63.

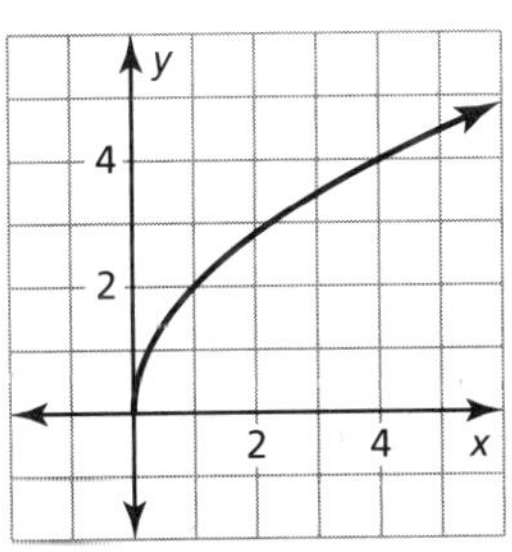

range: $y \geq 0$

64.

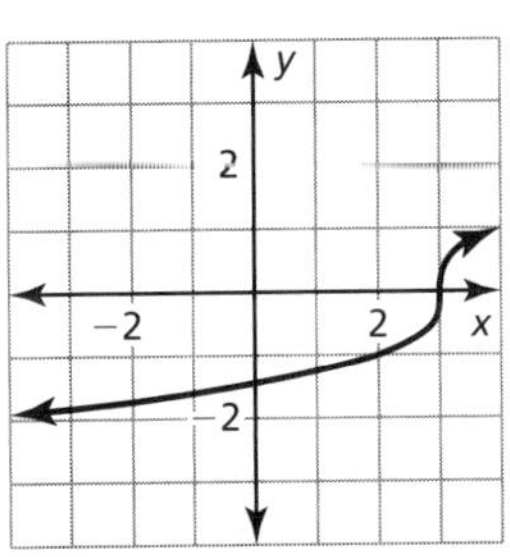

translation 3 units right

65.

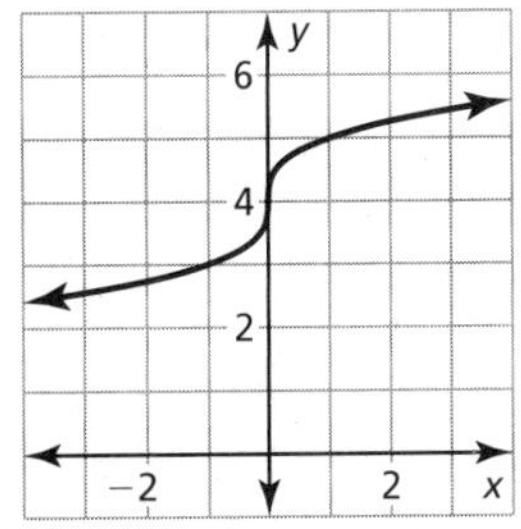

translation 4 units up

66.

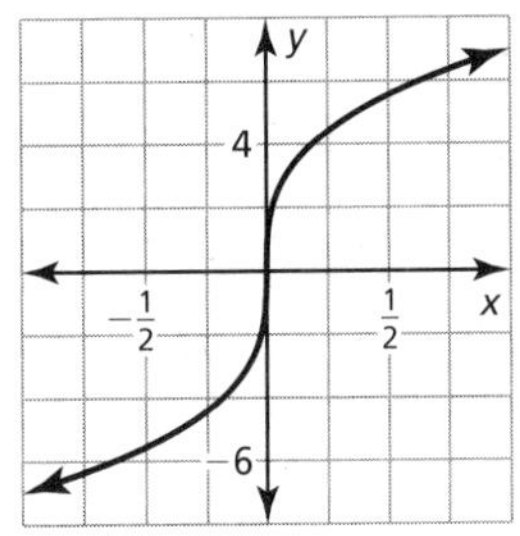

vertical stretch by a factor of 7

67.

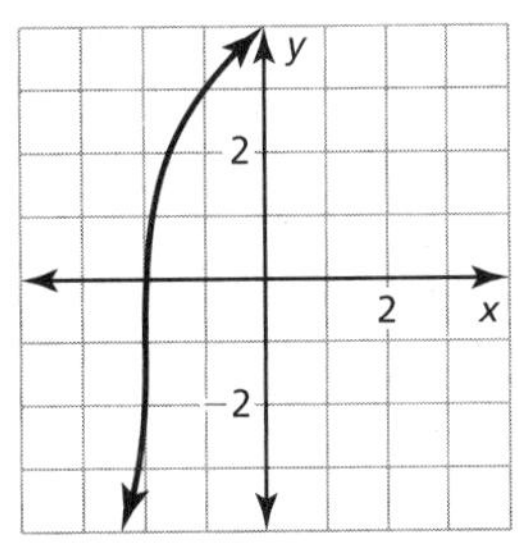

vertical stretch by a factor of 4 followed by a translation 2 units left and 1 unit down

68.

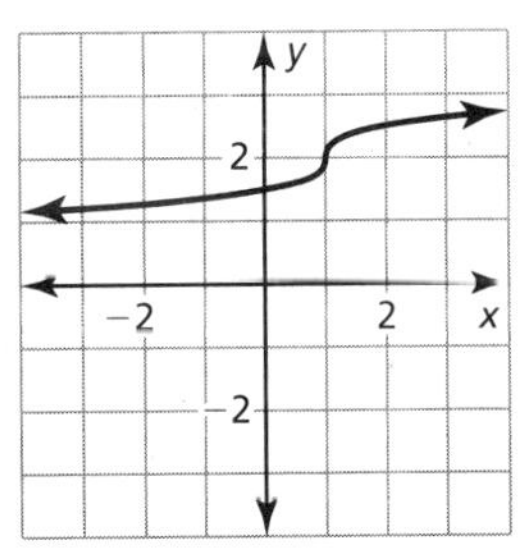

vertical shrink by a factor of $\frac{1}{2}$ followed by a translation 1 unit right and 2 units up

69.

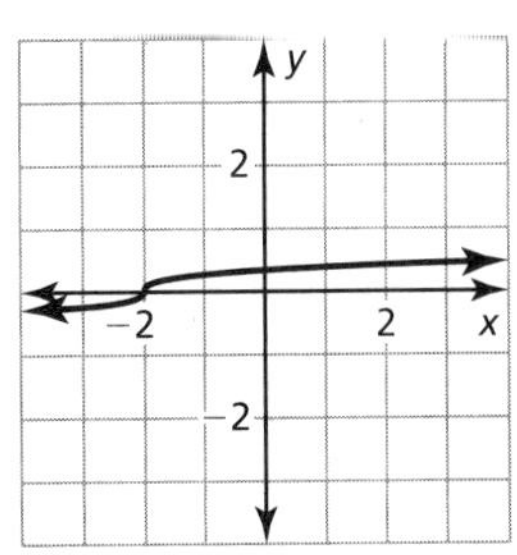

translation 2 units left and a vertical shrink by a factor of 0.25

Answers

70. $x = 25$ **71.** $x = 36$ **72.** $y = 49$

73. $w = 11$ **74.** $r = 102$ **75.** $x = 1$

76. $x = 2$ **77.** $x = 4$

78.

Input	6	5	4	3	2	1	0
Output	−2	−1	0	1	2	3	4

79.

Input	−9	−5	−1	3	7	11	15
Output	5	2	−1	−4	−7	−10	−13

80. $\cos\theta = -\frac{3}{5}$, $\tan\theta = \frac{4}{3}$, $\csc\theta = -\frac{5}{4}$, $\sec\theta = -\frac{5}{3}$, $\cot\theta = \frac{3}{4}$

81. $\sin\theta = -\dfrac{6\sqrt{37}}{37}$, $\tan\theta = 6$, $\cos\theta = \dfrac{\sqrt{37}}{37}$, $\csc\theta = \dfrac{\sqrt{37}}{6}$, $\sec\theta = \sqrt{37}$

82. $\sin\theta = \dfrac{\sqrt{55}}{8}$, $\tan\theta = -\dfrac{\sqrt{55}}{3}$, $\csc\theta = \dfrac{8\sqrt{55}}{55}$, $\sec\theta = -\dfrac{8}{3}$, $\cot\theta = -\dfrac{3\sqrt{55}}{55}$

83. $\sin\theta = -\dfrac{2\sqrt{6}}{5}$, $\cos\theta = \dfrac{1}{5}$, $\tan\theta = -2\sqrt{6}$, $\csc\theta = -\dfrac{5\sqrt{6}}{12}$, $\cot\theta = -\dfrac{\sqrt{6}}{12}$

84. $A \approx 34.8°$, $C \approx 91.2°$, $c \approx 21.0$

85. $C \approx 26°$, $a \approx 13.9$, $c \approx 8.2$

86. $A \approx 32.8°$, $B \approx 41.4°$, $C \approx 105.8°$

87. $A \approx 74.7°$, $B \approx 27.3°$, $c \approx 21.3$

Chapter 10

10.1 Start Thinking

mean = 75

Score	Frequency
50–59	2
60–69	3
70–79	8
80–89	5
90–99	2

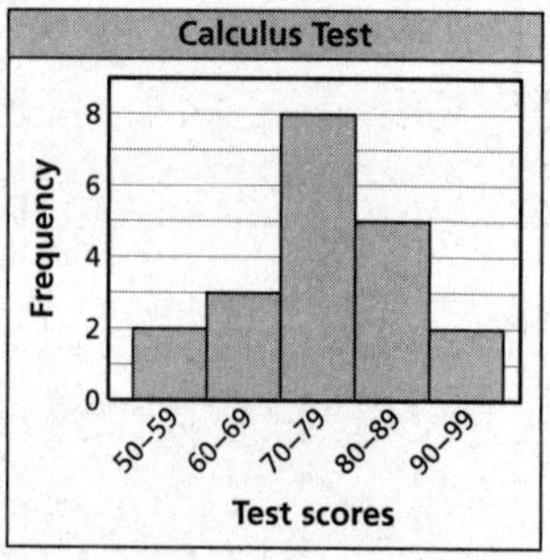

Sample answer: The histogram appears to have a reasonably symmetric distribution about the mean value of 75.

10.1 Warm Up

1. mean = 18.1, median = 17.5

2. mean $= \dfrac{855}{7} \approx 122.14$, median = 120

3. mean $= \dfrac{177}{40} \approx 4.43$, median = 4.6

10.1 Cumulative Review Warm Up

1. $g(x) = -|x - 1| + 3$

2. $g(x) = -\frac{1}{2}|x| - 5$

10.1 Practice A

1. 50% **2.** 47.5% **3.** 0.15% **4.** 99.85%

5. 47.5% **6.** 68% **7.** 47.5% **8.** 16%

9. a. 81.5% **b.** 16%

10. The probability that x is from 15 to 27 was found instead of the probability that x is at least 15.

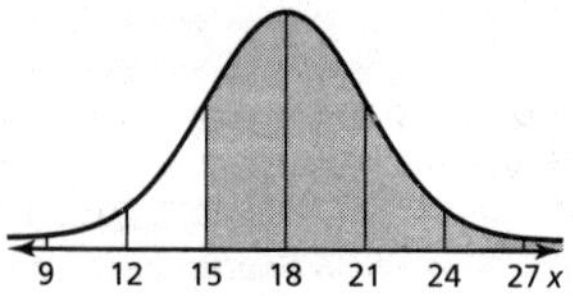

The probability that x is at least 15 is 0.84.

10.1 Practice B

1. 16% **2.** 5% **3.** 97.5%

4. 83.85% **5.** 13.5% **6.** 97.35%

7. 84% **8.** 2.5%

9. a. 95% **b.** 2.5%

10. The probability that x is from 18 to 24 was found instead of the probability that x is at most 24.

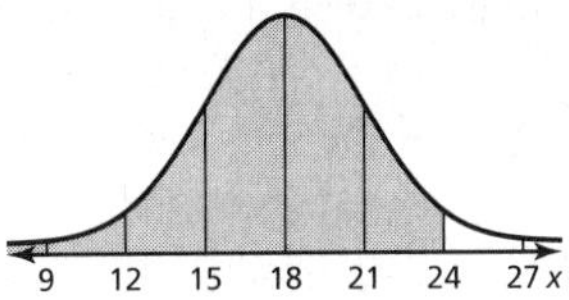

The probability that x is at most 24 is 0.975.

10.1 Enrichment and Extension

1. $S_{k2} \approx -0.178,$ so the distribution of banana prices is approximately symmetric.

2. $S_{k2} \approx 0.381,$ so the distribution of test differences is approximately symmetric.

3. $S_{k2} \approx 1.160,$ so the distribution of mail is skewed right.

10.1 Puzzle Time

FISH AND SHIPS

10.2 Start Thinking

The theoretical probability is $\frac{1}{8}$; *Sample answers*:

Some students may find that it takes only one toss of the three coins to get all three to land heads up. However, as a group they should notice that the average number of tosses is somewhere around eight. The experimental probability will quite likely differ from the theoretical probability. However, as the number of tosses increases (sample size increases), the two probabilities should get closer together.

10.2 Warm Up

1. The sample space is the set of all 52 cards.

2. The sample space is the set given that includes all possible color combinations with two spins; {Red-Red, Red-Blue, Red-Yellow, Red-Green, Blue-Blue, Blue-Yellow, Blue-Green, Blue-Red, Yellow-Yellow, Yellow-Green, Yellow-Red, Yellow-Blue, Green-Green, Green-Red, Green-Blue, Green-Yellow}

3. The sample space is the set of all possible responses and may include, yes, no, and maybe.

4. The sample space is the set of all possible responses and may include, yes, no, and sometimes.

10.2 Cumulative Review Warm Up

1. $(4, 0)$

2. infinitely many solutions of the form $(x, x - 8)$

3. $(-2, 1)$ **4.** $\left(4, -\frac{3}{4}\right)$ **5.** $(4, 3)$ **6.** $(0, 0)$

10.2 Practice A

1. sample; There are more than 25 bags of popcorn.

2. population; degrees of every (all) employees

3. The population is all students at the university, and the sample is the 1641 students that were surveyed. Of the students surveyed, 479 of them do not know the name of their college's mascot.

4. The population is all people in the United States, and the sample is the 1000 households with at least one child that were surveyed. Of the households surveyed, 874 of them have two or more computers.

5. parameter; it is referring to all of the football players on the team

6. statistic; it is referring to only the surveyed utility bills

7. a. The maker's claim is most likely true.

b. The maker's claim is most likely false.

10.2 Practice B

1. population; every student in the school

2. sample; more than 80 people could access the website

3. The population is every employee in the office building, and the sample is the 648 employees that were surveyed. Of the employees surveyed, 147 of them ride the subway to work each day.

4. The population is all homeowners in the State of Florida, and the sample is the 2500 homeowners that were surveyed. Of the homeowners surveyed, 1145 of them have switched their homeowner's insurance policy to a different company within the last 3 years.

5. statistic; it is referring to only the surveyed hockey players

6. parameter; it is referring to all the sold tickets

7. a. The maker's claim is most likely false.

b. The maker's claim is most likely true.

Answers

10.2 Enrichment and Extension

1. a. Type I: Decide garlic is effective in reducing tick bites when it really is not effective; Consequence: People would consume more garlic but not have the protection against tick bites and could develop Lyme's disease (or some other disease carried by ticks); Type II: Decide garlic is not effective in reducing tick bites when it really is effective; Consequence: People would not use the garlic (a nontoxic treatment) to prevent tick bites, but have to use a toxic repellant.

   b. *Sample answer*: Type I is more serious because people could potentially develop a serious disease from tick bites.

2. a. Type I: Say that there are too many defects when in reality there are not; Consequence: You would stop the assembly process when there is no problem, wasting production time; Type II: Say that there are not too many defects when in reality there are; Consequence: You would continue with the assembly process, but have many defective products, which could cause you to lose money.

   b. *Sample answer*: The factory owner might consider Type I more serious because he loses production time for no reason, which could cause him to lose more money than the defective products.

3. a. Type I: Decide that the shop is not meeting standards, when it is; Consequence: The shop owner could be justly fined; Type II: Decide the shop is certified as meeting standards when it is not; Consequence: Substandard cars are polluting the environment.

   b. *Sample answer*: Either could be argued as worse depending on whether you work for the shop or care about the welfare of the environment.

4. a. Type I: The software seems like it helps (and the professor buys it) even when it really did not; Consequence: The students are wasting their time using the software, and it is a waste of money; Type II: The software doesn't seem to help (and the professor doesn't buy it) when it would have helped; Consequence: The students would be missing out on software that helps them.

   b. *Sample answer*: Type I might be more serious, because the money could have been better spent elsewhere.

10.2 Puzzle Time

ON A MAP

10.3 Start Thinking

1. *Sample answer*: Surveying 50 random students at your school may not give the school board good information in regards to allocating funds for a robotics team. Depending on the total number of students in the school, the 50 surveyed may be a very small percentage or a very large percentage of the entire school population. The school board could get better information by surveying all the students to determine exactly how many (what percentage) of the students are actually interested in participating in robotics.

2. *Sample answer*: If the Student Government Association wants to know if families would be interested in participating in a school carnival, they should probably survey the parents as well as the students.

10.3 Warm Up

1. *Sample answer*: Because the survey only includes pet owners, it is quite likely that they would want hotels to accept pets. It might be better to survey hotel guests in general rather than pet owners who are likely biased in favor of pets.

2. *Sample answer*: Shoppers in the mall are quite likely to be biased toward extended mall hours so that they can shop longer. A better option would be to survey shoppers along with mall employees and store owners, as well as people living in the neighborhood around the mall.

3. *Sample answer*: Students would most likely be biased in such a way to encourage a decreased homework load. Surveying teachers, parents, administrators, and students might give more reliable results.

4. *Sample answer*: Because the survey includes individuals inside an athletic club, they are probably more likely to exercise. Surveying a larger and more diverse group would give more reliable results.

10.3 Cumulative Review Warm Up

1. 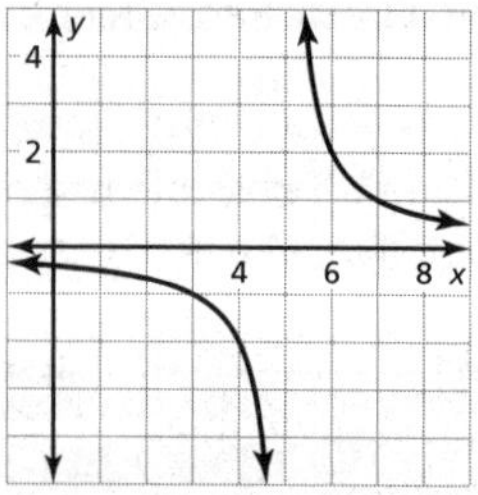

2. 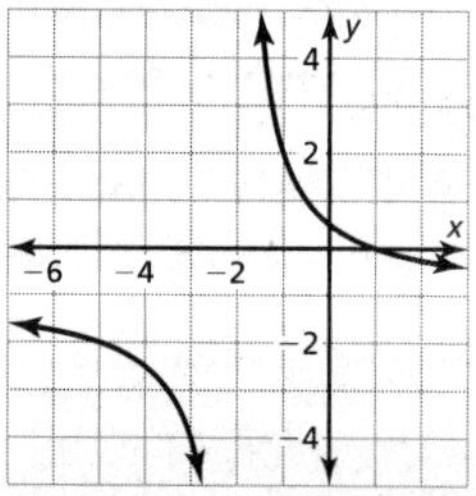

Answers

3. 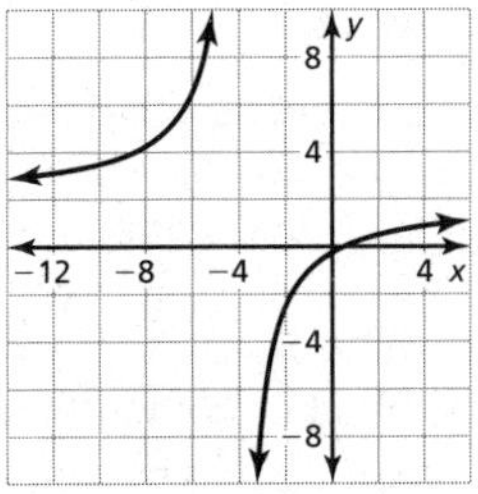

10.3 Practice A

1. self-selected sample
2. convenience sample
3. systematic; These customers shop at a locally-owned hardware store, so they would probably want the store to remain opened for business.
4. systematic; Members with last names that begin with the later letters of the alphabet do not get an opportunity to voice their opinion, and some of the first 20 members with the same last names might be related.
5. unbiased; The sample represents the population.
6. unbiased; The students are randomly selected.
7. A sensitive question is being asked. This type of question is better asked anonymously.
8. By saying "could be toxic," one would feel guilty if he or she did not agree. These words could be removed from the question and the individual could make his or her own judgment about lesions.

10.3 Practice B

1. cluster sampling
2. systematic sampling
3. cluster; The cluster was not randomly selected; People who own waterfront property will support lower taxes for those who own waterfront property
4. self-select; Those who are listening to the station probably prefer country music, so they are more likely to vote for country music
5. biased; Local fisherman are not likely to support the bill, because they would probably want to catch as many lobster as possible.
6. biased; Only one age group was surveyed.
7. People may hesitate to vocalize their agreement if they are standing next to employees at the plant. This question would be better asked anonymously.
8. The use of the words "far beyond" could influence someone to not disagree. The words could be removed.

10.3 Enrichment and Extension

1. Method 1: experiment; Method 2: survey; Method 3: observational study; Method 1; Method 2 and 3 depend on finding farms using the desired fertilizers, and different farms may use different amounts of fertilizer.
2. Method 1: experiment; Method 2: observational study; Method 3: survey; Method 1; Method 2 may not be random, and Method 3 relies on the drivers accurately remembering their battery usage.

10.3 Puzzle Time

TO UNWIND

10.4 Start Thinking

To determine whether a spinner is fair, you could examine its design and make sure each space has the same area and that the spinner actually spins freely over each space. You could also test the spinner by recording the results of 100 or 1000 (or more) spins to see whether each number/color came up about the same number of times. In a similar way, you could examine a coin to make sure that one side did not seem to be weighted. Then you could test the coin by recording the results of 100 or 1000 (or more) flips to see whether each side came up about the same number of times.

10.4 Warm Up

1. experiment
2. observational study

10.4 Cumulative Review Warm Up

1. $3^{1/3}$
2. 324
3. $9\sqrt{3}$
4. $\sqrt[6]{12}$
5. 2
6. $5^{1/2}$

10.4 Practice A

1. This study is a randomized comparative experiment. The treatment is the use of the computer for testing. The treatment group is the 150 students who test on the computer. The control group is the 150 students who test using paper and pencil.
2. This study is not a randomized comparative experiment because the people were not randomly assigned to a control group and a treatment group. The conclusion may not be valid. There may be other reasons why those who drank water had lower cholesterol.

3. This study is best done as an experiment. Half of the team is randomly assigned to yoga, and the other half is not.

4. This study is best done as an observational study because a time frame of 10 years is too long for an experiment of this type. After 10 years, the drugs may be obsolete. The effects of the drugs can be examined on those who have already taken them.

5. **a.** Muscle damage is more likely as one gets older. Thus, the breakdown according to age may affect the results. This study can be improved by just randomly assigning people to groups, regardless of age.

   b. The experimental design is good. However a potential problem may be that injuries can occur.

10.4 Practice B

1. This study is not a randomized comparative experiment because the people were not randomly assigned to a control group and a treatment group. The conclusion may not be valid. There may be other reasons why those who switched were more satisfied with the condition of their pool.

2. This study is a randomized comparative experiment. The treatment is the use of the recyclables containers. The treatment group is the 75 customers who used the recyclables containers. The control group is the 75 customers who used the recyclables bins.

3. This study is best done as an observational study because the organization cannot force someone to donate 20% to charity. Job satisfaction can be rated for those who already give 20% to charity.

4. This study is best done as an experiment. The rancher randomly chooses half the cows to receive the new feed, and the other half do not.

5. **a.** The experimental design is good. However, a possible potential problem may be that someone dislikes the taste of diet soda.

   b. A potential problem is the division based on exercise habits. Those who exercise regularly tend to eat less sugar, and may tend to drink more diet soda than the other group. Thus, their bodies may already be conditioned to the sugar cravings. This design could be improved by dividing the group randomly rather than based on exercise habits.

10.4 Enrichment and Extension

1. **a.** class size: small, medium, large; textbook: traditional, new

   b. average final grades for class

   c. 6

   d. The experiment could be blind or double-blind if the students and/or the researcher do not know whether the class is using the traditional or new textbook until after the experiment.

2. **a.** watering: daily, three days a week; fertilizer: traditional, organic

   b. average final heat rating for the plot at the end of the growing season

   c. 4

   d. The experiment could be blind if the researcher does not know which plant has been fertilized with each fertilizer until after the experiment.

3. **a.** dog food: new, old

   b. improvement in kidneys

   c. 2

   d. The experiment could be blind if the researcher does not know which food has been fed to each dog until after the experiment.

4. **a.** walking: using no device, with standard walker, with rolling walker

   b. cadence

   c. 6

   d. The experiment could be blind or double-blind if the walkers do not know what is being measured and/or the researchers do not know which walkers were required to respond to the signal until after the experiment.

5. *Sample answer*: Randomly divide the subjects into two groups of 50. Give 50 the inhaler and 50 the shot. Measure insulin levels three times a day. Compare the results.

10.4 Puzzle Time

WAVY

10.5 Start Thinking

The sample survey results show that 35% of the students have part-time jobs; *Sample answer*: Because the original prediction is 10 percentage points off, we may choose to change our prediction based on the sample survey. However, we may also decide to redo the survey with a greater number of students to make sure our sample was a good representation of the entire population.

Answers

10.5 Warm Up

1. 185.8
2. 4.8

10.5 Cumulative Review Warm Up

1. $x^5 - 15x^4 + 90x^3 - 270x^2 + 405x - 243$
2. $16x^4 + 160x^3 + 600x^2 + 1000x + 625$
3. $-128x^7 + 448x^6 - 672x^5 + 560x^4 - 280x^3 + 84x^2 - 14x + 1$
4. $x^4 + 4x^3y^2 + 6x^2y^4 + 4xy^6 + y^8$
5. $64a^6 - 576a^5b + 2160a^4b^2 - 4320a^3b^3 + 4860a^2b^4 - 2916ab^5 + 729b^6$
6. $x^{24} - 8x^{21}y^4 + 28x^{18}y^8 - 56x^{15}y^{12} + 70x^{12}y^{16} - 56x^9y^{20} + 28x^6y^{24} - 8x^3y^{28} + y^{32}$

10.5 Practice A

1. a. The sample mean is 9.3, so an estimate for the population mean would be 9.3.
   b. The sample proportion is approximately 0.233, so an estimate for the population proportion would be 0.233.
   c. The sample proportion is approximately 0.267, so an estimate for the population proportion would be 0.267.
2. not very confident; The sample mean is an estimate of the population mean.
3. a. Based on your survey, 0.43 is just outside the margin of error, so you can conclude that the claim is probably incorrect.
   b. Based on your survey, 0.43 is within the margin of error, so you can conclude that the claim is probably correct.
   c. 95% of the time, a sample proportion should lie in the interval from 0.29 to 0.57.

10.5 Practice B

1. a. The sample mean is 5.825, so an estimate for the population mean would be 5.825.
   b. The sample proportion is 0.325, so an estimate for the population proportion would be 0.325.
   c. The sample proportion is 0.275, so an estimate for the population proportion would be 0.275.
2. not very confident; The sample mean is an estimate of the population mean.
3. a. Based on your survey, 0.45 is outside the margin of error, so you can conclude that the claim is probably incorrect.
   b. Based on your survey, 0.45 is within the margin of error, so you can conclude that the claim is probably correct.
   c. 95% of the time, a sample proportion should lie in the interval from 0.31 to 0.59.

10.5 Enrichment and Extension

1. *Sample answer*: Should the school start a foreign language club?
2. *Sample answer*: yes; The type of club is specific and it is an easy yes or no answer.
3. The entire student population.
4. *Sample answer*: ask 100 students chosen randomly by ID number
5. *Sample answer*: no; The sample is random.
6. *Sample answer*: yes; The sample is random.
7. *Sample answer*: yes; The question may need changed to, "Would you join a foreign language club if it were offered?"
8. *Sample answer*: $\pm 10\%$; $p \pm 10\%$ where p is the percent of students who responded "yes"; The results became more accurate when more people were surveyed because the margin of error decreased.

10.5 Puzzle Time

A SWALLOW

10.6 Start Thinking

Sample answer: The company might produce a small amount of the sports drink in the suggested size bottle and see how the sales compare to the original size over a given period of time. They could do the same thing with a label change. They could test a small market with just a change in the label. And finally, they could produce a small amount of the sports drink with both a change in size and labeling. By comparing the sales within similar markets, they could determine if the label change or the change in size, or both, affect the sales in a positive or negative way.

10.6 Warm Up

1. mean of the control: 14.79; mean of the treatment: 15.02; difference: 0.23
2. mean of the control: 59.6; mean of the treatment: 55.4; difference: -4.2

Answers

10.6 Cumulative Review Warm Up

1. 1
2. $\tan^2 x$
3. $2\csc x$
4. $\cos x$
5. $\sec x$
6. $\cos x - 1$

10.6 Practice A

1. a. 3.625

b. 5.375

c. 1.75

d. 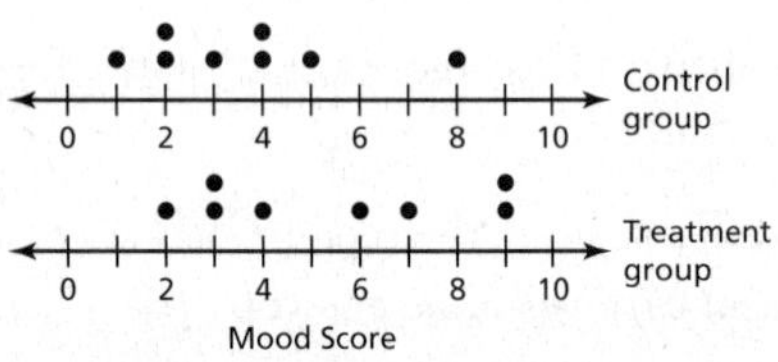

e. The plot of the data shows that the two sets are fairly symmetric. The control group may have a possible outlier, but it is a valid data item. So, the mean is a suitable measure of center. The mean score of the treatment group is 1.75 higher than that of the control group. It appears that mood therapy might be slightly effective, but the sample size is small and the difference could be due to chance.

f. *Sample answer*:

	Mood Score							
Control Group	5	9	2	3	3	7	8	4
Treatment Group	1	3	4	2	6	9	2	4

g. *Sample answer*: 5.125

h. *Sample answer*: 3.875

i. *Sample answer*: –1.25

10.6 Practice B

1. a. 3.35

b. 2.425

c. –0.925

d. 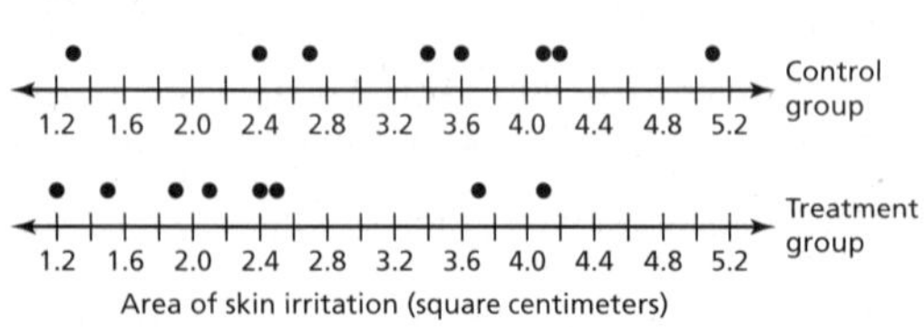

e. The plot of the data shows that the two sets are fairly symmetric and have no extreme values. So, the mean is a suitable measure of center. The mean score of the treatment group is 0.925 lower than that of the control group. It appears that photo therapy might be slightly effective, but the sample size is small and the difference could be due to chance.

f. *Sample answer*:

	Area of Skin Irritation (square centimeters)							
Control Group	4.1	2.5	5.1	1.2	3.4	3.7	2.7	1.3
Treatment Group	4.2	2.1	3.6	2.4	1.5	1.9	2.4	4.1

g. *Sample answer*: 3

h. *Sample answer*: 2.775

i. *Sample answer*: –0.225

10.6 Enrichment and Extension

1. a. *Sample answer:*

	Total Yield of Tomato Plants (kilograms)				
Control Group	2.4	0.7	1.7	1.4	1.6
Treatment Group	1.4	1.4	1.4	2.4	1.4

	Total Yield of Tomato Plants (kilograms)				
Control Group	1.5	1.5	1.8	1.9	1.9
Treatment Group	1.8	1.9	1.3	0.7	2.4

b. *Sample answer:*

	Total Yield of Tomato Plants (kilograms)				
Control Group	1.8	1.8	1.6	1.2	1.9
Treatment Group	1.9	1.2	1.7	1.4	1.4

	Total Yield of Tomato Plants (kilograms)				
Control Group	2.0	1.9	1.4	0.9	0.7
Treatment Group	1.5	1.3	1.7	0.9	-

c. *Sample answer:* The experimental difference of the means is -0.03 using bootstrap and about -0.08 using jackknife.

Answers

2. a.

	Depression Score			
Control Group	47	46	43	47
Treatment Group	45	47	39	47

	Depression Score			
Control Group	39	40	46	47
Treatment Group	47	42	40	49

b.

	Depression Score			
Control Group	40	39	41	39
Treatment Group	40	46	47	46

	Depression Score			
Control Group	45	43	42	49
Treatment Group	47	45	43	-

c. *Sample answer:* The experimental difference of the means is 0.125 using bootstrap and about 2.61 using jackknife.

3. a.

	Circumference (inches)			
Control Group	34.8	34.6	33	33
Treatment Group	31.4	33	35.4	33.4

	Circumference (inches)			
Control Group	38.2	38.2	32.6	33.4
Treatment Group	38.2	34.6	31.4	35.4

b.

	Circumference (inches)			
Control Group	33.4	32.6	35.2	34.8
Treatment Group	33	34.6	35.2	35.4

	Circumference (inches)			
Control Group	33	33.4	33	35
Treatment Group	34.6	31.4	32.4	-

c. *Sample answer:* The experimental difference of the means using bootstrap is -0.625 and 0.0 using jackknife.

10.6 Puzzle Time

ROCKET

Cumulative Review

1. mean: 12, median: 14, mode: 14

2. mean: 45, median: 46, mode: 46

3. mean 69.7, median: 68, mode: 80

4. mean: 44.1, median: 44, mode: 43 and 47

5. mean: 13.3, median: 16, mode: none

6. mean: 15.4, median: 15, mode: 14 and 15

7. 4.3; *Sample answer*: The typical data value differs from the mean by about 4.3 units.

8. 2.6; *Sample answer*: The typical data value differs from the mean by about 2.6 units.

9. 6.6; *Sample answer*: The typical data value differs from the mean by about 6.6 units.

10. 1.6; *Sample answer*: The typical data value differs from the mean by about 1.6 units.

11. 2.6; *Sample answer*: The typical data value differs from the mean by about 2.6 units.

12. 3.8; *Sample answer*: The typical data value differs from the mean by about 3.8 units.

13. inverse variation

14. direct variation

15. neither

16. neither

17. direct variation

18. direct variation

19. $y = \frac{7}{x}; \frac{7}{3}$

20. $y = -\frac{40}{x}; -\frac{40}{3}$

21. $y = -\frac{7}{x}; -\frac{7}{3}$

22. $y = -\frac{10}{x}; -\frac{10}{3}$

23. $y = \frac{56}{9x}; \frac{56}{27}$

24. $y = \frac{3}{2x}; \frac{1}{2}$

25. mean: 84.7, median: 85, mode: 71, 75, 77, 92, and 95

26. $\frac{1}{2x - 9x^4}$

27. $\frac{6 + 2x^3}{3}$

28. $\frac{x - 4}{x + 5}$

29. $\frac{x + 9}{x + 4}$

30. $\frac{x - 9}{x + 6}$

31. $\frac{x - 5}{x + 3}$

32. $\frac{y^4}{3x^4}$

33. $\frac{y^2}{6}$

34. $\frac{x^2 - 6x + 8}{x}$

Answers

35. $\dfrac{x+2}{x-10}$ 36. $\dfrac{x-2}{x+7}$ 37. $\dfrac{x+3}{x-10}$

38. $\dfrac{20}{7x}$ 39. $\dfrac{8+3x}{x-2}$ 40. $\dfrac{-7x}{3x+4}$

41. $\dfrac{-x^2+4x+18}{x^2+3x-40}$ 42. $\dfrac{3x^2-27x-68}{x^2-12x+11}$

43. $\dfrac{x^2+x+62}{x^2-11x-12}$

44. a. $P = 10x + 32$
 b. $A = 6x^2 + 39x + 63$
 c. 52 in.; 165 in.^{2}

45. $\sin\theta = \dfrac{3}{5}$, $\cos\theta = \dfrac{4}{5}$, $\tan\theta = \dfrac{3}{4}$, $\csc\theta = \dfrac{5}{3}$,
 $\sec\theta = \dfrac{5}{4}$, $\cot\theta = \dfrac{4}{3}$

46. $\sin\theta = \dfrac{3}{5}$, $\cos\theta = \dfrac{4}{5}$, $\tan\theta = \dfrac{3}{4}$, $\csc\theta = \dfrac{5}{3}$,
 $\sec\theta = \dfrac{5}{4}$, $\cot\theta = \dfrac{4}{3}$

47. $\sin\theta = \dfrac{3}{7}$, $\cos\theta = \dfrac{2\sqrt{10}}{7}$, $\tan\theta = \dfrac{3\sqrt{10}}{20}$,
 $\csc\theta = \dfrac{7}{3}$, $\sec\theta = \dfrac{7\sqrt{10}}{20}$, $\cot\theta = \dfrac{2\sqrt{10}}{3}$

48. $\sin\theta = \dfrac{4\sqrt{137}}{137}$, $\cos\theta = \dfrac{11\sqrt{137}}{137}$, $\tan\theta = \dfrac{4}{11}$,
 $\csc\theta = \dfrac{\sqrt{137}}{4}$, $\sec\theta = \dfrac{\sqrt{137}}{11}$, $\cot\theta = \dfrac{11}{4}$

49. $x = 12$ 50. $x = 7\sqrt{2}$

51. $x = 3\sqrt{3}$ 52. $x = 4$

53. *Sample answer:* $-272°$ and $448°$

54. *Sample answer:* $-665°$ and $55°$

55. *Sample answer:* $-\dfrac{11\pi}{4}$ and $\dfrac{5\pi}{4}$

56. *Sample answer:* $-\dfrac{7\pi}{6}$ and $\dfrac{17\pi}{6}$

57. *Sample answer:* $-300°$ and $60°$

58. *Sample answer:* $-\dfrac{10\pi}{3}$ and $\dfrac{2\pi}{3}$

59. $\dfrac{11\pi}{18}$ 60. $\dfrac{31\pi}{18}$ 61. $-240°$

62. $150°$ 63. $-\dfrac{\pi}{3}$ 64. $-135°$

65. $\sin 180° = 0$, $\cos 180° = -1$, $\tan 180° = 0$,
 $\csc 180° =$ undefined, $\sec 180° = -1$,
 $\cot 180° =$ undefined

66. $\sin 495° = \dfrac{\sqrt{2}}{2}$, $\cos 495° = -\dfrac{\sqrt{2}}{2}$,
 $\tan 495° = -1$, $\csc 495° = \sqrt{2}$,
 $\sec 495° = -\sqrt{2}$, $\cot 495° = -1$

67. $\sin\left(-\dfrac{\pi}{3}\right) = -\dfrac{\sqrt{3}}{2}$, $\cos\left(-\dfrac{\pi}{3}\right) = \dfrac{1}{2}$,
 $\tan\left(-\dfrac{\pi}{3}\right) = -\sqrt{3}$, $\csc\left(-\dfrac{\pi}{3}\right) = \dfrac{-2\sqrt{3}}{3}$,
 $\sec\left(-\dfrac{\pi}{3}\right) = 2$, $\cot\left(-\dfrac{\pi}{3}\right) = -\dfrac{\sqrt{3}}{3}$

68. $\sin\dfrac{\pi}{6} = \dfrac{1}{2}$, $\cos\dfrac{\pi}{6} = \dfrac{\sqrt{3}}{2}$, $\tan\dfrac{\pi}{6} = \dfrac{\sqrt{3}}{3}$,
 $\csc\dfrac{\pi}{6} = 2$, $\sec\dfrac{\pi}{6} = \dfrac{2\sqrt{3}}{3}$, $\cot\dfrac{\pi}{6} = \sqrt{3}$

69. $\sin(-120°) = -\dfrac{\sqrt{3}}{2}$, $\cos(-120°) = -\dfrac{1}{2}$,
 $\tan(-120°) = \sqrt{3}$, $\csc(-120°) = -\dfrac{2\sqrt{3}}{3}$,
 $\sec(-120°) = -2$, $\cot(-120°) = \dfrac{\sqrt{3}}{3}$

70. $\sin\dfrac{7\pi}{4} = -\dfrac{\sqrt{2}}{2}$, $\cos\dfrac{7\pi}{4} = \dfrac{\sqrt{2}}{2}$, $\tan\dfrac{7\pi}{4} = -1$,
 $\csc\dfrac{7\pi}{4} = -\sqrt{2}$, $\sec\dfrac{7\pi}{4} = \sqrt{2}$, $\cot\dfrac{7\pi}{4} = -1$

Answers

71.

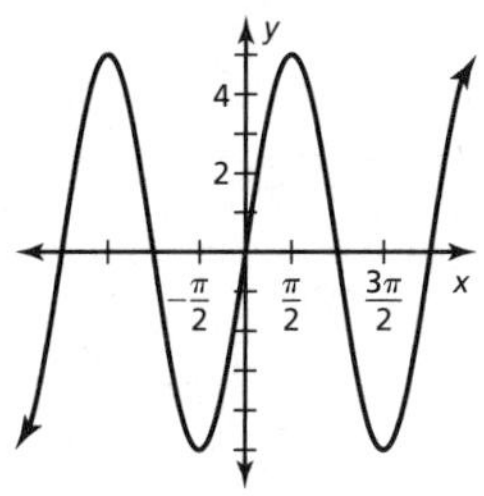

72.

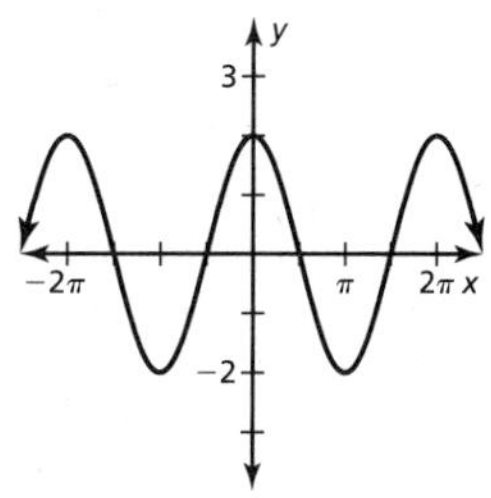

73.

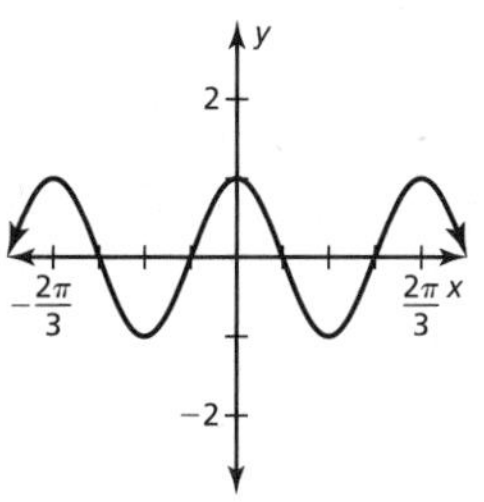

74.

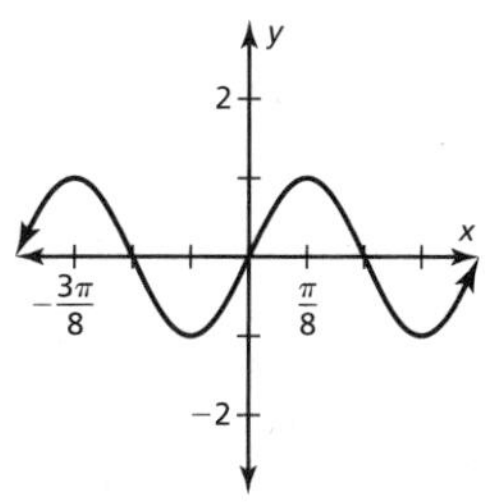

75.

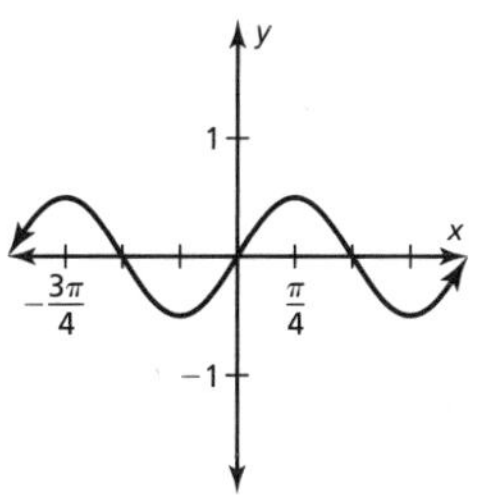

76.

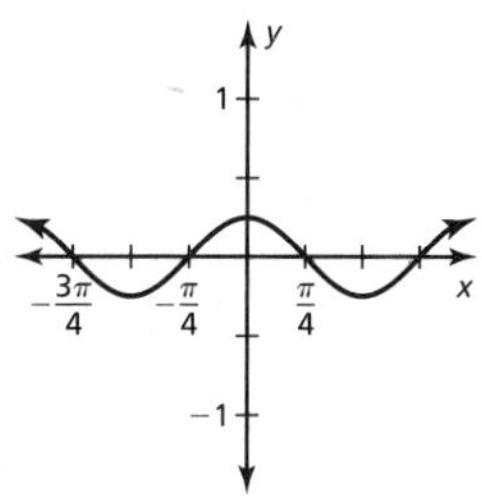

77.

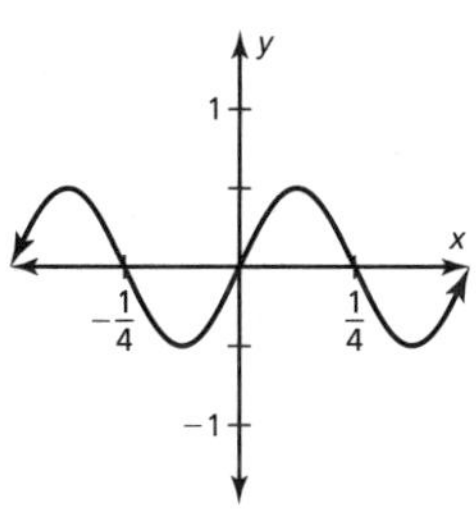

78.

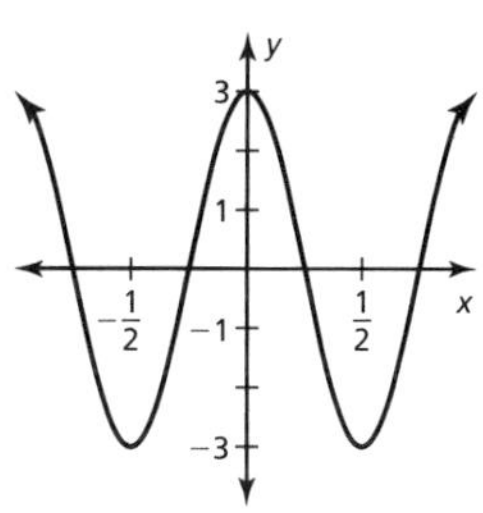

79.

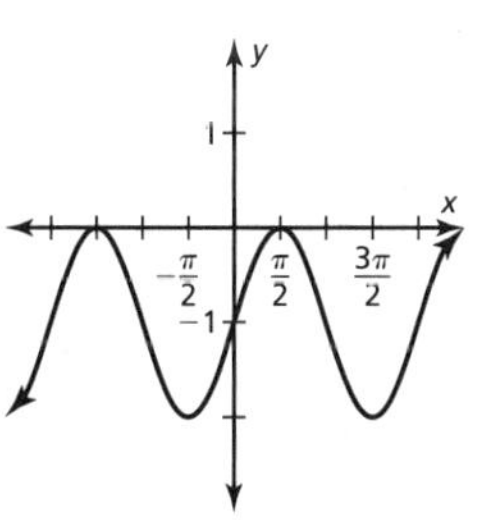

80.

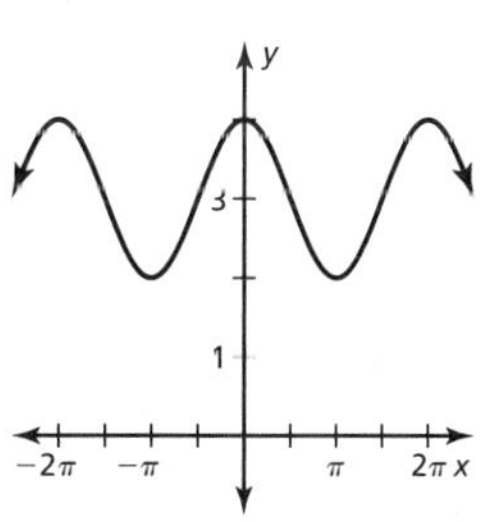

81.

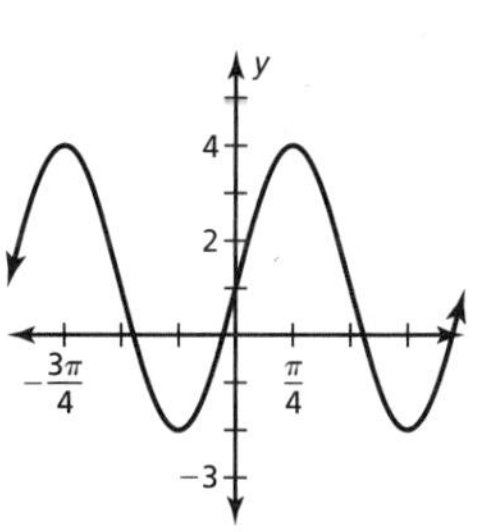

82.

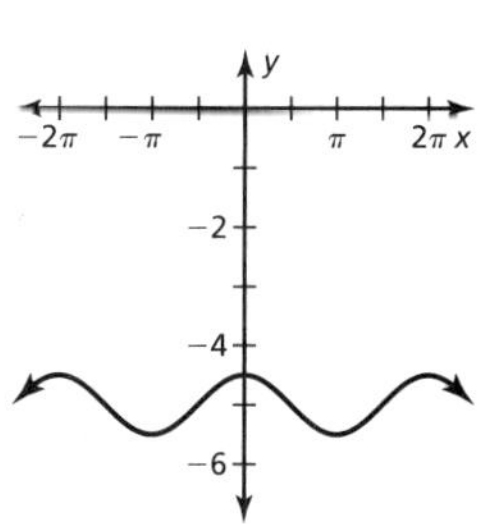

83. 0.34

84. 0.9985

85. 0.975

86. 0.9735

87. population: every student in the school district; sample: the 1230 high school students that were surveyed; The sample consists of 1144 students who feel the school should keep the music program and 86 students who do not.

88. population: every citizen in the town; sample: the 1000 citizens who were surveyed; The sample consists of 872 citizens who want to replace the street signs and 128 citizens who do not.

89. cluster sample; Students who take AP Calculus are more likely to have more homework than students who are not in AP classes. They are also likely to study more than students who are not in AP classes.

90. systematic sample; People who go to the gym are likely to have a different opinion about the importance of personal fitness than people who do not go to the gym.